41/6

THE MACHINERY
OF THE BODY

BY ANTON J. CARLSON AND VICTOR JOHNSON

THE MACHINERY OF THE BODY

FOURTH EDITION

THE UNIVERSITY OF CHICAGO PRESS

THE UNIVERSITY OF CHICAGO PRESS, CHICAGO 37

Cambridge University Press, London, N.W. 1, England

PREFACE TO
THE FIRST EDITION

This elementary introduction to the machinery that regulates the animal body is a precipitate of years of experience with the attempt to teach human physiology to college students at the University of Chicago. We have found most of these young people eager for as much comprehension of themselves as is possible even with little or no previous experience in science. Innate curiosity, fanned by personal interests, energizes the quest for understanding, particularly when demonstrations and experiments add clarity and conviction to the spoken and the written word. We realize that the printed page sometimes seems a feeble aid to understanding, except when written by the hand of a genius. And yet we offer this essay in the hope that college students and other lay readers may find it helpful and not merely another book.

We are convinced that even the beginner in the science of physiology should gain some familiarity with the wealth of experiments and rechecked observations on which our present generalizations are founded. And, now and then, he should be led on to the end of the path of present knowledge, so as to gain an appreciation of the need for further experimentation if the normal and abnormal body processes are to be better understood and more effectively controlled. This method of approach to a study of the principles of physiology not only proves a fascinating path to knowledge but also constitutes a safeguard against dogmatism and an aid to a development of that healthy skepticism which, in this field, recognizes no authority but experimental evidence and which perceives that our generalizations are tentative and subject to change when new facts are unearthed.

The compulsion to brevity in an elementary book tends either to presentations implying an unwarranted degree of certainty in our knowledge or to a mere recording of questioned facts and controversial theories. Conscious of both these mortal sins in the teaching of science, we have tried to avoid them. On the success of that attempt, we invite the candid criticism of our colleagues.

In the main the material is so presented as to be intelligible to the reader who has had no previous training in physics, chemistry, or biology. This more elementary material, under headings not starred, constitutes a consecutive story in itself. In addition, there are sections of the book whose headings are starred (*). These sections delve a little more deeply into the subject matter and constitute an immediately available source of more advanced reading, which may be used as the interests of the reader dictate. Footnotes are also in the category of advanced reading.[1]

Sound films have become an integral part of the teaching of elementary science at the University of Chicago. A number of illustrations in the text are taken from some of the films in the biology series. We are indebted to Erpi Classroom Films, Inc., and to Mr. James Brill and Mr. Melvin Brodshaug of that organization, for help in selecting and preparing appropriate scenes from the films.

For reading parts of the manuscript and for valuable suggestions we are indebted to Drs. Paul Cannon, R. W. Gerard (especially in the planning of chap. 2), Harold Gulliksen, N. Kleitman, A. B. Luckhardt, F. C. McLean, Herluf Strandskov, and Howard Swan— all of the University of Chicago. We are grateful to Dr. William Bloom and to Saunders and Company for permission to redraw a number of the illustrations from Maximow and Bloom's *Text-Book of Histology.*

We have received invaluable aid from Dr. Merle Coulter, whose long experience in the teaching of science and generous expenditure of time and energy have helped us at every stage in the planning and preparation of the manuscript.

ANTON J. CARLSON
VICTOR JOHNSON

CHICAGO, ILLINOIS
    August 1937

1. Division of the subject matter into relatively elementary material (not starred) and more advanced reading (starred [*]) was discontinued in the fourth edition, 1953, upon the advice of several readers.

PREFACE TO
THE SECOND EDITION

Advances in our knowledge and welcome criticism and suggestions from students and colleagues have prompted numerous changes in the interest of clarity and accuracy. We trust that the usefulness of the book will be enhanced by these changes, by the new chapter on reproduction, by the new material on alcohol, tobacco, and the viruses, and also by the new illustrations selected from the latest biology sound films made at the University of Chicago.

A. J. C.
V. J.

January 1941

PREFACE TO
THE THIRD EDITION

It has been gratifying to the authors to continue to receive numerous suggestions from investigators, teachers, and students for improving the usefulness of this volume. Many such suggestions have been employed in the preparation of this latest edition.

In the years elapsing since the previous edition research has steadily added to our knowledge of how the human body functions in health and in disease. World War II, while interfering seriously with the training of scientists, did stimulate important advances in the sciences of medicine and physiology. The more significant of these are incorporated into this revision, including our newer knowledge of blood transfusions and the use of blood banks in wartime and in peacetime, of nutrition and the vitamins, of the employment of radioactive elements in medicine, and of the control of infections with such new drugs as the sulfa compounds and penicillin. Suitable illustrations have been added, and numerous minor changes have been made.

A. J. C.
V. J.

May 1948

PREFACE TO
THE FOURTH EDITION

This fourth edition of *The Machinery of the Body* is the most comprehensive revision of the book since its original publication in 1937. There has been extensive reorganization of much of the material: a great deal of the subject matter in the former chapter on the cell has been incorporated into other sections of the book. The physiology of muscle and nerve cells, for example, has been integrated into a later chapter dealing in detail with muscles and nerves.

New material has been added on the scientific method; on the bodily injuries which result from the blast, the heat, and the radiation of atomic-bomb explosions; on the employment of radioactive isotopes in such physiological research as determining the life-span of red and white blood cells and in the control of thyroid-gland cancer; on the antibiotics and their more recently discovered limitations; on biological warfare; on the important new immunizing effects of gamma globulin in combating poliomyelitis; and on blood banks and blood donations.

This edition also includes our newer knowledge of the deranged physiology in pernicious anemia and the role of vitamin B_{12} in this disorder. The modern method of artificial respiration—more effective than older procedures—is described and newly illustrated. Included also are new presentations of vitamins, brain waves, glands of internal secretion and their hormones (including cortisone), effects of tobacco, traumatic shock, the nerve impulse, pregnancy tests, and emotional factors in disease.

These more important additions, as well as numerous minor alterations, are derived from the ever increasing understanding of physiology resulting from research. All the changes are intended to bring the book up to date and to increase its interest, accuracy, and clarity, while still preserving the primary object of the volume: to provide the reader with a basic understanding of how man functions in health and in disease, to indicate how we derive such information by experimentation, and to stimulate the reader to seek further knowledge.

There are new illustrations in the chapter on endocrine glands.

Upon the advice of several colleagues, the former practice of dividing the book sections into the more elementary and the relatively advanced has been discontinued.

The comments and suggestions of readers of *The Machinery of the Body* have been invaluable guides in revisions of the volume. In the present edition the authors are particularly indebted to the painstaking criticisms submitted by Drs. Fred E. D'Amour, of the University of Denver, Victor E. Hall, of the University of California at Los Angeles, and Eleanor M. Larsen, of the University of Wisconsin, who advised many improvements but are absolved from responsibility for such defects as remain.

A. J. C.

V. J.

June 1953

TABLE OF CONTENTS

LIST OF ILLUSTRATIONS xv

LIST OF TABLES xxi

1. THE SCIENCE OF PHYSIOLOGY 1

Physical and biological science—properties of living things—living and nonliving—virus—attitude of the biological scientist—physiological machinery—evolutionary adaptations to the environment—the scientific method—controls—experimentation—similarity of man and other animals—health and disease—the cell in physiology—structure and interrelationships—tissues of the body—studies upon cells

2. THE ORGANIZATION OF PROTOPLASM 22

Composition of protoplasm—organic and inorganic compounds—significance of water and salts—starches, fats, and proteins—role of enzymes in life—colloidal states—metabolism and energy—diffusion and filtration—permeability of cells—osmosis

3. BLOOD AND THE INTERNAL ENVIRONMENT 62

Evolution of large animals—role of the circulating fluid—the environment of cells—cells and liquid of blood—volume of blood—composition of plasma—water balance—clotting of blood—role of blood platelets—automatic repair of damaged vessels—delayed clotting—structure and function of red blood cells—destruction by hemolysis—transplantation of tissues—numbers of red cells in the blood—life-history—regulation of red blood cell formation—physiological constants—anemia—blood transfusions—cross-matching—blood types—varieties of white blood cells—defense against infection—repair of damaged tissue—fluctuations in number of white blood cells—the blood platelets—blood damage in atomic explosions

4. THE WORK OF THE HEART 126

Anatomy of the circulatory system—discovery of blood circulation—William Harvey—the evolution of the mammalian heart—chambers and valves—microscopic structure—graphic methods of study—contractions of the heart—refractory period—all-or-none response—electrical variations—cause of the automatic beating—sequence of events in a single beat—pressure changes—heart sounds—the electrocardiogram—disorders of the heart—control of strength of heartbeat—control of heart rate—physical, chemical, and nervous factors—the problem of inhibition—reflexes involving the heart

5. Blood Flow and Blood Pressure   165

Anatomy of arteries and veins—significance of capillaries—rate of
blood flow—slow and constant capillary flow—resistance to flow and
blood pressure—return of blood to the heart—blood flow to organs—
stopcocks—correlation of blood flow with tissue activity—chemical
and nervous influences—carbon dioxide and epinephrine—the vasomotor system—determination of arterial blood pressure—Stephen
Hales—the mercury manometer—cyclic fluctuations in pressure with
the heartbeat—influence of pumping of heart on blood pressure—
role of resistance to flow—nervous regulation of blood pressure—
physiological constancy of blood pressure—effect of blood volume
and gravity—the circulation in physical exercise—disorders of the
circulation—the lymph vessels and their functions

6. Mechanisms and Functions of Breathing   215

Energy liberation and life—history of the breathing apparatus—
gross and miscroscopic anatomy of the lungs—mechanics of inspiration and expiration—pressure changes in lungs and chest cavity—
quantity of air breathed—differences in inspired and expired air—
the breathing muscles and their motor nerves—control of breathing
by a nerve center in the brain—reflexes modifying breathing—chemical control of breathing—role of carbon dioxide—adaptations of
breathing rate to varied physiological conditions—use of oxygen in
exercise—artificial respiration—vocalization and other modifications
of breathing—influence of breathing on blood and lymph flow—exchange of oxygen between lungs and blood—transportation of oxygen in the blood—role of hemoglobin—carbon monoxide poisoning—
transportation of carbon dioxide—regulation of blood neutrality—
cell respiration—disorders of respiration—correlated adjustments of
circulation and respiration—physical training

7. The Work of the Alimentary Canal   264

Modification of food in digestion—relationship of the circulatory and
digestive systems in evolution—organs and glands of digestion—nature
of secretion—chemical digestion by saliva—gastric digestion—action
of bile and pancreatic juice—bacteria of the large bowel—reflex control of salivary secretion—chemical control of pancreatic secretion—
factors modifying gastric secretion—study of digestive movements—
peristalsis—swallowing—movements of the stomach and intestines—
regulation of swallowing by nerves—protection against entrance of
food into the lungs—nerves of stomach and intestines—control of
emptying of stomach—diarrhea and constipation—control of gastrointestinal movements by bulk—emptying of the rectum—appetite,
hunger, and thirst—absorption of food and water into the blood—
indigestion—ulcer and cancer—appendicitis—gallstones

8. The History of Foods in the Body   309

Absorption, storage, and utilization of carbohydrates and fats—
combustion of proteins—urea formation—interconversion of foods in

the body—animal heat—Lavoisier—measurement of heat production—indirect determinations—factors modifying heat production—
basal metabolic rate—influence of size, age, and sex—role of the thyroid gland—the thermostat of the body—regulation of heat production—control of heat dissipation—reflex regulation of body temperature—fever—diet and nutrition—energy requirements of the body—
quantity and kinds of proteins—animal and vegetable proteins—
water and salts—the vitamin-deficiency diseases—vitamins in foods
—roughage and variety in the diet—alcohol and tobacco—elimination of wastes—the work of the kidneys—machinery of urine secretion—derangements of kidney function

9. THE ACTION OF MUSCLE AND NERVE 367
Adjustment of animals to external environmental changes—skeletal
movements—mechanical changes—irritability—refractory state—
all-or-none principle—gradations of contractions—the muscle twitch
—tetanic contraction—muscle tone—energy for the contraction
process—role of combustion—muscle fatigue—effects of exercise—
visceral muscle—nerve stimulation—nerve impulse—speed of transmission—all-or-none law—refractory state—all nerve impulses
alike—nature of effective stimuli—membrane theory of nerve conduction—action and injury currents—polarization of the membrane
—Lillie's iron-wire model of nerve—the functions of the nerve sheaths
—degeneration and regeneration of nerve—the law of Bell and Magendie—autonomic nervous system—drug experimentation—inhibition
—chemical mediation of nerve influences

10. MECHANISMS OF CORRELATION—THE SPINAL CORD AND BRAIN . . 414
Integration of body parts—evolution of the central nervous system—
reflex action—complexity of spinal reflexes—role of the stretch reflexes—posture and the modulation of movement—characteristics of
conduction through the reflex arc—inhibition—reciprocal innervation of muscle groups—chemical theory of conduction across nerve
synapses—nerve pathways of the spinal cord—reflex centers and
pathways of the brain stem—the brain of man and lower animals—
structure of the cerebral hemispheres—surface configuration—localization of function—motor and sensory areas—interpretation of sensations—role of the cortex in behavior—nature of the learning process—conditioned reflexes—sleep—the functions of the cerebellum—
mental and emotional abnormalities—insanity

11. SENSORY MECHANISMS 466
The "five senses"—distance receptors—structure of the eye—the
lens system and image formation—adjustments of the eye to near and
far vision—nearsightedness and farsightedness—cataract—the functions of the iris—judgment of distance—stereoscopic vision—rods
and cones of the retina—night blindness—color vision—eye injuries
and infections—anatomy of the ear—the cochlea and its stimulation
by sound waves—defects of hearing—sense of balance and rotation—

body-righting reflexes—the semicircular canals—taste and smell—
the skin senses—pain—Weber's law of sensation

12. CHEMICAL CORRELATION: THE GLANDS OF INTERNAL SECRETION .   .   502
   The endocrine glands and their secretions—experimental studies—
   gland abnormalities in man—the pancreatic islets in diabetes—discovery of insulin—the cause of diabetes—defective function of the
   thyroid glands in adults—influence of basal metabolic rate—cretinism—goiter—iodine and the thyroid hormone—overactivity of the
   thyroid—indispensability of the parathyroid glands—tetany—calcium and muscle irritability—control of calcium balance by the parathyroids—discovery and action of epinephrine—physiological role of
   the hormone—the emergency theory—Addison's disease and the
   adrenal cortex—experimentally produced insufficiency—the cortical
   hormone—cortisone—ACTH—regulation of salt balance—reproduction and the ovaries and testes—the female reproductive cycle—
   chemical control by the ovaries—the follicular and luteal hormones—
   the endocrine activity of the testes—the posterior lobe of the hypophysis—control of growth by the anterior lobe—gigantism and
   dwarfism—control of gonad function—gonard-sitmulating hormones
   —secretin and gastrin—the liver—thymus and pineal body—nervous
   and chemical control of the endocrine glands—interrelations of the
   endocrine glands

13. BODY DEFENSES AGAINST DISEASE   576
   Emergency functions of the body machinery in disease—margins of
   safety—functions of pain—bacterial disease—natural immunity—
   prevention of bacterial entry into the body—inflammation—phagocytosis—lymph vessels and nodes—acquired immunity—Pasteur's
   experiments—artificial immunity—vaccination—chemicals employed
   in combating infections—biological warfare

14. REPRODUCTION AND EARLY DEVELOPMENT   599
   Cell growth and reproduction—cell volumes and cell surface—cell division—mitosis—vegetative and spore reproduction—sex—differentiation of gametes—reproduction in fish and frog—the land egg—
   mammals—the placenta—outmoded notions—germ layers—recapitulation—reworking old parts—birth—circulatory changes—infant
   nutrition

15. CONCLUSION   623

SELECTED REFERENCES   625

THE AUTHORS   627

INDEX   633

LIST OF ILLUSTRATIONS

1. ANIMAL AND PLANT CELLS 13
2. COMMON TYPES OF CELLS AND TISSUES 15
3. TRANSFORMATIONS OF LYMPHOCYTES IN TISSUE CULTURE 20
4. BEHAVIOR OF ELECTROLYTES IN SOLUTION 28
5. COLLOIDAL STATES 46
6. DIFFUSION 51
7. DIFFUSION IN THE LUNGS 52
8. PERMEABILITY OF CELL MEMBRANES 56
9. OSMOSIS 57
10. SWELLING AND BURSTING OF CELLS IN DILUTE SALT SOLUTION . . 59
11. OSMOSIS AND FILTRATION IN CAPILLARIES 60
12. AMEBA, HYDRA, AND FLATWORM 64
13. SEPARATION OF PLASMA AND FORMED ELEMENTS OF BLOOD . . . 65
14. DISTURBANCES OF THE FILTRATION-OSMOSIS BALANCE IN CAPILLARIES 71
15. SEPARATION OF SERUM FROM CLOTTED BLOOD 74
16. RELATION OF CESSATION OF BLOOD FLOW TO CLOTTING 75
17. MAMMALIAN RED BLOOD CELLS 82
18. OSMOTIC EFFECTS UPON RED BLOOD CELLS 85
19. AGGLUTINATION OF RED BLOOD CELLS 89
20. PIPETTE AND SLIDE USED FOR COUNTING RED BLOOD CELLS . . . 91
21. PHAGOCYTIC CELLS IN THE LIVER 92
22. BONE OF THE LEG SHOWING RED BONE MARROW 92
23. STIMULATING EFFECT OF REDUCED BLOOD-OXYGEN TENSION UPON THE
    RED BONE MARROW 97
24. ANEMIA 100
25. INTERACTIONS AMONG BLOOD GROUPS 107
26. SCHEMATIC REPRESENTATION OF A AND B SUBSTANCES IN RED CELLS 108
27. COMMONER KINDS OF WHITE BLOOD CELLS 116
28. OTHER VARIETIES OF WHITE BLOOD CELLS 116
29. AMEBOID MOTION 119
30. DIAGRAM OF THE CIRCULATORY SYSTEM 128
31. AN EXPERIMENT OF HARVEY ON BLOOD CIRCULATION 130
32. EVOLUTION OF THE FOUR-CHAMBERED HEART 131
33. THE VALVES OF THE HEART 133
34. RELAXATION AND CONTRACTION OF THE HEART, SHOWING VALVE ACTION 134
35. THE CONDUCTION SYSTEM OF THE HEART 135
36. THE KYMOGRAPH USED FOR RECORDING THE HEARTBEAT 137
37. THE REFRACTORY PERIOD OF THE HEART 139
38. THE ALL-OR-NONE LAW 140
39. ELECTRICAL CHANGES IN CONTRACTING HEART MUSCLE 141
40. PRESSURE CHANGES AND VALVE ACTION IN THE HEART 147
41. THE ELECTROCARDIOGRAM 148

42. EFFECT OF TEMPERATURE ON THE HEART 154
43. EFFERENT NERVES OF THE HEART 157
44. VAGUS INHIBITION OF THE HEART 158
45. REFLEX INHIBITION OF THE HEART 161
46. AUTOMATIC GOVERNOR OF THE HEART RATE 163
47. MECHANISMS SUPPLYING BLOOD TO ACTIVE MUSCLES 164
48. STRUCTURE OF ARTERY, VEIN, AND CAPILLARIES 167
49. RATE OF FLOW OF LIQUID THROUGH A SYSTEM WHOSE DIAMETER
    CHANGES   169
50. DIMINUTION IN PRESSURE IN A TUBE THROUGH WHICH WATER FLOWS 173
51. THE GRADIENT OF PRESSURE IN THE CIRCULATORY SYSTEM   . . . 174
52. VALVES IN VEINS 175
53. ADAPTATIONS OF BLOOD FLOW TO RATE OF TISSUE ACTIVITY . . . 177
54. EAR OF A RABBIT SHOWING BLOOD-VESSEL REACTIONS 181
55. THE MERCURY MANOMETER 184
56. DETERMINATION OF BLOOD PRESSURE IN MAN 187
57. EFFECT OF VAGUS STIMULATION ON ARTERIAL BLOOD PRESSURE . . 189
58. EFFECT OF CUTTING VAGI ON ARTERIAL BLOOD PRESSURE 190
59. EFFECT OF EPINEPHRINE INJECTION ON ARTERIAL BLOOD PRESSURE . 190
60. VISCOSITY, RATE OF FLOW, AND VESSEL CALIBER 192
61. THE VASOCONSTRICTOR CENTER AND NERVES 193
62. THE CAROTID SINUSES AND THEIR AFFERENT NERVES 198
63. CAROTID SINUS EFFECTS ON ARTERIAL BLOOD PRESSURE 199
64. REFLEX ACTION OF AORTIC DEPRESSOR NERVE 201
65. EFFECT OF HEMORRHAGE AND OF TRANSFUSION ON ARTERIAL BLOOD
    PRESSURE 204
66. BLOOD CIRCULATORY AND LYMPHATIC SYSTEMS 211
67. LYMPH VESSELS AND LYMPH NODES OF THE ARM 213
68. LYMPH NODE 214
69. ESSENTIAL FEATURES OF ALL ORGANS OF RESPIRATION 218
70. THE BREATHING PASSAGES IN MAN 219
71. METAL CAST OF AIR SPACES AND PASSAGES OF THE LUNGS 220
72. MICROSCOPIC APPEARANCE OF LUNG TISSUE 221
73. DIAPHRAGMATIC MOVEMENTS IN BREATHING 222
74. RIB MOVEMENTS IN BREATHING 222
75. MECHANICAL DUPLICATION OF BREATHING MOVEMENTS 225
76. PRESSURE CHANGES IN THE LUNGS 226
77. APPARATUS FOR RECORDING BREATHING MOVEMENTS 227
78. RECORDING THE INTRATHORACIC PRESSURE CHANGES IN MAN . . 228
79. INTRATHORACIC PRESSURE CHANGES 228
80. MECHANICS OF INTRATHORACIC PRESSURE CHANGES 229
81. VOLUMES OF INSPIRED AND EXPIRED AIR 233
82. NERVES OF BREATHING 235
83. REFLEX INHIBITION OF BREATHING 237
84. EFFECT OF CUTTING THE VAGI ON BREATHING 238
85. EFFECT OF CARBON DIOXIDE DIMINUTION ON BREATHING . . . 242
86. BACK-PRESSURE–ARM-LIFT METHOD OF ARTIFICIAL RESPIRATION . . 245

87. The Larynx and the Vocal Cords 247
88. Cells of Intestine and Alveoli Compared 251
89. Gas Tensions in Liquids 255
90. Tensions of Respiratory Gases in the Body 256
91. Curves of Dissociation of Oxyhemoglobin 256
92. Organs and Glands of the Digestive Tract 267
93. Gastrointestinal Organs of Man 268
94. Structure of the Intestinal Wall 269
95. Visible Changes in Secretory Cells 271
96. The Salivary Glands in Man 273
97. Digestion of Starch by Saliva 274
98. Digestion of Protein by Gastric Juice 275
99. Control of Salivary and Pancreatic Secretion 281
100. The Pavlov Pouch 283
101. Peristalsis in the Esophagus 285
102. Peristalsis in the Stomach 286
103. Tracing of Swallowing 287
104. Peristalsis of the Stomach 288
105. Churning Movements in the Small Intestine 289
106. Passageways for Food and Air 293
107. Hunger Contractions 300
108. The Vermiform Appendix of Man 307
109. Storage of Carbohydrate in the Liver 312
110. Chamber for Direct Determination of Heat Production . . 320
111. Apparatus for Indirect Determination of Heat Production . 323
112. Factors Modifying Heat Production 324
113. Normal Basal Metabolic Rates 327
114. Relationship of Caloric Intake and Output to Body Weight . 329
115. Body-Temperature Fluctuations in Turtle and Rabbit . . . 331
116. Body-Temperature Fluctuations in Man 337
117. Effect of Inadequate Protein Intake 339
118. Effect of Salt Deficiency in the Diet 342
119. Vitamin A Deficiency 344
120. Experimental Rickets 346
121. Rickets in the Child 347
122. Functioning Isolated Kidney 357
123. The Urinary System 358
124. A Secreting Unit of the Kidney 359
125. Tracing of Urine Flow from the Kidney 364
126. The Human Skeleton 369
127. Flexor and Extensor Muscles of the Arm 370
128. Changes in Skeletal Muscle Fiber during Contraction . . 371
129. The Staircase Phenomenon 374
130. Single Twitch of a Frog Muscle 376
131. Tetanus in Skeletal Muscle 377
132. Fatigue in Skeletal Muscle 383
133. Electrical Changes in Active Nerve 386

134. Nerve-Muscle Preparation 387
135. Ignited Powder Fuse and Conduction of the Nerve Impulse
     Compared 388
136. Isolation of Single Nerve Fibers 390
137. Polarization of Nerve Fibers 393
138. Transmission of the Nerve Impulse 394
139. Current of Injury and Current of Action 396
140. Electrotonus 397
141. Stimulation of Nerve by Anode and Cathode 398
142. Lillie's Iron-Wire Model of Nerve 399
143. The Sheaths of Nerve Fibers 401
144. Conduction in the Iron Wire 402
145. Spinal Cord and Nerve Roots 404
146. Cross-Section of Spinal Cord, Showing Nerve Roots . . . 404
147. Autonomic Nerves and Ganglia 408
148. Autonomic Innervation of Some of the Viscera 409
149. Effect of Drugs on the Vagus Nerve Endings 410
150. Evolution of the Central Nervous System 418
151. Essential Components of a Reflex 419
152. The Flexion Reflex in a Frog 420
153. Complex Reflex Behavior 422
154. Reciprocal Innervation of Antagonistic Muscles 430
155. Comparison of a Reflex Center and the Sinus Node of the Heart 432
156. Sensory and Motor Pathways between Brain and Periphery . 437
157. Nerve-Fiber Tracts of the Spinal Cord 438
158. Embryonic Development of the Human Brain 439
159. The Right Cerebral Hemisphere of Man 448
160. The Brain in Vertical Section 448
161. The Curve of Learning 455
162. The Curve of Forgetting 455
163. The Eyeball in Its Socket 468
164. Structure of the Eyeball 469
165. Image Formation by a Glass Lens and by the Eye 470
166. Ciliary Body, Iris, and Lens Attachment 472
167. Projection, as Demonstrated by Scheiner's Experiment . . 473
168. Errors of Refraction and Their Correction 475
169. The Optic Pathways 478
170. Stereoscopic Vision 479
171. Rods and Cones of the Retina 480
172. Demonstration of the Blind Spot 481
173. Structure of the Ear 484
174. The Organ of Hearing 486
175. The Nonacoustic Labyrinths 489
176. Sense Organs of Static Equilibrium 490
177. Equilibrium Organ of the Crayfish 490
178. Sense Organs of Dynamic Equilibrium 493
179. Sense Organs of Taste and Smell 495

180. Cold Spots on the Finger 499
181. Sensory Organs in the Skin 499
182. End Organ of Muscle Sense 500
183. Locations of the Glands of Internal Secretion 505
184. Gland Tissues of the Pancreas 509
185. Injection of Insulin 513
186. The Thyroid Gland 517
187. Effect of the Thyroid Hormone on Oxygen Consumption . . 518
188. Hypothyroidism in the Adult 519
189. Hypothyroidism in the Young 520
190. Normal and Cretin Rabbits 521
191. An Adult Cretin 523
192. An Unusually Large Goiter 526
193. Causal Relationships of Goiter to Hypo- and Hyperthyroid
     Function 527
194. Involuntary Bulging of the Eyes—Exophthalmos 527
195. The Parathyroid Glands 531
196. Parathyroid Tetany 532
197. The Adrenal Gland 536
198. The Genital System of the Human Male 544
199. Genesis of Sperm in the Human Testis 545
200. The Genital System of the Human Female 547
201. Follicle and Corpus Luteum in the Ovary 547
202. Pituitary Gigantism in Man 561
203. Experimental Dwarfism 562
204. Hormone Control of Milk Secretion 567
205. Control of the Pancreas by Secretin 568
206. Disease-producing Bacteria 580
207. Injury to the Skin Providing a Portal of Entry for Bacteria . 583
208. Ciliated Cells 584
209. Behavior of White Blood Cells in Defense against Infection . 587
210. Photograph of a Large Phagocyte 588
211. Prevention of Spread of Bacteria 589
212. Clumping of Bacteria by Antibodies 592
213. Cell Division in Nonnucleated Cells 602
214. Cell Division in Nucleated Cells: Mitosis 602
215. Human Sperm and Egg 605
216. Embryonic Membranes in Man and in the Land Egg . . . 606
217. The Placenta 608
218. Formation of the Nervous System from Ectoderm 611
219. Formation of the Neural Tube in the Pig 611
220. Pharyngeal Gill Slits and Arches 615
221. Embryonic Development of the Human and the Pig Embryos . 616
222. Birth of the Young 617
223. Circulatory Changes at Birth 619

LIST OF TABLES

1. COMPARISON OF THE RELATIVE CONCENTRATIONS OF CERTAIN COMMON SALT IONS IN SEA WATER AND IN THE BODY FLUIDS OF CERTAIN ANIMALS   30
2. MAIN CONSTITUENTS OF THE BLOOD PLASMA   72
3. DIFFERENCES IN COMPOSITION OF INSPIRED AND EXPIRED AIR . . . 234
4. SUMMARY OF THE MAIN FEATURES OF THE CHEMICAL FACTORS IN DIGESTION 276
5. EFFECT OF CONCENTRATION OF SALT ON THE ABSORPTION OF SALT AND WATER FROM THE INTESTINE 303
6. HEAT GENERATED BY THE BURNING OF FOODS 321
7. RELATIONSHIP OF SIZE OF ANIMALS TO RATE OF HEAT PRODUCTION . 326
8. REFLEX ADJUSTMENTS TO EXTERNAL HEAT OR COLD 335
9. PROTEIN REQUIREMENT OF A DOG WITH PROTEIN AS THE SOLE SOURCE OF ENERGY 340
10. THE MOST IMPORTANT VITAMINS, THEIR MAIN SOURCES, AND THE CHIEF VITAMIN-DEFICIENCY CONDITIONS 348
11. THE MAIN STRUCTURES INNERVATED BY THE TWELVE CRANIAL NERVES 440
12. EFFECTS OF UNDERACTIVITY AND OVERACTIVITY OF THE THYROID GLANDS 528
13. THE REPRODUCTIVE CYCLE IN THE HUMAN FEMALE 548
14. MAIN DERIVATIVES OF THE THREE EMBRYONIC GERM LAYERS . . . 610
15. THE RECAPITULATION PRINCIPLE 614

CHAPTER ONE

THE SCIENCE OF PHYSIOLOGY

I. BIOLOGY: THE STUDY OF LIFE
   A. Properties of life
      1. Living things metabolize   2. Living things grow   3. Living things reproduce   4. Living things become adapted to their environment   5. Living things are highly organized
   B. Living and nonliving

II. ATTITUDE OF THE PHYSIOLOGIST
   A. Mechanisms
   B. Adaptive significance

III. THE SCIENTIFIC METHOD
   A. Common errors of observation and reasoning
   B. Controls; statistics
   C. Experimentation

IV. THE CELL IN PHYSIOLOGY
   A. Discovery of the cell
   B. Structural features
   C. Tissues, organs, and systems
   D. Kinds of cells and tissues
      1. Epithelium   2. Muscle   3. Nerve   4. Connecting and supporting tissue
   E. Methods of study of the cell
      1. Chemical analysis   2. Microscopic examination   3. Micromanipulation   4. Isolated living cells   5. Tissue culture   6. Perfusion   7. Other methods

I. BIOLOGY: THE STUDY OF LIFE

Physiology is one of the *biological sciences*. Both the biological and the *physical sciences* deal with the objective phenomena encountered in nature. At first glance it would appear that the problems faced in these two divisions of science are fundamentally different. The materials they work with seem to be in distinct categories. The physical sciences deal with the nonliving matter and forces in the world and the universe. Biology is the science of life, and the objects of its study display certain properties which we tend to consider as peculiar to living things. These may be summarized as follows.

A. Properties of Life

1. *Living things metabolize.*—They take in materials from their surroundings, change them chemically, and convert them into new products. There exists in the living complex a dynamic equilibrium, involving the passage of a constant stream of materials and energy through the living system.

2. *Living things grow.*—From the materials metabolized is manufactured new life-stuff, possessing properties specific for the living, synthesizing agent. In growth, living substance produces more of itself.

3. *Living things reproduce.*—Characteristically, there is a progressive increase in the number of units constituting the entire individual and also the production of new individuals; a variety of methods for effecting these is encountered.

4. *Living things become adapted to their environment.*—Adaptation not only is the long-time process involving evolutionary changes extending over countless generations but is a matter also of constant readjustment to environmental changes by each individual. Living things are sensitive to changes in the environment and are capable of reacting and adjusting more or less successfully to them. Irritability to stimulation is a universal accompaniment of life.

5. *Living things are highly organized.*—Organization implies interdependence and interaction, controlling and subordinate parts, regulation and co-ordination, integration of constituent parts into a well-knit unit or whole. This feature of life is perhaps its most outstanding character, and its importance is stressed by the very name of the unit in biology—the *organism*.

In these general ways, then, the materials with which the biologist works seem to be different from those of the physical scientist. Yet it has been pointed out frequently that each of the characteristics of life mentioned is paralleled in one or another nonliving system. A candle flame has a metabolism of a sort—consuming wax and oxygen and liberating carbonic acid gas and water. It displays the phenomenon of movement and can produce more of itself in suitable environments. Clouds grow and multiply in number. Recording and measuring devices possess a kind of irritability, and machines and motors have a high degree of organization and integration. On the whole, however, it is difficult to mention any one physical system which possesses many of the properties of living things.

B. Living and Nonliving

We must bear in mind that, so far as present knowledge goes, there is no sharp line of demarcation between living and nonliving. The boundary between the two must be considered rather in the nature of a gradual transition, as is true with the division of biological objects, in turn, into plants and animals. The extremes are readily differentiated, but the existence of transitional phases must be recognized. There are realms in biology in which it is difficult to say whether the phenomena observed are those of a living or a nonliving system, as, for example, in the field of *virus* studies in bacteriology.

There is a disease of tobacco plants which produces a characteristic pattern on the leaves—tobacco mosaic disease. This disorder is somewhat like a bacterial infection, for one can infect a healthy leaf by placing upon it a minute quantity of material from an infected leaf. But ordinary microscopic examination of this infectious material reveals nothing; there are no bacteria, no organisms, no cells. One can even press out the juice from the material, pass it through fine filters, and infect healthy leaves with the filtered juice. We deal here with a tiny agent which seems alive, since, like the larger bacteria—admittedly alive—it grows, reproduces, and causes an infectious disease.

What is the infectious agent? It is called virus, or *filterable virus*, because it can pass through the pores of fine filters. Investigations have shown the tobacco mosaic virus to be an extremely complex chemical known as a *protein* (see p. 36). Solid crystals of the protein can be obtained from the infectious juice in which it is normally dissolved. Kept in a test tube, these crystals display no more of the properties of life than so many crystals of table salt. But a crystal placed on the healthy tobacco leaf produces the disease. Note that, as the disease spreads over the leaf, the virus protein grows and reproduces; *it manufactures more of itself*[1]—one of the prime characteristics of life.

Some biologists think that in virus protein we have one of the first stages in the origin of life: a transition form between living and nonliving, in which a number of life-properties are displayed by a pure chemical. Perhaps this represents a stage of evolution preceding the origin of the relatively more complex and microscopically visible cell, which is the unit of structure for most living things.

1. It is also suggested that perhaps the leaf manufactures the virus protein. But clearly the virus itself is the stimulus for this synthetic reaction.

Interestingly enough, other biologists say no; far from being an early stage in the origin of life, viruses are highly degenerate forms, with an immeasurably long evolutionary history. Such scientists emphasize the fact that viruses cannot live alone. They can live only in living cells. Virus protein in a test tube is dead. It springs to life only in contact with live cells. It is a *parasite*.

Now, it is common for parasites to degenerate in the course of evolution. The tapeworm, evolved from forms which had a digestive tract, has lost this system. This worm has no need to digest its food; it lives in its food, already digested by the host. Many parasites have no sense organs, even though their evolutionary ancestors did. They have no need for sense organs to detect food or enemies. Perhaps viruses are an extreme form of parasitic degeneration. Perhap they have lost not only digestive and locomotor and sensory mechanisms but even cell structure. The host cells which harbor them perform all the labors of living, leaving this degenerated molecular form of life only the task of reproducing itself.

What is the true answer? Are viruses the beginning of life—of evolution? Are they the end stage of a long evolutionary development? The answer lies in the future—in the laboratory and the mind of some scientist.

When investigative methods are considered, we find that biological and physical sciences are very closely akin. Though the materials he observes may be peculiar in certain respects, the biologist employs, in general, the same methods as the physical scientist. The approach in each case is experimental wherever possible. The same precautions are employed in making observations, and there is in both cases the same attempt to quantitate objective phenomena and discover significant causal relationships. There is the common aim of trying to understand nature and, if possible, to control it to suit the ends of man.

The working hypothesis of the biologist is that eventually the phenomenon of life will be explained in terms of physics and chemistry. He does not dogmatically assert that this will surely come to pass. He simply governs his scientific actions and devises and conducts his experiments as though this will be the final culmination of biological research. The justification for such a point of view is twofold. First, there is a certain amount of direct confirmation of the hypothesis. It has been shown time and time again that processes thought to be absolutely dependent upon living organisms can be duplicated in the test tube. Second, assumption of this hypothesis has led to very fruitful results. Even if ultimately it should be shown that life-phenomena are fundamentally different from the phenomena of chemistry and physics in a qualitative sense, nevertheless it is true that a better understanding of many life-phenomena has resulted from the adoption of this working hypothesis. So manifold are the purely chemical and physical reactions in living organisms that we now have in biology the daughter-sciences of *biochemistry* and *biophysics*.

II. ATTITUDE OF THE PHYSIOLOGIST

Biochemistry and biophysics are offshoots of the science of physiology, which seeks to explain the underlying machinery of the life-processes of the organism. They have resulted from the attempts of the physiologist to explain the functions of the body in terms of physics and chemistry. Biochemistry and especially biophysics are young sciences but are expanding rapidly and encompassing in their fields of investigation more and more of the phenomena once thought to be peculiar to systems possessing the indefinable property of life.

The adoption of the same attitude in physiology as seems natural in the physical sciences is not simply a matter of deciding to do so. It is rather a process of learning, and, like all learning, it is slow and painful. Though the point of view is presented at the outset, the acquisition of it by the student of physiology is for the most part a gradual process—an important by-product of the contemplation of many specific physiological situations.

A. MECHANISMS

Let us cite an example of the difficulty of approaching physiological problems from the same point of view as problems in the realms of pure physics or chemistry might be approached. A thermostat is a physical instrument by means of which the temperature of a room is kept more or less constant. Suppose that, quite ignorant of how a thermostat operates, you put this question to its owner: "How is it that, as soon as the temperature of this room falls, your gas furnace is turned on?" You may get this reply: "Why, that happens in order to keep the room warm and comfortable. You see, if this room were to get too cool, we'd all be uncomfortable; and, so, to prevent this, the furnace goes on at the right time." You would conclude that this man might know something about the utility of his thermostat but that he was quite ignorant of how it operates. You would say that, by his reply, the man imputes a sort of intelligence to a purely physical system. Of course, a reply such as that given would be so unusual that you might also conclude that the man either misunderstood your question or else was simple-minded, to say the least.

Yet students of physiology, and not always beginners in the field, will sometimes confuse, in exactly this manner, the *utility* of a physiological reaction and the physiological *machinery* for bringing it about. In explanation of why breathing is accelerated in exercise, this is sometimes said: "It happens because the active muscles need more oxygen or in order to rid the body of the excess of carbon dioxide produced by the muscles." This is an expression simply of the utility of the accelerated breathing. The physiologist must interest himself in the specific physiological machinery—nervous, chemical, physical—which is automatically set into motion by activity of muscles and whose end result is an accelerated rate of breathing. He is interested in *mechanisms*. By mechanism, then, is meant an automatically operating sequence of physiological events set into motion by some environmental change and resulting finally in a readjustment negating or counteracting the effects of the initial environmental change. The full significance of the term *physiological mechanisms* will become apparent only with the multiplication of examples in all that is to follow.

B. ADAPTIVE SIGNIFICANCE

It is, of course, quite legitimate to consider the utility to the organism of any physiological reaction. Such consideration, in fact, occupies an important place in biology in studies of the adaptation of an organism to its environment. Reactions which are useful to the organism may be of survival value in evolution. They represent evolutionary adaptations to the environment, and, as such, their usefulness is an important biological consideration. A useful, or "purposive," reaction, such as the withdrawal of a limb from an injurious stimulus, is of tremendous adaptive significance; and forms which failed to develop the adaptation were eliminated by natural selection in the struggle for existence. Again, the physiological response to loss of blood is an accelerated production of blood constituents— a reaction of obvious utility. Because this reaction is useful, it may be looked upon as an adaptation of animals to an environment in which loss of blood is a not uncommon occurrence.

But knowledge of usefulness and of evolutionary significance leaves unanswered the question of the physiological machinery involved. Exactly what is it about loss of blood which stimulates the blood-forming organs? What *makes* them increase their activity in such circumstances?

Furthermore, to say that a reaction occurs because it is needed dismisses the phenomenon from further analysis and closes the door to investigation and possible control. Only when the interlocking chain of reactions involved is understood do we appreciate the true significance of a physiological process, and only then are we enabled to interfere intelligently when the physiological machinery is damaged by disease.

Of course, the physiologist is not always successful. He frequently fails entirely to answer the question, "What is the mechanism here; what underlying physiological machinery is operating?" And he rarely answers the question completely. Usually, he must admit that here or there are cogs in the machine, events in the sequence of causal relationships, which he cannot at present explain.

III. THE SCIENTIFIC METHOD

Our dependable knowledge about man and the world is derived from precise observations and well-devised experiments. In some scientific fields, such as astronomy, the primary sources of our information are measurements, recording of natural phenomena, and mathematical calculations. In other scientific areas, such as physiology and medicine, it is easier to manipulate the forces at play in experimentation.

To devise an experiment which will provide an observer with a reasonably sound answer to the problem he seeks to solve requires special training and experience. With a little thought, certain of the errors of the untrained and inexperienced can be avoided.

A. COMMON ERRORS OF OBSERVATION AND REASONING

An error may result from mistaken diagnoses. If cancer is diagnosed and treatment instituted, and the patient survives and is well ten or twenty years later, he may conclude that the treatment cured the cancer. Such errors are common in testimonials of patients offering support to therapeutic claims of many useless drugs. It is assumed that the original diagnosis was correct. It may not have been, particularly if the diagnosis was made by one who is untrained or careless. The aphorism is apt: "Many *diagnoses* have been cured."

Another common reasoning error in everyday life is that of *Post hoc, ergo propter hoc*, which is the Latin for "After this, therefore because of this": because event *B* follows event *A* in time, *A* is the cause of *B*. An example: "I fell asleep on the beach; soon after I awoke, my skin became fiery red." The conclusion that sleeping causes sunburn is no more absurd than many we hear or read every day regarding bodily ailments and the effectiveness of alleged remedies. Query into the claim that "taking these red pills cured me" may reveal that the patient was actually helped *after* taking the red pills but that he also had a vacation during the pill regime. Which was responsible for the improvement, the pills or the vacation?

To avoid the *post hoc* error, one must be on the alert for possible causal factors for the results observed other than that which may appear most prominent.

B. Controls; Statistics

A well-devised experiment seeks to avoid *post hoc* errors by means of *controls*. Two sets of observations are made: one on the experimental group and another on the control group. In testing the effectiveness of a chemical agent against an infection, it is necessary to do more than inject the chemical into some of those who are infected. There must be controls who have the same infection, are in the same environment, and are in every conceivable respect managed identically with those treated with the chemical, including psychological factors: the controls must even *think* they are being treated with the chemical being tested. They must receive injections of some inert substance which looks like the chemical under study. Those who are assessing the therapeutic efficacy of the chemical should indicate which subjects are improved and which are not while they are still completely unaware of which individuals received the treatment and which were controls. Subjected to such study, with the employment of appropriate controls, many cures or remedies would be found to be no more effective than tap water.

False conclusions may still be reached by an experimenter in physiology and medicine despite extreme care in so selecting subjects for study that the control group and the experimental group are as nearly identical as possible. There is a great variability in living material and in the responses of living things to various stimuli. This difficulty may often be overcome by the employment of considerable numbers of subjects. If a drug upon injection into one or a

few dogs causes an elevation of blood pressure, it is unsafe to gen-
eralize that this drug always or nearly always causes an elevation of
blood pressure. Perhaps by accident the few dogs employed were not
representative of dogs in general. Many subjects must be used, many
experiments repeated, before conclusions are safe.

Difficulties in interpreting results may still remain, even when
thousands of subjects are used, as in certain promising experiments
testing the efficacy of a protein fraction of the blood in protecting
against *poliomyelitis*, or infantile paralysis (see also pp. 592–93).
Special mathematical techniques employed in *statistics* may be required to determine whether the differences observed in the experimental and the control groups are probably significant or probably
due to chance.

C. EXPERIMENTATION

The physicist who would learn about electricity must study electrical energy *in action*. True, he wants also to know about the physical
and chemical structure of the battery which can generate electricity,
but he really gets down to work when he closes the circuit and begins
to observe what the *flow* of electricity is and what it does. In biology
much information is derived from studies of the structure and chemical composition of the dead cells or organs of animals. It may even
be possible, sometimes, to form some idea of how these structures
function in the living animal. But any clear notion of how the body
and its parts do their work in life must come from studies of *living*
material. We may as well expect a physicist to work out the principles of electron flow, or induced currents, or motors, or dynamos
from a study of the parts of a battery as for a physiologist to discover the nature of the movements of the stomach, and all the many
factors affecting it, from observations upon the dead organ. To learn
about electricity, the physicist must study the battery at work. In
order that one may learn about the life-activities of the body, the
living body at work must be studied.

This necessity imposes upon the investigator a responsibility
which the physicist escapes. The zeal for knowledge about life must
never cause an infraction of the inviolable rule of the student of
physiology: *Inflict no pain upon the living subjects of your experiments.*

Anesthesia makes this possible. General anesthetics like ether owe
their great usefulness to the fact that they so affect the brain as to

cause unconsciousness and insensibility to pain, without seriously interfering with most other bodily functions. The internal organs carry on; the machinery of breathing and the pumping of blood are unimpaired. Anesthesia has been a great boon to mankind in making extensive surgery possible and in reducing the pains of childbirth. Anesthesia is a boon to the investigator as well, for it enables him to study the internal organs of animals in action, without infliction of pain. Any experiment involving pain or even a great degree of discomfort is done only on animals anesthetized in exactly the same way as human beings undergoing a surgical operation. The misguided and misinformed people who oppose animal experimentation have done incalculable harm to mankind, although the good sense of the people at large has been discrediting and disarming these fanatics to an increasing and gratifying extent.

The student and investigator in physiology can also be reasonably sure that, in general, what he learns about animals from his experimentation will be applicable to man, in whom his main interest centers. The common evolutionary heritage of man and of the higher animals strongly suggests their essential identity. More concrete evidence comes from the experiments themselves. Patient investigation reveals that, in their function as well as in their structure, the organs and systems of man and the other higher animals are fundamentally alike. It was chiefly from the dog, but also from other vertebrates, that man learned the function of his own pancreas and, equally important, how his pancreas might cease to function properly, and how to combat the deleterious or fatal consequences of such malfunctions.

There is an intimate relationship between physiology, which deals with the normal organism, and medicine, which deals with disease. Disease is largely abnormal physiology—a breakdown of the body machinery. An understanding of normal processes is obviously indispensable for the understanding and control of abnormal function by the physician. The physiologist, in turn, learns a great deal about the normal organism from studies of the defects produced in man by disease. In disease nature performs experiments—destroying this organ, crippling that function—from which we may often learn much about the normal activities of the body. Because growth was retarded in people whose thyroid glands were small and abnormal in appearance, it was possible to postulate that at least one function of the normal thyroid glands is the control of growth. Further experimentation on animals fully confirmed this hypothesis and produced much additional knowledge. Thus, throughout this volume, we shall consider many instances of the abnormal because not only are they of interest in themselves but they also help us to understand the normal operation of the living machine.

IV. THE CELL IN PHYSIOLOGY

Why does the mouth "water" when food is chewed? By what mechanisms is saliva poured into the mouth just at the time when food is present to be digested? Experiment reveals that a reflex is responsible, involving the co-ordinated action of a number of living units. Food in the mouth mechanically or chemically stimulates tiny sensitive structures in the tongue called *taste buds*. "Nerve messages" are sent through the nerves of which the taste buds are the sensory terminations. In the brain these messages—more properly termed *nerve impulses*—are relayed to the nerves of secretion, which transmit them to the salivary glands. The gland is stimulated to manufacture and secrete saliva into a duct which carries the juice into the mouth (see Fig. 99, p. 281).

To know this is to know a great deal more about salivary secretion than simply that there is a copious flow of the juice when it is needed. But it is far from knowing the ultimate details of the mechanism. Still unexplained are the activities of the various units constituting the reflex. An understanding of the whole process must await elucidation of how the parts function.

For example, we may ask: What sort of chemical or physical reaction occurs between the foods in the mouth and the parts of the taste buds? What is the nature of the activity initiated here? How does such activity produce nerve impulses in the nerves of the system? Why does this chemical stimulation affect taste buds especially rather than other tissues or other nerve endings in the mouth? What are nerve impulses? Just what chemical, thermal, or electrical changes go on in a nerve which make up the self-propagating disturbance called the nerve impulse? How fast does the impulse travel? By what means is the nerve impulse relayed from sensory nerve to secretory nerve? What physical, chemical, and structural changes occur in a nerve during the period when it is recovering from the effects of transmission of an impulse? What is the nature of the stimulating action of nerve impulses upon the secretory components of the salivary gland? What is secretion? How is it possible for the salivary glands to manufacture products which no other part of the body can manufacture? By what means is the manufactured product secreted into the duct? And, finally, what chemical, physical, and physicochemical reactions are going on in all these structures during the resting stage, maintaining them in readiness for activity?

These are only a few of the questions that we might ask about this single relatively simple process. And all these questions pertain to the unit parts of the reflex. Similar questions will arise repeatedly. In the main we shall be concerned with the functions of organs and systems and the mechanisms for correlating the functioning of the systems into the smoothly operating, well-integrated total organism; with the manner in which the circulatory system is interrelated with the skeletal muscle system and the nervous with the digestive system, etc. But in all this, in the last analysis, we shall be confronted with the problem of the unit part—the *cell*—and how it does its work.

Needless to say, many questions we may raise about cell physiology remain unanswered so far. And, trite though the expression is, it is nevertheless often true that the answer to one question raises a dozen new problems. No matter how much we may learn about the organism or the cell, for a long time to come we shall be confronted with that most fundamental question of biology: "What is life itself?" This question, also, must be answered ultimately in terms of cell physiology by investigation into the activities of cells.

A. The Discovery of the Cell

Over a hundred years ago (1838–39) two German biologists, Matthias Schleiden and Theodor Schwann, first clearly described the cell as the structural unit of plant and animal organisms. The name *cell* was used because the first work (Schleiden's) dealt with plant structure, in which the heavy *cellulose* wall of the unit was prominent. Although the name is also used for the units of animal structure, these are not so clearly cell-like in appearance. They are globular units, possessing a cell boundary or thin membrane rather than an actual cell wall.

It would be impossible to overestimate the importance of this discovery to the sciences of biology. The whole structure of modern zoölogy, botany, embryology, genetics, anatomy, physiology, pathology, and medicine is laid upon this foundation. The cell is the unit of structure, of development, and of function, both normal and abnormal.

B. Structural Features

The general structural plan of the cell proper is approximately the same throughout the whole of the plant and animal kingdoms. Except in bacteria and certain primitive plants, there is an internal differentiation of structure common to all: the division of the cell into the *cytoplasm* and *nucleus* and the presence within the cytoplasm of such *cell organs* as, for example, the green *chloroplast* of plants. Associated with differentiation of structure is also a specialization of function, and we find specific cellular structures concerned predominantly with one or another of the manifold life-activities of the cell. The chloroplast manufactures food for the plant; the

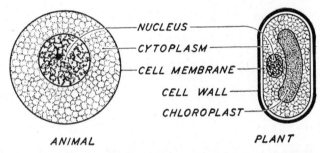

NUCLEUS

CYTOPLASM

CELL MEMBRANE

CELL WALL

CHLOROPLAST

ANIMAL   PLANT

Fig. 1.—Structural plan of animal and plant cells. Animal cells lack the cell wall and the chloroplast, which contains the green pigment, *chlorophyll*.

nucleus controls growth and cell division. In the most primitive of unicellular forms—the blue-green algae and the related bacteria—there is almost no differentiation of internal structure or function. Apparently, the cell as a whole engages in the activities of life, the entire substance of the cell participating in a rather diffuse manner.

In both plant and animal cells the living bit of material is bounded by the *cell membrane*, whose nature and properties will be considered at length later. In addition to the cell membrane, plant cells are encased in the heavy covering, or cell wall, already described.

In unicellular forms a single cell constitutes the entire organism. The evolutionary transition from single-celled plants and animals through colonial forms to the true multicellular organisms is largely a matter of specialization of cells, structurally and functionally, with the development of complex interrelations and interdependence of one type of cell upon another.

C. Tissues, Organs, and Systems

Groups of cells of a similar structure which perform specialized functions constitute what are called *tissues*. Specialization of the cells of multicellular forms in evolution, then, is partly a matter of tissue differentiation. In the bodies of vertebrates, and even in the invertebrate forms, several kinds of tissues are found.

A further specialization, or organization of the various tissues into *organs*, is met with in the higher plants and animals. The heart, for example, consists of several kinds of tissues built into a structural and functional unit. The liver, kidney, stomach, and glands are examples of organs, each of characteristic shape and internal combinations of several tissues.

Finally, we have the organization of organs into unified *systems*. The digestive system, for example, is composed of many organs: the gullet, the stomach, the intestines, and all the associated glands. The circulatory system includes the heart and the blood and lymph vessels. The nervous system includes nerves, the spinal cord, and the organs of the brain.

D. Kinds of Cells and Tissues

The various types of cells constituting the tissues of the body possess the general features common to all cells; but each kind has, in addition, certain structural modifications associated with specialization of function (see Fig. 2).

1. *Epithelium—the covering tissue.*—This is made up of cells which may be thin and flat, or tall and columnar, or any shape intermediate between the two. Epithelial tissue may be a single layer in thickness, or it may be composed of several layers of cells. It is the covering and lining tissue. It covers the exposed surfaces of the body—the skin, the eyes, the lips; it also lines the cavities, tracts, vessels, and ducts in the interior of the body. It lines the digestive tract, the blood vessels, the abdominal and chest cavities, the ducts of glands, as well as making up the secretory cells of glands. In any case, the characteristic thing about epithelial tissue is that one of its surfaces is usually "free," exposed either to the exterior of the body or to the cavity of a hollow structure.

2. *Muscle—the contracting tissue.*—There are three varieties of muscle tissue: the "striped" (or *striated*) muscles, which move the skeletal parts; the "smooth" muscles of the internal organs and blood

vessels; and the muscle of the heart. Striped muscle is composed of bundles of long, thin, cylindrical fibers, arranged parallel to one another, varying in size from 0.01 to 0.1 mm. in thickness and from a few millimeters to several centimeters in length. The fibers are the cells. Rather, each fiber is an organization of several cells whose substance (*sarcoplasm*) has fused into a single unit; each fiber is

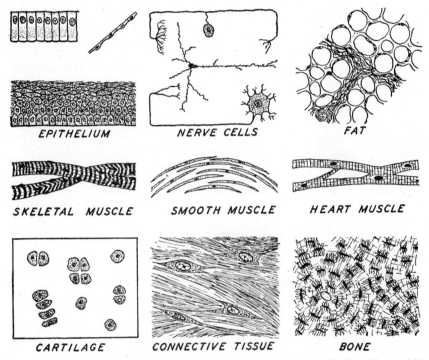

EPITHELIUM   NERVE CELLS   FAT

SKELETAL MUSCLE   SMOOTH MUSCLE   HEART MUSCLE

CARTILAGE   CONNECTIVE TISSUE   BONE

FIG. 2.—The common types of cells and tissues in the human body. The fat globules within the fat cells, the gristle in which cartilage cells are imbedded, and the hard material of bone are shown unshaded. Bone cells (with fibrous processes) are in black. Note the branching of the heart muscle fibers.

multinucleate. The term "striated," or "striped," is derived from the fact that under the microscope one sees alternate light and dark striations, arranged at right angles to the long axis of the fiber. It is also called *skeletal* muscle because, by means of attachments to bones, contractions of these muscles move the parts of the skeleton.

"Smooth" muscle is so named because of the absence of cross-striations. It is, therefore, also called *unstriated* muscle. The cells comprising it are spindle-shaped; the long axes of adjacent cells are parallel to one another, so that a contraction of each cell results in a shortening (and thickening) of the group of cells constituting the tissue.

The heart or *cardiac* muscle somewhat resembles striated muscle in appearance, having similar light and dark striations. In its mode of action it is intermediate between striated and smooth muscle, as will be indicated later.

3. *Nerve—the conducting tissue.*—The characteristic structural feature of nerve cells, or *neurones*, is the extension of long, thin processes, called *dendrites* and *axones*, from the main body of the cell. Bundles of these extensions make up the nerves of the body and the nerve pathways in the brain and spinal cord. Many variations of this general structural cell plan are to be found in various parts of the nervous system.

4. *Connecting and supporting tissue.*—Under this heading is grouped a variety of cells having a common embryonic origin. These tissues are characterized by the large amount of extracellular, or nonliving, materials they contain. The *connective tissue* proper is made up of spindle-shaped cells, individually resembling somewhat the smooth muscle cells. These cells have the peculiar property of manufacturing long fibrous strands, which come to make up the bulk of the tissue. The fibers are interlaced with one another, giving the tissue a tough, fabric-like consistency. The tissue possesses also a degree of elasticity, which varies with the numbers of elastic fibers included in it. Connective tissue, as the name implies, connects the cells of the body to one another and binds together the tissues of all the organs. It is to be found almost everywhere in the body. It binds the nerve and muscle fibers of nerve trunks or muscles into compact bundles and holds together the cells of the internal organs.

If all the other tissues of the body could in some manner be dissolved away, leaving only connective tissue, the body as a whole as well as the many organs of its interior would retain their characteristic shape or outline.

Bone and *cartilage* ("gristle") are living tissues, composed of living cells which are capable of manufacturing the extracellular rigid components[2] of the tissues. It is these cellular products which give bone and cartilage their properties of rigidity and strength, upon which depend their more obvious functions.

Fat tissue is composed of specialized cells which have the ability to take up fat and store it—a single globule within the interior

2. In the case of bone, this consists largely of calcium and phosphorus compounds.

of each cell. The bulk of the cell is simply inert fat, with the living components, including cytoplasm and nucleus, compressed to a thin covering wall surrounding the fat droplet. This gives the cell the appearance of a signet ring when seen under the microscope. Numbers of fat cells are bound together into a more or less well-knit structural mass by the ubiquitous connective tissue. Fat tissue is found especially under the skin, and much of it is located in and about a number of the internal organs.

Finally, there are other cell groups, also of *mesodermal*[3] origin, in which the constituent cells are not bound together into a closely knit tissue. Examples of these are the blood and lymph cells, and cells possessing ameboid properties, which wander about the tissues of the body engulfing foreign particles. These latter, as we shall see, help defend the body against certain infectious diseases by engulfing and destroying the invading bacteria.

E. METHODS OF STUDY OF THE CELL

1. *Chemical analysis.*—Much has been learned by studying the chemistry of living or dead cells and tissues. Samples of tissues can be subjected to chemical analysis before, during, and after activity, and cell activities can often be related to the properties of the chemicals themselves. The products of living cells, such as secretions or waste gases, may be analyzed, moreover, and much of the physiology of the cell clarified thereby. A knowledge of the chemistry of the secretion of the thyroid gland, for example, has contributed notably to our understanding of the physiology of this gland. Closely allied to the chemical analysis of cells and their products is the attempt to synthesize those products artificially. Many cell components and products have now been manufactured in the test tube.

2. *Microscopic examination.*—The microscopic structure of cells and tissues may be studied[4] by making what are called "sections." Tissues are subjected to chemicals which quickly kill the cells and "fix" them, preserving the structural relationships within cells and between cells fairly faithfully. The tissue is then imbedded in some

3. Early in embryonic development it is possible to detect a differentiation of the multiplying cells into three groups or layers. The *ectoderm* is the outer layer, from which the skin, for example, is derived. The *endoderm*, or inner layer, gives rise to the lining of the alimentary canal and its glands. Developing from the *mesoderm*, or in-between layer, is the great bulk of the body of adult man, including muscle, bone, blood, and blood vessels (see also chap. 14).

4. This branch of anatomy is called *histology.*

such material as paraffin, which lends a certain rigidity to the preparation, so that it can be cut into slices. The slices are so thin
that they are more or less transparent to bright light, enabling one
to make out structural details under the microscope. Finally, the
sections are stained. They are treated with mixtures of dyes which
serve to accentuate internal structural differences. By this technique
it is incidentally also possible to learn something about the chemistry
of cell components. Knowing the chemical composition and properties of a dye, and observing what parts of the cell are stained by the
dye (i.e., what parts of the cell react with the specific components of
the dye), we may deduce certain chemical properties of cell parts
too small to be isolated easily and subjected to direct analysis. Staining reactions show, for example, that the nucleus is less alkaline than
the rest of the cell.

3. *Micromanipulation.*—In this technique extremely fine needles
are inserted into the interior of the cell under the microscope. Manipulation of the needle shows, for example, that the external surface
of the cell is relatively rigid, of a jelly-like consistency; that the
nucleus, likewise, is bounded by a membrane structure; and that the
cytoplasm is mostly of a semifluid consistency, through which the
point of the needle passes with but little distortion of cell structure.

By manipulation it is also possible to dissect the cell and to demonstrate that, when the nucleus is removed, the cell soon dies. Or, if a
cell is cut in two in such a way that the intact nucleus remains in
one of the parts, that part is capable of repairing the wound and of
continuing its existence. The portion removed from the nucleus
always dies.

4. *Isolated living cells.*—Some cells, like those of the blood, can
easily be removed from the body and observed in the living state
under the microscope. Ameboid movements of the white blood cells
and the engulfing of particles by them have been studied in this way
at great length.

However, cells removed from their native environment soon decrease their activity and eventually die unless special precautions are
taken. The cells should be maintained at approximately body temperature, although somewhat lower temperatures are suitable for
certain studies of isolated cells. Ordinarily, such cells will die much
sooner than in their normal habitat, for a number of reasons. Nutriments are ultimately exhausted from the artificial environment.
Even before this, death will result from the accumulation of deleteri-

ous waste products. Or, again, bacteria may gain entrance to the fluid medium containing the cells and cause their death.

5. *Tissue culture.*—By a special technique, however, these eventualities can be prevented or delayed. "Cultures" of tissues, with living, growing cells, can be made by removing cells from the body, guarding against bacterial infection, and placing them in sterile nutrient media in sterile containers. The proper temperature is maintained, and at regular intervals bits of the tissue culture are transferred into a freshly prepared medium, so that the cells obtain a continuous supply of nutrients and are never subjected to a harmful accumulation of waste products.

By means of such repeated transplantations to fresh media, tissues have been kept alive for long periods. At the Rockefeller Institute in New York City, cells were kept which had descended through countless transplants from tissue removed from an embryonic chick heart years earlier. These cells (which were of the connective-tissue variety) were kept alive. They grew and multiplied and carried out all their normal activities, apparently, many years after the death of the chick from which the tissue was originally taken.

At any time the culture could be placed under the microscope, and the growth and transformations of the living cells observed. Photographs could be taken, or bits of the tissue could be fixed and stained and sectioned, from time to time.

An example of tissue-culture experimentation less spectacular, but no less important than the cultures of the chick heart, is the work of the late Dr. A. A. Maximow, of the University of Chicago. He made cultures of certain white blood cells called *lymphocytes* (see p. 115). Observing them day after day, he found that they changed their shape, eventually becoming indistinguishable from connective-tissue cells. They not only resembled connective-tissue cells but gave all the staining reactions of such cells. Like them, they actually produced typical connective-tissue fibers. The whole culture of blood cells was ultimately transformed into typical connective tissue. Here was a demonstration that, at least under certain conditions, lymphocytes can transform themselves into connective-tissue cells.

This striking discovery points to a possible function of the lymphocytes in the healing of wounds. When skin tissues are destroyed by a wound, for example, the break is closed largely by connective tissue in the healing process. Scars of all kinds are composed chiefly of connective-tissue fibers manufactured by connective-tissue cells.

Most of this tissue arises from pre-existing connective tissue which invades the area of destruction from the adjacent regions. But, in the healing of wounds, lymphocytes are often found in abundance in the vicinity. Their function here is not definitely known, except that the work described suggests that they contribute to the healing process. Perhaps physiologically, just as in tissue culture, these special white blood cells change into connective-tissue cells and aid in the formation of scar tissue.

Tissue culture is especially adapted for the study of cells of compact organs, where individual cells cannot be isolated without

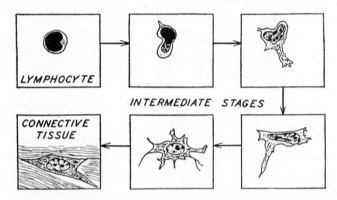

Fig. 3.—Successive stages in the transformation of lymphocytes (certain white blood cells) into connective tissue in *tissue culture*. It is suggested that these changes may also occur in the body when scars are formed in the healing of wounds.

damage to them. Cells from the liver or heart or the nervous system, for example, as well as abnormal cells from tumors and cancers lend themselves to such studies.

6. *Perfusion.*—Somewhat akin to this, at least in principle, is a technique in which whole organs are removed from the body, and a nutrient liquid, even blood, is artificially circulated (*perfused*) through the blood vessels of the preparation. Some organs survive for many hours or even days under such treatment. Analyses of the composition of the ingoing and outcoming liquids furnish much indirect information about the chemical activities of the cells of the organ: what materials are used up by the cells and what wastes or other substances are produced. Even the beating heart has been subjected to such treatment with fruitful results.

7. *Other methods.*—A vast array of significant data has been secured, and much cell physiology elucidated, by subjecting cells to

various stimulating and depressing agencies and noting, measuring, and recording the results. This has been done extensively on nerve and muscle tissue. Unfertilized eggs of vertebrates have also been artificially stimulated to commence cell division and grow, essentially as though the egg had been fertilized by a male reproductive cell. Full-grown, apparently normal frogs have been produced in this way.

In much of this work very elaborate techniques are necessary, involving the use of complicated physical apparatus. The various devices for measuring electrical changes in active cells include the most sensitive instruments for measuring electrical potentials known to science, and some of them require expert physicists for their construction and operation.

CHAPTER TWO

THE ORGANIZATION
OF PROTOPLASM

I. CHEMICAL COMPOSITION
   A. General chemical considerations
   B. Atomic components
   C. Water
   D. Inorganic salts
   E. Carbohydrates
   F. Fats
   G. Proteins
   H. Enzymes
   I. Radioactive chemicals

II. PHYSICAL STATES OF PROTOPLASM
   A. Internal organization
   B. Colloids
   C. Sol and gel states
   D. Electrical charges

III. ENERGY RELATIONSHIPS
   A. Metabolism and energy
   B. Sources of energy

IV. MOVEMENT OF MATERIALS THROUGH CELLS
   A. Diffusion
   B. Filtration
   C. Permeability of cell membranes
     1. Size of molecules 2. Fat solubility 3. Electrical charges. 4. Unknown factors
   D. Osmosis
     1. Nature of the phenomenon 2. Osmotic equilibria in the body

This chapter is concerned with certain basic chemical and physical concepts of protoplasm, cells, and membranes which will be encountered again and again in later chapters dealing with organs, systems, and the body as a whole. Some readers will prefer to skip this chapter, temporarily, and to refer back to its sections as the need arises later in the book. In any case, it will soon be appreciated that some knowledge of these phenomena is indispensable for an understanding of the machinery of the body.

I. CHEMICAL COMPOSITION

A. General Chemical Considerations

The science of chemistry deals with the composition and especially the transformations of matter—its properties and interactions. Most of the familiar physical objects about us are mixtures of simpler materials, despite their appearance of structural homogeneity. Soda water, for example, is a mixture of solid, gas, and liquid—of sugar, carbon dioxide, and water. The mixture is very intimate, but it is a mixture nevertheless; and any tiny portion which we might examine contains small units of pure water, pure sugar, and pure carbon dioxide gas. It is often possible to subject mixtures to various treatments which separate the ingredients from one another. Each of the constituent parts will be found to have definite properties which distinguish that material from all others.

But we find that eventually any further subdivision of the ingredients results in a loss of their characteristic properties. When this point is reached, our materials have been reduced to what is called the *molecular* state. Thus in the physical world there are molecules of water, of sugars, of salts, of carbon dioxide, etc. A drop of pure water or a crystal of pure salt or sugar contains nothing but water or salt or sugar, as the case may be.

Pure substances of this sort are called *compounds*. Most molecules are compounds, which implies that, even though they are "pure" substances, these also are combinations of still simpler components. These further subdivisions of matter are called *elements* or *atoms*. Or, speaking synthetically rather than analytically, we find that atoms are combined in nature to form molecular compounds. A molecule of table salt is a chemical union of one atom of sodium and one atom of chlorine.

Chemists have agreed upon shorthand symbols for atoms. Sodium is designated by the abbreviation Na (from *natrium*, the Latin word for sodium) and chlorine by Cl. This scheme makes it possible also to indicate the atomic composition of molecules. Table salt, or sodium chloride, is NaCl. Of course, some molecules may contain more than one atom of a kind, which can also be indicated in the molecular formula. A water molecule is composed of two atoms of hydrogen plus one of oxygen, and its molecular formula is H_2O. Some molecules are relatively large and contain many atoms. Grape sugar, for example, is made up of six atoms of carbon, twelve of hydrogen, and six of oxygen, and its formula is $C_6H_{12}O_6$.

Shorthand expressions called *equations* indicate interactions between atoms and molecules. The formation of sodium chloride from sodium and chlorine is indicated thus:

$$Na + Cl \rightarrow NaCl .$$

The formation of two molecules of water from two molecules of hydrogen and one molecule of oxygen is written:

$$2H_2 + O_2 \rightarrow 2H_2O .$$

Notice that in this equation there are as many total atoms of each kind in the reacting molecules as in the product formed. Both at the start and at the conclusion of the reaction there is a total of four atoms of hydrogen and two of oxygen. No atoms have been lost, and none formed, in accordance with the principle of conservation of matter. An equation which takes this into account is said to be balanced.

One of the most fundamental reactions in biology is that which occurs in green plants, in the union of six molecules of water and six of carbon dioxide gas, to give one molecule of glucose and six molecules of oxygen gas. The balanced equation for this reaction of *photosynthesis* is:

$$6CO_2 + 6H_2O \rightarrow C_6H_{12}O_6 + 6O_2 .$$

The reverse of any of these reactions—namely, a breakdown of a substance into its component parts—can also be indicated:

$$NaCl \rightarrow Na + Cl$$
$$2H_2O \rightarrow 2H_2 + O_2 .$$

The reverse of photosynthesis also occurs in nature in the process called *cell respiration:*

$$C_6H_{12}O_6 + 6O_2 \rightarrow 6CO_2 + 6H_2O .$$

Reversible reactions such as this are indicated with a double arrow:

$$C_6H_{12}O_6 + 6O_2 \rightleftarrows 6CO_2 + 6H_2O .$$

Chemically, this reversible reaction is of the *oxidation-reduction* type. The reaction to the right is an oxidation, being a union of oxygen with the carbon atoms of the sugar. The carbon of the sugar is said to be oxidized to carbon dioxide, water being a by-product. Photosynthesis, the reaction to the left, is a reduction, in which the carbon atoms of the CO_2 lose oxygen in the synthesis of the sugar molecule.

Obviously, there are far more kinds of materials familiar to us (mixtures) than there are pure molecules. Likewise, the number of kinds of molecules exceeds greatly the number of kinds of atoms or elements. Of the latter, there are fewer than one hundred known to chemists. All the diversified complex materials of the world, then, are combinations of these relatively few atoms or elements.

B. ATOMIC COMPONENTS

The substance of which cells are made is called *protoplasm*, which literally means the "formed thing of first importance." It would seem reasonable to expect that protoplasm, possessing the unique property of life, might be made of peculiar, or at least of rare, elements. Yet we find protoplasm to be composed of the very same atoms which are the most common in the nonliving materials of the world. No atom peculiar to living things is known. Carbon, hydrogen, oxygen, nitrogen, phosphorus, sulfur, sodium, chlorine, potassium, calcium, iron, iodine, and magnesium are the most common elements of protoplasm.[1]

These facts were not always appreciated. The obvious though erroneous view that protoplasm is peculiar in its chemical constituents was held for a long time. Even after many of the reactions of protoplasm and its constituents were known, they were called "organic associations" rather than "chemical reactions." Early chemical studies served to accentuate the differences between organic associations and the chemical reactions of the nonliving world. Even today chemistry divides its subject matter into the two main divisions of organic and inorganic. But whereas in the past organic chemistry was looked upon as a study of the properties and reactions peculiar to the constituents and products of protoplasm, today it is more broadly defined as the chemistry of the carbon compounds.

The gradual change in point of view of chemists dates from the historic work of Friedrich Wöhler in 1828. Previous to this it was believed that organic compounds could be produced only by living cells. *Urea*, a waste product of cellular activity found in the urine of

1. Using chemical symbols, the atoms of protoplasm are as follows, arranged approximately in the order of the relative frequency of their occurrence: C H O P K I N S Ca Fe Mg Na Cl; which is to say, "C. Hopkin's Cafe is mighty good." Being a crude memory scheme, this is all to be taken with a "grain of salt."

vertebrates, was thought to be such a compound. Wöhler, however, succeeded in manufacturing urea in the test tube from simpler molecules, quite independently of the presence of life, or live cells, or products of live cells. From this time the notion developed that organic associations were chemical reactions basically similar to those of inorganic chemistry and that organic compounds were combinations differing in no fundamental way from the compounds of nonliving systems.

It was said of the eminent German organic chemist, Emil Fischer, that he had two ambitions in life: to make a lump of sugar and to lay an egg. The first aim he achieved. He synthesized various sugars in the test tube. His second aim, to produce life, has, of course, not yet been realized. But though life itself, or living protoplasm, has never been manufactured, many of the constituents and products of cells have been synthesized in the test tube. Examples will be mentioned from time to time in the following pages.

C. WATER

Of the compounds found in protoplasm, ordinary water is the most common. From 60 to 99 per cent of protoplasm is nothing more mysterious than water.[2] It was undoubtedly in a watery medium that the proper combination of compounds came together in the proper manner, in a proper environmental setting, to produce the first living protoplasm in the remote past of geologic time. Later, evolution of living forms occurred in water, and down to the present time we find water an indispensable constituent of living cells as well as a necessary component of the cell environment.

Apart from its evolutionary significance, and its ubiquitous distribution on the face of the earth, water owes its prominence in living systems to certain important physical and chemical properties. Solids, gases, and liquids of many kinds dissolve in water. When solids like crystals of sugar go into solution, they are broken up into single molecules, which ultimately become uniformly distributed throughout the watery medium. Many liquids besides water will dissolve substances, but water will dissolve more materials than any other known pure liquid. Water is the closest approximation we have to a universal solvent.

The importance of the phenomenon of solution is not only that

---

2. Much of this is thought to be "bound water" securely held to other molecules of the system by physical forces.

the dissolved materials become homogeneously distributed throughout the system but also that chemical reactions occur much more
readily between dissolved materials than between materials in solid
form. Chemical reactions are largely interactions between molecules,
atoms, or *ions* (see below), and such reactions will be facilitated
by any agency which reduces solids to a more finely divided state.
If sodium chloride crystals are mixed with silver nitrate crystals, no
measurable reaction occurs; but, if sodium chloride and silver nitrate
are intimately intermixed in the finely divided state occurring in a
solution, there will occur an almost instantaneous reaction between
billions of contiguous parts of molecules, with the formation of
silver chloride and sodium nitrate.

The significance of all this in protoplasm is, of course, that life
itself depends upon a great variety of intermittent or continuous
chemical reactions between protoplasmic ingredients. These are
made possible, then, partly by the dissolved state in which many of
the reacting agents are held in the cell and its immediate environment.

Other relationships between water and solid particles besides that
of solution are common in nature, both in living and in nonliving
systems. There is the simple mixture or *suspension*, such as that of
sand in water. The characteristics of a suspension are that the solid
particles are of relatively large size, even macroscopic, being composed of large numbers of molecules or masses of atoms. There is a
tendency also for the solid particles to separate out from the watery
phase whenever the particles are of a specific gravity different from
that of water. The particles sink to the bottom, or float to the top,
unless the mixture is constantly agitated. The suspended particles
need not be solids. They may be liquid, as in a suspension of ether
or oil in water.

A special type of suspension, quite different from that described,
and having properties very important to protoplasm, is the *colloidal
suspension*, which will be discussed at greater length later (see p. 44).

Another property of water of significance both in biology and in
inorganic chemistry is the manner in which it produces or makes
possible a splitting of certain substances into smaller components
called "ions." When sodium chloride goes into solution, little if any
of the material exists as whole molecules. Each molecule is split into
a sodium and a chlorine ion. Ionization, even more than simple solution, markedly affects the chemical reactivity of a compound. In

general, it is true that ions are much more highly reactive than the corresponding molecules.

Many of the properties of ions depend upon the electrical charges they possess. The sodium ion is positively charged, and the chlorine ion is negatively charged. Sodium ions are designated by Na^+, and chlorine ions by Cl^-. The formation of these ions in a solution of salt can be indicated by the formula

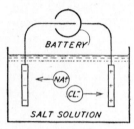

$$NaCl \rightarrow Na^+ + Cl^-.$$

Fig. 4.—Behavior of electrolytes in solution. Since like electrical charges repel one another and since unlike charges attract one another, sodium ions move toward the negative pole of a battery, and chlorine ions move toward the positive pole.

It is because of these charges that solutions containing ions are able to conduct electricity. When the positive and negative poles of a battery are dipped into a salt solution, the sodium ions are attracted to the negative pole, where their positive charges are neutralized. The chlorine particles migrate similarly to the positive pole. Molecules which ionize in water are called *electrolytes* because of this capacity of their ions to conduct electricity.

The properties of electrolytes are in large measure determined by the kinds of ions they yield in solution. Three great classes of electrolytes are differentiated on this basis. *Acids*, as a group, possess their peculiar properties, including their sour taste, because in solution they yield positively charged hydrogen ions. Hydrochloric acid, for example, ionizes thus in water:

$$HCl \rightarrow H^+ + Cl^-.$$

A second group, the *alkalies* or *bases*, always yield negatively charged combinations of oxygen and hydrogen, designated as OH ions. The alkali sodium hydroxide ionizes in this way:

$$NaOH \rightarrow Na^+ + OH^-.$$

The degree of acidity or alkalinity of a compound depends upon the extent to which it ionizes in water—the extent, in other words, to which the molecule yields H^+ or OH^- in a solution. Some compounds yield both H^+ and OH^- in solution. An important group of protoplasmic compounds is of this nature. To a slight extent, water itself dissociates into both these ions:

$$HOH \rightarrow H^+ + OH^-.$$

The third class of electrolytes are the *salts*, which ionize into particles neither of which is H^+ or OH^-. Sodium chloride is an example:

$$NaCl \rightarrow Na^+ + Cl^-.$$

Many of the important properties of cells, such as permeability, irritability, and nerve conduction, appear to be related to the presence of electrical charges upon cell membranes, which, in turn, depend upon the presence of electrolytes in and about the cells.

Finally, water can absorb more heat from warm surroundings with less change in its own temperature than almost any other known substance. This is important in living systems because temperature changes markedly influence chemical reactions and, therefore, physiological reactions as well. Indeed, the temperatures at which most cells can exist and function at all are rather limited, covering a range of some 100° F. Because of the abundance of water in protoplasm, then, it is better able to maintain its temperature within this range, even though the external temperature may fluctuate over a still wider range.

In mammals and birds, whose internal temperature normally fluctuates by only a few degrees, a number of physiological mechanisms supplement this physical effect of water. In these higher vertebrates the effect of water in buffering the body against extreme and sudden changes of temperature is considerably less significant than the physiological temperature-regulating devices.

D. INORGANIC SALTS

A variety of inorganic salts is found dissolved in the cells and body fluids. In the latter, ordinary table salt is one of the commonest. It has been found that in the kinds of salts present, as well as in their relative concentrations, the body fluids are remarkably like sea water.[3] It is reasonable to suppose that, with the evolution of protoplasm itself in the sea, these salts became an integral part of protoplasm and that, when multicellular forms arose, the body fluids incorporated into the organism were simply modified sea water taken from the immediate environment. When sea forms invaded the land in the Cambrian period, they carried with them in their body fluids an environment for the cells very similar to sea water.

A comparison of the relative concentrations of the main salt ions in sea water and in the body fluids of animals (see Table 1) reveals

---

3. Especially true of the positively charged ions of salts.

a striking similarity. So far as these common salt ions are concerned, it would be approximately correct to state that the body fluids, even of land animals, including man, have the composition of dilute sea water.

Present in low concentration—about 0.9 per cent by weight[4] in the body fluids of mammals—and relatively simple chemically, the salts nevertheless exert a profound influence upon cell activity. We might expect this to be true, inasmuch as, through countless eons of geologic time, protoplasm and cells have adjusted themselves to a specific salt environment, both as to kind and as to relative concentration.

If, for example, the calcium-ion concentration of cells or body fluids is diminished, a heightened irritability results. Cells such as those of skeletal muscles begin to twitch spontaneously. In mam-

TABLE 1

COMPARISON OF THE RELATIVE CONCENTRATIONS OF CERTAIN
COMMON SALT IONS IN SEA WATER AND IN THE
BODY FLUIDS OF CERTAIN ANIMALS

	Na	K	Ca	Cl
Ocean water..............	100	3.6	3.9	180.9
Marine invertebrate (limulus)..	100	5.6	4.1	186.9
Cartilaginous fish (dogfish)....	100	4.6	2.7	165.7
Blood serum of man..........	100	6.7	3.1	128.8

mals a considerable lowering of the blood calcium leads to convulsions and death. Corresponding to the importance of calcium to cell activity, we find that elaborate mechanisms have been evolved for the maintenance of suitable concentration of these ions in the cell environment (see chap. 12).

A proper balance of sodium, calcium, and potassium ions is indispensable for the normal action of the muscle cells of the heart. In a pure sodium chloride solution a frog heart soon stops beating in a state of relaxation. With sodium and calcium present, but no potas-

4. I.e., 0.9 gm. in 100 gm. of liquid. The *metric system* of measurement is used in science:

Length: 1 meter (about 39 inches) = 100 centimeters
1 centimeter (about 0.4 inch) = 10 millimeters
Volume: 1 liter (about 1 quart) = 1,000 cubic centimeters
1 cubic centimeter = 1,000 cubic millimeters
Weight: 1 kilogram (2.2 pounds) = 1,000 grams
1 gram = weight of 1 cubic centimeter of water
Temperatures: zero degree centigrade is the freezing point of water (32° Fahrenheit) and 100 degrees centigrade is the boiling point of water (212° Fahrenheit).

sium, it stops in a state of maintained contraction. Too much potassium stops the heart in relaxation. Only when these three ions are present in the proper relative concentrations will the heart continue beating normally. (For *osmotic* properties of salts see p. 58.)

E. CARBOHYDRATES

The three great classes of organic compounds in cells are the *carbohydrates, fats,* and *proteins.* The carbohydrates are composed entirely of carbon, hydrogen, and oxygen. Relative to the fats, there is more oxygen in the molecule, and the ratio of the number of hydrogen atoms to oxygen atoms in a molecule is 2 to 1, as in water. In this respect, carbohydrates may be considered as watered or hydrated carbon, from which relationship they derive their name.

Three general kinds of carbohydrates are distinguished, on the basis of the complexity of the molecule. First there are the *simple sugars* (or *monosaccharides*), whose formula is $C_6H_{12}O_6$. Grape sugar, or *glucose,* is an example. There exist in nature several simple sugars of exactly this formula, each having in its molecule the same number of atoms of the three characteristic ingredients. Though similar to one another in certain general respects, there are definite minor variations in chemical properties. Why should this be, when all of them have the same (so-called "empirical") formula? The differences in properties are apparently due to differences in the manner in which the atoms are arranged in the internal structure of the molecule. Similar situations are frequently encountered in organic chemistry. Ordinary grain alcohol, for example, and one kind of ether have certain properties in common, as one might surmise from the fact that the formula for each is C_2H_6O. Yet there are certain definite distinctions in chemical properties which are related to differences in arrangement of the constituent atoms.[5]

5. These differences are indicated in what is called the *structural formula* of the compound. Thus the formula for one kind of ether is

$$H-\underset{\underset{H}{|}}{\overset{\overset{H}{|}}{C}}-O-\underset{\underset{H}{|}}{\overset{\overset{H}{|}}{C}}-H$$

and for alcohol

$$H-\underset{\underset{H}{|}}{\overset{\overset{H}{|}}{C}}-\underset{\underset{H}{|}}{\overset{\overset{H}{|}}{C}}-OH$$

Compounds having the same empirical formula but different internal structural arrangements of atoms are called *isomers.* There are, then, several isomeric simple sugars having the same

Second, there are the *double sugars* (or *disaccharides*). Each molecule of a double sugar is composed of two molecules of simple sugars combined into a single molecule, usually with the loss of one molecule of water. We could indicate the formation of a double-sugar molecule from two molecules of a simple sugar or sugars in this way:

$$2C_6H_{12}O_6 \rightarrow C_{12}H_{22}O_{11} + H_2O \ .$$

The differences in properties of various double sugars are due, again, to differences in the arrangement of the constituent atoms, which here means differences in the kinds of simple sugars of which they are composed. One double sugar, for example, is composed of two glucose molecules in combination. Another, ordinary cane sugar, is composed of a molecule of glucose combined with a molecule of fructose.

Finally, in the *compound sugars* (or *polysaccharides*) there are great numbers of simple sugars combined to form a single large molecule. These are the starches, which are of two general types—the plant starches and the animal starches, or *glycogen*. The compound sugars vary, then, partly because of the kinds of simple sugars which enter into their makeup. In addition, they vary as a result of differences merely in the total number of simple sugars they contain. The number of simple sugars in most compound carbohydrates is unknown, though the kinds of sugars may be known. The formula for starch or glycogen would then be $(C_6H_{10}O_5)_n$, where n represents an unknown number.

Glycogen is a storage form of carbohydrate in animals, comparable to the starch in plants, and is particularly abundant in liver and muscle cells. In a sense, it is an ideal storage form, for its molecule

---

empirical formula but with different structural formulas. Two common isomers of this sort are glucose and fructose:

```
      H  H  H  H  H
      |  |  |  |  |
  H—C—C—C—C—C—C—H    (glucose, or grape sugar)
    |  |  |  |  |  ||
    O  O  O  O  O  O
    |  |  |  |  |
    H  H  H  H  H


      H  H  H  H     H
      |  |  |  |     |
  H—C—C—C—C—C—C—H    (fructose, a component
    |  |  |  |  || |       of cane sugar)
    O  O  O  O  O  O
    |  |  |  |     |
    H  H  H  H     H
```

is more "compact" than would be a collection of an equal number of simple-sugar molecules, containing, as it does, less water than the latter. Furthermore, it is very readily converted into a glucose compound, which apparently is a necessary preliminary to combustion. Starch and glycogen are stored fuel, laid down when glucose is present in abundance and available as fuel when glucose from other sources is less abundant.

Though their chief role seems to be that of a fuel for cells, some carbohydrates are also built into the structure of protoplasm; to a certain extent they serve as building material as well as fuel.

For some poorly understood reason, also, carbohydrate in the form of glucose is an indispensable constituent of the blood. The concentration of glucose in and about cells in mammals is normally about 0.1 per cent by weight. No particular disturbance of cell activity occurs from a simple, uncomplicated increase in this quantity. But, if this level is appreciably reduced by any means whatsoever, there is a sharp increase in the irritability of certain nerve cells in the brain. These are thrown into activity "spontaneously"; that is, by a multitude of very slight environmental changes which ordinarily are too small to stimulate normal cells. Spontaneous muscular twitchings may occur, which may develop ultimately into convulsions. The individual loses consciousness and may die.

This indispensability of sugar reminds one of the importance of the salts in the cell environment. There is, however, an important difference. The salt ions directly influence the irritability of nerve and muscle tissue, so that a living muscle removed from the body of an anesthetized animal will become highly irritable and will twitch spontaneously if the muscle is bathed in a solution containing an improper salt balance. But, in a proper salt environment, these same isolated muscles show no abnormal behavior even in the complete absence of glucose. The muscular twitchings and convulsions of an animal with a low blood-sugar concentration are abolished by severing the nerves connecting the muscles with the brain or spinal cord. Apparently the low blood sugar increases the irritability and activity of certain nerve cells in the brain, which then produce involuntary muscle contractions through nervous connections with those muscles.

The reason for this dependence of the intact animal—actually certain brain cells—upon glucose is far from clear. At any rate, there are elaborate mechanisms in the body concerned with the maintenance of the proper glucose concentration in the body fluids. These mechanisms involve a number of organs, including the liver, the nervous system, and at least two glands of internal secretion—the pancreas and adrenals.

In a disorder of the pancreas,[6] which is one of the sugar-regulating organs, the sugar content of the body fluids falls to dangerously low levels; and, unless the condition is artificially controlled, convulsions, unconsciousness, and death result. In another defect of pancreatic function, *diabetes*, there is an elevation of the sugar content of the blood. The injection of *insulin* (see chap. 12) corrects this condition. But, if an excessive dosage of insulin is given to control the disease, the glucose concentration of the body may reach a dangerously low level. Eating candy or sugar will restore the sugar level to normal and may prevent a fatal outcome.

F. FATS

The true fats are also composed solely of carbon, hydrogen, and oxygen, though their molecular structure is decidedly different from that of the carbohydrates. Correspondingly, the properties of fats are also quite distinct. An essential structural difference between fats and carbohydrates is that in a molecule of fat there is far less oxygen in proportion to the carbon and hydrogen. A common animal fat (*tripalmitin*) has the formula $C_{51}H_{98}O_6$. Chemical studies show that a fat molecule can be split into two kinds of smaller molecules, called *glycerol* and *fatty acids*. Each molecule of fat is made up of one molecule of glycerol and three molecules of fatty acid attached to the glycerol.[7] The significance of this structural

---

6. Overactivity of the *internal secretory* portion.

7. The formula of glycerol is $C_3H_8O_3$. The fat whose empirical formula is $C_{51}H_{98}O_6$ (tripalmitin) is constructed as follows:

feature of fats will become more apparent in a later section. There it will be seen that during digestion the fat of the food must be broken down by the digestive juices into its component parts— glycerol and fatty acids—before it can be absorbed into the system.

The fats, like the carbohydrates, serve as fuel. Together, these two furnish most of the energy expended in the living body. Fat also is stored in the body in the fat cells, as described earlier. There are certain notable differences between the carbohydrates and the fats as fuel. In the first place, the fats yield over twice as much energy or heat, per gram of substance, as do the carbohydrates. It is fuel of an even more concentrated form than the carbohydrates. This makes fat a more economical form in which to store food reserves, at least as regards bulk.

One might raise the question, then, why all the food stores of the body are not in the form of fat. A more acceptable question to a biologist, who accepts nature as it is, would be: Of what advantage is it to have some of the reserve food stored as carbohydrate and some as fat? The answer to this question brings out another important difference between carbohydrates and fats.

Stored glycogen can be converted into glucose, and thus be made available for immediate combustion, much more quickly than the fat stores can be drawn upon. In mammals the glycogen stores are actually drawn upon even in the intervals between meals. During digestion and absorption of carbohydrate the immediate excess in the body is stored, chiefly in the liver and muscles. Then, only an hour or so later, when carbohydrate is no longer entering the system from the intestines, the stores are drawn upon. Consequently, glucose is always available to the cells. If fasting is continued for some days, or if there is a period of deficient intake of food, when the glycogen stores near depletion, only then will the reserves of fat be drawn upon to any considerable extent.

An apt comparison has been made to illustrate this point. Stored glycogen is like a checking account, easily drawn upon. The available balance fluctuates relatively rapidly. Stored fat is rather like a savings account, where the balance remains at a more nearly stationary level, to be called upon only when the checking-account balance becomes low. This analogy may be extended, for, just as the reserves in a checking and a savings account are interchangeable, so also are the fats and carbohydrates.

Carbohydrates can be transformed into fats in the body. Also, fats (or at least certain components of the fat molecule) can be converted into glucose. The first of these statements is simply a more sophisticated expression of the homely bit of knowledge that starches and sweets are fattening. This conversion is, of course, an important factor in animal husbandry. Hogs are fattened by feeding starch-rich corn. In terms of the bank analogy we should say that, when the checking account reaches a certain maximum, any additional incoming funds are transferred to the more permanent savings account.

To a greater extent than carbohydrate, the fats and related fatlike compounds play an important role in the structure of all cells. The protoplasmic bounding membrane of the cell contains fatlike compounds, as do also intracellular membranes like that which bounds the nucleus. Special structures, like the sheaths about nerve fibers, also contain fatlike substances in abundance.

Deposited in great quantities just under the skin, fat functions as insulation. It is a poor conductor of heat and, in this location, is very effective in conserving the body heat. This is especially important in cold environments. Polar and marine mammals are well protected with a thick layer of fat (blubber) under the skin. Human females often resist the cold better than the traditionally superior male for the reason that women are generally more amply equipped with an insulating layer of fat than are men. The fat deposited under the skin, present even in thin people, provides a degree of protection against mechanical injury to the deeper body parts.

In the course of this discussion of fats, the terms "true fats" and "fatlike compounds" have been used. The true fats are those which are composed of a combination of glycerol and fatty acids. The fatlike compounds are similar to this but contain other molecules besides the fatty acids and glycerol. As the terminology suggests, the latter compounds, though of a somewhat different composition and displaying somewhat different properties, have certain features in common with the true fats.[8]

G. PROTEINS

The name *protein* literally means "of first importance." These compounds are always found in protoplasm and apparently are more intimately bound up with the integrity and life of the cell

8. The terms *lipins* for true fats and *lipoids* for fatlike substances are usually used to designate these related compounds.

itself than are either carbohydrates or fats. Throughout our discussion of physiology, justification for this statement will become more and more apparent.

From this one might expect a priori that proteins are extremely complex compounds. The molecules are very much larger than fat molecules and even exceed in size the rather large molecules of glycogen and the starches. Hundreds or even thousands of atoms enter into the formation of a single protein molecule. The great bulk of the molecule is composed, again, of carbon, hydrogen, and oxygen; but, in addition, there are atoms of sulfur, phosphorus, and especially nitrogen,[9] which latter may be considered the characteristic element of proteins.

Chemical analyses reveal that the protein molecule can be split into simpler components. Just as a single starch molecule is composed of many molecules of simpler sugars, so the protein molecule is a combination of many molecules of what are called *amino acids*.[10] There are known at present more than twenty different amino acids, constructed atomically upon the same general plan but differing in certain details of internal structure and of physiological significance and function. Much of our knowledge of protein chemistry and physiology has been clarified by studies upon the properties of these constituent parts.

A striking situation in nature is the extremely large number of different native proteins. A great variety of proteins is found in any one individual. In man the proteins in the cells of the thyroid gland, skeletal muscle, and liver, for example, all differ from one another.

9. Hemoglobin has been assigned the formula

$$C_{3032}H_{4816}O_{872}N_{780}S_8Fe_4 .$$

In general, about one-half of the protein molecule (by weight) is carbon, one-fifteenth hydrogen, one-fifth oxygen, and one-sixth nitrogen.

10. The structural formula of the smallest and simplest of the known amino acids is as follows:

$$\begin{array}{c} H \\ | \\ H-C-NH_2 \\ | \\ COOH \end{array}$$

The other amino acids are more complex structurally, but common to all of them is the $H-C-NH_2$ group, or radical. The NH_2 group is called an amino group in organic chemistry,

|
COOH

and compounds possessing the COOH complex are acid in nature, by virtue of ionization in which the positively charged H ion dissociates from the remainder of the molecule, which then is negatively charged. Hence the name "amino acid."

Again, certain differences are apparent from individual to individual within a species; some of the blood proteins of one man may differ slightly but significantly from those in the blood of another man. Finally, every species of animal (or plant) has some proteins characteristic of itself; the proteins of the muscle cells, kidney cells, or blood of man differ more or less from the proteins of corresponding tissues of other vertebrates.

The degree of divergence from one another in the structure and properties of proteins of different species varies with the degree of intimacy of the evolutionary relationships of the species involved. The more closely related evolutionary groups have, in general, more closely related proteins. Organisms farther removed from one another in the scale of evolution have proteins less similar to one another. The blood proteins of a man, for example, may be the same or only slightly different from those of another man. They are closely allied to those of the blood of apes. Greater divergence exists in the case of man and lower mammals, while the proteins of the blood of man are distinctly different from those in the body fluids of invertebrates. These relationships have been very useful in studies of evolution, adding important confirmatory data to evolutionary affinities established by other means. The blood proteins of marine and land mammals, for example, are more similar than are those of whales and other marine vertebrates which are not mammals.

We see from this that the number of proteins in animals and plants is extremely large. And, from all the available evidence, differences in proteins are largely due to differences in the numbers and kinds of constituent amino acids, as well as the internal arrangement of these amino acids with respect to one another. A likely analogy has been drawn between the structure of a protein molecule and the typographical composition of a paragraph of print.[11] Let the twenty-six letters of the alphabet represent the approximately equal number of different amino acids. Printed paragraphs vary from one another entirely on the basis of the numbers and arrangements of the constituent letters. Though only twenty-six distinct letters are available to the typesetter, an extremely large number of distinct combinations of letters into paragraphs is theoretically possible, especially as a given letter may occur more than once.

---

11. The hemoglobin molecule has been estimated to contain approximately five hundred amino acid molecules. Some proteins are thought to contain as many as two thousand amino acid molecules.

More than fats or sugars, the proteins enter into the actual composition of protoplasm. No cell is known which lacks protein as an important structural component.

A complete elucidation of protein chemistry will go a long way toward an understanding of life-activities in cells. Much progress has been made in the artificial synthesis of proteins in the test tube by linking together various pure amino acids. Relatively long chains have been constructed,[12] but so far, at best, the chains of amino acids are much shorter than those of a naturally occurring protein. Words or short sentences rather than paragraphs have been synthesized. In all cases, of course, the manufactured molecules are not alive.

Though of greatest importance as structural components of protoplasm, the proteins are also capable of being burned and liberating heat and energy. Proteins, or rather some of the amino acids, can be converted into glucose in the body by a chemical rearrangement of the atoms.[13] This conversion frequently occurs before proteins are burned in the body. A direct consequence of this capacity of the body for the conversion of protein into carbohydrate is that the energy of part of the protein molecule may be stored for future use in the form of glycogen.

But protein as such is not stored in any great quantity. Certainly, there is far less of it in strictly storage form than there is of fat or carbohydrate. Yet, in times of prolonged fasting, the proteins of the protoplasm itself are burned. When the glycogen stores near depletion, and the fat deposits become exhausted, the proteins are broken into amino acids and burned. In starvation, then, the body cells literally burn themselves slowly, obtaining from the combustion the energy indispensable for a continuation of the life-processes. Obviously, the successful management of such an abnormal contingency is limited. Sooner or later, continued withdrawal of proteins from protoplasm will injure the cells irreparably. As soon as this state is reached in cells upon whose activities the integrity of the organism depends, permanent damage is done, and, ultimately, death is the consequence.

By way of summary, we may say that the carbohydrates, fats, and proteins are organic compounds which comprise the great bulk of the materials in cells and necessary for cells. The relatively simpler

12. Including twenty or more amino acid molecules.
13. This includes the splitting-off of the NH_2 component of the molecule.

carbohydrates and fats play a lesser role in the chemical composition of protoplasm but are of greatest significance as sources of energy. Large reserves of these compounds are stored in the body in various cells, the carbohydrates being stored in a readily accessible form. The fat stores are in a more permanent, less readily accessible form, though more concentrated, as regards energy content per unit weight. Proteins, by far the most complex of the three, are of greatest importance as building materials for the construction of protoplasm, although the proteins also may be burned. The body is capable of a certain amount of interconversion of these materials. The most notable exception to this is the inability of the animal body to convert either fats or carbohydrates into amino acids or proteins, at least to any significant degree. A limited conversion of simpler substances into some amino acids apparently can occur in the animal body. The main interconversions occurring in animal may be schematically represented in this manner:

$$\text{Proteins} \rightarrow \text{Carbohydrates} \rightleftarrows \text{Fats}.$$

H. Enzymes

Enzymes, whose chemical composition is unknown in most cases, are substances indispensable for the life of the cell. A few of the enzymes which are secreted by cells have been prepared in an apparently pure, crystalline form. These seem to be protein in nature. At any rate, if the enzymes are not themselves proteins, they are ordinarily so intimately associated chemically with proteins that treatment of enzyme-containing products in such a way as to destroy the proteins also destroys the enzyme to a parallel extent. Heat, which coagulates and precipitates proteins, always destroys enzymes.

Some enzymes contain atoms of the heavy metals, such as iron, in their molecules. It is for this reason that compounds like cyanides, which combine with the heavy metals, also inactivate enzymes and, with the destruction of enzymes, cause death. Enzymes directly contribute extremely little if any energy to the cell. Yet energy-liberating processes and chemical reactions in general practically cease in cells when the enzymes are destroyed.

Enzymes are *catalysts*. In general chemistry catalytic agents of many sorts are well known. A catalyst is any agent which hastens the attainment of an equilibrium point in a chemical reaction. If the reaction is one of extensive breakdown of a compound, a catalyst markedly hurries the consummation of the breakdown. If the reaction is synthetic, a suitable catalyst will accelerate the synthesis. The catalyst itself, however, does not enter into the composition of the products of the reaction, so that only very small amounts of it disappear and require replenishment. Almost as much of it as was present at the outset remains after the completion of the reaction or the attainment of an equilibrium point in a reversible reaction. In this way a given quantity of catalyst may be used over and over again, and only small amounts of it are adequate to facilitate chemical changes in a large quantity of reacting materials.

Enzymes are organic catalysts. They act in the manner described for catalysts in general but are produced by cells, are generally distributed in living systems, and are indispensable for the catalysis of the chemical reactions occurring in protoplasm.

The importance becomes apparent when we consider that cellular compounds are of themselves very inert materials. To break down a protein molecule into amino acids or a starch molecule into simple sugars in the test tube requires treatment with strong alkalies or acids. Even then the reactions are very slow. In the alimentary canal these reactions go on quite rapidly and smoothly at relatively low temperatures, without the aid of strong acid or alkali. Enzymes are responsible for this. Again, in the test tube, sugars or fats oxidize immeasurably slowly, even in an abundance of oxygen, unless they are subjected to temperatures far above those compatible with life. Yet oxidation of these compounds at ordinary body temperature is going on all the time in cells. Certain enzymes make this possible.

And so it is with the innumerable chemical reactions of the cell and the organism. Without enzymes the constituents of cells are inert, nonreactive, dead. In the presence of suitable enzymes the cellular contents enter into a host of reactions. Because of enzymes the cell becomes the busiest of chemical laboratories, in which occur all those complicated interrelated reactions of oxidation, breakdown, reduction, synthesis, etc., which make the cell and life itself the dynamic thing we know it to be.

Enzymes are produced by cells. They are manufactured from ingredients which the cells abstract from their surroundings. Sometimes the enzymes are liberated by the cell to perform their functions somewhere outside the cell. The digestive enzymes of higher animals are of this sort. Manufactured in the digestive glands and passed into the alimentary canal through ducts, they exert their catalytic action upon the foods in the stomach or intestine. Other enzymes remain within the cell. Such enzymes include those which facilitate the intracellular oxidation of foodstuffs, or the synthesis of the compounds of protoplasm.

The enzymes are remarkably specific in their action. An enzyme which catalyzes one chemical reaction will usually be without effect upon others. Not only are there distinct and specific enzymes for the breakdown of proteins, others for fats, and still others for carbohydrates, but there may be more than one enzyme involved in the breakdown of any one of these. In the digestion of carbohydrate, for example, the enzyme of the saliva is capable only of starting the breakdown of the starch molecule. Another enzyme, secreted by another gland, is required to catalyze the completion of the process down to the simple-sugar stage. It is by the combined effect of the two, acting successively, that starch is digested to simple sugars.

It is believed that, for perhaps every chemical reaction occurring in protoplasm, there is a corresponding enzyme. Many of them are known at present, and new ones are being discovered almost every year.

There seems much justification for the belief that in enzymes and enzyme action may be found the ultimate explanation of life itself. Yet the enzymes themselves are not alive. This was not always appreciated. It was believed, for example, that the *fermentation* of sugar by yeast cells depended upon the immediate presence and activity of the living cells. In 1897 it was demonstrated (by Eduard Buchner) that fermentation can go on in the absence of yeast cells, provided that there is present a nonliving compound produced by yeast cells and extracted from them. This compound, of course, is a fermenting enzyme. This discovery is of considerable historical significance in biology and was also responsible for the term "enzyme" itself, which literally means "in yeast."

But, while enzymes are nonliving cellular products, and some of them have been prepared in crystalline, probably pure, form, and their chemical nature in many instances is fairly well known, no one has yet been able to synthesize an enzyme artificially, despite numerous efforts to do so.

Anticipating material to be discussed later, we may pose the question of the relationship between enzymes, hormones, and vitamins and their respective mode of action. *Hormones*, we shall see, are products manufactured by animal cells, and *vitamins* are compounds widely distributed in living matter. Vitamin and hormone action bears some resemblance to enzyme action, in that these materials contribute no energy to cells. Yet they are indispensable for normal cell activity and the integrity of the organism and, in some cases, are necessary for life itself. Furthermore, hormones and vitamins are highly effective in extremely small quantities. Further than this in the establishment of a common mode of action, or a fundamental underlying principle, we cannot go at present. Our knowledge of the chemistry of some of the hormones and vitamins is well advanced, and in many cases the structural formula is known, and artificial synthesis of the product has been accomplished. Such chemical knowledge has afforded no clue as to any common principle or mode of action.

I. RADIOACTIVE CHEMICALS

There have been beneficial by-products of atomic-bomb research of far-reaching importance to physiology and medicine. A great variety of chemical substances can be made radioactive, emitting rays not unlike those of radium. Such chemicals—*radioactive isotopes*—seem to retain all the normal properties of the original nonradioactive material and commonly enter into the same metabolic processes of the body as their precursors.

Thus, radioactive iodine, like ordinary iodine, when fed, rapidly accumulates in the thyroid gland. Its presence and amount in the gland can be determined by measuring the radiation emanating from the neck with a *Geiger counter*. Employment of this procedure is common and useful in detecting derangements of the thyroid gland involving faulty absorption and metabolism of iodine by the thyroid gland.

"*Tagged*" *atoms* or molecules of this kind—relatively easy to detect in the various parts of the body by detection of the radiation—are invaluable aids in studying many aspects of metabolism. For example, let us pose the question, "How much time elapses between the intake of iron in the food and the incorporation of that iron into the hemoglobin of the red blood cells?" The employment of radioactive iron provides an answer. Only a few hours may elapse between the feeding of radioactive iron compounds and the appearance of radioactivity within the red blood cells, indicating a remarkably rapid utilization of iron.

A new technique has been developed for indirectly tagging chemical substances which cannot be tagged directly. Certain plant products employed as drugs in the treatment of human disease which cannot be made radioactive directly can be grown in an environment containing radioactive carbon in the carbon dioxide of the air. In photosynthesis this radioactive carbon dioxide is incorporated into the substance of the plant, including the drug extracted from the plant. The absorption of the drug from the intestines, the parts of the body in which it might be especially concentrated, its excretion, etc., can be determined by means of the telltale tagged radioactive carbon atoms in the drug.

An example of the use of radioactive substances in combating certain kinds of cancer is discussed later in this volume (see p. 529).

II. PHYSICAL STATES OF PROTOPLASM

A. INTERNAL ORGANIZATION

In the preceding discussion protoplasm has been dissected and its constituent molecules separated from one another and further analyzed in turn. The properties and composition of the fragments have been examined in some detail. But, if we should crush and grind and extract a living cell without losing a single one of its constituents, we should obviously be dealing with a nonliving collection of chemicals. Or if, on the other hand, we were to concoct a mixture of all the known cell constituents, in the correct proportions, we should not have a living cell, or even protoplasm. We should still lack the precise orderliness of internal interrelationships and the chemical, physical, and structural integration of the constituent parts of the cell. Protoplasm is something more than the sum total of the chemical properties and states described. What further can be said of the organization of the foregoing materials in protoplasm?

B. COLLOIDS

In the state of the proteins in protoplasm lies a secret of protoplasmic organization of far-reaching consequence. Protein is present in what is called a *colloidal suspension*. Colloidal states are not peculiar to proteins, or even to living systems, but are common in inorganic systems, where many of their properties have been investigated.

Colloids have certain properties in common with true solutions. The suspended colloidal particles are distributed homogeneously throughout the water. Colloidal particles, like ions or molecules in solution, do not settle out.

There are, however, certain important distinctions in the case of colloidal suspensions of protein which explain much about the organization and properties of protoplasm and cells. Colloidal particles are much larger than the ionic or molecular particles disseminated throughout a solution. Estimates place the size of colloidal particles at 0.000001–0.0001 mm. in diameter. The latter size would be just at the lower limit of visibility under the ordinary microscope. This, of course, is far larger than a sodium ion, or even a molecule of glucose. It is recalled that protein molecules are relatively very large. Some of the larger of these are about 0.0001 mm. in diameter. Each protein molecule probably constitutes a discrete colloidal particle in protoplasm.

To say that colloidal particles are "large" is strictly relative. In terms of the size of a cell they are extremely small. A consequence of this is that within a cell there is a tremendously large total surface area of the colloidal particles. Intracellular reactions are believed to occur essentially at the surfaces of the colloidal particles, just as reactions between any two inorganic reagents occur only at the points of contact, or adjoining surfaces of those compounds. Consequently, the distribution of the cell proteins into these tiny particles and the multiplication of reactive surface area thereby play a significant role in determining the rapid rate at which intracellular reactions are known to occur. Some of the protoplasmic poisons are thought to exert their deleterious action by covering the highly reactive surfaces of the colloidal particles.

C. SOL AND GEL STATES

One characteristic peculiar to colloidal suspensions is the ability to undergo distinct changes in consistency or rigidity. The very same colloidal suspension may, at one time, be as fluid as water and, at another time, have the consistency of jelly. A well-known example of this is the case of gelatin, familiar to everyone. Gelatin is a protein which forms a colloidal suspension in water. When the system is warm, its consistency is that of water; when cooled, it jellies. The fluid condition is called the *sol state*, and the semisolid condition the *gel state*. The two states are readily reversible. We can change the system from the sol to the gel state, and back again as often as we wish, simply by cooling or warming, as the case may be. This is not merely a matter of freezing and thawing, as might be guessed from the example given. The temperature changes are too small to freeze or thaw the water of the system. Furthermore, it is possible to produce similar reversible changes in a colloidal suspension quite independently of changes of temperature. There is no change in the chemical composition of the suspension. The change is one of rearrangement of the relationships between the solid particles, or *solid phase*, and the watery portion, or *watery phase*, of the colloidal system.

The nature of the change from gel to sol or the reverse is not well understood in terms of just what is happening to the colloidal particles. One suggestion is illustrated in Figure 5. The first diagram

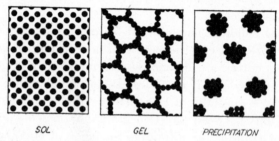

FIG. 5.—Colloidal states. In the fluid sol state the particles bounce about freely. In the gel state the particles probably form a continuous network, rendering the system jelly-like. When colloidal particles clump together into larger masses, they "settle out" by gravity, or precipitate.

represents the sol state. In this the solid particles (*black*) are separated from one another by water (*white*) throughout. The solid phase is said to be discontinuous. The watery phase is continuous; that is, one could trace an imaginary course from any one part of the suspension to any other part along a watery pathway. The second diagram represents the gel state. Here the solid particles are contiguous throughout, and water is entrapped in little lakes within the meshwork. Here the watery phase is discontinuous, and the solid phase continuous. In both cases note that there has been no change either in the size or nature of the solid particles or in the number of particles per unit volume of the suspension.

Whether this is a true picture cannot be said with certainty. It would explain nicely the physical characteristics of fluidity or relative rigidity. It is readily seen that one system would pour like water, whereas the other would tend to be semisolid. We know also that tremendous pressures are required to force the water from a colloidal gel, which would be expected if a gel is constructed as described.

At any rate, sol and gel states, and the transformations of one to the other, are common in the colloidal state of protoplasm. Plasma membranes, as well as the network of membranes extending through the cytoplasm, are colloidal gels. The bulk of the cytoplasm is largely in the sol state. The gelled protoplasm of the surface of the cell and the intracellular membranes probably maintain the structure of the cell. If the surface membrane is mechanically disrupted by microdissection, the more fluid interior, or at least a portion of it, tends to escape. If the environmental conditions are suitable, the escaping sol is converted into the gel state. In this way the membrane is restored, and the cellular "wound" is healed. Sol-to-gel and gel-to-sol transformations in cells are probably also involved in ameboid movement and perhaps in muscle contraction, as well as in other cellular activities which involve rapid changes in shape or contour of cells.

The clotting of blood is largely such a phenomenon. Suspended in the fluid portion of the blood are colloidal particles of a special protein (called *fibrinogen*). Normally, this is in a sol state. But under certain conditions, such as the shedding of blood or injury to a blood vessel, mechanisms are set into operation which convert this protein into a special kind of gel. The fluid blood is thereby changed into a semisolid clot.[14] The utility of this in controlling blood loss is obvious. More details of its mechanism will be discussed in chapter 3.

For completeness, and also for clarity, Figure 5 includes a third diagram. This illustrates what occurs when colloidal particles clump together, or *agglutinate*. The large clusters do not remain in suspension but settle out as a precipitate. Heating produces such precipitation of some colloids. In protoplasm this is a nonphysiological process.[15] It occurs in abnormal situations, such as heating, or treatment with some of the poisons, and produces death of the cell.

D. Electrical Charges

Some of the properties of colloids and of colloidal cell membranes are apparently related to the presence of electrical charges upon the particles. It is thought that these charges are partly responsible

14. This particular sol-to-gel transformation is irreversible.

15. By *physiological process* is meant a normally occurring process. *Nonphysiological* refers to an abnormal process.

for the state of diffusion of the particles throughout the watery phase, for, in any given medium, all the colloidal particles possess like electrical charges. Either all are negatively charged or else all are positively charged. Since like charges repel one another, the charged colloidal particles would be held apart. Treatment of many colloidal suspensions in such a way as to neutralize the charges causes the particles immediately to clump together into large clusters and settle out of the suspension. Obviously this destroys the colloidal organization, and treatment of protoplasm in this manner so deranges its internal structure that life-activities cease at once.

III. ENERGY RELATIONSHIPS

A. METABOLISM AND ENERGY

The term *metabolism* has been applied to the sum total of the chemical reactions that occur within cells or within the organism. Our discussion so far has clearly indicated that many reactions take place continuously. Most of these can better be understood later, and one or another of them will be referred to and discussed in subsequent chapters. At present only some of the main features of the chemical reactions of protoplasm will be presented.

Roughly, metabolic processes are divided into two main categories: those in which the compounds of cells are synthesized, or built up, called *anabolism;* and those in which chemical breakdown or decomposition occurs, called *catabolism.* Both usually proceed at all times. Anabolism is obviously involved in any increase in the total bulk of protoplasm, as in growth or reproduction, where new protoplasmic compounds are being synthesized. But it goes on even independently of growth. In the life-activities of most cells there is a continuous destruction, a steady disintegration, an uninterrupted catabolism of protoplasm. There is a certain amount of wear and tear, just as there is upon a mechanical system like a machine. A machine does not grow, yet there is a steady wearing of its parts, and, as long as it continues to run, repairs must be made and worn-out parts must be replaced. Protoplasm likewise replaces its "worn-out" parts, and proteins and other compounds continue to be formed as long as life lasts. In the living organism, of course, the machine itself does the replacing of its own parts. Protoplasm is constantly manufacturing more of itself.

All these processes require energy. Growth, repair, synthesis, and construction of new protoplasm require energy just as truly as energy in some form is required to manufacture and replace the parts of a car. Energy is likewise necessary for the maintenance of the dynamic equilibrium in cells called "life"; for the preservation of protoplasmic and cellular structure, such as the cell membranes; and for the activities of the "resting" cell.

B. Sources of Energy

Ultimately, this energy comes from the sun. In photosynthesis green plants bind the energy of sunlight into carbohydrates, from which, in turn, the plant can manufacture both fats and, with the addition of nitrogen compounds, proteins. Starches, fats, and proteins contain the stored energy of sunlight, which is drawn upon in the life-processes of protoplasm. Only by a chemical shattering of the energy-containing molecule is its energy made available to cells.

Cells accomplish this by the process called *respiration*. Broadly speaking, respiration is an oxidative phenomenon, a combustion process. Sugar, as was pointed out earlier, may be oxidized by cells in the following reaction:

$$C_6H_{12}O_6 + 6O_2 \rightarrow 6CO_2 + 6H_2O + Energy .$$

The important feature of this reaction, from our present point of view, is not the chemical end products it yields but the liberation of energy it entails. For this reaction, obviously, free oxygen must be available to the cell and must be present within it as such in the dissolved state. This process is called *aerobic* (meaning "with air") to distinguish it from a second kind of energy-liberating process.

Some cells derive energy from foods in the absence of free oxygen, in which case oxidation of the kind described above cannot occur. It is well known in chemistry that oxidation can occur *anaerobically* —independently of free, or molecular, oxygen. Oxygen may be obtained from some other compound which has atomic oxygen in its molecule, or the oxygen from one part of a molecule may oxidize the rest of the molecule, as in the following reaction:

$$C_6H_{12}O_6 \rightarrow 2C_2H_6O + 2CO_2 + Energy .$$

In this instance two atoms of carbon of the glucose have been oxidized to carbon dioxide by oxygen obtained from the rest of the molecule. The carbon in the alcohol molecules (C_2H_6O) has been reduced.

Precisely this reaction occurs in some living cells and is called *fermentation*. The significant feature here, again, is that the reaction yields energy. It is a molecule-shattering process which liberates the energy bound within the molecule in essentially the same way that oxidation with free oxygen yields energy. In fermentation, however, the energy yield per molecule of glucose is less than in respiration. The reason for this is clear. Some of the energy originally contained in the glucose molecule still remains bound in the relatively large alcohol molecule at the end of the reaction.

Reactions of this type enable cells to obtain energy and to survive even in the absence of free oxygen. The degree to which cells can do this varies considerably. Mammalian cells can obtain energy anaerobically only for limited periods of time. Yeast cells can continue living and growing and reproducing indefinitely, fulfilling all their energy requirements by fermentation of glucose into alcohol and carbon dioxide in the complete absence of free oxygen. The relatively small energy yield per molecule of glucose is compensated for by the breakdown of more molecules. A yeast cell is capable of fermenting approximately its own weight of glucose per minute.

Some cells (certain bacteria) are able to obtain energy only anaerobically and actually perish in the presence of free oxygen. In any case, specific enzymes must be present to catalyze the reactions, whether they be aerobic or anaerobic.

IV. MOVEMENT OF MATERIALS THROUGH CELLS

The never ceasing chemical activity of the cell necessitates a constant entrance of materials into it and passage of other materials out. Oxygen and glucose, for example, enter the cell more or less constantly, and waste products of metabolic activity leave. In other instances materials are moved through cells, as in the case of the respiratory gas oxygen, which passes from the air passages of the lungs through the cells of the blood-vessel walls into the blood, and again through similar cells, as it leaves the blood on the final stage of its journey. Likewise, glucose present in the intestinal cavity traverses the cells of the intestinal lining and the *capillary*[16] walls before it arrives at the cells where it is to be used.

Much of this movement of molecules is of a purely physical nature, often observed and studied at length in nonliving systems. *Diffusion*, *filtration*, and *osmosis* are all involved.

16. The tiniest of the blood vessels are called *capillaries*. Their thin walls are but one cell in thickness.

A. Diffusion

Diffusion depends upon the fact that particles (ions or molecules) in solution are in a continuous state of rapid motion, resulting in frequent collisions. If a lump of sugar is dropped into water, molecules from the surface of the sugar will immediately go into solution; these molecules, by virtue of their motion, tend to spread themselves throughout the solvent. At first the concentration of dissolved sugar molecules in regions immediately adjacent to the undissolved remnant of sugar is greatest. In areas of the solvent most distant from the dissolving solid the concentration of dissolved molecules is least. In intermediate localities an intermediate concentration

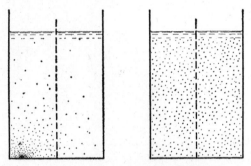

Fig. 6.—Diffusion. *Left:* Migration of sugar molecules away from a dissolving lump of sugar. *Right:* Establishment of equilibrium, when the membrane (*dotted line*) permits the passage of sugar molecules (see also Fig. 9).

obtains. At any rate, there is a *gradient* of decreasing concentration from the lump of sugar to the portion of the solution most distantly removed from the solid sugar. This would also be true just at the moment that the last bit of sugar is dissolved.

But, after a time, the movements of the frequently colliding molecules result in an equalized distribution throughout the entire solvent. The concentration of sugar in any unit volume becomes equal to the concentration of sugar in any other unit volume of the system. Thus there has been, during the whole process, a migration of molecules from regions of high concentration, where collisions are frequent, to regions of lower concentration. This movement of molecules is called "diffusion." We shall encounter it again and again in many physiological systems.

Diffusion occurs just as described, even though the watery medium is separated into two parts by a mechanical partition, provided the partition is *permeable to* (i.e., "permits the passage of") the dissolved diffusing molecules.

In unicellular organisms, or multicellular forms of the simpler types, exchange of materials between interior and exterior of the organism is accomplished in part by diffusion. Molecules used up by cells are usually less concentrated within the cell than in the environment and will diffuse into the cell, provided the plasma membrane

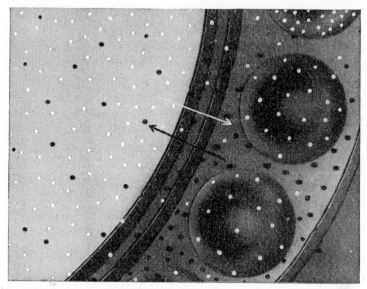

Fig. 7.—Diffusion in the lungs. Oxygen (*white dots*) diffuses from lung air sacs into lung blood capillaries (*as indicated by white arrow*) because free oxygen is more concentrated in the air sac than in the blood. Carbon dioxide (*black dots*) diffuses in the reverse direction (*as indicated by black arrow*) because it is more concentrated in blood than in the air sacs. Oxygen entering the blood by this physical process is transported by the red blood cells. The flow of blood is upward in this picture. (From the sound film, *Mechanisms of Breathing.*)

does not bar the way. Likewise, molecules produced in the cells will necessarily be more concentrated there than in the environment and, unless they are chemically bound to the protoplasm, will tend to diffuse from the cell. Thus a direct exchange of materials between the living cells and the external environment is effected.[17]

Similar activity is involved in the metabolic exchanges which take place between the cells of the large mammalian body and the fluids in their immediate environment. But, in addition, we must also

17. Of course, this is not the only mechanism of exchange. Phagocytosis by the ameba, ciliary movement in the gullet of *Paramecium*, and action of the contractile vacuoles are modes of interchange of materials independent of diffusion.

explain the transport of materials to the vicinity of the cell and the movement of wastes away from the cell. Many physiological processes are involved, but, here too, diffusion plays a significant role. Consider, for example, the manner in which oxygen gas gets from lungs to protoplasm and the waste carbon dioxide traverses the reverse route, eliminating the circulation of the blood from consideration for the moment. Dissolved oxygen passes through lung capillary walls, from the lung cavities into the blood. The concentration of free oxygen atoms is higher in the lung cavities than in the blood, and the cells through which it must pass in order to negotiate this journey are freely permeable to it. Likewise, oxygen diffuses from the interior of capillaries to the fluids bathing the tissues and finally into the cells themselves, where the concentration of free oxygen is very low. At no time in this movement are there barriers to diffusion in the form of cell membranes impermeable to oxygen. Carbon dioxide molecules produced in cells by oxidation move in exactly the reverse direction, driven to do so by a decreasing gradient of molecular concentration, from cells to tissue fluids to blood to interior of the lung cavities. (See chap. 6 also.)

B. Filtration

Filtration also is a purely physical phenomenon, producing a movement of materials across and through cells. It is dependent upon differences in mechanical pressure on the two sides of the cell involved. Its nature can be illustrated by the following. Suppose we have a quantity of water contained within a closed bag of linen of fairly fine mesh. The bag, in turn, is immersed in water. If the fluid in the bag is compressed by mechanically squeezing the bag, there will be a movement of water from the interior to the exterior of the bag, provided the mesh is not so fine as to render the bag "watertight." The mechanical elevation of pressure within the bag simply squeezes fluids out. And, if in this fluid there are dissolved ions or molecules which also can pass through the meshes of the bag, these likewise will be squeezed out with the watery solvent into the surrounding water.

In mammals (as well as in vertebrates and even some invertebrates) there is an exactly corresponding situation, with the liquid blood at a pressure above atmospheric, separated by the cells of the capillary blood-vessel walls from the fluids bathing the tissue cells, which are at atmospheric pressure. These cells permit the passage not only of water but also of dissolved ions and certain of the dissolved molecules. Consequently, there is an apparently continuous squeezing of the water of the blood, along with some dissolved constituents, out of the blood stream. In this manner dissolved oxygen and nutriments and other materials used in activity are transported from the blood to the cells. Occurring as it does simultaneously with diffusion, it is probably a more effective means of transport, inasmuch as it moves materials more rapidly.

Obviously, such migration of metabolic products can occur only where mechanical pressure differences exist, such as in the system of blood under pressure plus tissue fluids. Of itself, it causes only a one-way movement and cannot explain the transport of materials from cells into blood against the mechanical pressure gradient. Another mechanism is involved here (see p. 60).

C. PERMEABILITY OF CELL MEMBRANES

We have stated that, in diffusion and filtration, dissolved particles move across cells and through cell membranes only when those membranes afford no obstruction. It is characteristic of cell membranes, however, that they do obstruct the passage of certain particles. Some particles can readily traverse membranes, in which case we say that the membrane is permeable to them. In other instances the way is barred; the membrane is impermeable. There is the striking phenomenon of "selective" permeability, in which a membrane "permits" passage to certain materials and "refuses" passage to others. Not all cell membranes are alike in this respect. Ions, for example, readily pass through or between the cells of the walls of the finest blood vessels but penetrate through the plasma membranes of some tissue cells only with difficulty if at all. One consequence of this is the difference in concentration of some ions inside and outside the cells. Inside red blood cells there is considerable potassium and little if any sodium; outside these cells there is many times as much sodium as potassium. If it were not for the selective permeability of the red cell membrane, the ions, moving as freely as water across it, would be equally concentrated inside and outside the cell.

What accounts for selective permeability? Why is it that small ions cannot cross some cell membranes, but certain larger molecules, such as those of glucose or gycerol, can?

We have been using certain objectionable terms, such as "selective," "rejection," and "admit" or "refuse" passage, which might imply the operation of some living force or conscious agency, something peculiar to living cells, something not encountered in the inorganic world. But, of course, nonliving membranes displaying "selective" permeability, or (a better term) *semipermeability*, are well known to the chemist, and the biologist seeks to explain cell-membrane permeability on the same physical basis as does the chemist. However, neither biologist nor chemist has been entirely successful in this.

1. *Size of molecules.*— Is the size of the molecule involved an important consideration? We might guess that perhaps there are minute pores in semipermeable membranes large enough to admit water but too small to admit larger particles. This seems to be the case for particles above a certain size. For, in general, intact protein molecules (the largest molecules we know) cannot penetrate cell boundaries, while the smaller amino acid molecules can.

2. *Fat solubility.*—But, on the other hand, among the smaller particles relative size seems to play a lesser role. An important factor here is the solubility of the molecule in fat. If molecules of approximately equal size are compared, it is found, in general, that those molecules which are readily soluble in fat enter or leave cells more readily. A large fat-soluble molecule may penetrate the cell more readily and quickly than a much smaller molecule not soluble in fat. This we relate to the fact that fats and fatlike materials constitute an integral part of the structure of the cell, particularly of the cell membrane. We may tentatively regard the cell membrane as a mosaic of gelled protoplasm, in which some of the blocks are missing, leaving minute interstices, and some of them composed of fat or fatlike materials. It is thought that the fat-soluble molecules enter the cell by first dissolving in the fatty portions of the membrane (see Fig. 8).

It is also found that, if such fat-soluble molecules are present in excess, they may dissolve out enough of the fats of the cell membrane to disrupt the cell. The general anesthetics and narcotics, such as ether and chloroform, which act alike upon single cells or multicellular plants and animals, are fat soluble and fat solvents. They enter freely through the fatty membranes and finally may dissolve them away, causing disintegration and death of the cell.

3. *Electrical charges.*—In some instances the electrical charges almost universally present upon cell surfaces seem important in de-

termining permeability or impermeability of certain cell membranes to ions, which likewise possess electrical charges. Thus a cell like a red blood corpuscle, with a positively charged surface membrane, would repel positively charged particles and thereby prevent their passage, while negatively charged particles or uncharged particles even larger than ions might readily enter or leave the cell. Of course, even a migration of negative ions into the cell could not occur unless there was a simultaneous migration of equal numbers of negatively charged particles in the opposite direction.

4. *Unknown factors.*—Though many phenomena of cell-membrane permeability can be explained on the basis of molecular size, fat solubility, and electrical charges, others cannot. Cells in many loca-

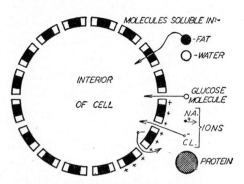

tions display specific permeability discriminations which cannot as yet be explained on known physical or chemical grounds. Sodium chloride can normally penetrate the cells of the intestinal lining. Some other salts, such as magnesium sulfate, cannot, or at least do so much more slowly. Similarly, amino acids freely pass through these cells in the process of absorption of digested foods. The very closely related but injurious compounds, the *amines*, which are produced by the intestinal bacteria, apparently penetrate the intestinal

FIG. 8.—Permeability of cell membranes. Molecules the size of glucose are thought to penetrate cell membranes through interstices too small to admit passage to larger water-soluble molecules or to protein molecules. Large fat-soluble molecules may pass through fatty portions of the cell membrane (*shown in black*). Plus-charged ions are repelled from the plus-charged cell surface. Minus-charged ions may enter or leave the cell under special conditions.

wall with somewhat greater difficulty. Some unknown properties or activities of the cells seem responsible for these important discriminations.

It is probable that active metabolic processes of cells contribute to the maintenance of different concentrations of certain substances on the two sides of the cell membrane (i.e., inside and outside the cell). According to this view, the membrane is not merely a passive structure, permitting some substances to pass through and rejecting others. In addition, a kind of "metabolic pump" apparently operates to determine the relative concentrations of materials adjacent to living membranes.

So the problem of cell-membrane permeability must be left in this state for the present. It is partly explained on physicochemical grounds and partly one of the unsolved puzzles of cell activity to be explained by future investigators. In any case the importance of the phenomenon in the life of the cell and the organism is universally recognized.

D. Osmosis

1. *Nature of the phenomenon.*—Osmosis, in inorganic systems or in cells, is a direct consequence of membrane semipermeability. Sup-

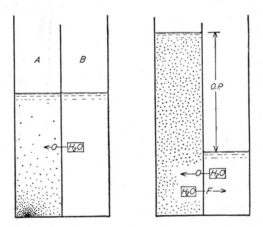

FIG. 9.—Osmosis. A membrane impermeable to sugar prevents the dissolving sugar molecules from diffusing into chamber B. Water migrates from B to A, and equilibrium is established (*right*) when the osmotic flow of water (O) is equalized by the filtration of water (F) in the opposite direction (see also Fig. 6).

pose the chamber illustrated in Figure 9 to be divided into two compartments, A and B, by a membrane permeable to water but impermeable to sugar molecules. A small lump of sugar is placed in compartment A. The sugar dissolves, and dissolved particles diffuse throughout compartment A only. Ultimately, there is an equal distribution of sugar molecules throughout all parts of A. Part B still contains only distilled water.

At the same time, osmosis also occurs. The level of the fluid on side A rises and that on side B falls. The reason for this movement of water is obscure and presents a problem to the biologist and to the physicist and chemist as well. One hypothesis states that in pure water all the molecules are in constant motion but that in a solution many water molecules adhere to, or are *adsorbed* on, the molecules of dissolved substance. In our example such water molecules, bound to the large sugar molecules, cannot move through the membrane. This leaves fewer freely diffusible water molecules on the *A* side than on the *B* side, where any water molecule is free to penetrate the membrane. Consequently, more water moves from *B* to *A* than from *A* to *B*.

Exactly the same movement of water from *B* to *A* would occur if sugar molecules were present on both sides of the membrane, provided the concentration were greater on side *A*. Thus, as in the case of diffusion, the movement of molecules (in this case, of water) has been such as to tend to make the concentration of dissolved molecules alike in all parts of the system. Such a state is never actually reached, however. Before this occurs, the level of the water in *A* becomes stationary at a fixed height above the level in *B*.

This height is determined by the difference in molecular concentration in chambers *A* and *B*. If no sugar is present in *B*, it depends upon, and is a measure of, the total sugar concentration in *A* alone. The height (*O.P.*) to which the water rises in *A*, expressed in millimeters, is a measure of the *osmotic pressure* of the solution in *A*.[18]

Why, we may ask, does the water rise only to a limited height? Why does not the movement of water continue until *A* is infinitely dilute, or until all the water in *B* has passed into *A*? The reason for this may be seen if we consider the action of gravity on the fluid in *A*. The gravitational pull is such that the fluid toward the bottom of *A* is subjected to a rather high pressure—a hydrostatic pressure exerted by the column of water above. This pressure clearly exerts an action tending to force fluid back through the membrane from *A* to *B*. It tends, that is, to cause a filtration of water against the osmotic flow of water. The greater the height (*O.P.*), the greater is this filtering force, or filtration pressure. Equilibrium is reached, and the level of *A* remains constant, when the filtration pressure on side *A*, owing to gravity, causes as much water to leave *A* as is caused to enter *A* by osmosis.

2. Osmotic equilibria in the body.—Osmosis operates in exactly this way in cell-membrane systems. Cell membranes generally are relatively impermeable to salts, or at least are less permeable to salts

---

18. Ordinarily, osmotic pressure is expressed in millimeters of mercury instead of water.

than to water. If cells are placed in a solution of salts whose concentration is a little greater than that within the cell, water passes out of the cells by osmosis, and they shrink in size. If they are now placed in solution of proper salt osmotic pressure, they fill out and resume their normal size and appearance. If cells are placed in a solution whose salt content is less than that within the cell, water passes from the solution into the cell by osmosis; and, if the process is observed under the microscope, one may see the cell swell and increase in size. When the difference in concentration inside and outside is sufficiently great, the influx of water will soon disrupt the cell membrane. The cell appears to burst and is destroyed (Fig. 10).[19]

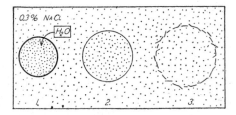

Fig. 10.—Swelling and bursting of a cell when placed in a solution of salt (*black dots*) of a concentration less than the salt concentration inside the cell.

Osmosis and cell destruction by weak salt solutions or pure water are not only of theoretical interest but also of practical importance both to the physiologist and to the physician. When living tissues are removed from the body of an experimental animal for physiological studies, it is necessary that the fluid which moistens or immerses the preparation have the proper salt osmotic pressure. If pure water or too weak a salt solution is employed, the cells of the tissue or organ are injured or destroyed, and the preparation dies.

Any cells to be observed in the living state, then, whether they be isolated cells of the blood or components of an isolated tissue or organ, must be moistened with a salt solution of a concentration equal to that of the blood or the interior of the cells, which in mammals is about 0.9 per cent. A 0.9 per cent sodium chloride solution is, therefore, sometimes called *physiological salt solution*.[20]

Likewise, when quantities of liquids are injected into an animal,

19. The term applied to this cellular disintegration is *cytolysis*, literally, a "solution"of the cell or a "loosening"—i.e., a freeing—of the cell contents.

20. It is also called *isotonic* salt solution. A solution weaker than this is *hypotonic;* a stronger solution, *hypertonic.*

either for experiment or for the treatment of human disorders, they must be of nearly the same osmotic pressure as the blood. Solid compounds of all sorts must be dissolved in physiological salt solution, for injection purposes, if cell injury is to be avoided.

Despite its name, "physiological salt solution" is not truly physiological. It is physiological with regard to its osmotic pressure and its sodium and chlorine ions, which are the most abundant inorganic ions in the body fluids. It is not physiological partly in that it lacks other ions, such as calcium and potassium, which, we have seen earlier, are necessary as constituents of the environment for normal activity of cells.[21]

In our description of the nature of osmosis we referred to the phenomenon of filtration and saw that equilibrium was reached when filtration and osmosis became equalized. A parallel situation obtains in the water balance existing in the body between blood and the tissue fluids. The pressure of the blood appears to force water continuously through the walls of the small blood vessels, along with such dissolved blood constituents as encounter no resistance to passage through the cells of the vessel walls. If this process operated unopposed, all the fluids of the blood would soon be filtered out of the blood vessels. In the blood, however, there are certain proteins in colloidal suspension. These are able to penetrate the normal vessel walls only with difficulty, if at all. In the liquids outside the blood stream there are also these proteins but in distinctly lower concentrations. That is, under normal conditions the protein osmotic pressure of the blood exceeds that of the tissue fluids. As a consequence, there is apparently a continuous

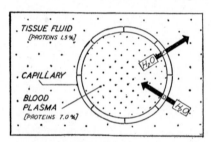

FIG. 11.—Osmosis and filtration in capillaries. The tendency for water to be filtered out of capillaries (F) is approximately equalized by the osmotic influx of water into capillaries (O) produced by the greater concentration of protein molecules (*black dots*) inside the vessels. In any single capillary, filtration may exceed osmosis, or vice versa. But in a given group of capillaries the total quantity of water passing out by filtration is nearly equal to that re-entering by osmosis.

21. In Ringer's solution this defect is partially corrected. It contains sodium, potassium, and calcium ions in concentrations equal to those occurring in body fluids. Many other solutions imitating the composition of the body fluids have been prepared and are used in special experiments. One contains glucose, as well as Na, K, and Ca, and others have even more nearly a true physiological composition. For most work 0.9 per cent sodium chloride solution or Ringer's solution is adequate.

influx of water from the tissues into the blood, by virtue of the higher protein osmotic pressure of the blood.

These two forces seem to be approximately equal. The filtration of water out of the blood stream is, on the average, approximately equal to the osmotic pull of water into the blood vessels.[22] This not only serves to maintain an approximate constancy of the blood volume but also provides a kind of circulation of water, with its nutriments or wastes, as the case may be, from blood to cells and back again into the blood.

22. That this is not a perfectly balanced dynamic equilibrium is shown by the fact that some of the water filtered through the capillaries re-enters the blood by way of the lymph vessels (see chap. 5).

CHAPTER THREE

BLOOD AND THE
INTERNAL ENVIRONMENT

I. Evolution of the Circulation

II. Solid and Liquid Components

III. Blood Volume

IV. The Plasma
  A. The inorganic salts
  B. The blood sugar
  C. The plasma proteins
    1. Viscosity   2. Water balance   3. Fibrinogen
  D. Special plasma substances

V. The Clotting of Blood
  A. Stoppage of flow; exposure to air
  B. Initiation of clotting
    1. Contact with injured cells   2. Contact with foreign surfaces: platelet disintegration
  C. Substances involved
    1. Clot-inducing material   2. The role of calcium ions   3. Fibrinogen and thrombin   4. Prothrombin
  D. Defibrination
  E. Complexity of the machinery
  F. Repair of damaged vessels
  G. Defective clotting

VI. The Red Blood Corpuscles
  A. Visible features
  B. Hemoglobin
  C. Stroma framework
  D. Hemolysis
    1. Osmotic equilibria   2. Fragility   3. Action of fat solvents   4. Action of biological agents   5. Species specificity
  E. Life-cycle of the red cell
    1. Red-cell counts; hemoglobin determination   2. Formation and destruction   3. Span of life   4. Correlating mechanisms   5. Physiological constants
  F. Anemia
    1. Hemorrhage   2. Bone-marrow defects   3. Nutritional deficiencies   4. Pernicious anemia

VII. Transfusion of Blood
  A. Early difficulties.
  B. Safe blood transfusions
    1. "Reactions" to transfusions   2. Cross-matching   3. Blood groups
    4. The Rh factor
  C. Value of blood transfusions
  D. Transfusions in war and peace

VIII. The White Blood Cells
  A. Kinds of cells
    1. Granular white cells   2. Agranular white cells   3. Varieties of granular and agranular white cells
  B. Origin and life-span
  C. Functions
    1. Defense against infection; phagocytosis   2. Controlling mechanisms
    3. Tissue repair
  D. Fluctuations in the white count
    1. Normal variations   2. Effect of infections: leucocytosis   3. Leucemia

IX. The Blood Platelets

X. Blood Damage in Atomic Explosions

I. EVOLUTION OF THE CIRCULATION

The evolution of circulatory systems goes hand in hand with the development of body bulk. In the case of unicellular forms, food materials, oxygen, and the water are contained in the immediate environment of each cell individual. For particles of such materials to reach any part of the individual only a relatively small distance need be traversed. Similarly, waste products formed in the life-activities of protoplasm need only travel, at most, a distance equal to the radius of the cell to be eliminated entirely from the organism. Even in the simpler multicellular forms a similar situation obtains. *Hydra* is multicellular, but the body structure is such that nearly every body cell is adjacent either to the exterior of the animal or to the large *gastrovascular* cavity (see Fig. 12). Wastes from any part of any cell are therefore passed to the exterior, or into the large body cavity directly, by the very cell in which the waste is produced. Food and oxygen in the environment lie at a distance not exceeding one cell diameter away from the protoplasm of any part of the organism.

The limitations of a body plan of this kind, where every cell is in contact with the external source of food and depository of wastes, are obvious. At no point can such an organism advantageously be more than two cells in thickness. With the development of mesoderm in forms like the present-day flatworm a greater diversity of tissues is made possible. From the mesoderm are developed such structures as muscle and bone, which make possible a more complex individual, with more diversified behavior, capable of invading a greater variety of environments.

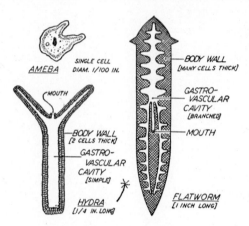

But with this increase in body bulk it becomes necessary to provide for a ready exchange of materials between cells lying deeply within the organism and the exterior. An early solution of the problem is seen in the evolution of the flatworm type of animal: The gastrovascular cavity is branched in such a way that the terminal ramifications of the system penetrate to all parts of the body. Food taken into the cavity is circulated to all parts of the branching system by the movements of the body and need travel but a short distance from some part of the cavity to reach its cellular destination. A system of tubes for elimination of wastes also ramifies through all parts of the body. Thus, again, any cell of the body is close to the exterior in the sense that near every cell is an end branch of the gastrovascular cavity and of the excretory system. This scheme seems adequate for the flatworms, but it would be a cumbersome and inadequate plan for a still bulkier animal.

In animals thousands of cells in thickness we encounter an entirely different device. The gastrointestinal tube, from which food enters the organism, is not branched. We find instead a many-branched system of tubes containing a circulating fluid, which reaches all parts of the body. Some one of these vessels now courses in the immediate vicinity of each cell. The system serves as an intermediary, transporting material between the cell and the source of food (stomach, small intestine), or oxygen (lungs), and the waste-eliminating organs (lungs, large intestine, kidneys). All these organs, in turn, open to the exterior of the animal.

The cells of the interior of the body are now spatially removed from the external environment by many intervening layers of cells. Only

FIG. 12.—Simple animals in which food is distributed to all parts of the organism without the aid of a blood circulatory system.

via the blood is communication possible. When considering the cells themselves, then, "environment" comes to have a new meaning. It refers to the conditions in the intercellular fluids immediately adjacent to the cells, or the *internal environment*, a concept of far-reaching significance. Obviously, the blood is a large factor in determining its nature. Foods and oxygen are brought to it by the blood, and wastes are removed from it in similar fashion. So the blood is more than simply a vehicle for transportation. It is an agency which preserves, in the internal environment, the conditions for normal cell activity.

This concept of the internal environment should be kept in the foreground not only during our consideration of blood but also in all that is to follow about the machinery of the body, for, in the words of Claude Bernard, "all the vital mechanisms, however varied they may be, have only one object [i.e., result], that of preserving constant the conditions of life in the internal environment."

II. SOLID AND LIQUID COMPONENTS

Blood is a convenient "tissue" to study because it is so readily obtainable. Even human blood can be had with only the slight inconvenience of insertion of a needle into a vein or, in smaller quantities, by lightly pricking the skin. Further, blood so obtained is not injured, and the blood constituents can be observed in their normal state. Many of the principles applicable to cells in general and many of the relationships of cells to their immediate environment have been worked out in studies on blood. Observations of this sort are more difficult to make upon compact tissues like liver or kidney, where isolation of cells is almost certain to entail some damage to those cells and where the tissue fluids which constitute the immediate internal environment of the cells are scant and difficult to obtain.

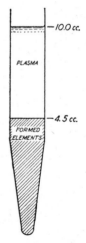

Fig. 13.—Separation of the formed elements (cells) of blood from the liquid portion by gravity.

Grossly, blood appears to be a homogeneous, red, viscous fluid. But, microscopically, one can see that it is composed of discrete particles suspended in a watery fluid. The particles, called the *formed elements*, consist of the *red blood cells*, the *white blood cells*, and the *platelets*. These may be separated from the fluid portion, or *plasma*, simply by allowing blood, treated to prevent clotting, to stand in a tube. The formed elements are slightly heavier (sp. gr., 1.090) than plasma (sp. gr., 1.030), and they slowly sink by gravity.[1] This difference in specific gravity is insufficient to cause settling of the formed elements from plasma in the circulation where blood is kept in continual agitation.

III. BLOOD VOLUME

The total volume of the whole blood in the living body can be estimated in several ways. A known quantity of a relatively innocuous dye is injected into a vein. A few minutes is then allowed for the dye to become well mixed with the blood. At this time, and before appreciable quantities of the dye have had time to escape from the blood stream, a measured quantity of blood is drawn and its dye content determined. From such figures the total blood volume is easily calculated. For example: Supposing 1.0 gm. of dye is injected, and time allowed for distribution of this throughout the circulating blood. Then, if a 5 cc. sample of blood is found to contain only 0.001 gm., we calculate that the 1.0 gm. of dye must be distributed throughout 1,000 times 5 cc., or 5,000 cc. On the assumption that our sample contains an average concentration of the dye, and that none of the dye has left the blood stream, this figure (5,000 cc.) is taken to be the total volume of circulating blood.

This and related methods based on the same principle indicate that in mammals the blood makes up about one-thirteenth of the total body weight. Individuals weighing 65–75 kg. (143–65 lb.), then, would have about 5–6 liters (5.2–6.3 qt.) of blood.

As a matter of fact, neither of the assumptions made in the application of the dye method holds strictly. The dye is not uniformly distributed throughout the whole blood, and some of the dye does pass out of the blood stream.

IV. THE PLASMA

The formed elements make up about 40–50 per cent of the volume of whole blood; and plasma, 50–60 per cent. These figures vary somewhat, even in health, with temporary physiological changes in the water content of the blood. If one perspires profusely, the cells are temporarily more concentrated, and the volume percentage of

1. In laboratory practice it is customary to hasten the separation by *centrifuging* the blood. A centrifuge whirls a tube of blood about in such a way that centrifugal force quickly drives the heavier formed elements to the bottom of the tube, leaving the supernatant clear or straw-colored plasma to be pipetted off.

plasma is reduced. If considerable water is drunk, for a short time there will be a larger proportion of plasma. In any case, the fluctuations are slight and temporary, and the relative proportions of cells and fluid remain fairly constant.

Approximately 90 per cent of the plasma consists of water, in which many substances are dissolved or suspended. Obviously, the plasma at one time or another contains every product which tissue cells use and obtain from the outside and also all substances produced by cells which are transported to other organs to be used, in turn, by them or excreted from the body. In addition, other materials are found, all of which contribute in some way or other to the maintenance of the relative constancy of the internal environment of cells. Plasma itself tends to maintain a remarkably constant composition despite the fact that materials are being added and removed at many points. Entering the blood, for example, are oxygen in the lungs, absorbed water and foods from the intestines, hormones from the glands of internal secretion, metabolic wastes of many kinds from all cells, and foods such as glucose from storage depots. Leaving the blood are carbon dioxide into the lungs, wastes into the large intestine, nutriments for all cells, excess foods into storage regions, oxygen into the tissues, and water and other substances into secretory organs such as the liver, kidney, sweat glands, and digestive glands.

It is not entirely correct to say that the composition is maintained *in spite* of these (and other) factors. Rather, the constancy is in large measure a result of complex interactions between them. They are so interrelated and so regulated that any change in blood composition calls forth readjustments of these factors, with a resultant restoration of the balance, so that there is preserved in the plasma and the internal medium that constancy of composition indispensable for life.[2]

A. The Inorganic Salts

The role of inorganic salts in this respect has been discussed in chapter 2. Present in roughly the same proportions found in sea water, certain of the salt ions have constituted the environment of cells all through evolutionary time, from primitive unicellular marine life up to the mammals, including man. Through these ages the evolutionary possibilities of cells have been limited by the chemical and physical composition of the blood and tissue fluids.

2. Regulation of the alkalinity of the blood is discussed in the chapter on breathing, p. 258.

When, therefore, that environment is appreciably altered as to chemical composition, the cells are unable to function properly. Too little calcium, too much potassium, too little sodium—all affect cellular and organ activity profoundly.

The kinds of salt ions found are chiefly chloride, bicarbonate, carbonate, phosphate, and sulfate; and sodium, calcium, potassium, and magnesium. Chloride and sodium ions are the most abundant. Although for the most part the salts are present in ionic form, some seem to be combined securely with proteins. This is especially true of part of the calcium. The inorganic salt concentration in mammals is about 0.9 per cent by weight;[3] over half of this is common table salt in solution. This salt concentration is the major factor in giving the plasma an osmotic pressure just equal to that within the blood cells and cells of other tissues, so that a proper water balance between cells and internal environment is maintained (see pp. 58–59).

A part of the salts also represents simply materials in transport to tissues where they are used, as exemplified by part of the calcium and the phosphates, which are utilized by bone cells for the construction of the nonliving inorganic constituents of bone, giving bone its characteristic hardness and rigidity. A large part of bone consists of calcium phosphate and calcium carbonate compounds. Calcium ions are also indispensable for the clotting of blood.

Again, the blood iodides are of importance in manufacture of the thyroid hormone in the thyroid gland, and iron is used in the manufacture of the pigment in red blood cells—*hemoglobin*.

B. The Blood Sugar

In health the concentration of glucose in the plasma is maintained between the limits of about 0.08 and 0.14 per cent, averaging about 0.1 per cent. We shall see that many mechanisms contribute to the maintenance of this *physiological constant*, correlating the addition of glucose to the plasma from storage depots and from the foods in the intestines, with removal of glucose from the blood into the cells generally, where it is oxidized; into the liver and muscles, where storage occurs; and into the kidneys, which may excrete it.

Important as a source of energy in transport, glucose also plays a part as an indispensable constituent of the internal environment, as we have already seen. A reduction in its concentration to 0.04 per

3. The salt concentration of the plasma and body fluids of reptiles and amphibia is about 0.7 per cent.

cent or less may produce muscle twitchings, convulsions, unconsciousness, and death (see pp. 33–34).

C. The Plasma Proteins

The plasma contains 7–9 per cent by weight of proteins of several
kinds. They are present in a colloidal sol state. Our first conjecture
about them might be that they represent digested proteins which
have been absorbed as amino acids and resynthesized to proteins
by the cells of the intestine, or in the blood, and are in the process of
being distributed to the tissues for use in construction of protoplasm.
Such a guess proves to be quite unfounded. They are not increased
in amount during digestion and absorption of a mixed meal, although at that time appreciable increases occur in the concentration
of blood glucose, fats, and amino acids. Nor is their concentration
diminished significantly in fasting of even long duration. Of course,
ultimately, these proteins must be traceable back to ingested foods,
but this is not their immediate source. What their history is between
their absorption and their appearance in the blood stream, we do
not know as yet. The available evidence indicates that much of the
plasma protein, certainly the fibrinogen, is manufactured in the
liver. The *antibodies* of the blood, active in combating infections
(see pp. 591–92), are found in the protein portion of the blood.

1. *Viscosity.*—On the other hand, some of the functions of the
plasma proteins are well understood. They are responsible in large
part for the viscosity of the plasma, which markedly affects the
arterial blood pressure (see chap. 5). At this time it is sufficient to
note that either an increase or a decrease in the blood viscosity im-
pairs the circulation, proportionally to the degree of deviation from
the normal. In extensive loss of blood, or *hemorrhage*, the fluids
which may be artificially injected to replace the lost blood often
fail to restore an adequate circulation, partly because their viscosity
does not approximate that of the blood plasma—a condition difficult to fulfil in actual practice except by the transfusion of blood.

2. *Water balance.*—The plasma proteins apparently play an important role in maintaining the water balance between blood and
tissue fluids and in establishing a sort of local circulation between
blood capillaries and tissue cells, as described in chapter 2. The fact
that the proteins of the tissue fluids and of the plasma appear to
be identical suggests either that these proteins are produced in
every organ of the body or that they do pass through the capillary

wall, though very slowly. But the further fact that their concentration is normally less in the tissue fluids than in the plasma indicates that, unlike most of the other plasma constituents, the proteins do not *readily* penetrate the capillary walls. This difference in concentration tends, by osmosis, to balance the filtration of water into the tissue spaces.

The manner in which this mechanism operates may be better appreciated by observing what happens when either factor is changed. A priori, we might guess that, under at least two distinct sets of conditions, there would be an excessive passage of water from the blood and an accumulation of it in the tissues. These are prolonged, persistent elevation of the filtration pressuer (i.e., blood pressure) in the capillaries and a reduced protein concentration of the blood. In fact, either of these conditions may occur abnormally, and either produces an excessive collection of fluids in the tissues, known as dropsy, or *edema*. The first occurs in circulatory defects in which there is interference with the flow of blood through veins. If the work of the heart and the pressure of blood in the arteries continue approximately normal, obstruction of veins tends to raise the pressure of the blood in the capillaries. The filtration pressure, that is, exceeds the plasma protein osmotic pressure.

The second condition mentioned obtains in certain diseases of the kidney, in which the damaged kidneys permit the plasma proteins to pass into the urine, where they are not found normally. Consequently, the blood-protein concentration falls, perhaps to 5 or even to 4 or 3 per cent. If the filtration pressure remains normal but the plasma protein osmotic factor is thus reduced, edema is the consequence.[4]

This water-balance mechanism is also of importance in the recovery from hemorrhage. An immediate consequence of extensive blood loss is a general reduction in blood pressure throughout the circulatory system (see p. 204), including the capillaries. As there is no immediate change in the concentration of plasma proteins, the tissue fluids should pass into the blood at an increased rate. After a hemorrhage such movement of water actually occurs. The lost blood, at least so far as the plasma volume is concerned, is replaced

---

4. It is freely admitted that other factors may also contribute to the edema under these abnormal conditions. For example, in venous obstruction the capillaries may be injured because of a reduced oxygen supply, or by overdistention, increasing the permeability of the capillary wall.

by drawing upon what may here be considered the reserve fluids of the tissues. These relationships, as well as the usual normal balance, are schematically represented in Figure 14.

3. *Fibrinogen.*—This protein, comprising only about one-twentieth of the total proteins,[5] has been studied extensively as an important factor in the clotting of blood. The origin of this protein alone of all the plasma proteins has apparently been satisfactorily determined. The evidence indicates that it is manufactured in the liver. This conclusion is based upon studies in which the fibrinogen content of the blood of an animal is experimentally depleted. Subsequent regeneration of the protein is found to be delayed by experimental damage to the liver but not to other organs. Moreover, the extent of the delay in reappearance of fibrinogen is proportional to the extent of the liver damage. Confirmatory evidence comes from the fact that in man liver damage by disease is sometimes accompanied by a deficient fibrinogen content of the blood; and, in general, the more severe the damage, the greater is the deficiency of fibrinogen.

Fibrinogen is as important to the body as is clotting of the blood. The basis of the clotting mechanism is a change in the colloidal state of fibrinogen—from its normal sol state in the circulating blood to the gel state characteristic of the clot.

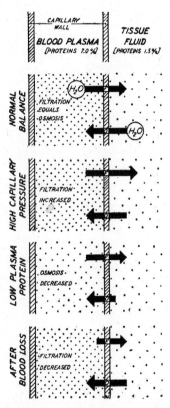

Fig. 14.—Disturbances of the normal balance between filtration (*F*) and osmosis (*O*). Black dots represent protein molecules. Lengths of arrows indicate rates of water movement.

D. Special Plasma Substances

Finally, the plasma contains a great variety of organic compounds, including hormones, enzymes, clotting elements not yet referred to, antibodies, urea, uric acid, creatin, creatinin, and many others—all of special significance, origin, function, and fate. Doubtless, also, the

5. Or about 4 parts in 1,000 of plasma.

TABLE 2

MAIN CONSTITUENTS OF THE BLOOD PLASMA

	Constituents	Origin	Significance	Fate
	Water	Food*	Aqueous medium necessary for life of all cells	Excreted by kidneys, lungs, sweat glands
Proteins (7-9%)	Fibrinogen	Liver	See blood coagulation, p. 74	Some used up when blood coagulates
	Albumin and globulin	Liver(?)	Maintain blood viscosity and osmotic pressure Some are antibodies	?
Nonprotein nitrogenous bodies	Urea Uric acid Creatinin Creatin Ammonia salts	Products of breakdown of proteins and related substances	Waste substances in transport to excretory organs	Excreted by kidneys
	Amino acids	Food Tissue breakdown	Food in transport	Used as building stones for new proteins; some excreted by kidneys
	Phosphatides	Food. Some can be synthesized in body	Cephalin important in blood coagulation Building stones of cell membranes	?
Nonnitrogenous bodies	Sugar (0.085%)	Food Storage depots in liver, muscles (glycogen)	Food in transport Provides medium to which some cells are accustomed	Burned to carbon dioxide and water, giving energy Excess stored in liver, muscles, or excreted by kidneys
	Fat	Food Fat depots of body	Food in transport	Burned to carbon dioxide and water Excess stored in fat depots
	Cholesterol	Partly from food	Large quantities present in nervous tissue and adrenal glands	Part excreted in bile
	Lactates	Glycogen breakdown especially in muscle contraction	Product of sugar breakdown	Burned to carbon dioxide and water Reconverted to glycogen; excreted by kidneys

* All constituents of body come ultimately from substances taken by mouth (of mother in case of embryo) or into lungs.

TABLE 2—*Continued*

MAIN CONSTITUENTS OF THE BLOOD PLASMA

Constituents		Origin	Significance	Fate
Inorganic salts (0.9%) Chlorides Carbonates Bicarbonates Sulfates Phosphates Iodides } of {	Sodium Potassium Calcium Magnesium Iron	Food Storage depots	Found in all cells, are necessary for life Provide medium to which cells are accustomed; proper function of all tissues depends on proper balance of these Maintain osmotic pressure of blood Calcium necessary in blood coagulation Iodine important in proper thyroid function	When present in excess, excreted by kidneys and colon
Special substances	Enzymes	?	?	?
	Adrenal hormones Thyroxin Insulin Hypophyseal hormones Sex hormones Parathyroid hormone	Adrenal gland Thyroid gland Pancreas Hypophysis Gonads Parathyroid glands	Hormones (chemical messengers) regulate and co-ordinate cell, tissue, and organ activity. (For details see chap. 12.)	Some (the excess?) appear in the urine
	Antibodies	?	Are proteins Act on bacteria, bacterial toxins, and foreign proteins	Appear in urine and other secretions
Blood gases	Oxygen (19 cc. in 100 cc. blood)	Diffuses into blood in lungs	Carried mainly in loose combination with hemoglobin inside red blood cells	Diffuses into tissues Used in tissue oxidations
	Carbon dioxide (50 cc. in 100 cc. blood)	Diffuses into blood from tissues	Waste substance produced in tissue oxidations. Carried mainly as sodium bicarbonate	Diffuses into lungs and is exhaled
	Nitrogen	Diffuses into blood in lungs	Inert. Present only as dissolved nitrogen	Diffuses into lungs and is exhaled

plasma contains compounds whose nature and function are as yet unrevealed to investigators. It has been impossible so far to duplicate all the properties of plasma in any artificially prepared complex of known ingredients of normal plasma. It has been impossible, that is, to prepare a synthetic medium which will support the growth of cells outside the body in tissue cultures. All culture media must include some plasma or tissue fluid, containing compounds whose nature is unknown and whose indispensability as an environmental constituent has so far defied analysis. Table 2, listing the commoner plasma constituents, emphasizes the complexity of the fluid.

V. THE CLOTTING OF BLOOD

In view of the importance of the circulating fluid, it is not surprising to find that elaborate mechanisms have been evolved which guard against its loss should a blood vessel chance to be ruptured. One of nature's devices in invertebrates has been to produce *spasms*[6]

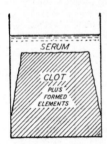

FIG. 15.—Separation of fluid serum from clotted blood.

of ruptured vessels, which serve to pinch off the opening. In vertebrates the same end is served by the *coagulation* or clotting of blood, which everyone has observed in his own blood escaping from a ruptured vessel. When blood is drawn from a vein into a beaker, it remains liquid only a short time. It is converted into a semisolid gelatinous mass, or clot, in some four to eight minutes.[7] If clotting blood is examined under the microscope, threadlike or needle-like processes are seen to appear. As they increase in length and number, they form an entangled interlacing network called *fibrin*. Between the meshes the formed elements (red and white blood cells) and some of the fluid become entrapped in the solidifying mass. As it solidifies, the clot also shrinks, squeezing out from its interstices a straw-colored liquid known as *serum*, which collects above the clot and retains its fluid consistency indefinitely.

The predominance of red cells gives the clot its color, but they do not otherwise contribute either to the formation or to the final internal structure of the clot. They are simply caught in the meshwork. It is possible, by the expedient of washing the clot in running

6. Strong contractions of the vessel walls.

7. This interval is referred to as the "clotting time." Essentially, clotting is a transformation in the colloidal state of one of the plasma proteins (fibrinogen) from the sol to the gel condition (see pp. 45–47).

water, to free it of red cells, in which case it retains all its former properties except the red color. Also, perfectly typical clotting occurs in blood freed of red and white cells. Coagulation is a phenomenon of the plasma rather than of the whole blood.

A. Stoppage of Flow; Exposure to Air

What causes clotting? One might guess that it is initiated either by exposure to air or by a stopping of the normal movement of the blood. But if blood is drawn with proper precautions into a suitably prepared vessel, clotting will be prevented or greatly delayed, even though the blood is freely exposed to air. The same experiment indicates that stopping the flow is not the important factor. We know also that there are regions in the body where quantities of blood may normally remain stationary for hours at a time without clotting. Further evidence is provided by the experiment of tying off a segment of vein with two ties, or *ligatures*, occluding the vein at points an inch or two apart. Though all

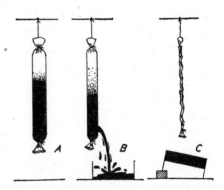

Fig. 16.—Stationary blood in a segment of vein removed from the body remains fluid a long time, the formed elements separating from the plasma by gravity (*A*). Allowing the blood to escape by cutting the vein (*B*) is soon followed by clotting, so that the vessel into which the blood has escaped can be inverted without the blood flowing out (*C*).

movement of blood has ceased in the isolated segment, the blood remains fluid almost indefinitely. If the vein segment is later cut open, the escaping blood will flow out freely and then clot in a few minutes.

B. Initiation of Clotting

Two distinct factors seem capable of initiating the clotting process: contact of the blood with injured tissues or damaged cells or contact of blood with "foreign" surfaces possessing certain physical properties different from those of the smooth lining of the blood vessels, with which the blood is normally in contact.

1. *Contact with injured cells.*—If blood is drawn into a vessel whose inner surfaces are properly prepared, taking care that fluid from tissues necessarily injured in the experiment at no time comes into contact with the blood, clotting fails to ccur, or is greatly delayed. This fluid blood, even after hours, can then be made to clot within a

few minutes by adding juices compressed from almost any tissue. Something present in cells generally, liberated when they are injured, is able to initiate the clotting process.

2. *Contact with foreign surfaces: platelet disintegration.*—If blood is drawn carefully, so as to avoid contact with injured tissues, it will clot in the normal time if the container is glass. But, if the glass beaker is lined with paraffin, clotting will be delayed. The essential difference here seems to be the physical nature of the glass or paraffin surface to which the blood is exposed. It is not clear just what this difference is, except that water solutions "wet" a clean glass surface but not a paraffin or waxy surface, leaving it quite dry. In general, surfaces which are "wet" by water, on which a thin film of water tends to remain after most of the water is drained off, behave like glass as regards clotting. Surfaces which are not "wet" by water, in general, retard clotting.

Clotting can even be induced in the blood vessels when the blood is exposed to the proper surfaces. If a pin is stuck through a vein, a thin clot forms on the pin as the blood flows by. Or, if particulate matter of a suitable kind is injected, each particle soon comes to be covered with a thin clot.

Just why this happens is not clear. But, if the process is carefully observed under the microscope, it will be seen that certain of the formed elements of the blood—the blood platelets—collect upon surfaces like glass and quickly disintegrate. At a paraffined surface this occurs only very slowly. The platelets seem to be involved in these surface relationships, and their disintegration apparently initiates clotting. This can be verified in other ways. If blood is drawn into an ice-cold vessel, and is itself immediately cooled sufficiently, clotting does not take place, and examination reveals that the platelets have remained intact. As soon as the blood is again warmed, the platelets quickly disintegrate, and clotting occurs. Why platelets respond in this way to temperature changes is not known, but the important consideration here is that the clotting seems to be dependent upon platelet disintegration. Various chemicals are also known which delay the breakdown of platelets. They all retard clotting.

C. Substances Involved

1. *Clot-inducing material.*—If blood is drawn so as to prevent both the entrance of tissue juices and the disintegration of platelets,

clotting can be induced by adding a little extract of platelets, a small amount of the fluid in which platelets have previously disintegrated. Apparently, then, platelets and tissue cells alike yield clot-inducing materials when injured. Chemical analysis of cells and platelets, and of the clot-inducing substances they liberate, reveals that similar materials are involved in the two cases. These substances are collectively called *thrombokinase*.[8]

We now begin to see how the clotting mechanism is automatically set into motion in nature. A vessel is ruptured by an injury. The blood flows out into the injured tissues. Thrombokinase liberated by the damaged cells initiates clotting. Further thrombokinase is liberated by the platelets, which disintegrate as they come into contact with the abnormal surfaces of the injured tissue. But this explains only the beginning of the process. What other materials besides thrombokinase are involved, and how do they interact to produce the final product—the fibrin clot?

2. *The role of calcium ions.*—It is easy to demonstrate that clotting does not occur in the absence of calcium ions. If a pinch of crystalline sodium oxalate is shaken with some freshly drawn blood, the salt goes into solution. It reacts at once with the calcium of the plasma, and insoluble calcium oxalate is formed and precipitates from the solution. Blood so treated remains liquid indefinitely. Any agent is an *anticoagulant* which so removes calcium ions from the plasma, or even converts the calcium from the normal ionic form to its condition in an un-ionized compound like calcium citrate. If at any subsequent time, even hours later, calcium ions are furnished by placing a bit of calcium chloride into the blood, a normal clot will form in a few minutes.

3. *Fibrinogen and thrombin.*—That the protein fibrinogen is also necessary in blood coagulation can be shown by removing it from fluid blood—by precipitating it, for example.[9] Blood so treated fails to clot. It has also been possible to isolate fibrinogen from the blood and to experiment with it alone. Another agent important in clotting, called *thrombin*, can be extracted from the serum remaining at the conclusion of the clotting process. If a bit of purified thrombin is added to a purified colloidal suspension of fibrinogen, solidification or fibrin formation occurs in typical fashion. This is apparently the

8. This term literally means "clot hastener." There are indications that one of these substances is identical with a material called *cephalin*, whose chemical formula is known.

9. Or by defibrination (see p. 78).

essential reaction of clotting: the conversion of fibrinogen to fibrin in the presence of thrombin. Strangely enough, though calcium ions are indispensable for clotting, this particular reaction proceeds in their complete absence.

4. *Prothrombin.*—Since fibrinogen is always present in the plasma, we might safely predict that thrombin itself is not present; otherwise, blood would not remain in a fluid state in the vessels. Examination of normal plasma reveals an absence here of thrombin as such. The evidence indicates it to be present in an inactive form called *prothrombin*.[10] Thus initiation of clotting consists in a conversion of prothrombin into the active form—thrombin. It is in this reaction that calcium ions and thrombokinase participate.

Prothrombin has been isolated from the blood plasma. Calcium ions are always present there. Only when thrombokinase is liberated by injured tissues or platelet disintegration is the prothrombin activated to thrombin. This, in turn, converts fibrinogen, also always present in plasma, into the fibrin clot. The whole clotting process actually consists of two successive reactions, which may be summarized as follows:

$a)$ Thrombokinase + Ca^{++} + Prothrombin $\rightarrow$ Thrombin

$b)$ Thrombin + Fibrinogen $\rightarrow$ Fibrin

D. Defibrination

We have previously seen that removal of fibrinogen from plasma or whole blood prevents coagulation. A convenient method of accomplishing this is by *defibrinating* the blood. Freshly drawn blood is stirred rather vigorously with wooden sticks, for example. In a few minutes a network of elastic fibers—the fibrin clot—forms upon the sticks and can be withdrawn, leaving the blood lacking in fibrinogen. It has been converted into the fibrin formed upon the sticks. The remaining blood, now called "defibrinated blood," stays fluid in-

10. The existence of biological agents in an inactive form is known elsewhere in the body. The hormone *secretin* is found in the intestinal lining in an inactive state called *prosecretin*, which is activated to secretin proper by the action of hydrochloric acid (see p. 280). Similar situations are seen in the case of some of the digestive enzymes, which are secreted in an inactive form and are able to effect digestive action only after activation by some other component of the gastric or intestinal juice.

The prothrombin of the plasma probably originates from the liver and bone marrow and perhaps also from the platelets, some of which disintegrate in the normal circulating blood.

definitely unless new fibrinogen is artificially added. Since red and white blood cells remain unchanged, defibrinated blood is convenient to use in studies upon these cells, unhampered by further coagulation.

The relationships of defibrinated blood and various other fractions of the blood to one another, as well as the terms used to designate them, may be clarified by the following:

Whole blood   = Formed elements *plus* Plasma

Serum   = Plasma *minus* Elements of clotting (essentially fibrinogen)

Defibrinated blood = Whole blood *minus* Elements of clotting

Defibrinated blood = Serum *plus* Formed elements

E. COMPLEXITY OF THE MACHINERY

The clotting machinery seems inordinately complex, involving intricately interrelated reactions of several substances derived from different sources. Even so, our account is somewhat simplified. Evidence is presented by Howell and others that still further substances are involved. According to this evidence, even the inactive prothrombin is not present as such in normal plasma but is bound by an anticoagulating substance called *antiprothrombin*. The role of thrombokinase is thought to be the formation of a union with antiprothrombin, freeing the prothrombin, which is then converted to active thrombin by calcium ions. In this theory there is no modification of the concept of what finally occurs, namely, the conversion of fibrinogen to fibrin by thrombin. In addition, *antithrombins*, as distinct from antiprothrombin, have also been detected in the plasma.

That the clotting process is so complex is not surprising when we consider the exacting requirements it should meet. The reactions evolved for converting the fluid blood into a solid should be such that they seldom or never occur when the blood circulates—thereby perhaps causing death—but are always set off when blood vessels are ruptured. These severe conditions are usually fulfilled by the clotting machinery. We see here a sure-fire device for the plugging of at least the smaller ruptured vessels which is set into motion by the very agent which ruptures the vessels. The injury tears the vessel and liberates thrombokinase. Thrombokinase initiates clotting.

F. Repair of Damaged Vessels

The clotting mechanism not only serves to help plug torn vessels but also often operates to forestall rupture of vessels. It is an adaptation not only for the filling of an open gap in a vessel but actually may prevent the development of a leak, or it may prophylactically strengthen and repair a weak spot in a vessel wall. A number of infectious organisms attack and destroy blood-vessel walls. One of these is the organism which produces syphilis.[11] In its growth it may destroy the tissues of the arteries. This, of course, involves cell destruction, since arteries are made of cells. In addition, platelets accumulate at the roughened spot on the wall and disintegrate. A local liberation of thrombokinase and prothrombin occurs, and a clot forms at the site of thinning and weakening of the vessel wall. This serves as a temporary repair device. However, the firmest of clots is at best rather tenuous, and the immediate strengthening of the wall is none too effective. But in time, unless the local damage continues at too rapid a rate, the clot becomes invaded by connective-tissue cells from adjacent portions of the arterial wall and is converted into tough fibrous connective tissue, which may be quite as strong as the original tissues were.

The student of evolution appreciates the time-honored adage, in paraphrase: "It is a good wind, indeed, which blows no ill." That adaptation is good indeed which does not also carry with it something of a threat to the organism. The clot which forms upon a vessel wall, called a *thrombus*, may extend across the cavity of the vessel and completely occlude it. This is often an adaptation to an extreme situation, where the damage to the wall has been great, or, at any rate, thrombokinase liberation extensive. Blood can now no longer pass through the vessel. If the vessel happens to be one which supplies blood to some structure vital to the organism, and if no other near-by vessels also supply that structure, the outcome is fatal. The dilemma faced by evolution here is difficult; but, on the whole, the automatically initiated intravascular clotting, or thrombus formation, in all cases of vessel-wall injury, is a positive adaptive value. Many perfectly normal individuals have thrombi at one point or another in some vessel or on a vessel wall and go through life without ill effects.

Thrombus formation holds a threat to life in another way. A clot

11. Of course, the organism of syphilis may grow and do damage elsewhere also, including especially the brain and spinal cord.

which has formed in a vessel not indispensable for the nourishment of an organ or the life of the individual may break loose, be carried off in the blood, and lodge finally in a vessel which is essential to life. Any occluding particle, of whatever nature, which is carried to the site of the occlusion by the moving blood is called an *embolus*.

G. DEFECTIVE CLOTTING

In any condition in which there is a deficient quantity of any of the clotting elements of the plasma, the mechanism of coagulation is impaired, and even slight wounds may cause excessive and even fatal hemorrhage. The number of blood platelets is sometimes so reduced in certain infectious diseases and in other conditions as to render the emergency liberation of thrombokinase inadequate for clot formation. Liver damage, as from chloroform or phosphorus poisoning, may reduce the fibrinogen content of the blood to a dangerous level.

Prothrombin deficiency sometimes occurs in newborn children or occasionally in adults. The adequate production of prothrombin depends partly upon a sufficient intake and absorption of a certain dietary constituent—vitamin K—which is not itself converted into prothrombin, apparently, but is necessary, perhaps as a catalyst, for the proper manufacture of prothrombin in the body. Bile, which the liver secretes into the alimentary canal (see p. 94), seems to be necessary for the absorption of vitamin K into the blood, so that a deficiency of bile may produce a deficiency of the vitamin in the blood and tissues even though there is plenty of vitamin K in the intestine. Delayed clotting of this origin can often be corrected by injecting the vitamin or giving bile by mouth.

Among the conditions causing defective coagulation, *hemophilia* has attracted considerable attention and stimulated much investigation. It is a familial disease, affecting mainly males, directly transmitted to the affected male only by the maternal parent.[12] It has played a significant part in history, for it has afflicted some of the royal houses in Europe. The (previous) royal house of Spain was affected, and the last czarevitch of Russia was a victim, inheriting the defect through his mother. Reports have it that much of Rasputin's evil influence upon the empress and emperor was due to his alleged healing influence upon the young czarevitch.

All the various factors involved in coagulation have been studied

---

12. It is thus a sex-linked, recessive character.

in attempts to solve the riddle of hemophilia. There is no deficiency of calcium, fibrinogen, prothrombin, or numbers of platelets. It has been shown, however, that the platelets of hemophiliac blood fail to disintegrate quickly when blood is shed, and the consequent thrombokinase liberation is inadequate. What stabilizes the platelets and what might be done to correct the condition are problems which only further investigation can solve.

VI. THE RED BLOOD CORPUSCLES

A. VISIBLE FEATURES

The red blood cells, or *erythrocytes*, are the most numerous of the formed elements, each cubic millimeter of human blood containing four and a half to five million. Human erythrocytes are biconcave

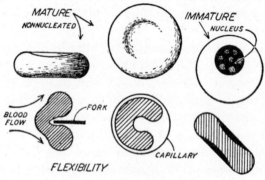

FIG. 17.—Mammalian red blood cells. Note how a cell may be bent when it strikes a fork in a vessel or twisted as it is forced through a narrow capillary, after which it again resumes its previous shape.

disks, a little less than 0.008 mm.[13] in diameter. They are normally of such uniformity in size that histologists frequently use them as handy units of measurement, including a red cell in a drawing to indicate the size of the cells in any tissue, relative to this nearly uniform red-cell size. They appear to be perfectly homogeneous, although there are indications that some internal structural differentiation of parts exists.

The striking visual characteristic of mammalian red cells is the absence of a nucleus. This at once raises the question, "Are these cells really alive?" Evidence has been presented that the life of the cell is in some way bound up with the nucleus. Certainly, these red cells are alive during the early stages of formation before they pass

13. About 1/3,200 inch.

into the circulating blood, for at that time they possess typical nuclei. At this developmental stage they must be considered alive. And, indeed, even the mature red cells of vertebrates other than mammals are nucleated. But in man and other mammals the nucleus is lost before the cell becomes a functioning unit.[14] It may, therefore, be more accurate to refer to these structures not as cells but as *corpuscles*—"little bodies." Perhaps the relatively short period of time during which red cells course through the blood stream—about four months, on the average—is in some way related to this absence of a nucleus.

B. HEMOGLOBIN

The most interesting chemical entity in the cell is hemoglobin, which is a protein (*globin*) in combination with an iron-containing pigment (*hematin*). It constitutes 95 per cent of the solids[15] of these cells. Each 100 cc. of blood contains about 15 gm. of hemoglobin. The main functions of the red cells are carried out by means of this substance, which, incidentally, also confers upon blood its red color. This pigment has the capacity for combining spontaneously with oxygen when free oxygen is present in relative abundance in the environment. The union is a loose one, so that, when the surroundings contain little or no free oxygen, the oxygen breaks apart from the molecule and, by physical diffusion, passes into the oxygen-poor regions.[16] These phenomena will be taken up in more detail with the mechanisms for transport of the respiratory gases in the blood (chap. 6).

C. STROMA FRAMEWORK

The peculiar shape of the red corpuscles suggests a definite internal structural organization, in spite of the apparent homogeneity of the cells. In no vertebrates are the cells perfectly spherical, though

14. There is just the suggestion that the loss of the nucleus is of service to the animal economy by increasing the capacity of the cell for carrying hemoglobin—the most significant constituent of the cell. In nucleated red cells the region of the cell immediately adjacent to the nucleus is relatively poor in hemoglobin. How the nucleus brings about this condition is quite unknown; but, granting its existence, we can see how a cell without a nucleus could possess a higher concentration of hemoglobin throughout.

15. Strangely enough, the red cells of the liquid blood have a higher proportion of solids (nearly 40 per cent of the cell) than muscle (with 25 per cent).

16. Another more complex activity—that of assisting in the carriage of carbon dioxide and in the prevention of too much acidity developing in the blood—the red cell also carries out largely by virtue of its hemoglobin.

they are suspended freely in a watery medium. Again, their remarkable elasticity indicates the existence of some flexible framework.
The cells can be observed to become twisted and bent into a variety
of shapes as they are forced through blood channels of a sufficiently
small diameter. But, just as soon as a wider channel is reached, they
resume their previous shape. Or a cell may be seen to strike a fork
in a vessel and be bent considerably by the stream on either side and
then slide on and again appear quite normal.

The structural framework of the corpuscle is called the *stroma*. It
is colorless; but, when it is stained, it presents the same shape as the
intact cell. So it is probably more than merely an envelope surrounding a mass of hemoglobin. Apparently, it sends extensions down into
the depths of the cell, forming an interlacing network extending
through the corpuscle and determining its shape, flexibility, and
elasticity. It is made up chiefly of proteins and of fatlike materials
(*phospholipins* and *cholesterol*). This structural framework, with the
nucleus, seems to comprise the entire cell in its more immature form.
At any rate, it is only in later stages of development that hemoglobin
makes its appearance in the maturing cell.

Just what may be the relationship of hemoglobin to stroma is not
clear. Some sort of loose chemical combination between the two
probably exists. Whether this be true or not, stroma and hemoglobin
are very readily separable experimentally and may be studied independently.

D. HEMOLYSIS

1. *Osmotic equilibria.*—The experimental separation of stroma and
hemoglobin—called *hemolysis*—is a specific example of the general
cellular phenomenon of *cytolysis*, in which the cell is disrupted and
its contents dispersed. Hemolysis is cytolysis of a red cell, conveniently observed because the chief constituent of the red cell is
colored.

When red cells are placed in a solution of salts found normally in
blood plasma, or simply table salt, made up in a total concentration
of 0.9 per cent (i.e., physiological salt solution), the cells maintain
their structural integrity for a long time. This concentration of 0.9
per cent represents also the total salt concentration within the red
cells. If now we dilute the solution to about 0.6 per cent (a *hypotonic*
solution), we can observe under the microscope that the cells become
spherical; they swell because water is drawn into the cells by osmosis

through the semipermeable surface. If the surrounding fluid is diluted still further, the larger of the swollen red cells suddenly disappear from view under the microscope. Actually, what has happened is that the cells rather suddenly lose their hemoglobin. Inasmuch as their visibility is dependent upon this pigment, they now become invisible. Though apparently the entire cell disintegrates, we know that this does not actually occur, for, if we stain the preparation suitably (e.g., with methylene blue), the cells again become visible. What we now observe is simply the framework or stroma of the cell, without the hemoglobin. The invisibility of the unstained stroma of the corpuscles has given them the name "ghosts."

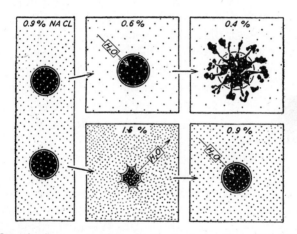

Fig. 18.—Osmotic effects upon red blood cells, showing swelling of the cells and *hemolysis* when they are placed in weak salt solutions and shrinking when they are placed in strong salt solutions. Dots represent salt particles. Hemoglobin is shown in black.

In shape the ghosts closely resemble the intact cell. In osmotic hemolysis the loosely joined stroma-hemoglobin complex is apparently broken, the character of the corpuscle membrane is altered by the distension of the cell with water, and the hemoglobin escapes from the cell. When the salt concentration in contact with the corpuscles is reduced to approximately 0.4 per cent, practically all the cells become hemolyzed; and the hemoglobin molecules, in the manner of all colloidal particles, become dispersed homogeneously throughout the system. If a test tube of hemolyzed red cells is allowed to "settle," or if it is centrifuged to hasten the settling, the invisible ghosts collect at the bottom of the tube; and the hemoglobin remains distributed throughout the fluid of the tube indefinitely, giving it a transparent pink or red coloration.

On the other hand, suppose that the cells are exposed to a salt solution more concentrated than that within the cells—somewhat over 1 per cent (a *hypertonic* solution). In this case the red cells can be seen to decrease in volume and to shrivel, so that the surface appears wrinkled.[17] As long as the cells are kept in this solution, they will retain the shriveled appearance. They do not lose their hemoglobin, however; and, should the cells be placed again in salt of a concentration of 0.9 per cent, they will resume their normal shape and appearance and seem to be little if at all affected by this treatment.

These changes in shape and size of cells are all due to osmosis, as indicated in chapter 2. In the system discussed here we deal with solutions of salts of differing concentrations, separated from one another by the cell membrane, which we have seen to be practically impermeable to salts. Consequently, when the red cell is placed in a weak salt solution, water enters the cell by osmosis, causes it to swell, and may hemolyze it. When the red cell is suspended in a more concentrated solution, it loses water by osmosis and shrinks to a smaller size.

2. *Fragility.*—The ease with which erythrocytes are hemolyzed by dilute salt solutions is frequently used as a criterion of the fragility of the corpuscles. Normal cells resist being hemolyzed when placed in solutions of a concentration just under 0.9 per cent, though, of course, they become more spherical. When the concentration is reduced to about 0.6 per cent or less, there is a hemolysis of some of the red cells, which are presumed to be weaker or more fragile than the others. If the blood cells, in general, are abnormal, so that their resistance to hemolysis by weak salt solutions is reduced, some of them become hemolyzed when exposed even to 0.8 per cent NaCl. This occurs in certain specific diseases, as when the ducts of the liver are obstructed, and toxic products accumulate in the blood stream. It has also been shown to occur in conditions of experimental *hyperthyroidism*. Determinations of the fragility of the red cell in this way may aid materially in the diagnosis of some diseases.

3. *Action of fat solvents.*—Substances which dissolve fats are generally cytolytic, by virtue of their action upon the cell membrane, dissolving out its fatty and fatlike constituents. When red cells are

17. Called *crenation*.

exposed to ether or alcohol or fatty acids in sufficient concentrations, a solution of the fatty elements of the membrane occurs. The complex of hemoglobin and stroma is disrupted, and again hemoglobin escapes from the cell. If sufficient fat solvent is present, the damage to the stroma may be extensive. Not only is the hemoglobin permitted to escape but the whole structural stroma framework of the cell may be shattered. Staining now reveals no ghosts. All that can be seen are a few shapeless remnants of the almost completely disintegrated cells.

The hemolytic action of ether is so great that even in prolonged surgical anesthesia hemolysis may occur in the blood stream, even though only a very low concentration of ether may be present. In such situations cytolysis of other tissue cells probably also takes place; and, if this involves cells vital to the organism, it would undoubtedly contribute to the fatal termination in those rare instances of overdosage with ether in surgery.

There seem to be individual peculiarities in the structure of the membrane in different kinds of cells, as regards the fatty components, for we find that cells of different tissues vary in their susceptibility to fat solvents. Chloroform, for example, is particularly harmful to liver cells in very small concentrations, and benzol injures especially the cells of the bone marrow.

4. *Action of biological agents.*—A distinction is sometimes made between hemolysis by biological agents (*hemolysins*) and by non-biological agents. The former is called *true hemolysis*, and the latter *laking*. Thus far we have considered only the phenomenon of laking, although the distinction in terms is rarely strictly observed. However, there is an interesting series of biological agents which cause hemolysis. The growth of malignant tumors may be accompanied by hemolysis, presumably from toxins liberated by the growth; but it is certain that the chief injurious effects of cancer are related to other actions than hemolysis. Certain specific bacteria cause hemolysis, and infectious agents like the malarial parasite, which live in the red corpuscles, cause a rupture and a liberation of the hemoglobin of millions of cells. Certain snake and spider venoms may also produce serious red-cell deficiencies through their hemolytic action.

5. *Species specificity.*—One of the most interesting forms of hemolysis, and a phenomenon of general biological significance, is disclosed by the following experiment: If the corpuscles of a rabbit

are exposed in a test tube to the body fluids or plasma of a dog, the rabbit cells can be observed first to cluster together instead of remaining dispersed. They are said to *agglutinate*. As the action continues, the agglutinated cells are then hemolyzed. The plasma of the dog apparently contains materials which clump and destroy the red cells of the rabbit. This agglutinating and lytic action is displayed to some degree whenever red cells of one species are exposed to body fluids of an animal of another species, and, rather remarkably, the more distantly the species are related phylogenetically (i.e., in the evolutionary scale), the greater, in general, is this destructive action. If only 0.04 cc. of eel's serum is injected into a rabbit, there will occur extensive hemolysis, as well as cytolysis of other cells, including especially nerve cells, and death ensues.

These phenomena are of fundamental biological significance. They indicate the existence of species differences in the chemical composition of the body fluids—differences proportional to the degree of evolutionary relationship between species,[18] no less striking than anatomical differences.

Practically these findings are also of great importance. It would be both useless and dangerous to transfuse blood from a sheep or dog into a human being to replace blood lost by hemorrhage, despite the fact that the hemoglobin of sheep's or dog's blood functions exactly as does the hemoglobin of human blood, for the injected corpuscles would be agglutinated or hemolyzed by the recipient's blood plasma. The agglutinated corpuscles might cause death by plugging millions of minute blood-vessel capillaries, which are of a caliber comparable to the diameter of a single red blood cell. In a similar way, any tissue or organ from an animal of one species will be cytolyzed—destroyed—upon its introduction into the body tissues of an animal of another species. Partly for this reason interspecies transplants are practically uniformly unsuccessful.

If this were not so, an entirely new field would be opened up in the control of various abnormalities, especially of the *glands of internal secretion*. We now know a great deal about the chemical composition of a number of *hormones*—those important agents upon whose continued production in the body life itself depends. And, from all our chemical and physiological information, it appears that the hormones of all mammals, and perhaps of all vertebrates, are quite similar in composition. The hormones of the thyroid gland of dog,

18. See discussion of protein specificity in chap. 2, p. 38.

rabbit, and man are practically identical in their action. As a matter of fact, in some human abnormalities where *endocrine* glands are underactive, it is possible in many cases to restore the individual toward normal by administration of extracts (containing the hormone) of glands of lower animals. An individual afflicted with underactivity of the thyroid need only eat the glands of cattle to be restored to a condition approximating normal. A diabetic, whose pancreas produces an inadequate quantity of hormone, may alleviate his condition by receiving injections of products of the pancreas of any convenient mammal. This type of control of endocrine-gland disease, while highly successful in many instances, entails the inconvenience of repeated administrations. A diabetic is not cured by insulin; his condition is merely improved as long as he continues to receive insulin injections. If he fails to continue the daily injections, his diabetes is still present, and death may supervene.

Therefore, because a cow's thyroid gland produces a hormone apparently identical in action with that of man, it would be highly desirable to be able to transplant the cow's gland into a man with defective thyroids. Letting the transplanted gland continue to grow in man, to manufacture its hormone, and to secrete it into the blood would be far more satisfactory to the sick human being and less expensive and inconvenient than the repeated daily ingestion of thyroid.

But, unfortunately, just as transfused (i.e., transplanted) blood cells from another species are destroyed, so also are the cells of any transplanted tissue. The blood and body fluids of man contain materials which destroy not only foreign red cells but also cells of any kind from another species. This method of controlling diseases of deficient endocrine glands cannot be used. Experimentally, interspecies

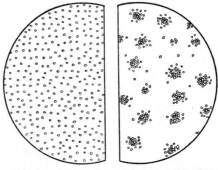

NORMALLY DISPERSED   AGGLUTINATED

Fig. 19.—Agglutination of red blood cells. Blood transfusion is dangerous or fatal if the transfused red cells are agglutinated by the plasma of the person receiving the transfusion. The clumps of cells plug the capillaries.

grafts have been attempted many times and with almost universal failure. Even if blood vessels grow into the transplanted tissue, so that its nourishment is adequate and its waste products drained

away, there is usually a gradual disappearance of the grafted
tissue. Cells in bits of the graft removed at intervals and examined microscopically can be seen to be undergoing solution and destruction. There is no evidence from any physiological experiment
that, even in a small percentage of cases, one can hope to make
successful interspecies tissue grafts which will grow and function
except perhaps in early stages of embryonic development. Human
beings have at various times subjected themselves to gland grafts.
This has been attempted especially in the case of the testis, whose
secretions are important for normal sexual activity. The only evidence for success in these grafts comes from the testimony of subjects who say that they "feel better" after the graft has been made.
Needless to state, such testimony is unscientific and essentially
worthless.

As a matter of fact, even transplants from one individual to another of the same species usually suffer the same fate. The body
fluids of some human individuals contain agents which seem to injure many of the cells of other individuals. This is of great practical
significance and must be borne in mind in every case of blood transfusion, which is really a form of transplantation. Careful tests must
be made, before blood is transfused, to determine whether the body
fluids or plasma of the recipient will clump the cells of the donor. If
incompatible blood is transfused, the donor's agglutinated cells may
plug a sufficient number of blood capillaries to kill the recipient.

E. LIFE-CYCLE OF THE RED CELL

1. *Red-cell counts; hemoglobin determination.*—One can rather
easily determine the number of red blood cells in a cubic millimeter
of blood. There are found to be approximately five million cells per
cubic millimeter in men and about four and a half million in women.
Some variation occurs with age, infants generally having a greater
number. Such large numbers are, of course, not counted directly.
And yet a fairly accurate determination can be made. Every clinical
examination of a patient usually includes a "red blood cell count"—
a determination of the number of red blood cells in a cubic millimeter of blood.

This is done by resorting to two devices: first, diluting the blood
by a measured amount and, second, counting the cells in only a
small part of a cubic millimeter (see Fig. 20). The dilution is carried
out by means of a pipette so marked that, when the blood is drawn

to the proper mark, and then diluted with a suitable fluid to the proper point, an accurately known dilution of the blood (usually two hundred times) has been made. Upon adequate mixing of blood and diluting fluid, a drop of the diluted blood is placed upon a counting chamber, which is scratched with very fine lines 1/20 mm. apart on the central part of the chamber. Additional rulings are present which are used in making white-cell counts. On each side of the ruled surface is a ridge of glass exactly 0.1 mm. high. If a cover slip is placed across the chamber resting upon the ridges, a distance of 0.1 mm. is present between the lower surface of the cover slip and the upper surface of the ruled counting chamber. If diluted blood fills

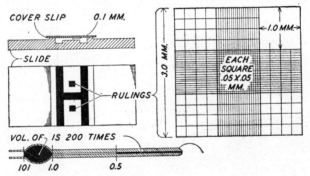

Fig. 20.—Pipette used to dilute blood and chamber or slide used to count the red cells. (See text for details.)

this space, we can see and also count the number of cells in any of the squares of 1/20 mm. by 1/20 mm.—actually in a volume of 1/20 mm. × 1/20 mm. × 1/10 mm. A simple calculation converts the number found in a volume of 1/4,000 cu. mm. of diluted blood into the number in a cubic millimeter of undiluted blood.[19]

The determination of the hemoglobin content of the blood is often of greater importance to physicians than the blood count itself. A person might have a normal red count, yet each cell might have less than the normal quantity of hemoglobin. If 100 cc. of blood should contain but 7 or 8 gm. of hemoglobin, instead of the normal amount of 15 gm., we say that the hemoglobin is 50 per cent of normal. Such an individual would be *anemic* and handicapped by defective oxygen transport, no matter what the red count might be.

19. E.g., if one counts 480 cells in 80 of the small (0.05 mm. × 0.05 mm.) squares, there will be an average of 480/80, or 6 cells per square. Multiplying 6 by 4,000 (since the 6 cells are in 1/4,000 cu. mm.) gives 24,000. Multiplying this figure by 200 (because the cell count was made on blood diluted 200 times) gives the figure 4.8 million, which is the number of cells in a cubic millimeter of the undiluted blood.

The hemoglobin content of blood can be determined by several methods, most of which depend upon the color of hemoglobin. The less hemoglobin that blood contains, the paler pink it is. The color of a known volume of blood, diluted by a measured amount, is compared with standard colors of varying shades of red or pink, which are made to correspond to the color of blood containing known quantities of hemoglobin.

2. *Formation and destruction.*—Of greater interest than the large numbers of cells is the relative constancy in number from day to day. True, the number of nerve cells in a cubic millimeter of nerve tissue, or the number of muscle cells in a unit volume of skeletal muscle,

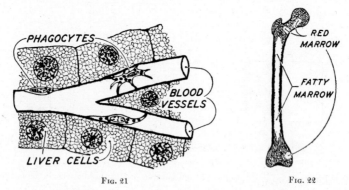

Fig. 21   Fig. 22

Fig. 21.—Phagocytic cells in the walls of the liver blood vessels. These cells engulf and destroy "worn-out" red blood cells.

Fig. 22.—Bone of the leg (femur) split open to show the large space in the shaft filled with fatty marrow and the more finely subdivided spaces at the ends of the bone filled with red marrow. The bone substance is shown black.

remains practically constant from day to day also, at least in the normal adult mammal. This is because few muscle or nerve cells are normally destroyed, and no new ones are produced. But red cells are continually being destroyed physiologically, and new ones formed— and at a tremendous rate. They are destroyed for the most part by the process of phagocytosis, mainly in the spleen and liver.

In these organs of man, as in *Hydra*, there are cells which, although attached to adjacent cells in a continuous tissue, and therefore incapable of ameboid locomotion, nevertheless are capable of engulfing particles. Such are certain specialized cells which are found at intervals in the inner lining of the walls of blood vessels, especially in the liver and spleen. Similar cells are found also in the *lymph nodes* (see pp. 213–14). As the blood (or lymph) flows past these cells, foreign particles in the lymph or blood may come in contact with them and be engulfed. These stationary *phagocytes* in the spleen and liver phagocytize and destroy red cells, which can be observed within the phagocytes in all stages of disintegration. Products of red blood cell breakdown can also be detected here chemically by the use of suitable stains.

Why the phagocytes of the liver and spleen should ingest some of the red blood cells and not others is quite unknown. Presumably, it is old, "worn," or injured cells which are destroyed, but so far there is no known criterion of age in adult red cells. As the blood is observed to circulate through the liver, all the cells may look alike, yet some travel quickly through the vessels and others go slowly, seem to adhere to the vessel wall, and then are phagocytized.

Since destruction proceeds always, and the total blood count remains nearly constant, we could postulate a continuous formation of red cells. This does not seem to occur in the circulating blood stream itself in the healthy adult. In mammals lack of a nucleus in the circulating cells precludes the possibility of *mitotic*[20] cell division. Actually, except in early embryonic life, the production of new red cells occurs outside the blood stream in the red bone marrow.

If a bone is split open and examined grossly, it will be seen that the bone is tubular. The hollow interior of the bony shaft is filled with a soft material. This is yellowish in color. But in the ends of the long bones, and pretty well throughout the interior of flat bones like the skull and vertebrae, there are streaks and patches of reddish tissue. Upon microscopic examination we see that the *yellow marrow* is made up of fat cells and constitutes simply a depot of fat storage. But, in the *red marrow*, red cells in all stages of development may be observed. There are the precursors of mature red cells—undifferentiated cells having little resemblance to the mature forms. Active cell division is observable among them. A nucleus is present, of course, even though the final product is nonnucleated in mammals. There are other cells intermediate in appearance between this parent-cell and the mature form, constituting a complete series. Only toward the end of this developmental series can hemoglobin be observed in appreciable amounts. This material seems to be added to the cell or manufactured by it late in its development. The hemoglobin content increases as the cell matures, the nucleus disappears,

20. Mitosis is the form of cell division characteristic of most cells. It is controlled by the nucleus (see chap. 14).

and at the end of the series we find the nonnucleated biconcave disks filled with hemoglobin, indistinguishable from mature erythrocytes of the circulating blood.

All this occurs in a connective-tissue network outside the regular blood stream in adult mammals. Normally, only the fully formed red cells pass into the blood stream. Why only the mature cells do this is not known. Only when red blood cell formation is occurring at an abnormally rapid rate do immature forms pass into the circulating blood.

3. *Span of life.*—Therefore, we know that formation and destruction are proceeding continuously. Can we tell anything about the rate of production and destruction and the factors controlling these activities? Can we tell how many red cells are being formed and destroyed daily? Direct observations cannot be made, but indirectly we can arrive at an approximation by several methods. By one of the older methods, the life-span appeared to be very short, perhaps from 10 to 30 days. By newer and probably more reliable experiments, it appears that red cells survive much longer, about 125 days.

The older method depends upon our knowledge of the ultimate fate in the body of the hemoglobin of the destroyed red cells. Hemoglobin disintegrates into a protein fraction, whose fate is unknown, and a nonprotein component—an iron-containing pigment called *hematin*. The latter material undergoes further breakdown. The iron fraction is retained in the phagocytic cells,[21] and much of it is apparently returned to the bone marrow to be used again in the manufacture of new hemoglobin.

A large portion of the hematin fragment is converted into *bile pigment* and is excreted by the liver. It is passed into the bile duct leading from the liver into the intestine (see Fig. 92, p. 267), finally to be passed from the body. Much of the coloring matter of the stools is due to bile pigments, which are still further altered chemically in their passage through the alimentary canal.[22]

Therefore, by determining the number of grams of bile pigments eliminated in the *feces* per day, one can calculate the grams of hemoglobin necessary to produce this, and thus also the number of red cells destroyed in its formation. Such studies have led to the con-

21. Iron can be demonstrated here by proper staining in histological preparations.

22. Therefore, when bile is prevented from passing into the intestines by a *gallstone* blocking the bile duct, colorless or "clay-colored" stools are passed, partly owing to the absence of bile pigments.

clusion that approximately one-tenth to one-thirtieth of all the red cells of the body are destroyed daily. Expressing the same result in another way, we may say that the average survival of a red cell is from 10 to 30 days.

It should be clear that this method of estimating the life-span of the red cell would be valid only if (1) *all* the hemoglobin which disintegrates appears as excreted bile pigment and (2) excreted bile pigment is derived only from disintegrated hemoglobin. There is evidence that the latter is not the case and that some of the excreted bile pigment comes from other sources than broken-down hemoglobin. If this is true, estimates of rate of red-cell destruction would be erroneously high and life-span low.

A newer method of approach was supplied by the employment of radioactive substances. Experimental animals or human beings can be fed substances known to be incorporated into the hemoglobin molecule in the body. Such substances, with radioactive "tagged" atoms in the molecule, can be prepared artificially. The radioactive substance is built into the hemoglobin molecule and the red cell. As red cells disintegrate, radioactive end products, readily detectable, appear in the excreta. The feeding of such radioactive building stones of hemoglobin results in a marked increase in radioactive waste excretion 100–140 days after the administration. This is interpreted as indicating that for this period of time the radioactive substances lay in the red cells and then were released and excreted when the red cells and hemoglobin were destroyed about 125 days later.

If the average life of the red cells is about 125 days, which other evidence also corroborates, it can be calculated that nearly two million cells are destroyed—and formed—per *second*.[23] Normally, this continues every day, every minute, every second, of our lives.

4. *Correlating mechanisms.*—What correlates these opposing activities, proceeding at a staggering rate and in anatomically separated structures—the liver and bone marrow—resulting in the remarkably constant number of circulating red cells? Are the organs simply "wound up" to keep going at the same pace? We know this is not true, for a variety of conditions can alter the rate of one or the other of these activities; but ordinarily the balance will soon be restored. The count may be maintained at a new level, but the activities are

23. In the 5 liters of blood of an average adult there are 25 *million million* red cells. If the cells survive 125 days, there are 200 billion destroyed per day, or over 2 million per second.

still interdependent. We must look here for some physiological chain
of events bridging the anatomical separation of the blood-forming
and blood-destroying organs—some automatically operating machinery which correlates the processes.

The discovery of perhaps a major factor in the mechanisms involved came in a roundabout manner from an experimental analysis
of the stimulating action upon the bone marrow exerted in high
altitudes. If an individual with a normal red blood cell count of five
million remains for about two or three weeks at an altitude of 14,000
feet above sea level, his blood count will gradually rise to about six
or seven million. Investigators turned their attention to the cause
of this. What is there specifically about high altitudes that stimulates red-cell production?

Any of the differences between the atmosphere at sea level and at
high altitudes might be suspected. In the latter environment there
is less dust, more ultraviolet light, a lower temperature, a lower atmospheric pressure, and less nitrogen and oxygen per 100 cc. of air.
Each of these should be tested to determine which is the causal
factor. Physiologists at the University of Wisconsin solved the problem by first subjecting rabbits, dogs, and rats at sea level to a
reduced total atmospheric pressure. An elevation of the red count
resulted. They then had the animals breathe air containing less
oxygen and more nitrogen than is present in the atmosphere. Though
the total pressure of these gases was at sea-level pressure, the same
red-cell rise occurred. Examination of the bone marrow revealed an
extensive increase in quantity of the red portion, and, microscopically, evidences of unusually rapid red-cell production were present.

It has long been obvious that a reduced oxygen content of the air
breathed leads to a reduced oxygen content of the blood. The Wisconsin investigators interpreted their findings as indicating that this
changed composition of blood exerts a chemical stimulus upon the
bone marrow. In further support of this conclusion, it is known
that abnormal conditions which prevent a proper oxygenation of
blood, such as interference with circulation through the lungs, may
also lead to a rise in the red-cell count.

How is it possible for the lack of a material to be stimulating?
How can a deficiency exert a chemical stimulation? It should, of
course, be no more puzzling to have stimulation by a deficiency than
by a surplus. In either case, the significant thing is the changed
chemical composition of the cell environment. A stimulus, by defini-

tion, is any environmental change, and no special difficulties are raised when the change is a deficiency rather than an excess of any environmental constituent. Cold, which is really a deficiency of heat, stimulates certain units in the nervous system and causes sensations all have experienced. More puzzling is the exact nature of the stimulus-response relationship in blood formation. Exactly how does this specific environmental change, namely, oxygen lack, exert its specific action, namely, accelerated activity, upon the bone-marrow cells?

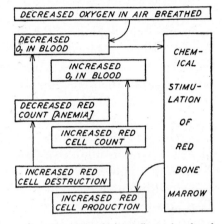

Oxygen lack is apparently not a general cell stimulant, although there are other cells in the body which respond to oxygen deficiency, such as certain nerve endings. But, in any case, the exact nature of the interaction between oxygen lack and the cell, and such intermediate mechanisms as might be involved, are unknown. Here is a link in a

FIG. 23.—Stimulating effect of reduced blood-oxygen concentration upon the red bone marrow. Note how any of the changes listed to the left tends to initiate an automatic adjustment in the opposite direction.

chain of mechanisms which has so far evaded satisfactory explanation.

The application of these findings to our problem is clear. That a lowered oxygen content of the blood stimulates the red bone marrow helps us explain the correlation of red-cell formation and destruction, which leads to the normal constancy of numbers. Suppose red-cell destruction should increase for some reason. A reduction in number of circulating red cells would be the immediate consequence. Since the oxygen of the blood is found almost entirely in the red cells in combination with hemoglobin, it follows that, if there are fewer red cells, there will be less oxygen per cubic centimeter of blood. But this is exactly the condition we have shown to be stimulating to the bone marrow, which quickly increases its output of cells. Consequently, there is an adjustment of the rate of production, corresponding to the accelerated rate of destruction. A similar sequence of events would prevent the normal bone marrow from ever "spontaneously" decreasing its activity if blood destruction remains constant.

On the other hand, a "spontaneous" abnormal overactivity of the bone marrow fails to cause a permanent increase in the red count, presumably because increased numbers of worn-out red cells would be presented to the phagocytes, and there would be a compensating increase in rate of phagocytic destruction.

Although other factors probably operate, this oxygen tension mechanism is important in maintaining the normal balance. By the same mechanism abnormal blood losses are also compensated for, as in hemorrhage or blood destruction by hemolytic agents. In all such cases the blood count is automatically restored to normal through stimulation of the bone marrow exerted by the reduced oxygen content of the blood. Examination of the bone marrow in these conditions usually reveals an abnormally great rate of activity.

5. *Physiological constants.*—Thus the relative constancy in number of red cells takes upon itself a far greater significance than the constancy in numbers of nerve or muscle cells. Here we have not a static condition of anatomical constancy but a dynamic equilibrium. It is an excellent example of a physiological constant. Many such constants will command our attention as we proceed,[24] and we shall see that a complete description of the mechanisms involved in their maintenance would encompass all of physiology.

F. ANEMIA

Anemia is an abnormality in which the red cells of the blood are reduced in number, or are deficient in hemoglobin, or both. Instead of a normal count of about five million cells, the anemic individual may have but four or three or even less than one million cells per cubic millimeter of blood, depending upon the severity of the anemia. Anemia is an abnormal state in which there is a breakdown of the formation-destruction balance which normally maintains the physiological constant of about five million.

The harm done to the organism in any anemia depends upon the chief function of the red cells, namely, that of transportation of oxygen. If there is a deficiency either of red cells or of hemoglobin, the quantity of oxygen supplied to the tissues generally is reduced, cell oxidations are hampered, energy liberation inadequate, and normal cellular function impaired. Nerves controlling muscles fatigue quickly, and, if the anemia is severe, the muscles function scarcely at all, and the patient is bedridden, able to carry on only the minimal

24. Recall also the state of physiological constancy of the plasma composition, p. 67.

energy-liberating reactions requisite for bare maintenance of life. When even this is no longer possible, death occurs.

We can conveniently separate the anemias into two or three groups: those due to excessive loss or destruction of red cells and those due to defective production by the bone marrow. Or a combination of both may be involved.

1. *Hemorrhage.*—Excessive destruction may result from a variety of causes. The presence of hemolytic agents of any kind in the circulating blood, or the actual mechanical loss of blood from the vessels, may produce a lowered red count. In either case, there is a compensatory reaction of the bone marrow. The reduced oxygen content of the anemic blood stimulates the red-cell-forming organs. Often, if the blood loss or destruction is not too excessive, the bone marrow is able to cope with the situation, and no anemia follows. As a matter of fact, a very slight loss of blood, as during a normal menstrual period in women, may so stimulate the bone marrow that the moderate deficiency is more than compensated for, and, despite the hemorrhage, the red count may be slightly above normal for a brief period. Here the bone marrow has "overcompensated." Physiological readjustments frequently overcompensate temporarily.

The ability of the bone marrow to replace losses is strikingly great. If fully a quarter of the blood is withdrawn at one time from a normal animal, it will be found that in a few weeks the red-cell count is back to normal. Lesser hemorrhages recurring chronically, as from an ulcer of the stomach, may also be fully compensated. But, if the bleeding is too extensive and too frequent, the bone marrow is unable to replace the cells as rapidly as they are lost, despite the fact that it has been stimulated to produce cells at a rate much in excess of normal. Anemia is the consequence.

2. *Bone-marrow defects.*—If there are no blood losses, and the rate of destruction is normal, an anemia might still result from underproduction of red cells because of defective bone-marrow function. Obviously, any agent or process which actually destroys the bone marrow would cause an anemia of this general kind. Benzol and lead are more or less destructive to bone-marrow tissue, and industrial workers who use these materials a great deal may absorb enough of them to cause a definite or even serious anemia. Likewise, tumors of the red bone marrow may destroy enough of the red-cell-forming tissue to produce anemia.

3. *Nutritional deficiencies.*—There may be inadequate red-cell

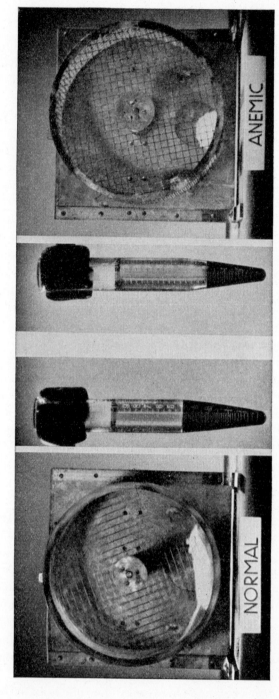

Fig. 24.—Anemia. The normal rat (*left*) is well and active, running vigorously in its revolving cage. The anemic rat is inactive, disinclined or unable to move about. The tubes show how the red cells settle out from the blood by gravity, leaving clear plasma above. Note that 10 cc. of anemic blood (*right tube*) contains only 2 cc. of cells, while 10 cc. of normal blood (*left tube*) contains 4 cc. (or often even more) of cells. (From the sound film, *Foods and Nutrition*.)

manufacture even if the bone marrow—the factory—is quite normal, provided the raw materials are not supplied. The growth and reproduction of the red cells obviously require an adequate supply of the proper compounds such as proteins and atoms such as iron to make both stroma and hemoglobin. A great many dietary and nutritional disorders and deficiencies prevent this from time to time.

The intake of iron, which is an indispensable atomic constituent of the hemoglobin molecule, has received considerable attention. There is no question but that diets deficient in iron cause anemia. The actual number of red cells may not be greatly reduced, but the hemoglobin content of each cell is. However, not very much iron is needed by the body each day. Recall that when red cells are normally destroyed, and the hemoglobin molecule disintegrated, the iron-containing fraction is not entirely lost. Much of it is retained by the liver and is later made available for use by the bone marrow in the manufacture of new hemoglobin. Dietary experiments have indicated that the daily iron requirement of man is about 6–10 mg. Usually, this amount is easily obtainable in an ordinary mixed diet, iron being fairly generally distributed in nature. Meat and eggs are especially rich in iron.[25]

However, during one period of a normal lifetime there is some danger of an inadequate iron intake. This is in early infancy, when milk is usually the sole item of the diet. Milk, otherwise an excellent food, is very poor in iron, and a prolonged milk diet will lead to an anemia on this account, in man or animals. The infant manages at this time, despite a rate of red-cell formation even faster than in adults, because during intrauterine life great stores of iron are obtained from the maternal blood and laid down in the liver. But, if weaning is unduly delayed, or excellent sources of iron, such as egg yolk, are not added to the milk diet rather early, the liver stores are depleted and an iron-deficiency anemia develops.

Other nutritional deficiencies, in which components necessary for the manufacture of cells in general are inadequate, may affect production of red cells as well. General undernourishment or deficient vitamin intake causes anemia. Sometimes the marrow fails to get the materials it requires even though the individual may eat a proper

25. Investigations at the University of Wisconsin indicate that a trace of copper is also necessary in the diet for normal hemoglobin manufacture. Apparently, copper is not itself a constituent of the hemoglobin molecule, but its presence seems necessary, perhaps in the role of a catalyst, for the adequate utilization of iron in hemoglobin synthesis.

diet. If there is a chronic severe diarrhea, or faulty digestion, there may be an inadequate absorption of the required materials.

4. *Pernicious anemia.*—Of special interest is the condition called *pernicious anemia*. It derives its name from the fact that prior to 1927 this condition was as uniformly fatal as inoperable cancer. The blood count decreases for a time, there might be a remission with a return of the count toward normal, and then a relapse more severe than the first attack. In each new attack the red-cell count goes lower, culminating in death after two to five years. Formerly, nothing could be done to stop this inevitably fatal course.

Some years ago a group of investigators at the University of California, under the direction of Dr. G. H. Whipple, became interested in experimental anemias produced in animals by repeated extensive hemorrhage. They observed the effects of various foods upon recovery from the anemia. They tried many foods quite empirically and at random, guided by no preconceived notions or hypotheses. Of all the articles they fed, they found that liver had the most striking effect. Of the dogs made anemic in the manner described, those fed liver recovered more rapidly than the others.

Now, of course, this experimental anemia was not pernicious anemia. The latter condition in man displays features which are quite distinct from most other anemias. The appearance of the red blood cells is abnormal, and serious changes in the alimentary canal and central nervous system occur. The dogs showed none of this.

Yet here was a case where red-cell production was in some way enhanced by the ingestion of a specific food, namely, liver. Why not try it, at least, upon patients with pernicious anemia? And so in 1927 two Harvard University physicians, G. R. Minot and W. P. Murphy, fed large quantities of liver to their patients daily. A striking recovery was effected. In the intervening years many investigators have confirmed these findings. A few days after beginning of liver feeding, the blood count commences to rise and in a few weeks approximates or even reaches the normal, and the subject is comparatively well. Note that liver does not cure the condition. When the liver intake is stopped, the anemia promptly returns, and, unless liver feeding is reinstituted, death is sure to follow.

Here the physiologist asks the question: "Does our knowledge of the normal physiology of red-cell formation and destruction contribute to a further understanding of the causes of pernicious anemia?" Apparently the primary difficulty in the bone marrow is with the formation of the stroma portion of the red cell. The cells produced are defective—abnormal in size and shape and excessively fragile. It seems that the manufacture of hemoglobin is not a primary defect. In fact, in pernicious anemia each cell contains more than the normal amount of hemoglobin. Yet there are so few cells that the quantity of hemoglobin in a cubic millimeter of blood is markedly reduced. Possessing these internal defects, the cells which are produced are destroyed more rapidly than usual, as evidenced by the fact that bile pigment, the end product of hemoglobin disintegration, is increased in quantity in the feces.

Can we learn anything about normal physiology from pernicious anemia? The findings with liver feeding suggest that this tissue contains a material necessary for normal red-cell production. What is the nature of this product? Liver has been subjected to treatment with various solvents, and it has been possible to extract from the whole tissue a product which contains practically all the material of liver effective in controlling pernicious anemia. When this material is injected, even in relatively small quantities, it controls the abnormality as well as does the feeding of whole liver.

Much has been learned of the chemical nature of the effective agent in liver. Actually, it is a vitamin (see pp. 343–48), called vitamin B_{12}, which has been prepared in pure crystals; the chemical structure of the substance has been partially determined.

The injection of infinitesimal quantities of vitamin B_{12} controls pernicious anemia. As little as one-millionth of a gram daily has been effective—a striking example of the powerful action of such biological agents as vitamins. But, if such small quantities of a substance occurring in many normal food substances can protect against pernicious anemia, why should this defect occur in anyone? Years of experimentation and investigation were required to provide the answer.

While very small quantities of *injected* B_{12} control pernicious anemia, much more must be given by mouth to be effective, unless gastric juice from a normal person is also administered. Gastric juice from a pernicious anemia patient is ineffective in this regard. Here is a striking and important difference between normal people and those with pernicious anemia. Apparently the former possess something in the gastric juice which is at least partially lacking in the stomach secretions in pernicious anemia—a something which enables the individual to avoid pernicious anemia on a small intake of vitamin B_{12} in the food. Presumably the substance in normal gastric juice facilitates the absorption of vitamin B_{12}.

From these observations there emerges the concept that the primary defect in pernicious anemia is probably the failure of the stomach (and possibly the intestine) to secrete enough of a substance necessary for the absorption into the blood stream of vitamin B_{12}, which is essential for the manufacture of normal red blood cells and the prevention of pernicious anemia. The condition can be prevented or corrected by feeding relatively large quantities of vitamin B_{12} or foods containing it, in which case there is absorption of the very small amounts required. Or these very small required amounts can be provided more directly by injection.

The fact that normal liver contains considerable vitamin B_{12} indicates that excesses of the vitamin absorbed by a normal individual are stored as a reserve safety factor in the liver.

Many questions still remain: What is the stomach secretion necessary for the absorption of vitamin B_{12}? What causes pernicious anemia patients to secrete too little of the substance? How does it assist in the absorption of vitamin B_{12}? How does this vitamin promote the formation of normal red blood cells and thus prevent the formation of faulty red blood cells characteristic of pernicious anemia? These and other questions remain for future investigation to answer. "Nature guards her secrets jealously." Only by persistent experimentation can these secrets be exposed, whether it be in the realm of pernicious anemia, cancer, or infantile paralysis. Even though we are fairly successful in controlling pernicious anemia, we know all too little of the basic causes of the abnormality. We know nothing about the prevention of its occurrence, and we cannot "cure" the disease, although we can so treat the patient that he can anticipate a relatively normal and useful life-span instead of dying within a few years of the onset of the condition.

VII. TRANSFUSION OF BLOOD

An organ or tissue operates very much the same in all normal human beings. Furthermore, each of us, in general, is equipped with more cells, tissues, or even organs than may be necessary for normal life; for example, man can get along with one eye, one ear, one kidney, one adrenal gland, or only part of his liver (see also "Margins of Safety," chap. 13). Therefore, it has been natural to attempt to replace tissues or organs lost in disease or accident in one man with

such surpluses as some other individual might possess and be willing
to donate. For various reasons, donations such as grafting and
transplantation of tissues have been unsuccessful, partly because of
the apparently insurmountable difficulty of maintaining the nourishment and life of tissues or organs during removal from the *donor* and
transplantation into the *recipient* and partly because of the difficulty
in re-establishing a life-sustaining flow of blood in the transplanted
tissue (see also "Species Specificity," p. 87).

A notable exception is the transplantation, or *transfusion*, of
blood. Undertaken with suitable precautions, this procedure is widely
used to replace blood which has been lost by injury or disease or in
a surgical operation.

A. Early Difficulties

The fact that man can lose considerable blood and still live and
recover completely indicates that he normally possesses a surplus of
this liquid and could spare some of it to an individual suffering excessive blood loss or blood destruction. Furthermore, as we have seen,
it is very easy to withdraw blood from a vein and reinject it into the
blood stream of someone else.

However, some of the early attempts at blood transfusion were
made not for the sound purpose of replacing lost blood but for such
fanciful reasons as the restoration of youth in the aged by the injection of blood withdrawn from a child. In one such account we
learn that the old man became ill, the child died, and the "physician"
fled. This was an early illustration of the dangers of indiscriminate
blood transfusions.

B. Safe Blood Transfusions

1. *"Reactions" to transfusions.*—If transfusions are made at random, from one human being to another, the recipient of the transfusion may display certain reactions in an important percentage of
the cases. Such reactions include muscle and chest pains, bloody
urine, collapse, and even death. Why should this be, since blood functions essentially the same in every normal human being? An obvious
approach to an answer is to observe under the microscope what
occurs when blood from one man is mixed with that from another.
Nothing significant in such studies is revealed unless the blood
samples examined are separated into their two main components,
the cells and the plasma, or, more usually, the serum.

In general, wherever a transfusion results in a reaction (i.e., an *incompatible* blood transfusion), it is found that mixing the donor's red blood cells with the recipient's serum or plasma causes the donor's cells to clump together in masses or even to be destroyed by hemolysis. Such an occurrence in the recipient's blood stream produces a blocking of innumerable capillary blood vessels, which are too tiny to admit clumps of red blood cells. A reaction—collapse or death—may be the consequence of such interference with blood flow in multiple microscopic areas. Such untoward effects almost never occur when preliminary testing reveals that the donor's red blood cells are not clumped by the recipient's serum.

2. *Cross-matching.*—Tests made upon thousands of individuals revealed many instances in which red blood cells of one individual are clumped (or *agglutinated*) by the plasma or serum of certain other individuals. Such tests are known as *cross-matching*. A transfusion is likely to be safe (i.e., *compatible*) only when cross matching reveals that the donor's red blood cells are not agglutinated by the plasma of the recipient. On the other hand, it is usually found to be of no consequence if cross-matching shows that the recipient's red blood cells are agglutinated by the donor's plasma. Although this effect takes place under the microscope, it apparently does not occur in the body. In practice, if cross-matching reveals that

> X's plasma agglutinates Y's red cells, and
> Y's plasma does not agglutinate X's red cells,

then it will usually follow that transfusion

> of X's blood into Y is safe and
> of Y's blood into X is unsafe.

Such cross-matching effects appear to be caused by chemical interactions between substances in the red blood cells of Y and in the plasma of X. The failure of Y's plasma to agglutinate X's red blood cells might be due to the absence of the appropriate substance either in X's cells or in Y's plasma.

3. *Blood groups.*—Cross-matching tests have shown that human beings can be classified into four main blood *types*, or *groups*, as regards these agglutination reactions. Some people have cells which are clumped by the plasma of everyone except those of the same blood group. Some have cells which are clumped by no one's plasma. Individuals of intermediate blood groups have red cells which are

clumped by the plasma of some but not all other people. A clarification of these relationships came at about the turn of the century, with the Nobel prize-winning researches of Karl Landsteiner in New York.

The results of these investigations are summarized in Figures 25 and 26, which indicate the significant properties of both the red blood

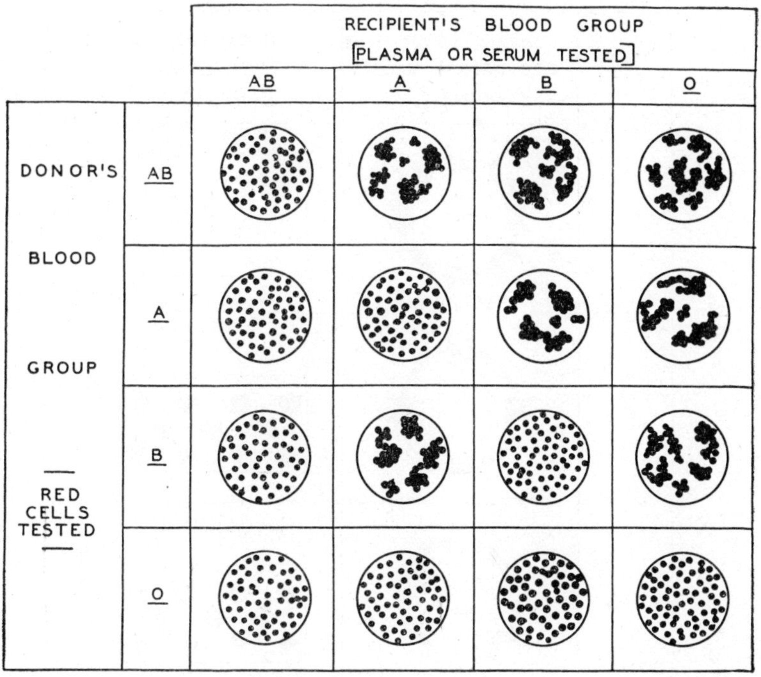

FIG. 25.—Interactions occurring when blood (i.e., red cells) of each of the four blood groups is mixed with the plasma or serum from each blood group. Circles are microscope fields. Black dots are red blood cells, which remain dispersed (as when cells from any group are exposed to serum from Group AB) or are agglutinated (as when cells from Group AB are mixed with serum from Groups A, B, or O). Note that, to identify the blood group of an individual, his red cells need be tested only with serum of Group A and of Group B.

cells and the plasma, as regards agglutination, for the four human blood groups: AB, A, B, and O. These symbols are derived from agglutination behavior of red blood cells (rather than the agglutinating properties of the plasma) of human beings, which behave, depending on the blood group of the individual, as if they possessed substance A (Group A), substance B (Group B), both substances A and B (Group AB), or neither substance A or B (Group O). As shown in Figure 26, which should be studied with the text, the plasmas of persons in the four blood groups also differ as regards the possession of chemicals causing red-cell agglutination. These chemicals are called substances *a* and *b*. Whenever plasma with substance *a* and cells with substance *A* are mixed on a glass slide, there is agglutination, as is true also when *b* substance plasma and *B* substance cells come into contact. Similarly, cells with both *A* and *B* substances

BLOOD GROUP	CHARACTER OF RED BLOOD CELLS [SCHEMATIC]	CHARACTER OF PLASMA [SCHEMATIC]
AB	A & B SUBSTANCES	
A		b SUBSTANCE
B		a SUBSTANCE
O		

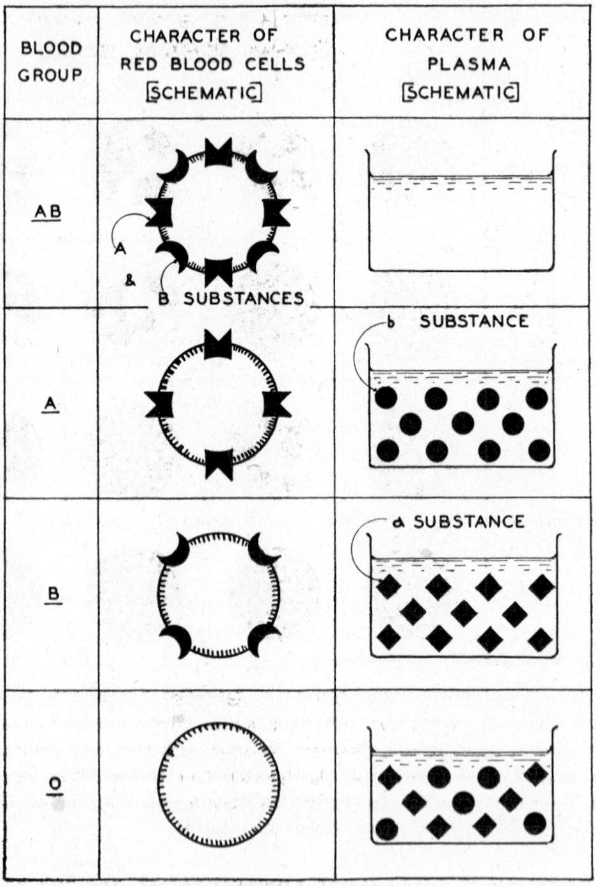

FIG. 26.—Schematic representation of *A* and *B* substances in red blood cells and of *a* and *b* substances in plasma in the four blood groups. Whenever plasma with *a* substance (Groups B and O) comes into contact with cells containing *A* substance (Groups AB and A), red cells are bound together: clumping or agglutination occurs. Similarly, plasma with *b* substance (Groups A and O) agglutinates red cells containing *B* substance (Groups AB and B). See also discussion in text, which should be read with repeated reference to this figure.

(Group AB) will be agglutinated by plasma with either *a* or *b* substances or both, and plasma with both *a* and *b* substances will agglutinate red cells with either *A* substance (Group A) or *B* substance

(Group B) or both (Group AB). Red cells containing neither A nor B substance (Group O) will be agglutinated by no plasma; plasma lacking both a and b substances (Group AB) will agglutinate no red blood cells. The apparent complexity of these statements is vastly simplified by a study of Figure 26, in which substances A and B and a and b are represented schematically in a manner calling attention to the property of substance a in plasma of fitting into or combining with substance A in adjacent red cells, thereby joining the red cells together into clusters. B and b substances are represented as fitting into each other but not fitting into the A or a substances.

Figure 26 will be understood when the reader comprehends why the following transfusions are usually *safe:*

> Group AB blood into a Group AB recipient
> Group A blood into a Group AB or A recipient
> Group B blood into a Group AB or B recipient
> Group O blood into any recipient

and why the following transfusions are usually *unsafe:*

> Group AB blood into a Group A, B, or O recipient
> Group A blood into a Group B or O recipient
> Group B blood into a Group A or O recipient.

Such understanding will also reveal why Group O individuals are called *universal donors,* whose blood can ordinarily be transfused into anyone safely, and why Group AB individuals (whose plasma lacks both a and b substances) are called *universal recipients,* into whom blood can ordinarily be transfused from individuals of any blood group without harm.

The thoughtful reader of the foregoing analysis will, by this time, have become puzzled about a problem which may be formulated specifically thus, regarding a transfusion from a donor of Group A into a recipient of Group AB: "Although it is clear that the donor's cells should not be agglutinated, why does not the plasma of the Group A donor, containing b substance, cause the agglutination of red cells of the recipient, with both A and B substances, and therefore still be harmful?" Theoretically, the latter should follow. But empirical observation, as is so often the case in science, fails to support the theory. Plasma with b substance (from a donor of Group A), when injected into a Group AB recipient, does not ordinarily lead to agglutination or cause reactions.

The explanation of this fact is not entirely clear. It is thought that plasma substances capable of causing agglutination in the test tube are ordinarily so diluted by the recipient's plasma as to become ineffective in agglutinating the recipient's red blood cells. In general, it is found that the character of the donor's plasma is unimportant as regards presence or absence of a or b substances. Ordinarily, it is the character of the donor's red blood cells, as regards presence or absence of A and B substances and their reactions with the recipient's plasma a and b substances, which determines whether a given blood transfusion is likely to be safe.

4. *The Rh factor.*—The discerning student will have noted, in the discussions of blood groups and transfusions, that caution was employed in statements that certain transfusion combinations were *relatively* safe or *usually* safe or *ordinarily* safe. Such qualifying adjectives are necessary because harmful reactions sometimes occur in transfusions entirely independently of the Group AB, A, B, or O relationships of the donor and recipient. Ordinarily it is entirely safe, for example, to transfuse the blood of a donor of one group into a recipient of the same group. However, if such transfusions are given repeatedly to the same recipient at appropriate intervals, reactions of increasing severity and danger may occur at successive transfusions in the case of certain individuals. Investigation reveals that, in the later transfusions, the recipient's plasma agglutinates and even destroys the donor's red blood cells, exactly as if the donor and recipient belonged to incompatible blood groups. Although the first transfusion was safe, and donor and recipient are apparently compatible, the recipient's plasma sometimes develops substances which agglutinate red cells formerly unaffected by that plasma.

This phenomenon resembles the appearance in blood plasma of substances harmful to certain bacteria, as a result of infection by those bacteria or of artificial injection of bacterial products into the body (see "Acquired Resistance to Infection," chap. 13). Just as such bacteria stimulate the body to produce antibacterial substances, transfusions sometimes seem also to stimulate the body to develop substances harmful to (i.e., produce agglutination of) certain kinds of red blood cells which were formerly safe to transfuse.

The red blood cells of certain individuals contain a substance or factor, apparently independent of A or B substances, which has been

called the *Rh factor* or substance. Determination of the existence of
this factor is accomplished by mixing human red blood cells with the
plasma of the rhesus[26] monkey. Such plasma contains substances
(called *anti-Rh substances*) which agglutinate the red blood cells of
many human beings (about 85 per cent of the general population),
who are designated *Rh positive*, having Rh substance in their red
cells. Red blood cells of the remainder of the population are not
agglutinated by rhesus plasma; such individuals are termed *Rh
negative*, lacking Rh substance in their red cells. Fortunately, human
blood plasma does not normally contain anti-Rh substances, so that
Rh relationships may be disregarded in first transfusions. Furthermore, repeated injections of Rh-positive red blood cells into Rh-
positive recipients is ordinarily without harmful consequence. However, two or three transfusions of Rh-positive blood (i.e., red blood
cells) into Rh-negative recipients causes the latter to form anti-Rh
substances which appear in the plasma and may cause agglutination
reactions upon the third or fourth or later transfusion of Rh-positive
blood. Therefore, when repeated transfusions are employed, and
when the recipient is Rh negative, it is necessary to employ donors
who are Rh negative as well as being of the appropriate AB, A, B,
or O blood group.

Another serious and usually fatal consequence of Rh incompatibil-
ity is occasionally encountered in newborn infants who are born with
a severe anemia and commonly do not live long. This condition is
closely related to the phenomena described in the preceding paragraphs. In this special kind of anemia the mother is always Rh negative (i.e., lacks Rh substance in her red blood cells) and the infant
is Rh positive, which trait was inherited from an Rh-positive father.
Why should Rh-negativity in the pregnant woman be deleterious to
the red blood cells of an Rh-positive infant before and immediately
after birth, since neither Rh-positive nor Rh-negative individuals
normally contain the damaging anti-Rh substance in their plasma?
The answer lies in the fact that during pregnancy the Rh-negative
mother develops these anti-Rh substances in detectable quantities
in her plasma, just as does an Rh-negative recipient of repeated
transfusions from an Rh-positive donor. Of course, the blood of the
fetus[27] does not actually flow into the mother's blood (see p. 608),

26. The term *Rh* is derived from "rhesus."

27. The term *fetus* is applied to the unborn young in the later stages of development in the
uterus.

but apparently either a few red blood cells or some of the Rh substance in the red cells of the fetus do enter the maternal blood stream, stimulating the mother to produce anti-Rh substances which appear in her plasma. These are virtually harmless to the woman, since they react only with the Rh substance which is absent from the red cells of the Rh-negative mother. However, the maternally manufactured anti-Rh substances apparently pass on into the fetal blood stream, where they react with the Rh substances in the red cells of the Rh-positive fetus, causing agglutination and such destruction of the fetal red cells that a severe anemia results. Persisting until birth, the anemia is sufficiently severe to be fatal in most instances.

A knowledge of these relationships has made feasible a measure of control. It is possible to detect a potentially dangerous combination of an Rh-negative mother and an Rh-positive fetus before birth. Every expectant mother should have her blood tested for the Rh factor. If she is in the Rh-positive majority, there need be no concern. But if she is Rh negative, tests upon her plasma should be made at regular intervals. The gradual development of anti-Rh substances in her plasma is almost certain evidence that the developing fetus is Rh positive, an inheritance from an Rh-positive father. Awareness of these relationships and, in a given pregnancy, knowledge of the threatening combination of an Rh-negativity of the mother's blood and the development of anti-Rh substances in the mother's plasma do not enable the physician to correct the condition. He can only anticipate the birth of a severely anemic infant. But he can have available an adequate supply of Rh-negative blood (which is also Type O) for transfusion immediately after birth. Unfortunately, such transfusions are too seldom lifesaving.

It might be argued that the condition described could be avoided if there were no fertile marriages between Rh-negative women and Rh-positive men. This conclusion is entirely correct; however, there is general agreement that such a ban is entirely unjustified. Since about 15 per cent of females (or males) are Rh negative, and 85 per cent of males (or females) are Rh positive, it follows that about 13 per cent of all marriages are of a combination of an Rh-negative wife and an Rh-positive husband. The operation of heredity is such that by no means all the offspring of such combinations are Rh positive. More fortunately, it is found that even the potentially dangerous combination of an Rh-negative mother and an Rh-positive fetus only rarely culminates in the kind of infantile anemia described. In the overwhelming majority of instances normal infants (as regards the Rh factor) result, presumably because the Rh substance of the fetal red cells fails to reach the maternal blood or stimulate the production of anti-Rh substance by the mother or because such anti-Rh substance as the mother produces fails to reach or influence the Rh-positive red blood cells of the infant. But, whatever might be the reason, the chances of producing defective anemic offspring are sufficiently remote as to constitute no absolute ban against the marriage of Rh-negative women and Rh-positive men.

C. Value of Blood Transfusions

The danger to life in the loss of blood from an injury or operation or at childbirth is not primarily in the loss of the formed elements of the blood but in the physical effect of the diminished volume of circulating liquid upon the blood pressure. Man would be somewhat weakened, but he would not be prostrated and would certainly not immediately succumb, if he were to lose half his red cells or white cells while still maintaining a normal total volume of circulating blood. But he would almost surely die from a sudden loss of one-half the volume of blood. The reason is the direct relationship of the blood volume to blood pressure (see pp. 204–5). A reduction in blood volume tends to reduce the blood pressure, as is the case in any elastic tube containing liquid under pressure, whether or not that liquid is moving. An excessive lowering of the blood pressure is fatal partly because the normal physical resistance to the circulation of the blood is too great to be overcome by a low blood pressure. The resulting faulty circulation causes damage and even death.

The value of a blood transfusion after a hemorrhage, therefore, lies primarily in its restoration of the blood volume, of the blood pressure, and of the blood circulation. Indeed, extensive use has been made of transfusions of plasma, since plasma is about as effective as whole blood in restoring blood volume and blood pressure.[28]

D. Transfusions in War and Peace

Many of the technical and scientific advances of wartime have no direct applicability to peacetime life and its needs. On the other hand, the driving necessities of war do result in a few rapid advances which might take longer in peacetime. The development of drugs to

28. An advantage of plasma transfusions is that the blood group of both donor and recipient may be disregarded, as should be clear from the discussion in the latter part of the section on blood groups.

combat certain infections (e.g., the *sulfa drugs* and *penicillin*) were hastened during World War II (see pp. 594–96). Great strides were also made in the recognition and control of tropical infections and of emotional states preventing adjustments to severe stress.

The employment of blood transfusions is in a similar category. Never before were so many transfusions and administrations of blood plasma given in a comparable period of time. Never before did so many people donate blood for such a purpose.

A major factor in the medical contribution to winning the war was the organized provision of large volumes of blood to replace that lost in war wounds and in the emergency surgery they necessitated. The organization of blood collections and the establishment of blood banks were as common as the monetary institutions whose terminology they borrowed. Processes for the safe preservation of blood and its fractions were developed on a colossal scale, including the drying of blood plasma and reducing the bulk to be transported, which required only the addition of water before use.[29]

In the later stages of the war the blood-donor service was so effectively organized that blood drawn on the Atlantic seaboard and flown to Europe would be circulating in the veins of a wounded soldier on the battlefront the following day.

Blood banks, a wartime product, will continue to serve a valuable purpose in peacetime. Such stored blood and plasma will be useful not only in catastrophes involving many injured people but in the treatment of individual instances of burns or injuries necessitating immediate replacement of blood, without even the delay of securing a donor. In routine surgery, likewise, the rapid availability of blood or plasma for transfusion when an emergency arises is of tremendous value. The blood-bank system, responsible for the saving of hundreds and thousands of men during the war, will continue to serve vital lifesaving purposes in peacetime.

Although contributions of blood to a blood bank or the Red Cross may save the life of a recipient, they involve virtually no danger to the donor. A healthy donor probably takes less risk in giving blood than he normally takes a dozen times every day in crossing streets, riding streetcars, or crowding into elevators where someone may have a cold. All normal people have much more blood than they need, and the small amount in a single donation is soon replaced. It has been

29. A disadvantage of employing blood plasma as a blood substitute is that blood plasma, lacking red and white blood cells, is not quite so good a blood substitute as is whole blood.

found desirable, however, to provide supplementary iron to women who give repeated donations.

VIII. THE WHITE BLOOD CELLS

The white blood cells are less numerous than the red cells, a cubic millimeter containing about seven thousand of them.[30] They are semitransparent and are difficult to see unless they are stained. This applies to most of the cells of the body except the red cells, which are naturally colored. The white cells are devoid of hemoglobin and differ from the red cells in other important structural features as well. They are always nucleated, even in the mature form circulating in the blood.

A. KINDS OF CELLS

The white blood cells, or leucocytes,[31] lack the uniformity in structure and appearance of the red blood corpuscles. Several kinds of white cells are differentiated on the basis of structure, origin, and function.

1. *Granular white cells.*—These white cells are rather large, possess lobulated nuclei, and have numerous small granules in the cytoplasm.[32] The granular cells constitute about 70 per cent of all leucocytes.

2. *Agranular white cells.*—About 30 per cent of the leucocytes lack granules in their cytoplasm.[33] Most of these are lymphocytes, which are variable in size but smaller than the granular white cells and possess a single, slightly indented nucleus which makes up the bulk of the cell, leaving a relatively thin rim of clear cytoplasm between nucleus and cell membrane.

3. *Varieties of granular and agranular white cells.*—The granular leucocytes are of three varieties whose cytoplasmic granules differ in size and chemical composition. Almost all of them possess small granules whose chemical composition is such that they stain with neutral (i.e., neither acid nor base) dyes and are therefore called

30. White blood cell counts are made in much the same way as the red count, using the same principles of dilution, and counting the cells in part of a cubic millimeter on the same accurately ruled counting chamber described earlier (p. 91). In actual practice the blood is diluted less, and the diluting fluid hemolyzes the red cells, which would otherwise make counting difficult because of the far greater number of red cells. Also, the white cells are stained.

31. From the Greek *leukos*, "white," plus *kytus*, "cell."

32. The term *granulocyte* has been applied to these cells. Because of their lobed nuclei, they are also commonly called *polymorphonuclear* cells.

33. From this structural feature, the term *agranulocyte* has been employed for them.

neutrophilic granular leucocytes. A small fraction of the granular white cells have larger cytoplasmic granules than the neutrophilic cells. In some of these, called *eosinophilic* white cells, the granules stain with acid dyes and in some, the *basophilic* cells, the granules react with basic dyes.

The agranular leucocytes consist of the more common lymphocytes, already described, and the *monocytes*. The latter are very large cells with horseshoe-shaped nuclei and an abundant clear cytoplasm.

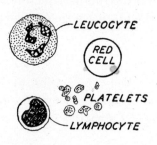
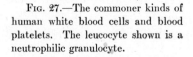
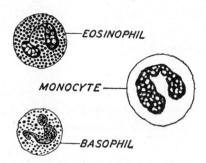

FIG. 27.—The commoner kinds of human white blood cells and blood platelets. The leucocyte shown is a neutrophilic granulocyte.

FIG. 28.—Less common varieties of white blood cells in man.

Summarizing, white cells or leucocytes are of the following varieties, occurring in approximately the percentage proportions shown:

a) *Granulocytes* (leucocytes with granular cytoplasm and lobulated nuclei):
    1. Neutrophilic granulocytes. . . . .   70 (normal range, 60–73%)
    2. Eosinophilic granulocytes.   3
    3. Basophilic granulocytes. . . . .   1 (or less)
b) *Agranulocytes* (leucocytes with clear cytoplasm):
    1. Lymphocytes (large and small) . . . . .   22 (normal range, 20–35%)
    2. Monocytes.   4

These proportions are determined by making what is called a *differential* white-cell count. A drop of blood is smeared in a thin layer on a slide, dried, and stained. Under a high-power (oil-immersion) microscope, a count is made of the actual numbers of each kind of cell in a total of two hundred or so white cells. From this, the relative proportions are easily calculated. Data obtained from differential counts are often of considerable value in the diagnosis of certain diseases in which the normal proportions are sometimes altered in a characteristic manner.

B. Origin and Life-Span

Though the circulating white cells contain nuclei, mitotic division does not occur in these cells in the blood stream. They grow and multiply outside the blood stream. Corresponding with structural differences, we find also a difference in the site of origin. The granular leucocytes[34] are formed in the red bone marrow in the same regions where red cells arise, and it is thought by some investigators of blood cells that these two kinds of cells, so distinctly different from each other in the mature form, arise from a common ancestral cell. As this cell grows in the bone marrow, it is thought that there are two main divergent lines of development in the daughter-cells, one culminating in the mature red cell, another in the granular leucocyte. Whether or not this is the case, it is clear that red corpuscles and granular leucocytes alike differentiate in the same tissue—the red bone marrow.

The lymphocytes develop in the *lymphatic* tissue, found in many regions of the body. The lymph nodes and the lymphatic elements of the tonsils and adenoids are tissues of this kind.[35]

It is clear from examination of the tissues which produce white cells that formation (and, therefore, destruction) of leucocytes continues unceasingly. Yet the normal life-cycle of leucocytes has resisted clarification. Different methods of investigation have yielded estimates of the life-span varying from less than an hour, which seems improbable, to three weeks.

The advent of radioactive atoms, a by-product of atomic-bomb research, has provided a promising tool of investigation. The feeding of a chemical compound which is used in the manufacture of white cells and which contains radioactive phosphorus leads to a detectable radioactivity in the white blood cells of man. Blood samples drawn on successive days and tests of the white blood cells for radioactivity reveal that radioactive white cells increase for a few days (as the fed material is incorporated into newly manufactured white cells) and then decrease in numbers as these radioactive cells disintegrate. The rates of these processes have led to the estimate that the average life-span of leucocytes is about thirteen days. If this is the average

34. This includes neutrophils, eosinophils, and basophils.

35. Most lymphocytes do not directly enter the blood stream from their site of origin. They are first freed into the lymphatic vessels and only later enter the blood as the lymph empties into the blood. The structure of the lymph nodes, lymphoid tissue, and the *lymphatic circulatory system* is described in some detail on pp. 211–14.

length of life of all leucocytes, admittedly some might live a much shorter time and some longer. Little is known of the fate of the great majority of the white cells. Presumably they disintegrate, and the breakdown products either are used again by the body or are excreted.

Only under special conditions, such as the combating of infections or the healing of wounds, do we know specifically about the fate of some leucocytes, as will be developed in the next section.

C. FUNCTIONS

Though fewer in number, the white cells are no less important than the red cells. When they are markedly reduced in numbers, as occurs in certain diseases, the individual becomes quite susceptible to infections, especially about the mouth and throat. Very great reduction in numbers (e.g., to 500 or 1,000 per cu. mm.) is fatal.

What is the function of these cells? Why are they indispensable to life? At present these questions can be answered only partially. We should expect the leucocytes to do work quite different from that of the red cells. Their appearance and internal structure is different; and, possessing no hemoglobin, they obviously do not transport oxygen. Likewise, we should expect the various kinds of white cells to have different functions.

1. *Defense against infection; phagocytosis.*—The best-known function of the commonest granular leucocytes depends upon their ability to engulf and digest particles, as is done by the ameba.

The ameba obtains its food by protruding protoplasmic extensions, which surround the food particle in such a way that the particle comes to lie within the unicellular organism in a little cavity where digestion takes place. This process, called *phagocytosis*, is also displayed by specialized cells of multicellular animals. In the inner lining of *Hydra*, for example (see Fig. 12, p. 64), there are cells which, though permanently attached to adjacent cells of the tissue, are able to protrude these extensions and to engulf food particles from the digestive cavity.

All through the evolutionary scale we find a persistence of this primitive food-getting reaction in specialized cells, even up to the mammals and man. There are cells which move about through the tissues of man and engulf and digest, ameba-fashion, foreign particles of many kinds, such as bacteria or dead cells or cell fragments.

Certain of the white cells, similarly, are able to migrate through the blood-vessel walls and to engulf bacteria and fragments of tissue cells.

In the immediate vicinity of infections by certain kinds of bacteria a remarkable series of reactions occur. When bacteria lodge in the deeper layers of the skin, they commence to destroy the tissues of the skin partly by means of toxic products which they liberate in their

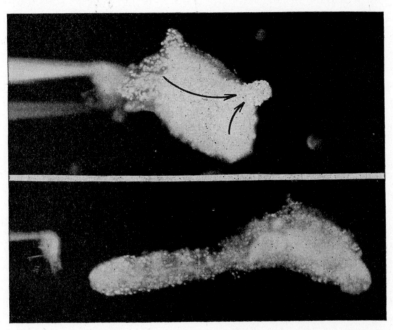

FIG. 29.—*Above:* Photograph of an ameba being touched (*on the left*) by a glass rod. This stimulus causes the animal to send out a protoplasmic extension (i.e., pseudopod or false foot) into which the substance of an animal flows (*indicated by arrows*). Thus (*below*) the animal moves away from the stimulating object. Amebae also engulf food by the extension of pseudopods. (From the sound film, *The Nervous System.*)

metabolism. Changes soon take place in the adjacent blood vessels. They dilate widely, leading a great quantity of blood into the infected region, producing the familiar reddening, and, because blood is warmer than skin, a characterized localized warmth. Quantities of fluid enter the tissue from the blood vessels, causing a swelling.[36]

In all this the blood leucocytes display a typical behavior. They seem to adhere to the blood-vessel wall in the injured area, and, by

36. These local changes, together with related phenomena, are collectively known as *inflammation* (see chap. 13).

ameboid movement, they move through the vessel wall in great numbers. Free from the circulating blood, they migrate toward the bacteria and phagocytize them. Fragments of local cells killed by the bacteria are also engulfed by the leucocytes. The ingested bacteria are usually killed and digested, but in the course of this process numbers of the leucocytes themselves may be destroyed by bacterial poisons. As they disintegrate, the leucocytes liberate their digestive enzymes, which, in turn, act upon other near-by dead cells and cell fragments. The net result is the local accumulation of blood and tissue fluids, digestive enzymes, dead tissue cells, living and dead leucocytes and bacteria, and cell fragments in all stages of disintegration. The whole conglomerate, thick, semiliquid mass is called *pus*. A relatively large collection of pus is known as an *abscess*.[37]

2. *Controlling mechanisms.*—What mechanisms control this behavior of the white cells? What makes them move through the vessel walls and approach and engulf the bacteria? The reaction seems to be of a primitive *tropistic* nature, in which the movements of the leucocytes are guided by certain bacterial products. The leucocytes appear to have evolved the reaction of migrating toward regions in which the concentration of certain bacterial products is highest. This automatically directs them to the bacteria themselves, just as an ameba is guided to its food by chemical tropisms. If we inquire further into just how certain chemicals are able to affect ameboid cells and stimulate them to an oriented locomotion, we find no clear answer. Tropisms of many kinds have been described in some detail, but their real nature remains obscure; the precise reaction between a specific chemical compound and the protoplasm is unknown. Nor, for that matter, can we say just what ameboid movement itself is, though we do call it a "primitive" form of behavior.

3. *Tissue repair.*—Even if the outcome of the warfare between leucocytes and bacteria is favorable, there is always more or less local damage to be repaired and tissues to be replaced. In the skin,

37. It is the neutrophilic cells which are involved in inflammation. The eosinophils seem also to function in defense. In invasion of the organism by certain animal parasites, these cells collect in great numbers in the infected areas, and they may be produced in abnormally large numbers in the bone marrow, so that more than the usual number are found in the circulating blood. This is especially true in infections by *Trichinella*, caused from eating infected pork which has not been adequately cooked. Also in asthma, and certain related conditions, the eosinophils of the blood increase in numbers, although their functions here are not known.

The functions of the basophils and the monocytes are even less well understood. The latter seem able to transform themselves into other kinds of cells under special conditions.

which has a great capacity to degenerate, there is considerable replacement by new skin cells. Other cells, like those of the liver or
kidney, are less capable of rapid replacement by cells of their own
type. Here, and to a lesser extent even in the skin, the destroyed
tissue is replaced by connective tissue. Connective-tissue cells are
distributed almost universally throughout the body; and, wherever
the damage may be, they soon fill the gap with their tough fibers.

It is at this repair stage of the reaction that lymphocytes are
thought to play a role. They are apparently incapable of ingesting
bacteria, and whether they are important in actual defense reactions
is not apparent. However, lymphocytes are frequently seen in great
numbers in areas of healing, or in regions of chronic infection, where
tissue destruction and repair are going on simultaneously. What do
they do here?

Earlier it was pointed out that, in tissue culture, lymphocytes are
able to transform themselves into connective-tissue cells, which, in
turn, form true fibrous connective tissue. The suggestion is that perhaps the lymphocytes undergo the same transformations in the body
and contribute to the connective-tissue formation in repair processes.
That this actually happens we are unable to determine with certainty. It by no means follows that, because cells behave in a certain
way in tissue culture, they will necessarily behave similarly in their
normal habitat in the body. Whether this occurs or not, it is still
quite possible that lymphocytes play some entirely unsuspected role
in defense against disease.

And, so far as we know, the lymphocytes and leucocytes as well
may have important functions quite independent of the emergency
situations described. But what these may be cannot be said with
certainty at present.

D. Fluctuations in the White Count

1. *Normal variations.*—In the level of the white-cell content of the
blood we encounter another example of a physiological constant.
Physiological constancy is relative. There are fluctuations, and the
normal is more often correctly expressed as a range, with certain upper and lower limits, than as a fixed, dead level. The normal range is
larger for the white cells than for reds. Counts anywhere from about
5,000 to 9,000 per cu. mm. are found in normal adult individuals and
are said, therefore, to lie within the range of the normal. Even in the
same person fluctuations of this magnitude may occur from time to

time. Attempts to correlate these changes with other normal physiological activities have not always been successful. It has been said by some, for example, that during digestion or muscular exercise or exposure to cold there is a temporary increase in numbers. Other investigators have been unable to confirm these findings. To establish relationships of this sort would be most important, for it might yield suggestions as to the possible functions of the white cells in the normal individual.

2. *Effect of infections: leucocytosis.*—There is no question about the increases in the white count which attend certain infections.[38] In pneumonia, appendicits, tonsillitis, and many other infectious diseases, the white count is elevated to 12,000, 15,000, 25,000, or perhaps even to 50,000. This we call a *leucocytosis*. With the aid of the white count, therefore, it is often possible for physicians to detect the presence of infections in internal structures like the appendix, which are hidden from view.[39] Within limits, the degree of the leucocytosis parallels the severity of the infection, so that counts taken at frequent intervals often give evidence as to whether the infectious process is increasing in severity or is subsiding.

The adaptive value of leucocytosis as a defensive reaction and its importance in the protection against disease are plain. When larger numbers circulate in the blood, more are available for action in the infected regions. Examination reveals that in most instances the additional cells are the phagocytic variety of leucocytes.[40]

The mechanisms of this adaptation are not thoroughly understood. Since there is no known extensive storage of mature granular leuco-

---

38. This is usually limited to infections by the type of bacterium known as the *coccus*. Other bacteria, including the *bacillus* of typhoid fever, may actually cause a reduction of the white count. This is called a *leucopenia*. Leucopenia, like anemia, will result also from extensive bone-marrow destruction, as in poisoning by benzol and other drugs, in overwhelming infections, or in degeneration of the bone marrow without apparent cause.

39. Obviously, the presence of fever and other signs of infection are of importance here also.

40. A normal differential count might give these figures, the total white count being 7,000:

Granular leucocytes (neutrophilic)   70 per cent—or 4,900 per cu. mm.
Lymphocytes. . . . .   22 per cent—or 1,540 per cu. mm.
Other white cells. . . . .   8 per cent—or   560 per cu. mm.

In an infectious leucocytosis with a white count of 35,000, the differential count might reveal this distribution:

Granular leucocytes (neu-
  trophilic). . . . .   94.0 per cent—or 32,900 per cu. mm.
Lymphocytes. . . . .   4.4 per cent—or 1,540 per cu. mm. as in normal
Other white cells. . . . .   1.6 per cent—or   560 per cu. mm. as in normal

cytes, it is clear that the increase represents an accelerated output by the leucocyte-forming system. Microscopic examination of the red bone marrow reveals an increased rate of mitotic cell division. But what is the mechanism of this? How does the presence and activity of bacteria in a localized region stimulate the distant bone marrow? We find here no such mechanism as that which operates in the case of red-cell production. This can be no question of oxygen lack. Nor does there seem to be an initial deficiency of any other sort which might serve as a bone-marrow stimulus. The leucocytosis is not preceded by a decreased white-cell count. It seems probable, from the available evidence, that specific bacterial products are responsible and that these diffuse into the blood stream and are carried by the circulation to the bone marrow, where they exert a specific stimulating action.

3. *Leucemia.*—Leucemia is an abnormal condition in which the white-cell count is greatly elevated, sometimes even reaching 500,000 cells per cu. mm. This condition is not to be confused with leucocytosis. The latter is a normal response to infections. In leucemia there is no known infection; it is an abnormality of the bone marrow itself,[41] in which the white-cell-forming elements multiply excessively rapidly for some unknown reason. Its cause is unknown, and it is usually not effectively controlled. Leucemia has no known adaptive value to the organism; in fact, it usually terminates fatally.

The white cells produced and freed into the blood are abnormal, showing great variations in size and internal structure and including many immature forms which normally do not enter the blood stream.

Leucemia possesses a number of features resembling cancer, notably the unrestrained growth of cells, the abnormal appearance of the cells, and the fatal outcome. It has been called "cancer of the blood."

IX. THE BLOOD PLATELETS

The third and last kind of formed elements are the blood platelets. They are roughly disk-shaped, far smaller than red cells, and show none of the special internal structural differentiation characteristic of cells. Their origin is obscure, though it is suggested that they too arise partly in the red bone marrow, because bone-marrow injury often markedly reduces their numbers. Some observations indicate

---

41. This would produce a *myeloid leucemia*, involving those white cells formed normally in the bone marrow. Leucemia might also be of the *lymphoid* type, in which the lymphatic tissues are overactive in the production of lymphocytes.

that they are also formed by phagocytic cells in the lungs. A platelet count is difficult to make because of the rapidity with which these bodies disintegrate in abnormal surroundings. Such counts as have been made indicate that the normal variations cover a wide range. Though averaging about 250,000 per cu. mm., counts anywhere from 200,000 to 600,000 have been considered normal.

Obscure though these bodies are in some respects, one of their functions—apparently the main one—is quite well understood. This has already been discussed in some detail in a previous section (p. 76) on the clotting of blood.

X. BLOOD DAMAGE IN ATOMIC EXPLOSIONS

The main direct injuries resulting from exposure to an atomic-bomb explosion are threefold: burns, blast injuries, and injuries from irradiation emanating from charged ionized particles. The first two of these are not particularly new kinds of injuries. The burns are no different from the burns produced by any other source of heat, but they may be extremely severe, depending upon the distance of the victim from the explosion and accidental or other shielding from the heat. The blast injuries are entirely mechanical, resulting from buffeting by intense pressure waves in the air. These are in no way different from the injuries which might be suffered from being struck by an automobile.

The third type of damage is different from ordinary bomb injuries, resulting from the release of intense radiant energy. The major effects are upon the formed elements of the blood. White cells and the white-cell-forming elements of the bone marrow are damaged first, so that there is an extremely low white blood cell count. As a consequence, there is an overwhelming bacterial invasion of the body. Apparently, in the normal person, white blood cells continually combat the many organisms that occur in the mouth, nose, throat, and elsewhere. When the total white cells are reduced, these organisms, unrestrained in their growth, become dangerous in numbers. Some of the bacterial invaders may be organisms ordinarily not harmful when the normal defenses of sufficient white blood cells are intact. Infected ulcerations of the skin, mouth, and throat result.

Blood platelets are destroyed, so that multiple hemorrhages may occur in many parts of the body. The red-cell-forming elements of the bone marrow are also injured, and severe anemia supervenes.

Attempts to alleviate the damage to the blood and blood-forming organs play a prominent role in caring for victims. Chemicals which counteract infections help to prevent the effects of reduced white cell numbers. Massive transfusions help to replace the deficiency in formed elements as well as replacing the blood fluids which are lost from the oozing burns.

It is of interest that, in an air explosion of a bomb, there is no significant contamination of the bombed area with radioactivity. The radioactivity is carried upward with the mushroom cloud, propelled by the blast and the heat, which, at the center of the explosion, is of the same order as the heat of the sun. The radioactive particles ascend to the stratosphere and are no longer harmful in the immediate vicinity. Consequently, there is virtually no danger from radioactivity, so that it is safe for rescuers to enter the bombed area even immediately after a bomb has exploded there.

Conditions are entirely different in an underwater explosion. Radioactivity in water droplets is scattered widely, so that a bombed area is badly contaminated and unsafe to enter for several days.

CHAPTER FOUR

THE WORK OF THE HEART

I. THE CIRCULATION OF THE BLOOD
   A. General anatomy
   B. The work of William Harvey
      1. Harvey's contribution   2. Harvey's experiments
II. STRUCTURE OF THE MAMMALIAN HEART
   A. Evolutionary development
   B. Chambers and valves
   C. Tissues
III. CONTRACTIONS OF THE HEART MUSCLE
   A. The graphic method of study
   B. The heart as an irritable tissue
      1. Long refractory period   2. The all-or-none response   3. Electrical variations
   C. Cause of the automatic beat
      1. What tissue is involved?   2. What stimulates this tissue?
IV. THE CARDIAC CYCLE
   A. Sequence of events
      1. The pacemaker   2. The conducting system   3. Pressure changes and valve action
   B. The heart sounds
   C. The electrocardiogram
V. DISORDERS OF THE HEART
VI. ADJUSTMENT OF THE HEART TO THE ACTIVITY OF THE BODY
VII. CONTROL OF OUTPUT PER BEAT
   A. The effect of metabolites
   B. The law of Starling
      1. Effect of stretch   2. Distention of the heart
VIII. CONTROL OF THE HEART RATE
   A. The effect of temperature
   B. Chemical factors
      1. Epinephrine   2. Carbon dioxide   3. The thyroid hormone
   C. The nerves of the heart
      1. Accelerator and inhibitory nerves   2. Continuous action of the vagus   3. The problem of inhibition
   D. The cardiac reflexes
      1. Nerve centers   2. The afferent nerves   3. The depressor reflexes
      4. The Bainbridge reflex

I. THE CIRCULATION OF THE BLOOD

A. General Anatomy

The heart is a muscular organ lying within the thorax, inclosed in a sac of fibrous connective tissue (the *pericardium*). In mammals it is completely divided by a partition into two parts, the so-called "left heart" and "right heart." Leading from the left heart is a large vessel, the *aorta*, which arches upward, backward, and then downward, extending to the lower abdominal cavity. All along its course it gives off *arteries*, which branch more and more profusely into smaller and smaller vessels, ramifying to all parts of the body. The very smallest of these vessels, the *capillaries*, whose walls are but one cell in thickness, and whose internal caliber is about that of a red-cell diameter, are diffusely distributed to organs and tissues everywhere.

The capillaries then unite to form tiny *veins*, which, in turn, join to form larger and larger veins. The veins of the lower portions of the body empty into the *inferior vena cava*, and the veins of the head and neck are tributaries of the *superior vena cava*. These two large venous channels empty into the right heart, completing what is called the *systemic circulation*.

From the right heart springs the *pulmonary artery*, which soon divides into two, one for each lung. Each pulmonary artery divides into smaller and smaller arteries and finally into the lung or pulmonary capillaries, which penetrate all parts of the organ. These, again, collect into larger and larger veins, finally forming the *pulmonary veins*, which empty into the left heart. This makes up the *pulmonary circulation*. Figure 30 is a schematic representation of these features of the circulation.[1]

B. The Work of William Harvey

So conditioned are we to thinking of the circulation of the blood that the concept is implicit even in a description of the gross anatomy of the circulatory system. About three hundred years ago, however, although much was known about the anatomy of the heart and the blood vessels and their distribution, the nature of the movement of the blood was unknown. At that time William Harvey demonstrated the fact that the blood moves in a continuous double circulation.

---

1. The circulatory changes which occur in the mammal at birth, when aeration of the blood in the lungs commences, are described in chap. 14.

His work marks the beginning of our accurate knowledge of the circulation.

1. *Harvey's contribution.*—Before Harvey's time many erroneous notions existed. Some thought that the blood moved forward and backward, in a sort of ebb and flow. Others, having no idea of any connection between arteries and veins (i.e., capillaries), believed that the arteries carried spirits and that the veins carried nutriments from

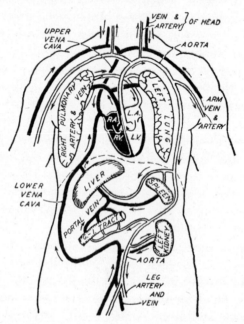

Fig. 30.—Diagram of the circulatory system showing the main blood vessels of man. The aorta and vena cava actually course along the midline of the body. In each organ the arteries divide into smaller vessels and capillaries (not shown). Vessels containing oxygenated blood are in outline. Vessels containing oxygen-poor blood are in black. Arrows indicate direction of bloodflow.

the heart to other tissues. Still others thought that the arteries contained fuliginous vapors, whatever they might be, and some even thought that the arteries contained air. Again, it was held that the blood moved from the right heart into the left through small holes in the partition between the two. Such holes, though invisible, *had* to be there.

Yet certain facts had already been demonstrated which were quite incompatible with these ideas. The existence of valves in the heart and in the veins failed to suggest to physicians that only a one-way movement of blood could occur. It had even been suggested that blood passes from the right to the left heart through the lung vessels, yet it was felt that some of it went directly through the partition between the left half and the right half of the heart.

Harvey's conclusions can be summarized as follows: All the blood moves in a continuous onward direction from right heart to left heart through the pulmonary circulation. All of it is pumped in one direction through the heart. All the blood moves in one direction through the arteries and continues its circulation on into the veins, which return it to the heart.

2. *Harvey's experiments.*—Harvey's observations were both anatomical and physiological. He made many dissections and experimented upon living animals of many kinds and upon himself. From the data collected he reasoned somewhat as follows.

Given the valves of the heart, heart contractions can force blood along in only one direction. There can be no ebb and flow. Pinching off both venae cavae causes an immediate collapse of the right heart. Stopping the pulmonary veins causes the left heart to collapse. On the other hand, occlusion of the aorta causes the left heart to swell with accumulated blood, and occlusion of the pulmonary artery causes the right heart to swell. Therefore, blood always enters the right heart through the venae cavae and leaves by the pulmonary artery. It enters the left heart from the pulmonary veins and leaves by the aorta.

Blocking of any artery causes a gradually increasing accumulation of blood on one side only—that toward the heart—and cutting an artery causes a spurting from only one cut end. The blood, therefore, must always flow through an artery in only one direction—away from the heart. Corresponding effects from blocking or cutting a vein indicate that in veins the blood flows only toward the heart. Furthermore, the valves of the veins prevent movement in the reverse direction.

Computation of the quantity of blood pumped into the arteries clearly indicates that it cannot simply be pumped to the tissues and there consumed. There is too much of it. It must somehow pass over into the veins, which themselves carry too much blood to be accounted for on any other basis than a connection with the arteries.[2]

Similarly, there is a minor or pulmonary circulation. The vessels

2. Harvey never saw capillaries. They were discovered four years after his death, in 1661, by Marcello Malpighi. Harvey had correctly inferred their existence.

here are quite adequate to explain the movement of all the blood
from the right to the left heart. There is no need to postulate imaginary openings directly connecting right and left hearts.

Finally, since it is possible to drain off the blood of the body by
opening any large artery or vein,[3] it must be true that, in the normal
course of events, all the blood must flow past any one point in the
system, over a period of time.

Only after years of careful research did Harvey publish his book
On the Motion of the Heart and Blood in Animals.[4] It won for him

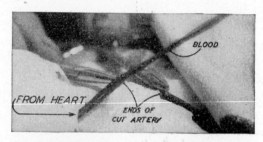

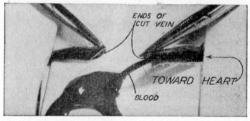

Fig. 31.—One of Harvey's experiments demonstrating the circulation of the blood. When
an artery is cut (*above*), blood spurts interruptedly in a direction away from the heart. When
a vein is cut (*below*), blood flows out steadily but more slowly in a direction toward the heart.
(From the sound film, *Heart and Circulation.*)

lasting renown as a physiologist and scientist of the first order, and it
cost him much of his medical practice. Sick people, especially those
with circulatory disorders, could no longer trust a physician with
such harebrained notions!

But the king of England stood by him and retained him as court
physician. Perhaps even this was due less to a true appreciation of
the intrinsic scientific worth of the treatise than to Harvey's eloquent
dedication of his work to the king, "the light of this age, and indeed
its very heart."

3. Not quite all the blood can so be drained from the vessels. The heart stops pumping
before all the blood is lost.

4. Published originally in 1628. This readable classic contribution to physiology and medicine is available at low cost (see "Selected References," following chap. 15).

II. STRUCTURE OF THE MAMMALIAN HEART

A. Evolutionary Development

The four-chambered heart of mammals (and birds) is the culmination of a long evolutionary development traceable through the various classes of vertebrates. In fish there is no division into right and left heart. The organ is little more than a straight muscular tube. The aorta carries the blood to the gills, where it is aerated. From

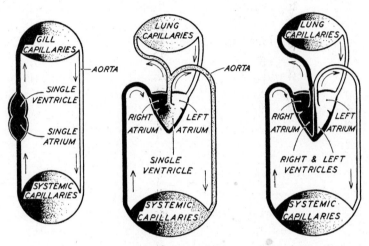

Fig. 32.—Evolution of the four-chambered heart. In fish (*left*) blood reaches the organs of the body ("systemic capillaries") under low pressure because it has already traversed the gill capillaries, where it is aerated. In amphibia like the frog (*center*), aerated blood (*unshaded*) is separated from unaerated (*black*) in the atria of the heart, but these mix in the ventricle, so that partially aerated blood (*dotted*) is sent both to the lungs and to the tissue capillaries. In mammals and birds (right) the complete separation of the heart into two halves makes possible the pumping of unmixed aerated blood to the systemic capillaries under high pressure. Arrows indicate the direction of blood flow.

here it passes, without first returning to the heart, to vessels which distribute it to all parts of the body. Only after this does it return again to the heart. In this arrangement blood is never pumped directly from heart to body tissues generally. It must first pass through the capillaries of the gills, where much of the impetus to flow imparted by the heart is lost.

In amphibia there is a beginning separation of the pulmonary and systemic circulations, with a corresponding partial division of the heart into a right and a left half, which makes possible the pumping of blood under high pressure to the body tissues. The single-chambered *ventricle* pumps blood into a vessel which soon divides into two.

One of these leads to the lungs, where aeration occurs. The other
carries blood to systemic structures. Blood from the lungs returns to
the *left atrium*, and from the systemic circulation to the *right
atrium*.[5] The aerated blood from the lungs and the venous blood
from the tissues, though separated in the atria, soon mix in the
single ventricular chamber.

In reptiles there commences a division of the ventricles also into
two chambers. In mammals and birds this is complete, and the heart
is divided into two separate halves. Each half, in turn, consists of
an atrium and a ventricle. The right atrium and right ventricle
contain unaerated blood; and the left atrium and left ventricle, aerated. There is no mixing of the two kinds of blood in the normal heart,
and systemic and pulmonary circuits are distinct and separate.

In early embryonic development the mammalian heart recapitu-
lates some of these evolutionary stages. The heart commences as a
pulsating tube and only reaches the mature condition later in development.

B. CHAMBERS AND VALVES

The atria, or *auricles*,[6] have distinctly thinner walls than the ventricles. The thicker walls of the ventricles are responsible for almost
all the pumping action of the heart. The walls of the left ventricle
are much thicker than those of the right. This we relate to the greater work done by the left ventricle. It pumps blood through the entire
systemic circuit, while the right ventricle has the easier task of
pumping blood only through the lungs, a much shorter distance.

On each side of the heart is a valve system, between atrium and
ventricle, which permits blood to flow only from the atrium to the
ventricle and which is closed automatically by blood starting to
move in the reverse direction. These are the left and right *atrioventricular* valves. Guarding each exit from the ventricles are also
valves of a somewhat different construction, called the *semilunar*
valves, from the fact that each is made of three half-moon-shaped
leaflets. The long margin of each leaflet is attached to the vessel wall,

5. Blood first enters the heart in the so-called *sinus venosus*, which is the thin-walled prominent first chamber of the heart in fish, amphibia, and reptiles. The sinus venosus empties into
the right atrium. In mammals and birds the sinus venosus has disappeared as a distinct chamber of the heart.

6. Strictly speaking, the term "auricle" refers anatomically to a small appendage of each
atrium. Sometimes, especially in physiology and medicine, "auricle" is used as a synonym of
"atrium."

the short margin being free. The aortic semilunar valves are located at the beginning of the aorta. They permit blood to flow from the left ventricle into the aorta, but they are closed by any reflux of blood in the reverse direction. The pulmonary semilunar valves lie at the beginning of the pulmonary artery and prevent backflow of blood from the pulmonary artery into the right ventricle. There are no true anatomical valves at the orifices of the left or right atria, where the veins empty into the heart. Figures 33 and 34 can give

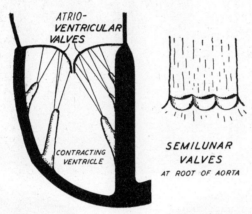

ATRIO-
VENTRICULAR
VALVES

CONTRACTING
VENTRICLE

SEMILUNAR
VALVES
AT ROOT OF AORTA

Fig. 33.—The valves of the heart. The atrioventricular valves are prevented from "turning inside out" when the ventricles contract by tough cords arranged as shown. The aorta (*right*) is slit open lengthwise and spread out flat (see also Fig. 34).

but an approximate notion of the arrangement of the chambers and valves of the heart. A clear picture can best be obtained by examining a dissected heart.

C. TISSUES

Microscopically, the heart is seen to consist mainly of the specialized kind of muscle called "cardiac muscle." The fibers branch and reunite extensively. As a matter of fact, the entire mass constituting left and right atria is really a single, long, twisted, multinucleated muscle fiber, branching profusely, with branches reuniting again, constituting a continuous mass of protoplasm. Similarly, the entire ventricular muscle mass is but one interlacing network of branches of a single multinucleated fiber. The muscle fibers are bound together with connective tissue into the characteristic shape of the organ. Connective tissue serves also to join the atrial with the ventricular muscle fibers. Muscle fibers connecting the atria with the ventricles seem to be lacking in some species. This discontinuity of the contractile tissue between atria and ventricles is an important consideration in an understanding of the mechanism by which contractions of atria and ventricles are synchronized.

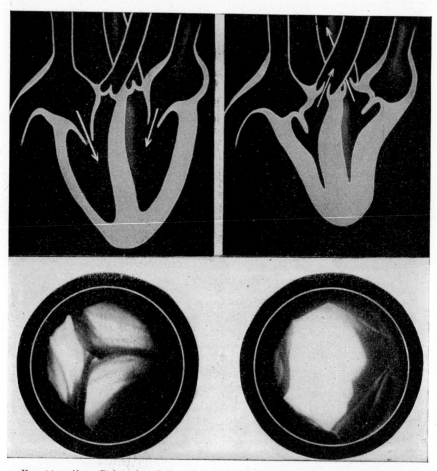

FIG. 34.—*Above:* Relaxation (*left*) and contraction (*right*) of the ventricles of the heart. In relaxation blood enters the ventricles (*white arrows*) through the open atrioventricular valves. At this time the semilunar valves are closed. Contraction of the ventricles forces the atrioventricular valves shut and opens the semilunar valves, through which blood is ejected (*white arrows*). *Below:* Photographs of the semilunar valves closed (*left*) and open (*right*), viewed from above. (From the sound film, *Heart and Circulation.*)

Nerve tissue is also present. There are nerve fibers which are terminations of the cardiac nerves—the nerves which originate in the central nervous system and innervate the heart.[7] In addition, whole

7. The *extrinsic* nerves of the heart.

neurones are also present, cell bodies and axones lying entirely within the organ.[8]

There is also a kind of tissue quite peculiar to the heart. It resembles embryonic heart tissue in appearance and is called *nodal tissue.*[9] It is most concentrated in special regions and is distributed in a manner which itself suggests its function. One such mass is located at the junction of the large veins with the right atrium. This is called the *sinoatrial*, or *sinus*, *node.*[10] Its fibers fuse with the true muscle fibers of the atrial muscle in which it is imbedded. In the lower part of the partition between the atria, another patch of this tissue is found, again blending its fibers with the adjacent atrial muscle fibers. This is called the *atrioventricular*, or *A-V*, *node.*

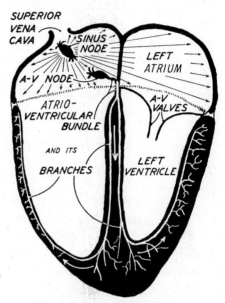

From the A-V node a bundle of nodal tissue, the atrioventricular or *A-V bundle*, passes downward to the ventricles. This bundle of persisting embryonic muscle, plus nerve tissue, is the important functional bridge between atria and ventricles. In the ventricles it

FIG. 35.—The conducting system of the heart. The beat originates at the sinus node (or primary pacemaker) and spreads through atrial muscle (*black arrows*) In this spread, the A-V node is activated and transmits an impulse to the ventricles along the A-V bundle and its branches. There is probably no continuity of muscle fibers from atria to ventricles in mammals.

branches profusely, its final terminations ramifying to all parts of the ventricular musculature. Its distribution is shown in Figure 35.

Blood vessels are also present in the heart walls. One might assume that the heart muscle is nourished by the blood within its chambers. This is true, however, only of an extremely thin innermost layer in mammals and birds. The muscle, especially that of the ventricles, is

8. The *intrinsic* neurones of the heart.

9. Also called *Purkinje tissue*, after its Czech discoverer, Johannes Purkinje.

10. It derives this name from its phylogenetic origin. It represents the vestige of the old sinus venosus, which existed as a distinct chamber in the heart of the lower vertebrates.

entirely too thick to receive adequate oxygen or nutriments by diffusion from the blood in the chambers. Like all other muscle, the
heart has blood vessels of its own. These are the *coronary* arteries,
arising from the aorta, which distribute their branches throughout
the muscle of atria and ventricles. Blood is pumped into these
arteries in exactly the same way that it is forced through any other
branch of the aorta. The arteries divide into smaller and smaller
vessels, finally of capillary size and structure. There are likewise
typical veins which drain away the capillary blood and which join
one another to form larger and larger veins. The venous blood is
emptied into the right atrium, after the fashion of venous blood from
any organ of the body.[11]

Finally, both the outer surface of the heart and the inner lining of
its chambers are covered with thin, flat, smooth, epithelial cells.

III. CONTRACTIONS OF THE HEART MUSCLE

A. THE GRAPHIC METHOD OF STUDY

Physiological observations and the results of experiments can
often be satisfactorily expressed by written figures or by drawings.
But especially in the case of organs like the heart, which move as
they function, it is convenient or even necessary to obtain an automatically inscribed record of the events in order to understand what
is happening. Often a graphic record of an experiment will give a
clearer and more accurate picture than pages of print. In the last
century a great many mechanical, electrical, and optical devices have
been invented for this purpose. An understanding of some of them
becomes indispensable for an appreciation of much of the material
in this volume.

The instrument called the *kymograph* is extensively used. This consists of a removable drum mounted upon an upright bar. A sensitive paper is mounted on the drum, and the drum is slowly rotated by
a spring clockwork. Continuous records can then be inscribed on the
moving paper. In most cases paper blackened with soot or smoke is
used, and the recording points of various instruments in contact with
the moving paper trace white lines.

The muscle lever is a device which transforms contractions of
muscle into mechanical movements of a lever, which movements can

11. An appreciable portion of the blood in the heart vessels (i.e., in the *coronary circulation*) makes its way into the chambers of the heart by numerous tiny veins which empty
directly into the ventricular chambers.

then be recorded on the smoked paper of the kymograph. It is used extensively in recording heartbeats. A muscle lever is of very light weight, with its fulcrum near one end. The long end of the lever contains the writing point, which rests lightly against the smoked paper with a minimum of friction. From the short end of the lever a string is stretched vertically to the apex of the heart of a frog, for example. The lever arms are of such difference in weight that the free apex of the heart is lifted, and the long axis of the heart lies in a vertical plane (see Fig. 36).

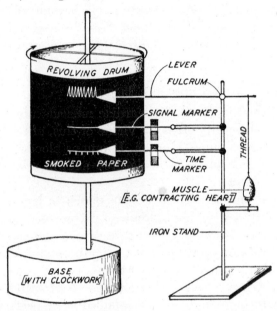

Fig. 36.—The kymograph being used for the automatic inscription of a record of a beating, isolated frog heart. (See discussion in text.)

In contraction the long axis of the heart shortens. This pulls upon the string, and the short arm of the lever is lowered. The writing point correspondingly moves upward. The extent of movement of the writing point depends not only upon the strength of contraction of the heart but also upon the relative lengths of the lever arms. The lever effects an amplification of the cardiac movements. Each contraction (*systole*) traces a vertical line upward upon the drum; each relaxation (*diastole*), a downward vertical line. If the drum is revolving and the smoked paper moving, clearly the record will be spread out horizontally. The upstroke and downstroke will inscribe a line resembling an inverted letter V. The faster the drum revolves, the greater the separation of the arms of the inverted V written with each cardiac cycle.

A complete tracing also includes a "signal" record and a time record. The signal marks the points of application and withdrawal of stimuli, for example. A time record is usually inscribed by an electrical device which traces a horizontal line interrupted at regular intervals of time (e.g., 5 seconds) by a short vertical line. The writing points of the lever, signal, and timer are all adjusted in a vertical line.

The usefulness of graphic recording will become clear only when many records like those in this book have been examined or, preferably, when the student has himself made such graphic records in the laboratory.

B. The Heart as an Irritable Tissue

Like other muscle tissue, the cardiac musculature possesses the property of contractility. It is by virtue of this, of course, that the heart performs its pumping action. It pumps blood from its chambers by contraction of its musculature—and, in relaxation, allows more blood to enter.

Like skeletal muscle, also, it is independently irritable. Contractions can be initiated by many kinds of extraneous stimuli—chemical, electrical, or thermal. Such stimuli can induce contractions independent of normal beats, or the stimuli can initiate heartbeats in a heart which has just stopped beating.

1. *Long refractory period.*—Heart-muscle contractions differ from those of skeletal muscle in certain qualitative respects which make this tissue particularly suitable for the study of certain general characteristics of muscle. The *refractory period* of the heart, for example, is much longer than that of skeletal muscle or of any other irritable tissue. Lasting about 0.1–0.2 second, which is about a hundred times as long as in skeletal muscle, the phenomenon is very easy to demonstrate, even with relatively crude apparatus. The refractory state continues for almost the entire period of systole. Thus a stimulus applied to a spontaneously beating heart will induce an early or premature contraction when applied during diastole. Stimuli applied early in systole, even if they are fairly strong, induce no early contractions, nor in any other way do they disturb the spontaneous rhythm. The heart at this time is almost absolutely refractory to stimuli. This period is considered to be a sort of recovery or restora-

tion stage of the heart tissues—a period in which there is a reversal of the chemical and physical changes which initiate contraction. Not until this reversal is well under way is the refractory period over, and the heart again susceptible to artificial stimulation.[12]

The long refractory period of the heart is an important physiological adaptation. Because of its long duration, it prevents the heart

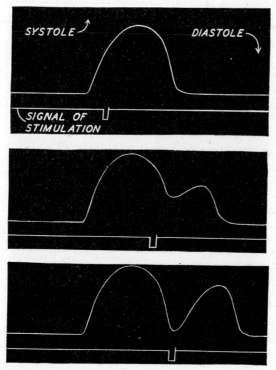

Fig. 37.—The refractory period. A stimulus applied to a spontaneously beating heart during systole is ineffective (*upper tracing*). A stimulus of the same intensity applied in diastole induces a new premature contraction. Note that a stimulus applied late in diastole (*lower tracing*) induces a greater response than one applied earlier in diastole (*middle tracing*). These records of a turtle heart were inscribed on a rapidly moving drum, using the apparatus shown in Fig. 36.

from maintaining itself in the contracted state, even though a continuous stimulus may be applied. A continuous stimulus will initiate a single beat and then become ineffective until the refractory period is over, by which time relaxation would already have commenced. Thus any abnormal, accidental, continuous stimulus would still cause a rhythmic response. The value of the long refractory period to the

12. See further discussion of the refractory period in chap. 9, p. 372.

organism is apparent when we consider that a prolonged, maintained contraction would be as ineffectual in pumping blood as would a cessation of the beat in complete relaxation.

2. *The all-or-none response.*—The heart is also the most suitable structure upon which to demonstrate the *all-or-none* principle. The ventricle of a frog heart can be experimentally isolated in such a way that its spontaneous beating ceases, but it retains the capacity to beat when artificially stimulated. If the amplitudes of the resulting contractions are measured, the following relationships can be shown

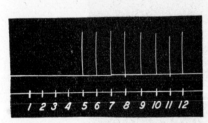

FIG. 38.—The all-or-none law. Kymograph record of contractions of a turtle ventricle induced by electrical stimuli of increasing intensity (*1–12*). The spontaneous beating of the heart has been arrested. Stimuli *1–4* were too weak to elicit a contraction. The response to stimulus *12* was no greater than to stimulus *5*, although the intensity of stimulus *12* was about twenty times greater than stimulus *5*.

Weak stimuli fail to elicit any response. By increasing the strength of the stimulus, a point is reached where the irritability threshold is exceeded, and each stimulus will induce a response of a specific magnitude. If, now, the strength of the stimulus is further increased, the amplitude of the contractions will *not* increase. Even if the stimulus exceeds the threshold strength twenty times, no contraction increase results. This all-or-none relationship is thought to be due to the fact that, when an irritable cell or unit, such as a nerve fiber, is adequately stimulated, it is "set off" completely. It reacts much as does a gunpowder explosion, where it is impossible to produce any sort of graduation of responses by grading the strength of the stimuli. It might also be compared to the discharge of a cartridge from a pistol. Here the stimulus is the pulling of the trigger, and any change in the strength of the stimulus, in the force with which the trigger is pulled, in no wise affects the speed or force of the response, that is, the explosion and subsequent projection of the bullet.

The phenomenon is so clearly shown by the heart ventricle because this structure is built from a single complex continuous fiber and behaves as such. Being a very large fiber, its contraction is so great that it can easily be measured and recorded. A skeletal muscle fiber is very small, and the movement resulting from activation of only one fiber can be measured only by highly refined techniques. Furthermore, skeletal muscles and nerves are made up of thousands of separate fibers, and gradations of stimulus strength produce gradations of response in the whole muscle or nerve because, with increasing strengths of stimulation, more and more fibers are activated.

It does not follow, in the heart or any other tissue, that, because varied intensities of stimulation do not affect the response, nothing else can. Factors modifying the condition of the heart muscle can and do modify the strength of its beat, just as changing the state of a train of gunpowder (as by dampening it somewhat) will alter the nature of the response.

3. *Electrical variations.*—In general, when tissue becomes active, it becomes electrically negative with respect to resting or inactive tissue. During contraction, that portion of a muscle tissue or muscle fiber which is active at a given instant is electrically negative with respect to adjacent inactive portions of the muscle, which either have just ceased acting or in the next instant will commence contraction. During activity, then, a muscle fiber becomes a minute battery, with positive and negative poles, the latter at the active region, the former at any inactive region. The electrical potentials[13] developed are physically identical with the potentials developed in any battery. If the poles of such a living battery are connected by a conductor, a current[14] flows. The current can be led through a suitable recording instrument which indicates the passage of the current and measures its magnitude.

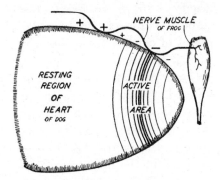

The potentials developed by the muscle tissue in the beating heart are large enough so that, if some other irritable tissue, such as a nerve, is touched to the heart at two different places, then, as the heart beats, and the activity reaches one point of contact with the nerve, the nerve will be stimulated, just as though an electric current from a battery had been applied to it (see Fig. 39). These

FIG. 39.—Electrical changes in contracting heart muscle. Each time the heart beats, the frog muscle contracts because the frog nerve is stimulated by the electrical currents developed by the contracting heart muscle.

electrical changes, when properly recorded, constitute an electrocardiogram (see pp. 148–49).

Though most readily demonstrated in muscle, and especially in

13. Called *action potentials.*   14. Called an *action current.*

heart muscle, these electrical phenomena are universal in active tissues. Active gland cells are negative with respect to inactive cells, and the portion of a nerve fiber which is conducting a nerve impulse at any given instant of time is negative with respect to resting regions of the nerve.

C. CAUSE OF THE AUTOMATIC BEAT

The automatic rhythmicity of the heart has received much attention. The heart is not entirely peculiar in this respect, for rhythmic responses are the rule in many organs and systems. The intermittent nature of the breathing movements is a good example, involving rhythmic contractions and relaxations of the muscles of breathing. But this is different from the cyclic action of heart muscle in the following important respect. If the nerves of the muscles of breathing are severed, the movements cease at once. Their contractions are entirely dependent upon rhythmic activation through their external or extrinsic nerves. The beating of the heart, on the other hand, continues even after all its nerves are cut. The rhythmicity and the automaticity are inherent in the heart itself. In fact, if the organ is completely removed from the body, it will continue to beat for some time. Nor is the integrity of the organ itself required for this automaticity; a bit of the heart muscle, cut off from the organ, may continue to contract rhythmically. Even in tissue culture, microscopic pieces of cardiac tissue sometimes continue to pulsate.

1. *What tissue is involved?*—What is the secret of this rhythmic action? What initiates this spontaneous automatic beating? It is found by isolating various portions of the heart that different regions possess differing degrees of automaticity. It is greatest in the atria, especially so at the sinoatrial node. It is least marked at the apex of the ventricles. This suggests that perhaps the nerve cells intrinsic in the heart itself inaugurate the beat, for nerve cells are most numerous at the junction of the large veins with the right atrium and least numerous at the apex of the ventricles. It would be desirable to remove the intrinsic nerve tissue of the heart to see whether this would stop the automatic beating. But in the vertebrate heart this cannot be done because nerve and muscle tissue are structurally too intimately interrelated. It is impossible to dissect out the nerve tissue without completely destroying the whole organ.

But it so happens that, in the heart of a certain invertebrate (*Limulus*, the king crab), nerve and muscle tissue are easily separable anatomically. All the nerve cells lie in a compact mass upon the surface of the heart and can readily be removed without damage to the muscle. When this is done, the heartbeat ceases at once. So there is no question but that, in the adult *Limulus* heart, the spontaneous activity is inherent in the nervous elements and not in the muscle itself. While it is sometimes dangerous to apply conclusions reached upon one species of animal to another species, yet we must at least admit the possibility that a similar situation might obtain in vertebrates. Some justification for this is had in the finding already mentioned—that the degree of automaticity in various parts of the vertebrate heart is roughly proportional to the quantity of nerve cells present.

Yet we find that the heart of the embryonic king crab commences its beat before any nerve cells are present. Furthermore, a culture of embryonic vertebrate heart muscle, which is free from nerve cells, goes on beating for some time.

In the king crab, then, the automaticity resides in the muscle itself in early embryonic life. Later this function is entirely taken over by the nervous elements which make their appearance, and the muscle proper loses its automaticity. In the vertebrate heart it is rather generally conceded that the automaticity probably resides in the heart-muscle tissue rather than in the nerve cells in the organ.

2. *What stimulates this tissue?*—Even if it be granted that the automaticity of the vertebrate heart resides in the heart-muscle tissue, the question would still remain: Is the tissue responding to some internal chemical metabolic activity of the protoplasm or to some external chemical stimulation? Are there environmental factors which are basically responsible for initiation of the beat, regardless of which tissue is first activated? Of this we can say but little. It is true that an isolated heart which has ceased beating might be stimulated to resume its beat by adding sodium ions to the medium. A heart which beats for a time in a calcium-free medium, and then stops, may also be made to beat again by adding calcium. Does this mean that calcium ions or sodium ions constitute the normal stimulus to the heartbeat? Or does it simply mean that, unless the proper ions are present, the conditions are unfavorable for activity? Do these ions actually stimulate, or do they merely furnish a suitable environment in which some other agent may cause the stimulation? If so, what is this unknown factor? Is it carbon dioxide, necessarily always present in living heart tissue and definitely known

to be the factor which stimulates the breathing machinery? Is it some hormone? Until these and other questions are answered, we must simply state that we do not know the primary cause of the heartbeat.

IV. THE CARDIAC CYCLE

A. Sequence of Events

Upon looking at a beating heart, one is inclined to agree with Harvey's first impression that "the motion of the heart was to be comprehended only by God . . . [because of] . . . the rapidity of the motion, which in many animals is accomplished in the twinkling of an eye, coming and going like a flash of lightning." But patient inspection reveals that the events occur in an orderly, regular sequence, which is more easily seen if the heart is beating slowly. There is a cycle consisting of contraction, or systole, and relaxation, or diastole, which includes the brief period of inactivity before the next systole. It can further be seen that the atria contract earlier in the cycle than the ventricles, the sequence being atrial systole, slight pause, ventricular systole, and, after corresponding relaxations, a repetition of these events.

1. *The pacemaker.*—Still closer examination shows that the right atrium beats a little in advance of the left atrium, though both ventricles contract together. Further details usually cannot be made out by direct inspection but can be ascertained by determining the manner in which the electrical changes spread over the heart. It is found that the electrical negativity, and therefore the contraction also, commences at the sinus node.[15] Further evidence of this is seen in the following experiment: If the sinus node is locally cooled, the rate of the heartbeat is diminished. Warming the node accelerates the rate. This is quite in accord with temperature effects generally, for cold usually slows physiological reactions and heat accelerates them. The important consideration at present, however, is that cooling or warming any other region of the heart fails to affect the rate. In fact, under all normal conditions, the rate can be changed only by stimulating or inhibiting the sinus node, at which spot the activity originates, whether this be done by heat or cold, by hormones, by metabolites, or by nerves which affect the rate of the beat.

---

15. In the lower vertebrates, where the sinus is still a distinct chamber—the sinus venosus—the beat originates in this chamber. The cycle here is: sinus contraction, right then immediately left atrium contraction, and—after a brief pause—ventricle contraction, with a short pause between successive cycles.

For this reason the sinus node is called the *pacemaker* of the heart. It sets the pace and starts the beat, and the contractions of the rest of the cardiac musculature come as a result of spread of activation from this region.

2. *The conducting system.*—In the atria the activation appears to spread directly through the muscle fibers. But the activity cannot continue to the ventricles in this direct fashion, because there is no true muscular continuity between atria and ventricles. The only means available for the stimulus to pass from atria to ventricles is through the atrioventricular bundle. Recall that this bundle of nodal tissue, composed of embryonic heart tissue and nerve fibers, originates in the atria, passes downward into the ventricles, and there branches profusely, penetrating to all parts of the ventricle. Presumably stimulated directly by the impulse coming through the atrial muscle, the bundle and its branches act much like a nerve, transmitting to all parts of the ventricle a stimulating influence much like an impulse in a peripheral nerve (see Fig. 35, p. 135). We know this to be true, for the atrioventricular sequence is disturbed when the bundle is damaged. This damage may be a result of disease, or it can be produced experimentally. If the bundle is sufficiently compressed mechanically, or is cut, it can no longer function; it can no longer conduct impulses. The ventricles either cease beating entirely (in lower vertebrates) or beat quite independently of the atria.[16]

A summary of the sequence of events in systole is as follows:

a) Beat initiated at sinus node
b) Adjacent atrial muscle excited; impulse spreads to all of atrial muscle, which contracts as it is stimulated
c) A-V node and bundle stimulated
d) Impulse rapidly transmitted to all parts of ventricles via nodal tissue
e) Entire ventricular muscle mass contracts practically at once

Relaxation of each chamber follows immediately upon completion of contraction, so that the atria are already relaxing at about the time the ventricles have begun contracting. After ventricular relaxation there is a brief period of inactivity of the whole heart. This completes the cycle.

16. The fact that the ventricles continue to beat at all after control by the atria is removed suggests that a pacemaker is also present in the ventricles. As a matter of fact, numerous regions in the atria or ventricles of mammals are capable of initiating contraction under abnormal conditions. These are called *secondary pacemakers*. They are normally held in abeyance under complete control of the *primary pacemaker* at the sinus node but become pacemakers in their own right when cut off from control by the sinus node, as in case of damage to the A-V bundle.

3. *Pressure changes and valve action.*—The beat of the heart and the mechanisms for synchronizing its parts are adaptations for pumping blood. We may now consider more directly what is occurring within the heart and how the blood is moved. It will be convenient to commence with the diastolic phase of the cycle. As the ventricles begin to relax, the pressure within the ventricles suddenly decreases. The moment this pressure falls below the pressure in the aorta or pulmonary artery, the semilunar valves are automatically forced shut, preventing backflow of blood. As the pressure in the ventricles continues to fall, it drops below the pressure in the atria, and the atrioventricular valves are pushed open. Blood then flows from the large veins through the resting atria into the ventricles. Therefore, a considerable quantity of blood is already present in the ventricles before the atria start contracting.

With atrial systole a final rush of blood is pumped into the ventricles. The ventricles themselves then contract, pressing on the blood within their cavities. This at once closes the atrioventricular valves, preventing escape of blood back into atria and veins. The pressure in the ventricles mounts rapidly as systole proceeds, and presently it exceeds the pressure in the large arteries. The semilunar valves are forced open, and a spurt of blood is ejected into the aorta and into the pulmonary artery. Meantime, atrial diastole has started, and the atria are filling passively with blood from the large veins, completing the cycle (see also Fig. 40).

B. The Heart Sounds

During each cycle characteristic sounds are produced by the heart. They can be heard by placing the ear against the chest over the heart or by leading the sounds to the ears through the tubes of an instrument called the *stethoscope*, the receiving end of which is placed over the heart. In each cycle two distinct sounds are heard, termed the "first" and "second" heart sounds. The first is low pitched, the second is sharper, louder, higher pitched, and of shorter duration. The only way to appreciate what these sounds are like is to listen to them. They roughly resemble the sounds of the syllables "lubb-*dup*."

The second sound is known to be due to vibrations set up by the sudden closure of the semilunar valves very soon after the beginning of ventricular diastole. Experimental injury to these valves modifies the sound, corresponding to faulty function. If they are slit open, for example, so that they do not close tightly in diastole, and blood therefore leaks back into the ventricles, the second sound is of rather a soft hissing character, called a *murmur*. Instead of the normal "lubb-*dup*," there is heard "lubb-*shhhh*."

This finding is of significance not only in indicating the cause of the second sound but also in detecting the existence of defective valves. If the valves are damaged by syphilis, for instance, the presence of the injury may be detected by the abnormality of the second sound.

The first heart sound is of more complex origin. It is mainly due to vibrations set up by closure of the atrioventricular valves at the

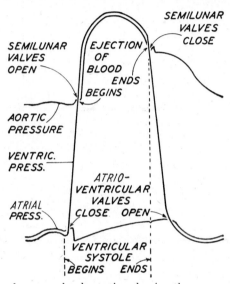

Fig. 40.—Pressure changes and valve action showing the sequence in which the heart valves open and close in a single cardiac cycle as the pressure in the ventricles rises in *systole* and falls in *diastole*.

beginning of ventricular systole; if these valves are damaged experimentally or by disease, the sound is greatly modified. However, some residual of sound still persists even though, for the moment, the flow of blood through the heart is stopped experimentally. During such time, of course, all valve action ceases. The second sound is eliminated, but some of the first sound remains. It is thought that, for the most part, this sound is caused mainly by vibrations set up by the closure of the atrioventricular valves, with a small component due to vibrations set up by the contractions of the muscle fibers. Even skeletal muscle fibers produce such vibrations, giving rise to sounds which can be heard by placing a stethoscope upon any contracting muscle.

C. The Electrocardiogram

The electrical charges (action potentials) developed by each beat of the heart can be demonstrated even at the surface of the body. At that instant of the cardiac cycle when the upper portion of the heart is active, either of the hands will be at a lower potential (more negative) than either foot because the hands are relatively closer to the active negative region of the heart than are the feet. The extremities of the body become the poles of a temporary true electrical battery. Though these potentials are smaller in intensity than those developed

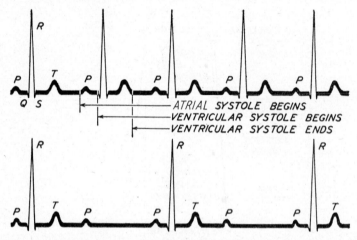

Fig. 41.—The electrocardiogram. In each normal cardiac cycle there occurs a series of waves shown above; the *P* wave corresponds to atrial contraction; the *QRS* and *T* waves correspond to ventricular contraction. When the conducting system of the heart is abnormal so that only every other atrial beat is followed by a ventricular beat, the tracing shown below is obtained.

on the surface of the heart itself, they are, nevertheless, of sufficient magnitude to be detected by sensitive electrical instruments. The *electrocardiograph* is such an instrument; it "picks up" and records the electrical variations developed in the hands or feet at each beat of the heart.

An electrocardiogram, which is a record of the deflections taken in this way, shows that a rather complex but characteristic series of oscillations or waves occurs during each cardiac cycle. Many experiments have enabled physiologists to correlate these deflections with the various mechanical events of the cardiac cycle. Figure 41 shows a typical electrocardiogram with the waves lettered in the conventional manner. The beginning of the initial wave (the *P* wave) marks the onset of atrial systole. The beginning of the QRS deflection marks the onset of ventricular systole, and the end of the T wave delimits the end of ventricular systole. That is, the P wave corresponds to the atrial beat; the QRS and T complex, to the ventricular beat.

Electrocardiograms have been very useful in the detection of certain irregularities in the heartbeat, such as the condition in which damage to the conducting tissue of the heart prevents every other atrial beat from being transmitted to the ventricles. The sequence of mechanical contractions in such an abnormal heart would be: atrium—long pause, atrium—ventricle, atrium—long pause, atrium —ventricle. The electrocardiogram would clearly reveal this state, by the sequence of the deflections: $P \ldots P—QRST;\ P \ldots P—QRST$. On the same principle, many other irregularities can be determined from the deflections of the galvanometer produced by the tiny potentials, of fleeting duration, at the surface of the body.

V. DISORDERS OF THE HEART

The heart is subject to a number of disorders which affect its efficacy as a pump. The contractile power of the muscle itself may be reduced by damage to the muscle from bacteria, chemical agents, or a diminution of the blood supply. A fairly common cause of death in adults is by occlusion of one or more of the arteries supplying the heart muscle with blood. The occlusion may be in the nature of a spasm of the blood vessel or, in *coronary occlusion*, an actual mechanical plugging by some solid object, usually a blood clot forming in a heart blood vessel. In either case the source of oxygen and nutriments is suddenly cut off from an area of the ventricular muscle. If this area is small, the "attack" may not be fatal. If an extensive area is involved, the pumping action of the heart is so impaired that death occurs in a few minutes. The extreme importance of this disease is indicated by the fact that it kills two and one-half times as many people as does cancer and is listed as at least a contributory cause of death on one-fourth of the death certificates of adult males.

The muscle may also be damaged in certain infections. The germs of scarlet fever or diphtheria, though they do not grow in the heart, produce toxins which reach the heart through the blood stream and are injurious to the heart tissues. In thyroid malfunction, also, an excessive production of thyroid hormone may be injurious to the heart muscle.

Overproduction of the thyroid hormone further affects the rhythm of the heart by its injurious effect upon the atrial muscle. The atrial fibers fail to contract in a co-ordinated fashion; and, instead of a normal beat, involving the whole atrium, we find that at any given instant many isolated areas are in a state of contraction, while others are undergoing relaxation, producing what is called *fibrillation* of the atria. Such fibrillation also results from a number of other disorders which are unrelated to thyroid-gland malfunction.

Defects of the valves constitute an important group of incapacitating or even fatal conditions. Mention has been made of syphilitic damage of the semilunar valves, in which a "leakage" from aorta into ventricles occurs during diastole. This infection may likewise cause a narrowing of the aortic opening and interfere with ejection of blood by the heart.

The atrioventricular valves, especially those of the left heart, may also become either defective, permitting leaks, or the orifice may become narrowed. The latter condition is a common aftereffect of rheumatic fever. Rheumatic fever itself usually does not imperil life when confined to the joints. But, long after the acute joint difficulty has subsided, defects of the left atrioventricular valve may make their appearance. It is thought that the bacteria of rheumatic fever do not actually grow upon the heart valves but that they injure the valves indirectly by means of harmful substances produced by the bacteria. One of the most important problems in medicine today is the control of rheumatic fever, and the prevention of the resulting heart disease, which is essentially a disease of youth and young adult life.

In much cardiac disease there is a remarkable compensatory reaction of the organ. Hampered in its function, the heart often enlarges. Its muscular content may increase, and frequently the resulting greater pumping power offsets the defect. The cause or causes of this "compensatory" enlargement are not known at present.

VI.   ADJUSTMENT OF THE HEART TO THE ACTIVITY OF THE BODY

The use of oxygen and nutriments by organs and tissues and the production of wastes continue unabated throughout life. The continuous circulation of the blood makes this possible. But there are also great variations in rate of activity of the various organs and of the body as a whole. In activity a skeletal muscle uses several times as much oxygen as is adequate for its resting metabolism. A gland which is secreting uses more oxygen and blood fluids than when it is resting. Correspondingly, circulatory adjustments occur which modify the blood supply, parallel with modifications in rate of tissue activity. Numerous complex mechanisms have been evolved as adaptations of the pump to these varying levels of tissue activity. A full appreciation of what these mechanisms are and how they operate can come only after a study of the circulatory system as a whole, of the actions of blood vessels, and of the factors controlling arterial blood pressure. At present we must confine our attention to the heart itself. We shall examine into the mechanisms which operate to increase or decrease its pumping action. Though, for the present, these are considered as more or less isolated mechanisms, it should be borne in mind that here we are investigating only a few cogs of the machinery set into motion by changes in tissue activity and varying requirements of organs for blood. As we go along, becoming more and more familiar with the operations of parts of the machine, we shall attempt to gear the parts together and to see how the whole smoothly running machinery works.

When necessary, the mammalian heart is capable of pumping seven or eight times as much blood as the quantity ejected during rest. It should be obvious that, in general, an increase in the total cardiac output per minute[17] can be effected in only the following ways: by increasing the rate of the beat, or the volume ejected per beat,[18] or by a combination of both.

VII. CONTROL OF OUTPUT PER BEAT

A normal human heart in bodily rest ejects about 60 cc. of blood from each ventricle per beat. Pumping at its maximum capacity, the heart can increase this to about 150–200 cc. That is, the heart is able to contract at least three times as forcibly as it does in times of rest. Here, apparently, is an outright contradiction of the all-or-nothing principle. Recall, however, that the all-or-none law states that the strength of the cardiac contraction is independent of the strength of the stimulus. But it *is* dependent upon other things, namely, upon the condition of the heart muscle. When the ventricles contract more forcibly and eject a greater quantity of blood, we cannot ascribe this to a more intense stimulation of the ventricular muscle.[19] It is due

17. This is called the *minute volume output*, or the minute volume, and is expressed in cubic centimeters.

18. This is called the *stroke volume*.   19. Via the A-V bundle.

rather to modifications within the heart muscle itself, which render it increasingly contractile. What sort of modifications might these be?

A. THE EFFECT OF METABOLITES

It has been shown that in the presence of a moderate excess of carbon dioxide the heartbeat is stronger. This, as well as other metabolites, so affects the heart muscle chemically that it contracts a little more strongly, even though the stimulus to the muscle remains of constant strength (see the "staircase" phenomenon, p. 374). The utility of this mechanism is apparent, for in muscular exercise there is a temporary increase in carbon dioxide of the blood from the exercise itself. This chemical change, then, may lead directly to more forcible beats.

B. THE LAW OF STARLING

1. *Effect of stretch.*—Perhaps the most potent physiological factor determining the stroke volume is the degree of stretch of the cardiac muscle just at the moment the beat commences. If we determine the strength of the contraction of heart muscle which is being stretched at the moment of stimulation, we find that it contracts far more forcibly than a lax, unstretched muscle. We shall see that this holds similarly for skeletal muscle. If we stimulate a moderately stretched muscle, it will contract more strongly and do more work than that same muscle stimulated when it is not under stretch, even though we employ stimuli of identical intensity in both cases.

2. *Distention of the heart.*—In the heart the main factor which modifies the stretch of the muscle is the quantity of blood within the chambers at the moment systole commences. The more blood contained within the chambers, the more, necessarily, will the muscle be stretched, just as the rubber of a balloon is under greater stretch when it contains more air. The greater also will be the ensuing contraction of the heart and ejection of blood. This relationship was discovered by the English physiologist, Ernest Starling, and is known as Starling's law of the heart. It states that the greater the volume of blood in the heart at the onset of systole, and the greater, therefore, the stretching of the heart muscle, the greater will be the quantity of blood ejected per beat. Of course, the law holds only within limits. Too great a stretch, from too extensive filling with blood in diastole, damages the muscle. Its contraction strength is reduced, and the cardiac output may be inadequate to support life.

The adaptive significance of this becomes clear when we learn that during muscular exercise the filling of the heart with blood during diastole is much faster and greater than in periods of rest. Exercise fills the heart more completely in diastole in ways to be discussed later, and this, by the operation of the Starling principle, induces a greater output per stroke.

As to what the precise mechanism is we do not know at present. Exactly how does stretching a muscle fiber, cardiac or skeletal, increase its contractile power? What internal change is induced in the muscle fiber by this physical change to modify its manner of response to stimulation?

VIII. CONTROL OF THE HEART RATE

A number of factors can modify the rate at which the heart beats. In bodily rest the rate in man is approximately seventy beats per minute. During physical exercise this may be increased threefold, with a consequently greatly increased volume of blood ejected per minute. What are these factors? By what mechanisms can the heart rate be modified? For convenience, we may group these mechanisms into thermal, chemical, and nervous effects.

A. THE EFFECT OF TEMPERATURE

In general, physiological reactions are accelerated by heat and slowed by cold. The rate of metabolic activities of all cells, and the speed of organ activity generally, change in this way with variations in the temperature, within limits. Here we see a parallel between living and nonliving systems, for many inorganic chemical reactions likewise are accelerated by heat.

It is for this reason that the so-called "cold-blooded" vertebrates are to a large extent at the mercy of the environmental temperature. The organism necessarily reacts sluggishly in the cold. With the development of mechanisms for maintenance of a constant body temperature, mammals and birds, as organisms, became independent of fortuitous changes in external temperature. Yet, even in these forms, there are slight changes in internal temperature and corresponding variations in rate of cellular, tissue, and organ activity.

The heart is quite like other muscle and nerve and ciliated epithelium and all other tissues in this respect. With the great changes in internal temperature in lower vertebrates, and much smaller variations in mammals and birds, there are always changes in heart rate. Figure 42, showing kymographic records of heart-muscle–lever tracings, illustrates experimental results of these temperature effects upon the turtle heart.

It should be emphasized that this is a purely local effect. The action is upon the heart itself, entirely independent of nerves. A frog or turtle heart removed from the body accelerates the rate of its beat when irrigated with warm saline solution and slows with cold.[20]

Of course, the rough proportionality of local temperature and the heart rate holds only within limits. If the temperature of any heart is lowered sufficiently, the protoplasmic organization of the fibers is

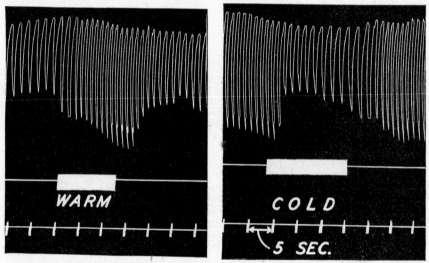

FIG. 42.—Effect of temperature on the heart. Application of warm salt solution (*left*) to a beating turtle heart increases the rate and strength of the beat. Application of cold salt solution (*right*) decreases its rate and strength. Upstroke in the tracing is systole. Tracings made with the apparatus shown in Fig. 36.

destroyed, and the beat stops forever. Likewise, if the temperature is elevated too greatly, the proteins are precipitated, the muscle loses its contractility, and the beat ceases.

Such great temperature changes are not normally encountered by the heart of mammals or birds. Severe exercise under normal conditions will elevate the body temperature only a few degrees Fahrenheit at most. Yet this is sufficient to increase the rate and, therefore, the minute output of the heart.

20. The effect is still further circumscribed within the heart itself, as we have seen (p. 144). Localized cooling or warming of the cardiac pacemaker at the sinus node produces exactly the same changes in rate as cooling or warming the entire heart. Temperature changes limited to any region of the heart, exclusive of the sinus node, are also without effect upon the rate.

Abnormal conditions which produce an elevation of the body temperature, or fever, also accelerate the heart rate.

B. CHEMICAL FACTORS

In this category we refer again only to agents which act upon the heart itself.[21] We exclude, for the present, chemicals which indirectly affect the heart rate and consider only those products which act independently of extrinsic nerves. These agents would be quite as effective, then, upon a denervated heart, or a heart beating outside the body, as upon the normal heart in the intact organism.

1. *Epinephrine.*—Epinephrine is a compound which is obtained from the adrenal glands in the abdomen by subjecting the glands to suitable solvents. If proper amounts of epinephrine are injected into an animal, very specific effects are produced upon the circulatory system, among which is an accelerated heart rate. Though these effects of epinephrine injection are undisputed, there remains some question as to the exact role of epinephrine in the body. But the best evidence seems to be that epinephrine is produced physiologically and that it is secreted in greater quantities under certain emergency or stress conditions, such as excitement and fear, and during severe muscular exercise. The "emergency" adaptive value of accelerated epinephrine secretion in such situations is thought by some investigators to depend in part upon the consequent increased cardiac rate and output,[22] attended by a corresponding increase in the blood supply of skeletal muscle.

2. *Carbon dioxide.*—Carbon dioxide is a by-product of the metabolism of all cells of the body and is harmful or even fatal if present in excess; it is a waste substance to be excreted from the body. But, in addition, it exerts important and even indispensable controlling influences upon several of the organs and systems. Nature has evolved mechanisms in which this by-product of metabolism is put to important adaptive use. We have already seen that carbon dioxide and probably other products of tissue metabolism, especially acids, increase the force of the heartbeat. The effect of these substances upon the heart rate is almost negligible, however, in the concentrations normally occurring in the blood. Apparently nature missed a bet here, for it would seem that the evolution of a

21. Actually upon the sinus node pacemaker.

22. Plus also an increase in the sugar content and the coagulability of the blood. Further discussion of the problem of epinephrine secretion and its probable emergency function will be found in chap. 12.

mechanism for the direct acceleration of the heart by carbon dioxide would be useful, since an increased rate of blood flow would carry the waste to the lungs faster whenever the substance is produced more rapidly. Only with a great increase in carbon dioxide or acid in the blood is there a pronounced effect on the rate of the heart: an apparently nonadaptive diminution.

3. *The thyroid hormone.*—*Thyroxin*, the hormone secreted by the thyroid gland, has a stimulating action upon the heart rate. Whether thyroxin is injected into an animal or powdered thyroid-gland substance eaten, or excessive amounts of the hormone are secreted in malfunction of the gland, the heart rate is accelerated. The importance of this mechanism in the normal individual seems open to question. The thyroid hormone stimulates the rate of oxidative reactions in all cells of the body, including those of the heart. The faster heart rate is thought to be secondary to the accelerated rate of metabolism which the heart shares with all tissues. Evidence for this view comes from the fact that, when the hormone is injected into the blood, there is not the immediate cardiac acceleration which would occur if thyroxin directly stimulated the pacemaker. It is only some hours later, after the general body metabolism has been elevated, that the heart rate is accelerated. Obviously, such a delayed reaction, whether direct or indirect, could be of no significance in the rapid circulatory changes and readjustments that are repeatedly being made in the organism from minute to minute.

C. The Nerves of the Heart

We know that skeletal muscles are absolutely dependent upon an intact innervation for their function. Physiologically, they can be stimulated only via their nerves; and, if these are cut or injured, the muscle is permanently paralyzed. This is not the case with heart muscle. The heart continues to beat in the complete absence of its extrinsic nerves; to a limited degree it may even adjust its rate and strength to the bodily needs by means of the thermal and chemical mechanisms we have just described. But in the normal intact animal the adjustments are effected mainly by the cardiac nerves. Though not indispensable for the actual beat, they nevertheless modify the rate of the beat. The efferent nerves are of two kinds. There are a pair of *accelerator* nerves and a pair of *inhibitory* nerves.

1. *Accelerator and inhibitory nerves.*—The accelerator nerves arise in the spinal cord in the chest region and reach the heart by a some-

what devious pathway. Artificial stimulation of these nerves[23] initiates nerve impulses, of course, and these impulses, carried to the heart, exert here a stimulating action. The heart beats faster. Even though a continuous stream of impulses is set up in the nerve, we obtain no resultant sustained contraction of the heart, such as would occur in skeletal muscle when its nerve is stimulated.[24]

The inhibitory nerves arise in the medulla of the brain, where their cell bodies lie. The fibers course downward in two rather large bundles called the *vagus nerves*. This name is derived from the same root as the word "vagabond." The vagus is indeed a vagabond nerve in the sense that it branches extensively and distributes its fibers to many of the internal organs. It "wanders" through the body, but in a far more orderly and regular fashion than the name implies.

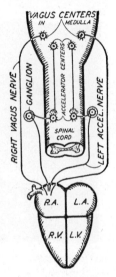

Passing downward through the neck in two large trunks, the vagus nerves give off cardiac branches. Artificial irritation of these nerves by any sort of stimulating agent diminishes the strength and rate of the heartbeat or may even stop the beat for a brief period.[25]

The utility of this double innervation is clear. The heart is thereby supplied with both an accelerating and a braking device, which admits of accurate and delicate control of the rate. Driving a car aptly illustrates the usefulness of such dual control. Imagine a car in which the only manner

FIG. 43.—The efferent nerves of the heart.

of changing its rate of speed is by means of the accelerator, by increasing or reducing the gas feed. The car would be far from ideally

23. In actual practice a nerve is usually cut before being experimentally stimulated, the cut ends being referred to as the *peripheral* and *central* ends. The peripheral end is the portion whose connection with the central nervous system is cut off. It is stimulated when efferent fibers are to be studied. The central end is the portion remaining in connection with the central nervous system and is stimulated when afferent effects or reflexes are being observed. Most nerve trunks are mixed, containing both afferent and efferent fibers.

24. The nerve impulses do not act upon the cardiac muscle, causing it to contract, but upon the pacemaker, stimulating it to act at an accelerated tempo. The beat is accelerated because successive activations of the cardiac muscle by the pacemaker now occur at shorter intervals than before.

25. The vagus fibers also terminate in the sinus node and exert their inhibitory effect on the heart rate by slowing the rate at which the sinus node activates the heart. The depression of the contraction strength is probably due to a direct action of these nerves on the heart muscle.

manageable. The possession of a braking device converts it into a far more easily controlled machine, whose usefulness is correspondingly greatly improved. The double innervation of the heart similarly results in finer, more quickly changeable adjustments.

2. *Continuous action of the vagus.*—In fact, the inhibitors exert a greater controlling action upon the heart rate than is apparent from the foregoing. Under physiological conditions, in man and many other animals, each vagus nerve is apparently in a continuous state of activity. Nerve impulses are continually passing down to the

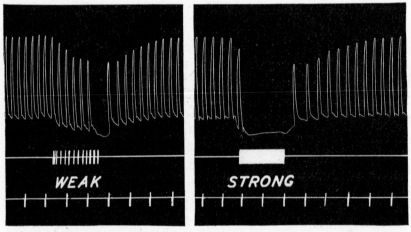

Fig. 44.—Vagus inhibition of the heart. Tracing of the contracting turtle atrium (upstroke is systole) showing a decrease in rate and strength of the beat from weak stimulation (*left*) of the peripheral end of the vagus nerve. Stronger stimulation (*right*) of the vagus stops the heart completely for a time. Time tracing shows 5-second intervals.

heart and exerting there a continuous slight inhibition of the pacemaker. The brakes are dragging a little, at all times. We know this to be true because, if the vagus nerves are severed in an animal, the heart rate accelerates. Nerve impulses continually arising in the medulla cannot now reach the heart, and it is said to have been released from its continued inhibition. Note that a direct consequence of this vagus activity is that acceleration of the heart may be affected physiologically by means of the vagus. Whenever vagus action is diminished, whenever fewer impulses traverse it to the heart, an acceleration results. This is like producing acceleration in a car traveling along with dragging brakes by reducing the braking action. This increases the car's speed independently of any change in the actual accelerator mechanism per se.

A continuous action of this sort is called a *tonic action*, or *tonus*. Many physiological examples of tonic action are encountered. Skeletal muscles are always engaged in a low degree of continuous activity, even at "rest." They are maintained in a state of partial contraction. Numerous other instances of the same phenomenon will be presented later. So the vagus nerves are tonically active, tonically inhibiting the heart. And increases in heart rate can be effected simply by a diminution of vagus tone, entirely independently of any effects of the accelerator nerves.

3. *The problem of inhibition.*—The action of inhibitory nerves presents a problem. The positive activation of a skeletal muscle by stimulating its nerve seems less puzzling, although much of the machinery involved is obscure. But, at any rate, we deal with something positive—nerve impulses—causing a positive change—contraction—in the muscle fibers. The nerve impulse travels because each active region of the nerve becomes itself the stimulus to the next adjacent region. When the termination of the fiber is reached, the impulse in some way passes on to the muscle fiber. There is a continuous transmission of activation, which in the nerve is a nerve impulse, and which in the muscle fiber is a contraction.

But how can we explain inhibition? How can nerve impulses cause a cessation or slowing of action of the structure innervated? How can a positive change like a traveling excitation in a fiber produce not excitation but inhibition at its termination? We know that the impulses in the vagi are of fundamentally the same nature as those in accelerator nerves. So far as the speed of propagation, or the electrical or chemical changes, or any other measurable features of nerve impulses are concerned, nerve impulses seem to be qualitatively nearly the same, no matter in what nerve they may travel. The impulses in the *cardioinhibitory* nerves are essentially the same as those in the *cardioaccelerator* nerves.

The cause of the diametrically opposite effects these impulses produce upon the heart must be sought elsewhere than in the nature of the nerve impulse. Apparently, the essential difference is dependent upon the chemical and physical organization of the fiber terminations in the heart and upon their relationship to the sinus node fibers. A nerve impulse reaching a vagus fiber termination induces certain localized changes which are not nerve impulses and which inhibit the sinus nodal tissue. Impulses arriving at accelerator terminations induce opposite local changes at the sinus node and stimulate it.[26]

D. The Cardiac Reflexes

So far, our account of the nervous control of the heart rate is rather fragmentary. We have seen that the cardiac nerves induce an acceleration or inhibition of the heart rate when they are artificially stimulated. Nothing has been said of the manner in which these nerves are activated physiologically. When we wish to observe the functions of nerves, it is most convenient to stimulate the trunks somewhere along their course—at some convenient place between the brain or spinal cord where they arise and the structure they innervate. Physiologically, nerves are never stimulated in this way. They are activated at one end only, and the impulse travels the full length of the fiber in only one direction. Nerves like the vagi or the accelerators, which are efferent nerves, transmitting impulses away from their origin in the brain or cord, are always activated, then, in the central nervous system.

1. *Nerve centers.*—The fibers of efferent nerves are protoplasmic extensions of nerve cells which lie in the central nervous system. The fibers from a whole cluster of nerve cells lying in the medulla make up the vagus nerve trunk. Such a cluster of nerve cells in the central nervous system is called a *nerve center.* So the fibers of any nerve, even those of skeletal muscles, have a corresponding center in the brain or cord. The center which gives rise to the cardioinhibitory fibers is called the *vagus center,* or the *cardioinhibitory center.* Likewise, in the medulla and the spinal cord, there are the cardioaccelerator centers. This account stresses rather the anatomical aspects of the center, but the "nerve center" is an important physiological concept as well.

Thus, any nerve impulses in the efferent vagus fibers must come from the vagus center, so that our problem of determining the cause of the impulses is carried back to the center itself. What influences the cardioregulatory centers (accelerator or inhibitory) to function? By what mechanisms are they stimulated to send impulses down the efferent fibers to effect adjustments in heart rate?

First, the centers may be influenced by nerve impulses reaching them from other parts of the central nervous system through con-

26. The nature of these changes at nerve terminations will be discussed at some length in chap. 9.

necting nerve fiber pathways. Certain activities of the brain influence the centers through nervous channels inside the brain or spinal cord and consequently modify the heart rate. Emotional states are predominantly effective. Excitement leads to a rapid heart rate, and extreme fear may cause cardiac slowing by affecting the cardioaccelerator or cardioinhibitory centers, respectively.[27] Many have experienced the cardiac slowing of fear. In the protected civilization in which most of us live (in peacetime) extreme fear is uncommon in waking hours, but during sleep intense emotion may be aroused in dreams. Cardioinhibition is a common result, and the dreamer awakes to notice a very slow pounding action of the heart. Of what possible adaptive value such a reaction could be is difficult to state.

Of at least equal importance are the reflex effects upon the cardioregulatory centers. The stimulation of the sensory nerves of many body regions affects the heart rate. Recall that sensory nerves or afferent nerves transmit impulses toward and into the central

nervous system. The impulses in these fibers originate at the peripheral terminations of the fibers. Such impulses travel in the afferent fibers to the brain or spinal cord, whence they are relayed through appropriate interconnecting nerve-fiber pathways to the cardio-regulatory centers. Here we see something of the physiological significance of nerve centers. They are essentially *reflex* centers, way stations in which afferent nerve impulses are shunted into efferent pathways, completing a reflex act. The centers may be likened to telephone exchanges, to which messages converge from many regions and from which they are sent out again through appropriate channels.

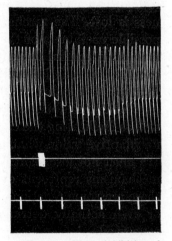

FIG. 45.—Reflex inhibition of the heart. A kymograph record of the beating frog heart shows a slowing of the rate caused by tapping (*indicated by signal*) upon the abdominal organs. Time in 5-second intervals.

2. *The afferent nerves.*—It is found that stimulation of almost any afferent nerve of the body can affect the heart rate. To prove such effects to be reflexes, it is only necessary to perform the experiment before and after cutting the efferent nerves, that is the

27. It is probable that emotional states also affect the heart rate and produce other circulatory changes by nonnervous mechanisms such as by causing increased epinephrine secretion.

vagi and the accelerators. If such cutting abolishes the action, we may conclude that it was reflex in nature and that the afferent nerve carried impulses to one of the cardioregulatory centers, whence they were immediately transmitted down the efferent fibers.

Two or three examples of such reflexes will suffice. If the abdomen of an anesthetized frog is exposed, and the heartbeats observed or recorded, it will be found that a gentle tapping upon the abdominal organs will cause a cardiac inhibition. The effect is abolished if the cardiac efferent nerves are severed. Here is a reflex started by mechanical stimulation of afferent nerves in the intestines. Impulses pass up to the medulla and stimulate the vagus center, and the heart is inhibited by impulses, in turn, passing down the vagus fibers. This reflex partially explains the efficacy of the "solar-plexus blow" in laying low a fistic opponent. The afferent nerves of the abdominal organs are stimulated by a sharp blow to the abdominal wall, and the heart is reflexly inhibited. It may beat so slowly that it pumps an inadequate quantity of blood to the brain, and consciousness is lost. The mechanism of this is clear, but, here again, the possible usefulness of the reaction is hard to see. Of what value is it to the organism to faint in the midst of battle? Is this an adaptation for inducing insensitivity to pain? Is it "nature's anesthesia"? Perhaps this reflex is of some value in combat by slowing the circulation, and therefore diminishing bleeding from wounds, as well as facilitating the clotting of the blood. Of course, there may be no adaptive value whatsoever. It may, indeed, be only a handicap. We must not lose sight of the fact that, though most physiological mechanisms represent evolutionary adaptations of positive value, there are some reactions in all animals which are either indifferent or even actually detrimental so far as value to the organism is concerned.

3. *The depressor reflexes.*—There is a pair of nerves, called the *depressor nerves*, whose fibers terminate in the arch of the aorta and in the ventricular muscle at the root of the aorta. They are afferent nerves and transmit impulses from the aorta into the medulla of the brain, whence they are relayed to the vagus center. Artificial stimulation of this nerve induces a slowing of the heart. The reflex is abolished by section of the vagus nerves; so we know that the vagus center and the vagus fibers to the heart are involved in the reflex. Granted that this reflex can be induced artificially, does it ever occur normally? Are the depressor nerves ever stimulated

physiologically? If so, how is such stimulation effected, and under what conditions does it occur? Answers to these questions must be ferreted out one at a time by experimentation.

It can be shown that, when the aorta is experimentally stretched from within, the depressor nerve is stimulated. Action potentials can be detected in the nerve, and the heart rate is reflexly slowed. Now, the only way in which the aorta can be stretched physiologically is by an elevation of the pressure of the blood within the aorta. If an excessive amount of blood is pumped into this vessel, we should expect the distention and stretching of the walls to initiate the reflex and slow the heart. We now begin to see a possible function of the reflex. Suppose that the heart were suddenly to increase its rate because of some irregularity. This would pump blood more rapidly into the aorta. The aorta would be distended, the afferent fibers of the depressor nerve would be stimulated, and the heart would be reflexly slowed. Its beat would tend in this way to be restored to the rate obtaining before the sudden change. Thus any abnormal acceleration of the heart rate automatically sets off machinery which, in turn, tends to slow the rate. The device

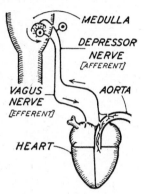

Fig. 46.—An automatic governor tending to maintain a constant heart rate. If the heart accelerates, more blood is pumped into the aorta. Stretching the aortic walls stimulates the depressor nerve, and there is a reflex slowing of the heart. Arrows indicate the course of the impulses in this reflex (see also Fig. 64).

acts much like a governor of a steam engine, which prevents the engine from running too rapidly.

Expressed otherwise, a governor acts to maintain a constant speed in the engine. The depressor-nerve mechanism likewise is an important factor in the maintenance of the heart rate at the normal level of about 70 during bodily rest. Any tendency for the heart to exceed this rate is soon automatically corrected.

A further discussion of this mechanism and its adaptive value to the organism will be found in the following chapter. At that time another mechanism of a parallel nature will be described—that of the *carotid sinus* and its nerves.

4. *The Bainbridge reflex.*—Figuring prominently in the cardiac acceleration of exercise is a reflex originating in the right atrium and great veins emptying into the atrium, for if these structures are de-

prived of their sensory nerves, there is a smaller increase in heart rate during exercise. Apparently any stretch of these structures, by an increase in the quantity of blood they contain, stimulates sensory nerve endings in the vessel walls. Afferent nerves (of the vagus trunk)

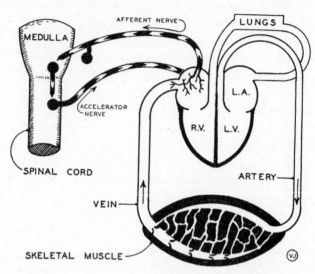

Fig. 47.—Schematic representation of some mechanisms automatically supplying large groups of muscles (as those of the legs) with more blood when they become active (e.g., in running). The sequence is as follows: contracting muscles pump blood faster through the valve-equipped veins; blood flows back to the heart more rapidly; the right atrium is distended; the stretch of the atrial wall stimulates afferent nerves; the accelerator nerve is stimulated reflexly; the heart rate accelerates, pumping blood in greater quantities; the active muscles are supplied with the extra blood without which the muscular work could not continue. Other mechanisms supplement these. Arrows indicate direction of blood flow or of nerve-impulse conduction.

transmit nerve impulses to the medulla and reflexly accelerate the heart. Exercise effects such stimulation by increasing the volume of blood returning through these great veins to the right heart (see Fig. 47 and pp. 175–76 for details of this mechanism).

CHAPTER FIVE

BLOOD FLOW
AND BLOOD PRESSURE

I. Physiological Anatomy of the Vessels
   A. The arteries and veins
   B. Capillaries
II. The Flow of Blood
   A. The circulation time
   B. Capillary flow
     1. Slow rate of flow 2. Constancy of flow
   C. Pressure of the blood
     1. Resistance to flow: friction 2. The gradient of pressure
   D. Blood flow in veins
III. Control of Blood Flow to Organs
   A. Stopcock action; constriction and dilatation of vessels
   B. Study of blood supply of organs
   C. Role of carbon dioxide and epinephrine
   D. Nervous factors
     1. Vasomotor nerves: tone 2. The problem of inhibition 3. Reflexes
IV. The Arterial Blood Pressure
   A. The work of Stephen Hales
   B. The mercury manometer
   C. Systolic and diastolic pressures
   D. Blood-pressure measurements in man
V. Adjustment of the Pressure to Varying Physiological Conditions
   A. The cardiac output
   B. Peripheral resistance
     1. Viscosity 2. Tonic action of vasoconstrictor nerves 3. The vasomotor centers 4. Action of carbon dioxide 5. Vasomotor reflexes 6. The modulator nerves 7. Constancy and change
   C. Blood volume
     1. Relationship to blood pressure 2. Splenic contractions
   D. Gravity
VI. Summary of the Effects of Exercise
VII. Common Disorders of the Circulation
VIII. The Lymphatic Circulation
   A. Lymph and lymph vessels
   B. Movement of lymph
   C. Functions of the lymphatic system

I. PHYSIOLOGICAL ANATOMY OF THE VESSELS

Throughout the vascular system all the vessels are hollow tubes of different diameters, ranging from the large aorta, an inch in diameter in man, to the microscopic capillaries barely large enough to admit a red blood cell. Variations in the thickness of the walls roughly parallel the variations in internal caliber, the wall of the aorta of man being about one-eighth inch in thickness, the capillary walls being of microscopic size. The internal structure as well as the thickness of the vessel walls are significantly different in various parts of the vascular tree. These structural differences are of considerable importance to the physiologist, who finds an intimate interrelationship here between structure and function.

A. THE ARTERIES AND VEINS

Much of the wall of the aorta and larger arteries consists of smooth muscle. The spindle-shaped cells are arranged circularly around the vessel. Contractions or relaxations of these fibers are capable of changing the caliber of the vessels.

The outermost coat of the arteries is made up largely of connective tissue, which also invades the muscular layer to a certain extent. Much of the connective-tissue layer consists of elastic fibers, which give the vessel its elasticity and distensibility, so important in the circulation. The connective-tissue layer renders the wall tough and resistant, so that, although it "gives" somewhat under high internal pressures, it resists rupture even by very high pressures.

The arteries and veins are lined with thin, flat, epithelial cells which always present a smooth surface to the moving blood.

In the vessel walls nerve fibers are also to be found. These are the terminations of the extrinsic nerves. Some of them are afferent fibers,[1] but for the most part they are efferent. They are the motor fibers which innervate the smooth muscle. We shall see that here, as in the heart, the nerve fibers are of two classes—excitatory and inhibitory. Stimulation of the former causes a slow contraction of the circular muscles; stimulation of the latter, a relaxation.

The extremely fine terminal branches of the arteries just before the capillary bed is reached are called *arterioles*, or "little arteries." They possess all the layers found in the larger arteries. Although each coat is thin, the muscle layer is relatively thick.

---

1. The depressor fibers of the arch of the aorta, for example.

The veins are structurally like the arteries except that their walls are thinner. They are rather easily collapsed, as can be demonstrated on the superficial veins of the skin. Figure 48 diagrammatically represents the structure of an artery and a vein and shows that the

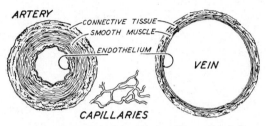

Fig. 48.—Structure of artery, vein, and capillaries. The walls of the capillaries consist of a continuation of the endothelial lining of the arteries and veins. The artery at the wrist has a diameter of over two hundred and fifty times that of a capillary.

caliber of a given vein is somewhat larger than the caliber of the corresponding artery. Capillaries are also shown.

B. CAPILLARIES

Connecting the arterioles with the veins is the capillary bed. As we pass from arterioles to capillaries, two significant anatomical changes are apparent. The walls now become exceedingly thin. The capillary vessels have lost all the coats of the arteries except one—the thin, flat, innermost cell layer. There is almost no connective tissue or no true muscle. For the most part the wall consists simply of a tubular extension of the continuous smooth lining of the arteries.

The branching of the vascular tree also is here more profuse than at any other part of the system. One arteriole breaks up into a number of capillaries whose individual diameters are only a little smaller than those of the tiniest arterioles.

Both of these anatomical features are of considerable functional significance. The thin walls we relate to the fact that in the capillary region the exchanges take place between tissues and blood. It is here that materials leave the blood vessels by filtration and diffusion and enter by diffusion and osmosis. Extremely little if any exchange of materials between tissues and blood can occur through the walls of the arteries and veins including even those of small caliber. This applies not only to the general tissue regions to which the blood carries materials needed in metabolism and from which wastes are drained. It also is true of the vessels of the lungs, where blood gives up carbon dioxide and takes up oxygen; of the intestines, where foods are absorbed; and also of the organs of storage or excretion. The capillaries are the only functional units, so far as any appreciable exchange of materials between blood and tissues is concerned; and the whole circulatory system—the heart, arteries, and veins, with all their controlling mechanisms—function "to serve the capillaries." The smaller caliber of the capillaries is of great significance here also. It is estimated that, in the capillaries of mammalian muscle, a cubic centimeter of blood is exposed to a total capillary wall surface of about 5,000 sq. cm.

The profuse branching at the point of transition from arterioles to capillaries is of related importance. The ultimate spreading network serves to distribute thin-walled capillaries to all parts of every tissue and organ. In the most active tissues, where metabolism occurs at a rapid rate, the capillaries are remarkably close together. In the diaphragm of the mammal the distance between adjacent capillaries is only about twice the diameter of the capillary itself. Expressing this in another way, a molecule of oxygen diffusing out from the exact center of a capillary, when it reaches the capillary wall, has already gone one-third of the total distance it must travel in order to reach the interior of muscle fibers farthest removed from a capillary. Less active tissues are less richly supplied with capillaries.[2] Even in muscle not all these capillaries are open at all times. Only during great activity of the muscle are all, or nearly all, of the capillaries filled with blood and functioning.

II. THE FLOW OF BLOOD

A. The Circulation Time

The rate of blood flow through the lesser or pulmonary circuit can be quite easily determined on an experimental animal like the rabbit. The *carotid artery* and the *jugular* vein, which, respectively, supply and drain the blood of the head, are exposed. A dye visible when mixed with blood, like methylene blue, is then injected into the jugular vein, and the time which elapses between the injection and visible appearance of the dye in the carotid artery is noted. In the interval the dye will have traveled the following course: jugular vein, superior vena cava, right atrium, right ventricle, pulmonary artery, lung capillaries, pulmonary vein, left atrium, left ventricle, aorta, and, finally, the carotid artery, one of the first branches of the aorta. This journey is made in about 10 seconds.

2. Fatty tissue is poorly supplied, and the lens of the eye is devoid of capillaries.

Measurements upon the major circulation time in the rabbit, using the same principle, reveal the fact that the longest circuit, such as that from the heart to the foot and back to the heart, requires only about 20 seconds. The total circulation time would be approximately the sum of these, or about 30 seconds. In man the total circulation time is about 60 seconds.[3]

B. Capillary Flow

From this we obtain an impression of great speed, of blood rushing along precipitously throughout its entire course. This view must be modified considerably, for great variations in rate of flow occur in different parts of the system. Measurements show that, the smaller the vessel, the slower the rate of flow in that vessel. This is dependent

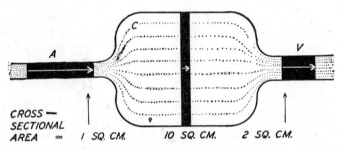

Fig. 49.—Rate of flow of liquid through a system whose diameter varies. Black areas represent equal volumes of liquid. The lengths of the white arrows represent rates of flow, which are 10 cm. per second in A, 1 cm. per second in C, and 5 cm. per second in V.

upon the fact that, when an artery branches, the total cross-sectional area of the branches is greater than the cross-sectional area of the original artery, even though, in branching, the caliber of each vessel is diminished.

1. *Slow rate of flow.*—How this influences rate of flow is shown in Figure 49. Suppose fluid to be moving from left to right through the system at a rate of 10 cm. per second in portion A. In any single second 10 cc. of fluid would then move into region C. In a corresponding second each 10 cc. of blood in section C, with its larger diameter, would move only 1 cm. to the right. At the same time 10 cc. would be forced into region V, where the linear rate of movement would be 5 cm. per second. There is, then, an exact inverse proportionality be-

3. Obviously, the time required for a systemic circuit depends upon which of the many routes is taken. It takes considerably longer for blood to go through a long circuit, such as to the foot, than through the short circuit of the vessel of the heart itself.

tween linear rate of flow and cross-sectional area. This would hold equally true if region C were divided into many small tubes, provided the *sum* of the cross-sectional areas of all the tubes is taken into consideration.

In the circulatory system the total cross-sectional area gradually increases as we go from aorta to arteries to arterioles, and the rate of blood flow correspondingly decreases. At the point where arterioles break into a profusion of capillaries, the area suddenly increases tremendously, with a sudden marked slowing of the rate of flow. The cross-sectional area of a given capillary bed is several hundred times the area of a large artery which gives rise to those capillaries. We therefore find the rate of flow in capillaries to be only a fraction of 1 per cent of the rate in large arteries.

Correspondingly, as the capillaries join to form *venules,* or "little veins," and these, in turn, unite into larger vessels, the total cross-sectional area diminishes, and the rate of flow increases again, as in section V of the mechanical model illustrated in Figure 49. The speed of flow in the large veins almost attains that of the large arteries but never quite equals it, for the caliber of a large vein is greater than that of the corresponding artery.

The adaptive value of this mechanism is clear. Blood is rapidly transported to and from the capillary regions, but in the capillaries the rate of flow slows down greatly, allowing time for exchange of materials between blood and tissues. The rate of flow in a medium-sized artery is about 250 mm. per second. In capillaries the rate is less than 1.0 mm. per second. That is, it requires nearly a second for a given red cell to pass through a single capillary 0.5 mm. in length.

2. *Constancy of flow.*—The flow of blood through capillaries differs from flow in arteries in another important respect. The heart pumps blood into the aorta only during systole. During the whole period from the end of one systole to the beginning of the next, while the semilunar valves are closed, most of the blood in the first few centimeters of the aorta is stationary. The onward flow here is intermittent, occurring mainly during contraction of the heart. Even in arteries a little farther removed from the heart there is a degree of intermittency of flow. When an artery is cut, the blood always escapes in spurts. As the capillary bed is approached, the intermittency becomes less and less marked, and in the capillaries themselves the flow is usually constant. The intermittent ejection of blood by the heart has become converted into a steady continuous flow in

the capillaries; as a result, exchange of materials between blood and tissues can proceed continuously. What is the mechanism of this?

It is practically entirely due to the elasticity of the arterial walls. If we were to construct a physical model of the circulatory system, making the walls of all the tubes and their branches of some such rigid material as glass, then an intermittent injection of fluid into the system would give an equally interrupted flow in the smallest glass branches. Fluid would flow through the smallest tubes only during the period of injection and stop entirely in the intervals between injections. But if the system was suitably constructed of some elastic material like rubber, the intermittent injections would be converted into a constant stream, provided also the injection pressure was suitable.

As the heart ejects blood into the aorta, the spurt of blood distends the elastic aortic wall. During cardiac diastole the stretched aortic wall clamps down again, squeezing blood forward, backflow being prevented by closure of the semilunar valves. This, in turn, distends the next segment of the aorta or artery, which presently recoils elastically. This gives rise to the traveling *pulse wave*, which can be felt in any artery.[4] The result is that, even during diastole of the heart, the blood is kept moving forward. As we progress farther and farther from the heart, the difference between rates of flow in systole and diastole becomes less and less, until no spurting at all is normally apparent in the capillary region.

This may also be expressed in terms of energy transformations. Motion of all kinds, including blood flow, demands the expenditure of energy. In systole of the heart energy is liberated, most of which is converted into energy of motion, and blood is pumped out. Some of the energy, however, is "expended" to distend the elastic aortic and arterial walls; that is, some of the energy liberated by the heart muscle is stored as potential energy of the stretched vessel walls. Then, when energy liberation by the heart temporarily ceases in diastole, the potential energy of the distended vessel walls is converted into kinetic energy. The vessel walls recoil elastically and push the blood along. Because the heart stores energy in the elastic vessels, a nearly constant source of energy is available to move the blood, although the heart itself yields energy only intermittently. We might compare this to the conversion of the intermittent move-

---

4. The rate of transmission of the pulse wave is 7–8 m. per second in the carotid artery. Compare this with the velocity of blood flow in that vessel—about 0.25 m. per second.

ment of winding a clock, into the constant movement of the hands, effected by storing energy of elasticity in the mainspring, which delivers energy to the moving parts in the intervals between succes-sive windings.

C. Pressure of the Blood

Mention has been made of the fact that the veins are more easily collapsible by externally applied pressure than are the arteries. This is due not so much to the thinner walls of the veins as to the low pressure of the blood within. If we measure the average pressure of the blood in successive regions of the vascular tree, we find that there is a continuous decrease from the heart through the arterial and capillary and venous regions. In the aorta close to the heart the average pressure is always highest; it is lower in the arterioles and capillaries and still lower in the veins. The lowest pressure is in the veins closest to the right atrium. At this point the pressure is about at atmospheric pressure or (in mammals) even lower. The rate at which blood escapes from a hole in a vessel demonstrates these differences very well; it spurts rapidly from an artery and flows much more slowly from capillaries or veins. Even large veins in the neck near the heart may bleed very little through a small hole in the wall.

1. *Resistance to flow: friction.*—This gradually decreasing pressure is again a purely physical phenomenon. It can be duplicated in a horizontal glass tube through which water is forced under pressure. Vertical tubes leading upward from the horizontal tube at intervals will register, by the level of water in them, the pressure at various points along the horizontal tube (see Fig. 50). The pressure near the pump is high, and it gradually diminishes with increasing distance from the source of the pressure. This is due entirely to friction. The energy of pressure is gradually dissipated as the heat of friction.

Without such a gradient of pressure, of course, the water would not continue to flow. It moves along because at any given point in the tube the pressure is higher than at any other point farther along. Similarly, the gradient of pressure in the blood vessels is primarily responsible for the onward flow of blood.

2. *The gradient of pressure.*—If at any point along the horizontal tube the friction becomes greater, the fall of pressure as the fluid flows through that region will be greater than the fall in any other part of the tube of equal length. Suppose, as in Figure 50, we split the horizontal tube into many smaller tubes of smaller caliber be-

tween points B and C. This greatly increases the frictional surface and, therefore, the dissipation of energy in overcoming the friction. Thus the fall in pressure between points B and C will be greater than the fall between A and B, or between C and D, even though all these points demarcate equally long segments of the tube. The steepness of the gradient of pressure-fall in any stretch of the tube depends

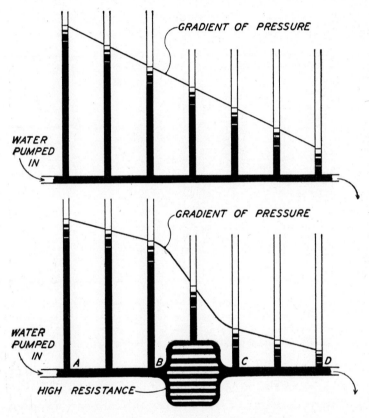

Fig. 50.—*Above:* Gradual diminution in pressure along a horizontal tube through which water (*in black*) is forced. *Below:* This drop in pressure is more rapid at regions along the horizontal tube where there is a high resistance to flow (i.e., between B and C).

upon the resistance to flow which the water encounters in that stretch. Note that, though the *pressure* continues to decrease from A to B to C to D, the *rate of flow* varies in a different manner. The linear speed of flow decreases as water flows from AB into the widened region beyond B. But then, even though the pressure continues to fall, the linear rate of flow accelerates as the water moves on into the *CD* segment, because of the previously described effect of changing the cross-sectional area.

Similarly, the downward gradient of pressure in the circulatory system is not uniform throughout. As fluid passes through portions where the most resistance is encountered, the rate of fall of the blood pressure is greatest. The strongest resistance is encountered in the arteriolar and capillary regions, where the vessels are of small caliber and where the total frictional surface area is largest. The rate at which the pressure falls on either side of this area (i.e., in arteries

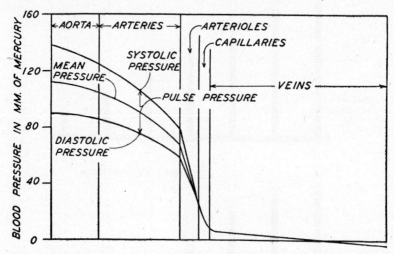

Fig. 51.—The gradient of pressure in the circulatory system, from aorta to arteries to arterioles and capillaries (where the gradient is steepest because of the high resistance) to veins, and back to the heart. Note that the pulse pressure (see p. 185) gradually diminishes, until in the capillaries the rate of blood flow is constant throughout each cardiac cycle.

or veins) is far less because here the friction is less. A certain amount of friction even in this area is responsible for what slight decline in pressure there is. Figure 51 illustrates this pressure drop and the rapid rate of the drop in the arteriolar and capillary regions.

D. Blood Flow in Veins

In the veins other agencies besides the pumping of the heart aid in the onward flow of blood. These are the skeletal muscle movements and the breathing movements. The veins are well equipped with valves (see Fig. 52) which permit movement of blood in only one direction—toward the heart. When a skeletal muscle contracts, it compresses the easily collapsible veins in its interior. This forces the blood from the veins in the direction of the heart. As the muscle relaxes, the veins are permitted to fill again with blood, which can come only from the capillaries because of the valves. In this way intermittent muscular contractions pump the blood of the veins forward in a manner mechanically much like the action of the heart itself. This mechanism is especially important in the return of blood to the heart from the legs against the action of gravity. It is a not uncommon experience to have fluid collect in the tissues of the legs (edema) when one stands quietly for a long time, or sits quietly on a long train journey or automobile ride with the inactive legs in a dependent position. This is much less likely to occur in walking, when muscle contractions facilitate the return of blood from the legs. And all of us have experienced the greater fatiguing effect of standing still, as compared with walking. In the former case the pumping action of the stationary muscles is minimal, and there is a sluggish flow of venous blood. Consequently, the wastes of metabolism, which contribute to fatigue, tend to accumulate in the muscles. Intermittent contractions and relaxations are necessary to move the venous blood along at a good speed.

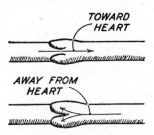

FIG. 52.—Valves in veins, permitting flow of blood toward the heart and preventing flow in the reverse direction.

The breathing motions also help to move the venous blood. In breathing, the volume of the chest is increased in inspiration, which serves to draw air into the lungs. Now, since the heart lies within the chest cavity, it also will be affected, and each inspiration will aspirate blood into the heart[5] from the veins in much the same way that it aspirates air into the lungs.

These actions are especially important during muscular exercise. At that time both the pumping action of the muscles and the aspirating effect of the accelerated breathing movements markedly increase. Consequently, blood is drained from the active muscles faster and flows back into the heart more rapidly. We can now gear together some important cogs in the animal machine. Recall the Bainbridge reflex, by which distention of the right atrium accelerates the heart rate, and the Starling principle, which states that, when the heart fills more completely in diastole, the resultant stretch of the heart-muscle fibers causes a more forcible contraction and the ejection of a greater volume of blood per heartbeat. In exercise, with an increased rate

5. Actually, into the thin-walled atria.

of filling and stretch of the heart, conditions are fulfilled for the operation of these mechanisms—for an increased rate and stroke of the heart. Consequently, the active muscles not only are drained of venous blood more quickly but are also supplied more swiftly with arterial blood. The very agent (muscle contraction) which produces a temporary physiological deficiency (i.e., requirement of more blood for the active muscles) is also instrumental in overcoming the deficiency by these mechanisms (see also Fig. 47, p. 164).

III. CONTROL OF BLOOD FLOW TO ORGANS

We have seen that the heart is capable of adjusting its output to the rate of activity of the body. When more blood is required, mechanisms are automatically set into operation which increase the rate and strength of the heartbeat and hasten the rate of the circulation. But note that this mechanism alone would increase the rate of blood flow to all the organs simultaneously. It allows for no differentially greater flow to one organ or system than to another. If the heart pumps more blood, all organs share in the general increased blood flow.

It is not common for all the organs of the body, however, to accelerate their activity at the same time. In muscular exercise more blood is required by the muscles; during digestion, by the digestive glands; during absorption, by the intestines; etc. To increase the blood supply of all organs when any one of them increases its activity would be very wasteful, and we find that, here at least, the body machinery is not prodigal. Not only can the output of the heart change but also there are mechanisms which alter the distribution of the blood to the various organs differentially, in accordance with their varying rates of metabolic activity.

A. Stopcock Action; Constriction and
Dilatation of Vessels

The nature of the mechanisms which effect such local changes in blood flow can again be illustrated in a physical system, like that shown in Figure 53. This consists of a pump, H, which regularly forces water into tube A and its branches to regions M and I. The water flows back to the pump through V, completing an artificial circulation. At points $C.M.$ and $C.I.$ in the tubes leading to M and I are stopcocks which are always only partly open but which may be widely opened or still further closed at will. Suppose also that it is possible to alter the rate of the pump or the length of its stroke. We may now ask what means would be available in such a system for increasing the quantity of water flowing into region M per minute. These we list:

   a) Increased output by the pump by either (1) increasing its rate or (2) increasing it force
   b) Closing still further the stopcocks at $C.I.$
   c) Opening widely the stopcocks at $C.M.$

In this simplified model of the circulatory system there are but two branches of the main tube leading from the pump instead of the many

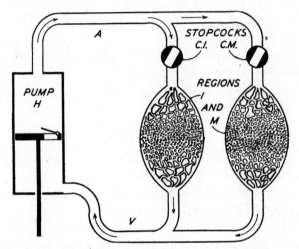

Fig. 53.—Schema to illustrate the adaptations of blood flow to the rate of tissue activity. An increased stroke and rate of the pump (*heart*) increases the flow to both regions I (*interior organs*) and M (*muscles*). Partial closure of stopcock $C.I.$ (*vasoconstriction*) and wider opening of stopcock $C.M.$ (*vasodilatation*) serve to shunt blood differentially through region M. Similar adjustments occur physiologically during muscular exercise.

branches of the aorta. There are but two local regions to consider instead of many in the body, yet the principles underlying the operation of the model are identical with those occurring in the body.

    The heart-pump analogy is obvious, and we have already dealt with the mechanisms which affect the volume pumped per minute. But what of the stopcocks? These are also to be found. They are not located in the main vessels to an organ, as the diagram suggests, but are present mainly in the tiny branches of those arteries—in the arterioles imbedded in the tissues and organs. Here they control the volume flow of blood into the capillaries.

The smooth muscle fibers encircling the arteriolar walls constitute the stopcocks. Always in a state of partial contraction, or tonus, they resemble partially closed stopcocks. An increase of the muscular tone, a further contraction of the circular fibers, diminishes the caliber of the arterioles and reduces the supply of blood to the capillaries springing from those aterioles. Such action is known as *vasoconstriction*, that is, constriction of the (arteriolar) vessels. A reduction of the circular muscle tone, a relaxation of the smooth muscles, opens the stopcocks widely. The arteriolar caliber is increased, and an increased volume of blood flows into and through the capillaries of that region. This is called *vasodilatation*. The latter not only permits blood to flow into the capillaries more rapidly but also fills capillaries which prior to the dilatation contained no blood. It opens previously empty, collapsed, nonfunctional capillaries, increasing the amount of blood available to all parts of the organ or tissue. Thus, by means of vasoconstriction or vasodilatation, blood can be shunted preferentially to this or that region of the body, depending upon its rate of activity.

B. Study of Blood Supply of Organs

Various means are available for detecting and measuring local changes in the supply of blood to a specific body region. By direct inspection blushing or blanching may be observed. Reddening indicates vasodilatation and the presence of more blood in the organ. Often it is also possible, either with the naked eye or more usually with the aid of a microscope, to observe the blood vessels contract or expand.

Again, it is sometimes feasible to measure changes in the total volume of the organ as an index of vascular changes. If a structure like the leg or the arm increases in volume as a result of an experimental procedure, it may be assumed that the increase is due to an increased blood content of the limb or to a dilatation of some of the blood vessels of the limb. This may be due to arteriolar vasodilatation, although a dilatation of the veins alone would produce a like effect. In such studies care must also be taken to avoid any other possible sources of volume change, such as loss of water by a gland which is secreting watery fluid. The volume decrease of the secreting gland might be erroneously interpreted as a vasoconstriction.

One of the most direct simple means of measuring these effects in the experimental animal is to collect all the venous blood which flows from an organ. This can be done with some small organs like the salivary glands. Into the vein which drains all the blood from the gland is placed a glass tube, or *cannula*, which diverts the blood from its usual course. The rate of blood flow through the gland can be determined by counting the rate at which the drops of blood escape from the cannula. A vasoconstriction in the arterioles supplying the gland will decrease the rate of flow of drops, and vasodilatation will increase the rate.

One of the earliest evidences of vasodilatation was studied by Claude Bernard, who, as a matter of fact, misinterpreted the results he obtained. He found that, when he cut a certain nerve in the neck of a rabbit (the *cervical sympathetic* nerve), the ear on the same side of the animal became warmer. If now he stimulated the nerve, the ear became cooler. He concluded that the nerve was thermogenic—that it regulated heat production in the ear. We know now that the temperature changes in this case are secondary to vasomotor effects.[6] Vasodilatation brings to the exposed surface more warm blood from the interior of the body, and the entire ear becomes warmer to the touch. Constriction cuts off the supply of warm blood, and the ear feels cold. By means of temperature changes, then, we can often obtain evidence of vasomotor changes in many exposed regions. Any exposed flushed part, in which vasodilatation is occurring, is warm to the touch.

C. ROLE OF CARBON DIOXIDE AND EPINEPHRINE

The mechanisms for effecting these vascular changes are chemical and nervous. Among the chemical effects, we find that carbon dioxide figures significantly. It apparently acts locally, producing relaxation of the circular smooth muscles and dilatation of the vessels. There is evidence, also, that other metabolic products of an acid nature act similarly. The mechanism is direct and effective. Whichever organ or tissue accelerates its activities is thereby assured an increased blood supply. The concentration of carbon dioxide will be greatly increased in the specific tissue which is especially active at any given time, and the vessels of that particular tissue are made to dilate by this compound.[7]

6. "Vasomotor" refers to vasodilatation or vasoconstriction.

7. Some difficulty arises in trying to see how metabolites can exert a dilating action on the arterioles, because presumably these waste products enter the blood stream in the capillaries and are at once carried *away* from the arterioles into the venules and veins. It is suggested that, since the arterioles as well as the capillaries are imbedded in the active tissue, some of the metabolites diffuse from the tissues partway through the arteriolar walls, reaching the circular muscle layer and effecting a relaxation.

Epinephrine produces different effects upon different blood vessels. Injections of quantities approximating those which may occur normally in the blood effect a relaxation of the arteriolar walls in the skeletal muscle system. At the same time the arteriolar muscles of the internal organs such as the stomach, liver, intestines, etc., are stimulated. There results a double adjustment ideally suited to facilitate blood flow through skeletal muscles. The vessels of the skeletal muscles dilate, and those of the abdominal regions constrict. Blood is shunted predominantly through the muscles and away from the internal organs. Recall that along with these vascular effects of carbon dioxide and epinephrine are cardiac effects also, which increase the minute volume output. All these effects add up in the same direction to increase the blood supply of the active muscles. Both the vascular and the cardiac effects of these chemicals are independent of the extrinsic nerves, acting equally well upon denervated vessels or heart.

D. NERVOUS FACTORS

1. *Vasomotor nerves: tone.*—The nervous control of the blood vessels involves efferents which are designated as *vasoconstrictor* or *vasodilator* nerves, depending upon whether they cause contraction or relaxation of the arteriolar muscles. The presence of both types of nerves is readily demonstrable, and, in general, the vessels of any specific organ are innervated by both kinds. Stimulation of a nerve of the neck, for example (the cervical sympathetic), causes a constriction of the vessels of the ear of the rabbit and of the lining of the mouth. The rabbit is frequently chosen to demonstrate this because the vessels themselves can be seen directly. They stand out as dark red lines when the semitransparent ear is held up to the light. The diminution in their caliber upon nerve stimulation is striking. Again, in the dog, stimulation of this same nerve causes a marked reduction in blood flow through the submaxillary salivary gland.

Vasoconstrictor nerves are usually in a state of continuous or tonic activity. The partial sustained contractions of the arteriolar muscles, to which we referred earlier as a state of maintained partial closure of the stopcocks, are dependent upon constant stimulation of those muscles by impulses in the vasoconstrictor nerves. Therefore, cutting the vasoconstrictor nerves should abolish this tone and should lead to a dilatation. Experimentally, we find this to be the case. Cutting the nerves to the vessels of a rabbit's ear results in an immediate dilatation, owing to a loss of tone of the circular muscles, which now no longer receive tonic vasoconstrictor impulses. This phenomenon is quite general in the vasomotor system and will be referred to again later as an important factor in the control of the arterial blood pressure.

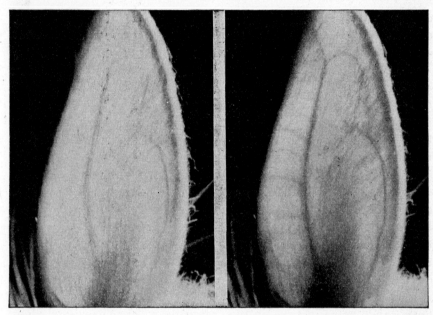

Fig. 54.—Ear of a normal rabbit (*left*). The blood vessels can scarcely be seen because they are maintained in a state of partial constriction by vasoconstrictor nerves. When these nerves are cut in an anesthetized animal, the vessels dilate markedly and stand out as dark lines (*right*). Then if this vasoconstrictor nerve is stimulated artificially (peripheral to the cut), the condition seen in the left picture is restored. (From the sound film, *Heart and Circulation*.)

When a vasodilator nerve of a group of arterioles is stimulated, the circular muscle fibers of the vessels relax. They relax to an even greater degree than the relaxation which attends cutting off of the tonic vasoconstrictor nerve impulses. If we cut the vasoconstrictor nerves of the salivary gland, a slight increase in blood flow through the gland occurs, reflecting the loss of vasoconstrictor tone. If now the vasodilator nerves of the gland are stimulated, the blood flow is still further increased. The dilatation greatly exceeds that of mere loss of constrictor tone.

2. *The problem of inhibition.*—The action of the vasodilator nerves

presents much the same problem that we faced in considering vagus inhibition of the heart. In vasodilator nerves we deal with impulses not essentially different from impulses in vasoconstrictor nerves, causing an inhibition of the structure innervated—the circular smooth muscle of the blood vessels. Somehow the nerve impulses of a vasodilator nerve produce some local change at the smooth muscle fibers which causes them to relax markedly. This change, of whatever chemical or physical nature it may be, is the exact opposite of the local change produced at the terminations of the vasoconstrictor fibers (see further discussion of inhibition, chap. 9).

3. *Reflexes.*—Of course, these efferent nerve fibers do not act as independent nerves. Their continuous tonic action, and changes in their action producing constriction or dilatation, must be traced back to the central nervous system and to reflexes. Here, as in the case of the cardiac efferents, we must consider the efferents in relation to nerve centers in the cord and brain as well as to various afferent systems. In the case of the vasoconstriction in the skin vessels upon exposure to cold, for example, we find that a reflex is involved. Cold stimulates sensory nerve endings in the skin. Impulses are transmitted up the afferent fibers to the central nervous system and, in a typical reflex fashion, are relayed back to the skin vessels by way of the vasoconstrictor efferents. The reddening of the skin in a warm environment is also essentially a reflex vasomotor phenomenon.

The operation of the vasomotor system and the vasomotor reflexes, and their functions in the organism, will be considered along with blood-pressure phenomena, with the factors which modify and control the arterial blood pressure.

IV. THE ARTERIAL BLOOD PRESSURE

The fact that blood escapes when a vessel is severed indicates that the pressure of the blood is greater than atmospheric pressure. The rate and force at which it escapes is a rough indication of this pressure, and we find, as pointed out in previous sections, that this varies considerably in different parts of the circulatory tree. It is greatest in the aorta and arteries and far lower in the veins.

A maintained high *arterial blood pressure* is required to force the blood through the small vessels, where the resistance is great. It must be high enough to force blood against gravity to the brain, which always requires a great deal of blood. Here, both frictional resistance encountered in the smaller vessels and gravitation must be overcome. The pressure must also be sufficiently high so that a residuum of pressure still remains in the blood beyond the capillary bed to drive the blood back again through the veins to the heart. Some of this return flow must also take place against gravity, as in the case of the return of blood from the legs; other mechanisms besides the pumping of the heart aid in this, as mentioned previously.

Furthermore, with a high pressure in the arteries, more rapid adjustments of blood flow are possible. A rather generalized vaso-constriction in a number of organs at once causes the highly compressed blood to flow quickly into unobstructed channels, into unconstricted vessels. Likewise, the high pressure insures an immediate inrush of blood into an organ when local vasodilatation occurs. In this respect blood pressure functions like the water pressure in our city mains. Because the water pressure is high, there is an immediate rapid flow of water when the stopcock of the faucet is opened.

A. The Work of Stephen Hales

Quantitative measurements of the pressure in the arteries were first made by the versatile English clergyman, Stephen Hales, in 1733. His experiments mark the beginning of quantitative studies upon the dynamics of the circulation, which have contributed so much to our understanding of the circulation in health and disease. Hales describes his experiment as follows:

> I laid a common field gate upon the ground, with some straw upon it, on which a white mare was cast on her right side . . . ; she was fourteen hands and three inches high; lean, though not to a degree, and about ten or twelve years old. [She was] . . . to have been killed as being unfit for service. Then laying bare the left carotid artery, I fixed to it towards the heart the brass pipe, to the . . . end of which a glass tube was fixed [vertically], which was twelve feet nine inches long. The blood rose in the tube . . . till it reached to nine feet six inches height.

From this it was possible to express the arterial blood pressure quantitatively. In Hales's experiment the pressure was such as to support the weight of a vertical column of water (or blood) 9.5 feet in height. Abbreviating this, we say that the pressure was 9.5 feet of water.

The level of the column does not remain constant but rises a little with each contraction of the heart and falls during diastole. We refer to these pressures as the *systolic* and the *diastolic pressures*. The average of the systolic and diastolic pressures is the *mean arterial pressure*. The difference between the two is called the *pulse pressure*.

B. The Mercury Manometer

Hales's device is rather cumbersome and does not lend itself read-ily to extensive experimental use. Today a modification of it is em-ployed, in which, instead of supporting a vertical column of blood, the pressure supports a column of mercury. The height of the column of mercury is only a fraction of that of an equal weight of water. As mercury is 13.6 times the weight of water, a column 1.0 cm. in height would be the equivalent of a column of water of 13.6 cm. As a result, a more manageable and less cumbersome procedure is available.

This modification involves the use of the *mercury manometer,* which consists of a glass U-tube, the bottom of which contains a quantity of mercury (Fig. 55, *A*). When both ends of the manometer

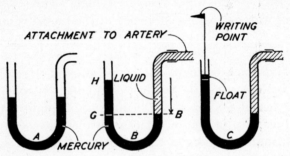

Fig. 55.—The mercury manometer. *A:* Condition with both arms of the U-tube open to the air. *B:* Condition with one arm of the manometer attached to an artery by rubber tubing. *C:* Manometer equipped with a float and a writing point for the inscription of a blood-pressure record.

are open to the air, the mercury columns are of equal height in the two arms. The surface of the mercury on both sides is being acted upon equally by the pressure of the air. If one arm is attached to an artery, there will then be an added downward pressure on the mer-cury on that side. As the level here falls, and the mercury is pushed into the open arm, a rise in the level in that arm will occur. Equi-librium will be reached (Fig. 55, *B*) when the downward pressure of the blood at point *B* is exactly equal to the downward pressure of mercury column *HG* at point *G*. The height of the column *HG* in millimeters is a quantitative measure of the pressure exerted at *B*, or the blood pressure. Blood pressure is expressed, then, in milli-meters of mercury, and we find that the average or mean arterial blood pressure in mammals is about 110–120 mm. Hg. By means of a "float" on the mercury of the open arm, and an attached writing point (Fig. 55, *C*), records of the arterial blood-pressure fluctuations may be inscribed upon the smoked paper of the revolving drum of a kymograph.[8]

C. Systolic and Diastolic Pressures

Fluctuations in arterial pressure occur with each heartbeat, which we correlate with the intermittent spurting of blood from a severed artery and with the arterial pulse. The fluctuations can also be noticed in the mercury manometer, the pressure rising with ejection of blood by the heart during cardiac systole and falling during diastole, when the semilunar valves are closed. We refer to these quantities as the systolic and diastolic pressures, respectively. More accurate measurements than are possible either with the Hales technique or the mercury manometer indicate that in mammals the normal systolic arterial pressure is about 120 mm. Hg., and the diastolic pressure about 80 mm. Hg. These are the values for the relaxed, resting, normal young adult. During each cycle of the heart the arterial pressure fluctuates between these points.[9] This fluctuation—the pulse pressure—normally measures about 40 mm. Hg. Many normal young adults have somewhat lower systolic, diastolic, and pulse pressures than those indicated. Excitement, exercise, and other normal states may appreciably increase these pressures, as will be seen presently.

8. It is of interest that the mercury manometer had already been invented when Hales made his observations. In fact, he himself employed the instrument in botanical research. Yet it seems not to have occurred to him to use it in his blood-pressure measurements. It was not until a century later that Jean L.-M. Poiseuille used the mercury manometer for this purpose. Also of interest is the fact that even such a simple device as the mercury manometer as we now know it represents the inventiveness of at least two men. It remained for a second investigator, Karl Friedrich Ludwig, to place a float on the manometer some years after Poiseuille used the plain U-tube.

These facts are not pointed out by way of disparagement but rather to indicate that often, in science, many years and more than one set of brains are required to make contributions which later appear minor and are taken quite for granted.

9. These fluctuations occur too rapidly to be registered accurately by a mercury manometer. Oscillations of the mercury do occur and enable us to determine the rate of the heartbeat from a mercury-manometer tracing. But the inertia of the heavy liquid is such that, before the mercury has had time to be pushed up to register the true systolic level, the heart has already gone into diastole. Similarly, before the diastolic level can be reached by the mercury, a new systole has begun. Consequently, the excursions of the mercury manometer with each cardiac cycle are only a few millimeters, instead of the true 40 mm. or so, of the normal pulse pressure. The more slowly the heart beats, the more time the heavy mercury has to follow the true intra-arterial pressure fluctuations. So when the heart beats slowly, the excusions of the manometer are greater, which should not be interpreted to mean that the heart is beating more forcibly. Even with a slow heart, the mercury manometer is useful chiefly for measuring and recording the mean arterial blood pressure and the heart rate.

The elasticity of the arteries and the action of the semilunar valves contribute significantly to the phenomenon of pulse pressure. As blood is ejected from the heart in systole, the rise in aortic pressure is reduced by the elasticity of the vessel walls. They "give," and the vessels distend, providing a greater vascular volume to accommodate the ejected blood. Likewise, during cardiac diastole, the pressure in the aorta and arteries is kept from falling rapidly because of the elastic recoil of the vessels, pressing upon the blood. Also, as the heart relaxes, the closure of the semilunar valves prevents the aortic or arterial pressure from falling to the low point reached within the heart itself in diastole, which is about atmospheric pressure.

As we pass from the aorta on to the arterioles, the pulse pressure becomes less and less, the effect of the arterial elasticity being to reduce progressively both the rise of pressure in systole and the fall of pressure in diastole. In the capillary bed the pulse pressure has completely disappeared. This is simply another way of stating what we have said earlier, namely, that there is a constant flow of blood in the capillaries instead of an intermittent flow, as in the aorta.

Therefore, in hardening of the arteries, when the elasticity of the vessels is decreased, the arterial pulse pressure is greatly accentuated. Likewise, damage to the semilunar valves, which permits leakage, will cause the diastolic pressure in the arteries to fall very low, approaching the intraventricular diastolic pressure. This again increases the arterial pulse pressure, often to such an extent that its abnormal magnitude can be detected by palpation of an abnormally strong pulse at the wrist. In this way, palpation of the pulse by the physician may aid in the detection of leakage of the semilunar valves.

D. Blood-Pressure Measurements in Man

The use of the mercury manometer as described necessitates exposure and opening of an artery and insertion of a cannula. Obviously, even this method is limited in its applications. It is suitable for work on experimental animals under anesthesia but cannot be employed on unanesthetized animals or on man. However, the mercury manometer can be used in an indirect method of measuring blood pressure.

The principle used is the measurement by a mercury manometer of the external pressure required just to collapse an artery. This pressure is applied to the main artery of the arm by inflating an airtight rubber cuff which is placed about the upper arm. The pressure in the cuff just sufficient to obliterate the pulse at the wrist approximates the systolic pressure.

Diastolic and pulse pressures may also be measured. Suppose we make measurements on a normal person whose pressures are 120 systolic and 80 diastolic. If the pressure in the cuff is raised to above 120 mm. Hg., no blood will flow into the lower arm and no pulse will be felt. If a stethoscope is placed upon the radial artery below the cuff, no sound will be heard. When the cuff has been deflated to a

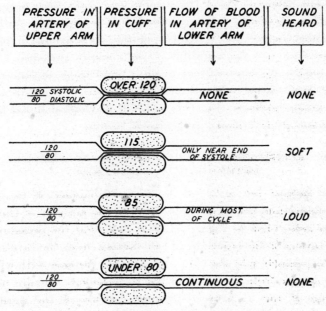

FIG. 56.—Schema illustrating the principles used in determining blood pressure in man by the indirect method. An artery and the compressing cuff (*filled with dots*) are represented in longitudinal section. Pressures are in millimeters of mercury. (See discussion in text.)

point just below 120, the blood in the vessels in the lower arm will remain stationary during most of the cardiac cycle. But at the height of each systole a small amount of blood will be forced past the cuff. This may be palpable in the artery of the wrist but becomes audible with a stethoscope first. The impact of the small spurt of blood upon the stationary column of blood sets up vibrations which are detectable as sound. As the pressure in the cuff is still further lowered, larger and larger spurts of blood pass the cuff at the height of systole, impinge upon and set into motion, for a brief time, the column in the lower-arm arteries. They set up more and more vigorous vibrations, which are heard as louder and louder sounds. Just before the cuff pressure is lowered to 80 mm. the sound with each impact will be maximal. At this time the arteries under the cuff are closed for only a brief period during each cardiac cycle—at the low point of diastole. When the cuff pressure is allowed to fall below 80 mm., blood will pass under the cuff throughout the entire cardiac cycle. The blood in the lower arm will always be moving, and there will therefore be no impact of moving blood upon a stationary column and no sound heard.

The point, then, at which the sounds first make their appearance, as the cuff is gradually deflated from a high pressure, is the systolic pressure, and the point at which the sounds are maximal in intensity, just before they become inaudible, is the diastolic pressure. Comparisons of results by this indirect method with those obtained simultaneously from direct measurements by means of a needle inserted into the artery indicate that this indirect method of blood-pressure measurement gives reasonably accurate results.

V. ADJUSTMENT OF THE PRESSURE TO VARYING PHYSIOLOGICAL CONDITIONS

Several factors are involved in the maintenance of the arterial blood pressure and the adjustment of it to the activity of the organism. If we refer back to Figure 53, it is clear that the pressure of the water in tube *A* is the resultant of two factors: first, the rate at which the fluid is pumped into the tube and, second, the ease with which the fluid can escape from the tube, or the resistance to onward flow into the regions *M* and *I*. Arterial pressure likewise is the resultant of these two factors: the rate at which blood is pumped into the arteries and the resistance to onward flow of blood into the capillaries and veins. The resistance encountered in the arteriolar and capillary regions we refer to as the *peripheral resistance*. Arterial blood pressure, then, depends largely upon the cardiac output and the peripheral resistance.

It is not to be concluded that, because more time will be devoted to these than to other factors, they are necessarily more important to the adaptations or the life of the organism. The volume of the circulating blood, for example, is at least as important as any other factor in maintaining a normal blood pressure (see pp. 204–5). However, the effect on blood pressure of the blood volume is normally of a more continuous, unchanging nature. But cardiac output and peripheral resistance are susceptible of rapid changes, and it is chiefly upon these that adjustments of pressure to the rate of body activity depend.

A. THE CARDIAC OUTPUT

No further discussion of factors which modify the cardiac output is needed here. Any increase in output, whether it be by increased rate or by increased volume ejected per stroke, elevates the arterial blood pressure. Increased production of carbon dioxide or epineph-

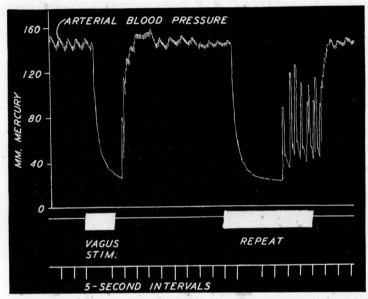

FIG. 57.—Kymograph record of the arterial blood pressure of an anesthetized dog, using a mercury manometer. Stimulation of the peripheral end of the vagus nerve causes a precipitous drop in pressure because the heart temporarily stops pumping blood. The heart cannot be inhibited permanently. As shown in the second part of the tracing, the heart recommences its beat (tall excursions) even though vagus nerve stimulation is prolonged. The small excursions at the beginning of the curve are due to heartbeats. The larger excursions at this point coincide with the breathing movements.

rine secretion, stimulation of the cardiac accelerator nerves, or an increase in return flow of venous blood to the heart, all elevate the arterial pressure, partly because of the resulting augmented cardiac output. Any diminution in rate or strength of the heart by whatever mechanism lowers the arterial blood pressure. Stimulation of the vagus nerve lowers the pressure by decreasing the heart rate (see Fig. 57). Reduced return of venous blood to the heart lowers the pressure by reducing the quantity of blood ejected by the heart per beat.

B. Peripheral Resistance

The peripheral resistance depends in large part upon the caliber of the arterioles, so that the problem of the role of peripheral resistance in blood pressure becomes largely a matter of the factors which control arteriolar vasoconstriction and vasodilatation. These factors have been dealt with briefly in connection with blood supply to organs. Any of the nervous or chemical factors mentioned which

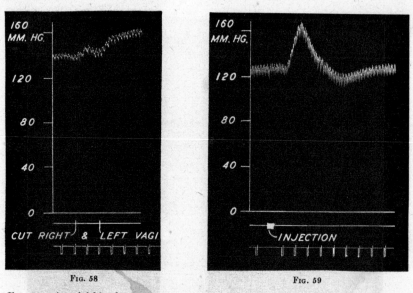

Fig. 58.   Fig. 59.

Fig. 58.—Arterial blood pressure curve of an anesthetized dog demonstrating the continuous inhibiting action of the vagus nerve on the heart. Cutting the vagus nerves (*indicated by signal*) releases the heart from the continuous brake action of the vagi. Its rate therefore increases, more blood is pumped into the arteries, and the blood pressure rises. Other factors are also involved in this rise. Time is shown in 5-second intervals.

Fig. 59.—Effect of epinephrine injection upon arterial blood pressure of an anesthetized dog. The increase in pressure is caused largely by the vasoconstriction induced by epinephrine increasing the peripheral resistance. In this experiment only a millionth of a gram of epinephrine was injected into a vein. Time is shown in 5-second intervals.

induced a widespread vasoconstriction will increase the peripheral resistance and elevate the blood pressure. Injection of epinephrine in adequate amounts, or the stimulation of a large vasoconstrictor nerve, will have such an effect. Any widespread vasodilatation is accompanied by a fall in blood pressure. Of course, if vasomotor changes occur in a highly restricted area, little or no change in the general arterial pressure may be produced. The vasodilatation which occurs in the salivary glands during secretion, for example, though it markedly increases the blood supply of those glands, opens up widely so small a proportion of the vascular pathways that the effect upon the total peripheral resistance is small or nil.

1. *Viscosity.*—The caliber of the vascular bed, especially of the arterioles and capillaries, determines the "external" resistance, which is dependent upon the friction of the moving blood against the vessel walls. Of great importance also is the factor of "internal" resistance to flow, or viscosity, which is an expression of the friction resulting from the sliding of molecules of fluid past one another in a moving stream. As blood (or any liquid) flows through a vessel (or any tube), the rate of flow is different at different points along the diameter of the stream. The external layer of fluid, in contact with the stationary wall, moves very slowly and encounters a certain external frictional resistance at the wall surface. The layer of fluid just within this outer layer moves a little faster than the latter and, consequently, encounters a certain frictional resistance of molecules of fluid moving past other molecules of the same fluid. As we go from the vessel wall inward, we find that the rate of flow becomes more and more rapid. Only in an innermost axial stream is there a column of fluid in which concentric layers are moving at the same rate. Here only, of the whole cross-section of the stream, are there no sliding of molecules past one another and no internal friction.

The viscosity of a fluid is an expression of the resistance of molecules of the fluid sliding past one another. The greater the viscosity, the greater is the internal friction, and the less readily the liquid will flow. If the fluid be sufficiently viscous, there will be no axial stream of appreciable cross-sectional area in which internal friction is absent. Figure 60 shows these relationships.

From this we can see that the important frictional effect of changing the vessel caliber in the case of a moving fluid of constant viscosity like blood is chiefly a matter of modifying the total internal friction. This is done not by any modification of the viscosity of the blood per se but by a diminution in the diameter of the frictionless axial stream of constant flow, as shown in Figure 60 (lower figures).

The importance of viscosity in maintaining the proper arterial blood pressure is seen when lost blood is replaced by a less viscous fluid such as physiological saline solution. After an experimental or accidental hemorrhage, if an equal volume of saline solution is injected to replace the lost volume, it may be impossible to restore the blood pressure to its former level, partly because the salt solution rapidly escapes from the blood vessels into the tissues, but partly also because of the relatively low viscosity of such a solution. Attempts have frequently been made to correct this defect of saline solution in combating the effects of extensive hemorrhages by adding substances to the solution which give it approximately the viscosity of blood.[10] One reason that blood itself is the best fluid to replace the lost volume after hemorrhage is due not only to the

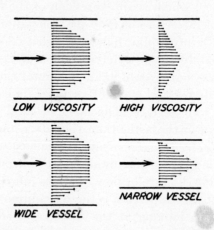

FIG. 60.—Relationships of viscosity, rate of flow, and vessel caliber. Arrows represent the pressure forcing liquid through various tubes (*shown in longitudinal section*). The lengths of the horizontal lines represent differing rates of flow at different points along the diameter of the tube. The two lower figures represent liquid of the same viscosity flowing through a wide and a narrow vessel.

oxygen-carrying red corpuscles it contains but also to its proper viscosity.

2. *Tonic action of vasoconstrictor nerves.*—The nervous factors controlling peripheral resistance need further elucidation at this time. Recall that many blood vessels are maintained in a state of tonic constriction by a continuous train of nerve impulses in the vasoconstrictor efferent nerves. Let us trace experimentally the origin and cause of these impulses. Cutting the vasoconstrictor nerves will cause a dilatation of the denervated vessels. If the cut nerve is a small one, and the ensuing dilatation therefore localized, there will be no effect upon the blood pressure. But, if the severed nerve is large, containing many tonically active fibers distributed to rather widespread areas, a definite fall in arterial blood pressure results. The

10. Blood owes its high viscosity largely to the formed elements. The viscosity of the plasma is due largely to the plasma proteins.

extensive dilatation reduces the total peripheral resistance sufficiently to produce this result. Blood pumped into the arterial system by the heart escapes more quickly through the widened arterioles and capillaries into the veins.

It is also found that, if the spinal cord is cut across in the chest region experimentally, or by accident in man, a rather general dilatation of most of the blood vessels of the body occurs. This is as extensive as if all the efferent vasoconstrictor nerves of the body below the chest region were cut. There is a simultaneous profound drop in blood pressure. No matter where this cut is made in the upper spinal cord, the same fall in pressure occurs.

However, if the spinal cord together with the medulla oblongata are disconnected from the rest of the brain above, by cutting across just above the medulla, no such marked vascular dilatation or fall in blood pressure occurs. In an animal of this sort, one may then cut away successive slices of the medulla from above downward. Quite suddenly, the removal of one thin slice of medulla will produce a great generalized vasodilatation and fall in blood pressure. We may conclude from this that the origin of the tonic vasoconstrictor impulses is in the medulla and that from here they pass downward through fiber tracts in the spinal cord, from which they ultimately emerge (chiefly in the thoracic and lumbar regions) to pass on into the

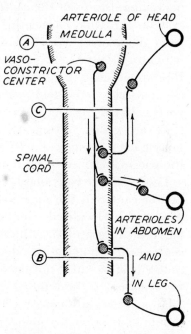

Fig. 61.—Nervous connections between the vasoconstrictor center in the medulla and vasoconstrictor nerves of various arteries. A cut across the medulla at A or across the spinal cord at B will not sever the connections between the vasoconstrictor center and the blood vessels. A cut across the spinal cord at C will completely remove the blood vessels from control by the vasoconstrictor center. Arrows indicate the course of nerve impulses from center to blood vessels.

peripheral vasoconstrictor nerves distributed to all parts of the body. Figure 61 shows these anatomical features diagrammatically. It is plain from this diagram that a cut at level A or at B will not appreciably affect the vasoconstrictor system. But a cut between A and

B will as completely remove the blood vessels from tonic vaso-constrictor control as will cutting the peripheral nerves themselves.

3. *The vasomotor centers.*—The cluster of nerve cells in the me-dulla, whence these impulses come, is called the *vasoconstrictor center*. The real explanation of tonic vasoconstriction and of the resulting constant high blood pressure lies in the continuous discharge of nerve impulses from the center through the efferent vasoconstrictor system. But what causes this sustained discharge of nerve impulses from the center? We shall see that, as in the case of the cardio-regulatory centers, this center also is affected by afferent nerves. But apparently the tonic action of the vasoconstrictor center is not entirely dependent upon such stimulation through afferent nerves. In this respect its action resembles that of the pacemaker of the heart, which continues intermittently to send stimuli to the heart muscle even though all extrinsic sources of nervous influence upon the pacemaker have been removed.

Less evidence is available for the existence of a dominating vaso-dilator center in the medulla. Generalized vasodilatation has been produced by localized stimulation of an area of the medulla, which has been interpreted as indicating the location at this point of a center controlling the vasodilator nerves of the body generally. This center is apparently not engaged in any appreciable tonic activity. It seems not to be sending impulses out over the peripheral vaso-dilator nerves in a constant stream. If such tonic action existed, it should be possible, upon cutting a vasodilator nerve, to observe the *opposite* of vasodilatation, that is, vasoconstriction. This does not occur.

Some of the vasodilator reflexes (e.g., those involving the erectile tissue of the genital organs) occur even though the spinal cord is cut across, removing all influences from the medullary centers upon the spinal-cord reflex centers. This suggests that the medullary vaso-dilator center exerts far less control over the spinal centers than is true in the constrictor system.

4. *Action of carbon dioxide.*—As we failed to determine the cause of the automatic rhythmicity of the cardiac pacemaker, so also are we at something of a loss to account for the automatic rhythmic nervous discharges from the vasoconstrictor center. Chemical fac-tors are undoubtedly of great importance here.

We know that a slight excess of carbon dioxide in the blood, and to a lesser extent other acid products, increases the action of the center. It is stimulated chemically and increases its vasoconstrictor discharges, and the blood pressure rises. On the other hand, when the carbon dioxide of the blood is abnormally reduced, the vasoconstrictor impulses are decreased in number, and the arterial pressure falls. This condition is easily produced experimentally in man by voluntarily breathing deeply and rapidly. The ensuing fall in blood pressure may be so marked that the circulation to the brain is impaired, and giddiness or even fainting may result. Here we see additional examples of uses to which the organism puts carbon dioxide, sometimes considered as merely a useless waste product. With less than a normal concentration of carbon dioxide in the blood, the activity of the vasoconstrictor center is impaired, the blood pressure falls, and the circulation is dangerously slowed.

We have seen that an excess of carbon dioxide, as in muscular exercise, increases the output of the heart and, in addition, acts directly upon the blood vessels of the tissues where it is produced, effecting a vasodilatation. This is supplemented by the stimulating action of the compound upon the vasoconstrictor center. There are produced a rather diffuse nervous vasoconstriction and an elevation of the blood pressure. However, those tissues in which the excess carbon dioxide is being produced are exempted from the generalized vasoconstriction, for in this region, and only in this region, the carbon dioxide is sufficiently highly concentrated to cause relaxation of the vessel walls.

Summarizing, we may say that in an active muscle carbon dioxide is produced in large amounts. Locally, it produces a vasodilatation. Some of it diffuses into the blood stream, slightly elevating the concentration of the compound here. Circulating through the body and through the capillaries of the vasoconstrictor center, it stimulates the cells of the center, increasing their vasoconstrictor discharges. When the stimulating effect of carbon dioxide upon the output of the heart is added to this, we see how effective is the mechanism for increasing the circulation through active muscles. Not only is more blood pumped by the heart per minute but the arterial blood pressure is elevated by the general vasoconstriction, and blood is shunted from the inactive regions and forced under increased pressure through the vessels of the active muscles, which alone are widely dilated. Later we shall encounter still further instances of the role of carbon dioxide in the animal economy and its indispensability to life itself.

5. *Vasomotor reflexes.*—Though the vasoconstrictor center is apparently able to maintain a great deal of its tonic action independently of nervous influences playing upon it, there is no doubt that its action is greatly affected by them. Such influences may reach the center from higher regions of the brain or from afferent systems and may induce either rather generalized or quite localized vasomotor changes. The vasomotor effects of certain conscious processes are well known. The sight of blood for the first time may arouse very disturbing emotions, and the vasomotor system may be so affected that a general vasodilatation and a profound drop in blood pressure occur, causing temporary anemia of the brain and fainting. Excitement may stimulate the constrictor center through nervous channels, elevating the blood pressure. Blushing and erection are further vasomotor phenomena which are induced by conscious processes, though these are sufficiently localized to prevent any fall in blood pressure.

The vasoconstrictor center is also influenced by impulses reaching it through afferent nerves. It may be activated or inhibited reflexly. We have already seen that the concepts "center" and "reflex" are inseparable. By "center" we mean "*reflex* center." The term "vasoconstrictor center" correctly implies that it is a region on which many afferent systems converge and from which numerous efferent systems diverge. There are several vasomotor reflexes, however, which apparently do not involve the medullary centers. Some of the reflexes mentioned earlier occur through the spinal cord and are not abolished by cutting across the spinal cord just below the medulla. Thus there are vasomotor centers in the spinal cord as well as in the medulla. The latter are generally of greater importance, however, and the majority of the reflexes, particularly those which affect extensive areas and involve definite blood-pressure changes, pass through the primary vasoconstrictor center in the medulla.

Afferent fibers[11] may influence the center in one of two ways: they induce either a rise or a fall in arterial blood pressure. Almost every afferent nerve in the body has fiber connections with the vasoconstrictor center, and stimulation of almost any afferent nerve may therefore affect the blood pressure. The more common effect is a stimulation of the center, inducing a reflex rise of blood pressure.

11. Afferent nerves whose stimulation causes a reflex vasoconstriction are called *pressor nerves*. Those which produce reflex vasodilatation are called *depressor nerves*. This terminology distinguishes these afferent nerves from the vasoconstrictor and vasodilator nerves themselves, which are all efferents.

This is particularly true of strong stimulation, which in the intact unanesthetized animal would be felt as pain. In general, stimuli strong enough to produce pain also reflexly raise the arterial blood pressure. The utility of this probably lies in the fact that the elevated blood pressure makes possible an immediate increase in flow to the skeletal muscles which are likely to become active (yielding carbon dioxide with its local vasodilating action) when painful stimuli provoke fight or flight.

Afferents which cause a reflex fall in blood pressure are also widely distributed, although they are often more difficult to demonstrate.

6. *The modulator nerves.*—Most afferent nerve trunks contain both pressor and depressor fibers; but, when the trunk as a whole is stimulated, the pressor effect usually predominates and masks the action of any depressors which may also have been stimulated. There are two notable exceptions to this in the cases of the *aortic depressor* and the *carotid sinus* nerves. These are purely sensory (rather, afferent) nerves; and, when they are stimulated, there always occur a reflex vasodilatation and a fall in blood pressure. These nerves have been studied extensively, and their important and interesting role in blood-pressure regulation has been fairly successfully elucidated.

The aortic depressor nerve is a branch of the vagus and, as its name implies, innervates the aorta. Its fibers ramify in the wall of the aorta just at its root, where it arises from the heart. The network of nerve-fiber terminations is so arranged as to be easily stimulated mechanically by distention of the aorta. In fact, with each beat of the heart, which forces blood into the aorta, the resulting distention stimulates these sensory endings, and nerve impulses can be detected passing up the depressor nerve. At each cardiac systole a volley of electrical action potentials appears. Whenever there is an elevation of the arterial blood pressure, whether it be from increase in the cardiac output or in peripheral resistance, the nerves are likewise stimulated.

Knowing what constitutes an effective stimulus to the depressor nerves, we may inquire further into the results of such stimulation. So we cut the depressor nerve and stimulate its central end in an anesthetized animal whose blood pressure is being recorded. Here we artificially produce afferent impulses exactly like those produced by distention of the aorta physiologically. The effect is an immediate slowing of the heart and a fall in blood pressure. The fall in blood pressure is only partly dependent upon the reflex cardiac slowing, for a fall in pressure still occurs even though the vagus nerves of the heart have been cut, in which case depressor nerve stimulation does not inhibit the heart. The fall in blood pressure must therefore be partly due to a vasodilatation, induced reflexly through the vaso-motor centers.

At the point at which the large artery carrying blood to the head (the *common carotid*) divides into the branch which supplies the external parts of the head (the *external carotid*) and the brain (the *internal carotid*), there is located a special structure called the *carotid sinus*. This is really a small, thin-walled swelling of the internal carotid artery just at the point at which it branches off from the common carotid. Its walls are richly supplied with nerve fibers which are the terminal sensory ramifications of the carotid sinus nerve. This, like the aortic depressor nerve, is purely afferent.

The physiological function of the carotid sinus is apparently exactly the same as that of the aortic depressor, as revealed by

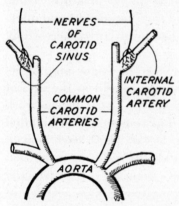

experimentation. One can perform crucial experiments here quite easily because the sinus, unlike the aorta, is located in a more accessible place in the neck and is off the course of the main blood flow. It is easy to vary the pressure within the sinus without directly affecting the pressure in other regions of the body. When this is done, we find again that a local elevation of internal pressure here causes a slowing of the heart and a drop in the general arterial blood pressure. A rapid volley of action potentials can be detected in the nerve at this time. If the nerve is cut, the general arterial blood pressure is unaffected by elevation of the pressure in the sinus. Thus we know that this is a reflex phenomenon, in which the carotid sinus nerve comprises the afferent side of the arc. We discover what efferent nerves are involved by the same principle of cutting nerves. If the sinus nerve is stimulated after the vagi are cut, cardiac slowing fails to occur. The efferent involved in the slowing was therefore the vagus. After the vagi are cut, however, the arterial

FIG. 62.—The carotid sinuses and their afferent nerves. These nerves, along with the aortic depressor nerves (see Fig. 46, p. 163), are called the *modulator nerves*.

pressure still falls when the sinus nerve is stimulated. Vasodilatation accomplishes this. Blood vessels whose efferent innervation has been destroyed fail to share in the dilatation, so we conclude that the efferent vasomotor fibers are also involved in the reflex.

A clear understanding of the significance of these nerves is possible only if we appreciate the fact that physiologically they are in continuous tonic activity. The normal arterial blood pressure distends the aortic arch and the carotid sinus sufficiently to initiate a con-

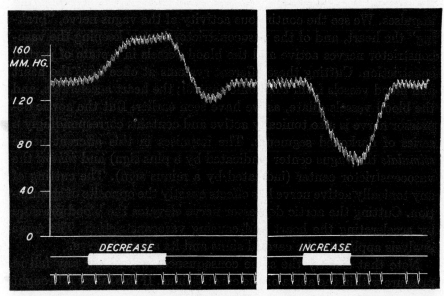

Fig. 63.—Effect of artificially decreasing and increasing the pressure within the carotid sinus upon the general arterial blood pressure of an anesthetized dog. A decrease of pressure (*left*) in the sinus causes a reflex vasoconstriction (and acceleration of the heart), producing an elevation of arterial pressure. An increase of pressure (*right*) in the sinus causes reflex changes in the reverse direction.

tinuous series of these depressor impulses. We have said that they can be detected directly with electrical instruments which measure the action potentials. This means that the vagus and vasoconstrictor centers are being continually bombarded by afferent impulses, which exert both a stimulation of the vagus center and an inhibition of the vasoconstrictor center.[12] Remembering a previous statement that the vagus center is also in tonic action, exerting a constant braking effect upon the heart, we see here an explanation of part of that tonic action. It is partly reflex, dependent upon stimulation of the

12. Or a stimulation of the vasodilator center. This will be discussed at greater length later.

vagus center by the tonically active depressor nerves of the arch of the aorta and the carotid sinus. Recall also that the vasoconstrictor center was shown to be in continuous activity. Here, however, the effect of the depressor nerves is to decrease that activity; that is, the tonic action of the vasoconstrictor center is being continually diminished by the afferent impulses of the depressor nerves.

Figure 64 (1) shows this situation for the aortic depressor nerve in diagrammatic form. The shaded segments represent nerve impulses; the spacing of the segments, the rate at which the nerve is conducting impulses. We see the continuous activity of the vagus nerve, "braking" the heart, and of the vasoconstrictor center, keeping the vasoconstrictor nerves active and the blood vessels in a state of partial constriction. Cutting these efferent systems at once frees the heart and blood vessels from this tonic control; the heart accelerates, and the blood vessels dilate, as we have seen earlier. But the aortic depressor nerve is also tonically active and contains correspondingly a series of darkened segments. The impulses in this afferent nerve *stimulate* the vagus center (indicated by a plus sign) and *inhibit* the vasoconstrictor center (indicated by a minus sign). The cutting of any tonically active nerve has effects exactly the opposite of stimulation. Cutting the aortic depressor nerve elevates the blood pressure by accelerating the heart and causing vasoconstriction. The same analysis applies to the carotid sinus and its afferent nerve.

Note that exactly the same consequences will follow any fall in pressure within the aorta or the carotid sinus. If this latter is brought about experimentally by simply pinching off the common carotid artery, cutting off the normal blood pressure to the sinus, an immediate reflex rise in blood pressure ensues.

But what does all this mean physiologically? Granted that these effects may be induced experimentally, how far are they applicable to the normal intact man or animal, and under what conditions and to what ends do the mechanisms operate normally?

First of all, they act as governors of the heart rate. They help to maintain the heart rate at that constant level which characterizes it as a physiological constant. Why does the heart of a resting individual continue beating rather evenly at the constant rate of about 70 per minute? If the rate should accelerate "spontaneously," more blood would be forced into the aorta and its branches, including the carotid artery and sinus. This distends them and stimulates the aortic depressor and sinus nerves, which, in turn, stimulate the

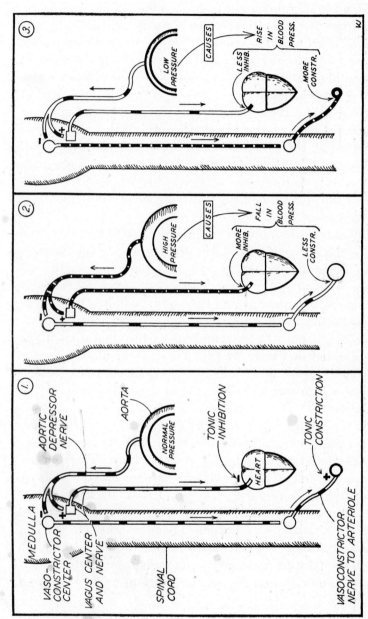

FIG. 64.—Schema of reflex action of aortic depressor nerve upon blood vessels and heart rate. Nerves are indicated by double lines. Nerve impulses are shown as shaded segments. Arrows indicate direction of nerve impulses. The spacing of the impulses indicates the frequency of impulses in the nerves. Plus signs indicate a stimulation; minus signs, an inhibition. The same reflex effects are produced by changes in pressure within the carotid sinus. Note that, in all cases, impulses in the depressor nerve inhibit the vasoconstrictor center and stimulate the vagus center. The extent of these two effects depends upon the number of impulses in the depressor nerve, which are moderately numerous in part 1, many in part 2, and few in part 3. (See discussion in text.)

vagus center, slowing the heart to its former rate. Likewise, if the heart should tend to slow down, the pressure would fall in these sensitive vascular areas. Both the tonic action of the nerves and the vagus inhibition of the heart would be reduced. Very soon the slowing heart would increase to its former rate. Any decrease from its constant rate, in this manner, automatically sets into action machinery which tends to correct the change.

In the second place, these mechanisms operate to preserve the physiological constancy of the blood pressure. When the arterial pressure tends to fall, there is a reflex vasoconstriction; when it tends to rise, a reflex vasodilatation is effected. In either case, a readjustment of the pressure to its former level is achieved. Perhaps a specific example of the operation of these two devices would make their significance more clear.

When a human being lies down, changing the axis of his body from the vertical to the horizontal position, gravity assists the heart to send blood to the brain. Rather, gravity no longer opposes the upward pumping of blood by the heart. Expressed otherwise, the pressure of the blood in the upper part of the body, including that in the arch of the aorta and the carotid sinus, is increased. Distention in these regions reflexly slows the heart and causes vasodilatation, in the manner diagrammed in Figure 64 (2). The carotid blood pressure falls to its previous level, and an adequate, yet not wasteful, circulation to the brain continues. Now consider what happens in the opposite situation.

A man who has been lying in a horizontal position, with an adequate blood pressure, suddenly stands upright. Gravity now opposes the action of the heart in pumping blood to the brain. The pressure in the carotid artery falls at once. So inadequate may be the pressure that insufficient blood is supplied to the brain, and "darkness before the eyes," giddiness, or even fainting, may occur. Of course, fainting itself corrects the situation and restores adequate brain circulation. But, if the resourcefulness of nature were exhausted in furnishing the organism with the fainting mechanism, man might be limited to the maintenance of the body in a horizontal position, or, at best, to a crawling existence.

If, as is usually true, we do not faint upon assumption of the upright position, we find that the sensation of giddiness, if it occurs at all, soon passes off. Here, the circulatory adjustment diagrammed in Figure 64 (3) has taken place. The reduced arterial pressure diminishes the constant stimulation of the aortic depressor or carotid sinus nerve, the vagus is less stimulated, and the heart accelerates. Also, the vasoconstrictor center is less inhibited, and vasoconstriction is increased. The net effect is an increase in blood pressure sufficient to offset the action of gravity. Reflexes of this sort are occurring again and again in the course of a normal day, along with the many changes in bodily posture.

In this discussion we have spoken of vasodilatation as a diminution of vasoconstriction. Active vasodilatation by positive stimulation of vasodilator nerves is also affected by sinus and aortic depressor reflexes. Thus we may look upon these afferent depressor nerves as *stimulating* the vasodilator center as well as *inhibiting* the vasoconstrictor center.

7. *Constancy and change.*—In these elaborate mechanisms we encounter an apparent contradiction. Here we have devices which serve to maintain a *constancy* of the heart rate and blood pressure. Yet, earlier, considerable time was devoted to study of those factors which *change* the heart rate and blood pressure. Actually, the constancy is maintained in the prevention of inadvertent, needless, wasteful, or extreme changes in the circulation. But when, as in exercise, active agents such as excitement, reflexes, epinephrine secretion, carbon dioxide production, etc., appear, the effect of these upon the circulatory system is such as to offset the leveling action of the machinery for maintaining constancy. The positive stimulation of the cardiac pacemaker by epinephrine, for example, may be so great as to overcome the effects of the aortic depressor and carotid sinus nerves. Unless such relationships existed, vascular changes of any sort, upon which muscular contractions and movements of the body are intimately dependent, would be impossible. Evolution has selected for the organism devices which serve to maintain certain constants but which are not so inflexible in their operation as to limit the organism to a motionless, vegetative existence.

We must think of the physiological constants in terms of cell environment. During rest this is maintained by a constancy in the cardiovascular system. But a continued maintenance of *constancy* in the cell environment in activity is dependent upon *changes* in the circulatory and other systems. Though heart rate and blood pressure are physiological constants in their own right, their constancy is subordinated to the maintenance of a constancy of more fundamental importance, namely, that of the cell environment.

C. Blood Volume

1. *Relationship to blood pressure.*—There is a direct relationship between the total volume of the blood and the arterial pressure, which is of a purely physical nature. In any closed elastic system of a given capacity, the more fluid it contains, the greater will be the pressure of that fluid. This applies whether or not the fluid is in motion. Thus, if any great diminution in blood volume occurs, the arterial pressure may fall to such an extent that an adequate circulation cannot be maintained. The immediate danger to life in a sudden

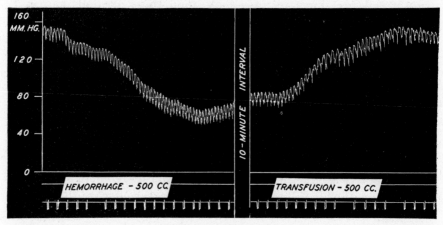

Fig. 65.—Effect of hemorrhage and of transfusion upon the arterial blood pressure of an anesthetized dog, recorded by a kymograph. Time in 5-second intervals.

extensive hemorrhage is not the loss of red cells but the fall in arterial pressure.

In such emergencies there is usually a compensatory generalized vasoconstriction, raising the arterial pressure of the blood which remains.[13] Artificial remedial measures are aimed at correcting the blood-volume defect at once, and fluids are administered extensively by mouth, under the skin, and, most effectively, directly into the circulation through a vein. Of course, a transfusion of blood itself would be the most efficacious procedure, for this would restore lost cells as well as fluid. But in an emergency, when considerable human

13. Effected partly by the modulator nerve mechanism (see p. 197) and probably also by epinephrine, which is apparently secreted in greater amounts at such times. Any abnormal increase in blood volume, which would elevate the pressure, tends also to be counteracted by a vasodilatation via the modulator nerve reflexes.

blood is not available for transfusion, salt solution of the proper osmotic concentration may elevate the pressure sufficiently to prevent death.

The importance of blood volume to life itself in the maintenance of arterial blood pressure stresses the importance of those physiological mechanisms which help to maintain the blood volume: the filtration-osmosis water balance between tissues and blood (see p. 70); kidney secretion, which diminishes when the blood volume and blood pressure are reduced (see p. 363); and the ingestion and absorption of water in the alimentary canal.

2. *Splenic contractions.*—The spleen contains a reserve supply of blood. In this organ there are cavities or sinuses, communicating with the capillaries, in which a considerable quantity of blood is stored, off the pathway of the circulating stream. In the splenic tissue a great deal of smooth muscle is to be found. Contraction of this muscle reduces the volume of the organ, squeezing quantities of the stagnant blood from its meshes into the circulating stream. The precise mechanisms which operate to cause the spleen to contract have not been completely worked out, although both nervous and chemical factors apparently operate. In some animals, during exercise and after hemorrhage, the spleen has been observed to contract. The resultant increased circulating blood volume serves to increase the supply of blood available for the active muscles, in the one instance, and helps to replace lost blood, in the other. Whether splenic contractions function significantly in man, in this way, has been seriously questioned.

D. GRAVITY

Gravity exerts a definite hydrostatic effect upon the circulation. Changes in the bodily posture and in the relationship of the body axis to the direction of gravitational pull alter the pressure of the blood in various parts of the body. In general, the pressure in the vessels of the highest parts of the body is less than that in the lower regions. The blood in the foot may be considered to derive its total pressure from two sources: the beat of the heart plus the weight of a column of blood equal to the distance from foot to heart. In the arteries, where the component of pressure due to the heart is great, the hydrostatic factor contributes a smaller percentage to the total pressure than it does in the veins or capillaries.

Nonetheless, when an individual suddenly assumes an upright posture, the pull of gravity upon the blood may, directly and indirectly, so reduce the pressure in the arteries supplying the head that the circulation to the brain is temporarily inadequate. The "conscious" activity in the brain may temporarily stop. Vision fails, "things get black before the eyes," and fainting may be the consequence. Note that here fainting may be of adaptive value. By this mechanism the long axis of the body is automatically restored to the horizontal position, and the resultant increased blood flow to the brain, no longer moving against gravity, prevents serious damage. When someone in a group faints, there is a universal impulse to frustrate this scheme of nature by frantic attempts to prop up the collapsed victim. Obviously, he should be permitted to lie down, and, if possible, he should be so placed that his head is lower than the rest of the body.

The fainting device is, of course, only a last resort. Usually other mechanisms come into play, and adjustments occur which maintain an adequate circulation to the brain, even though the posture remains upright (see p. 202). These mechanisms are ordinarily set into motion so quickly and effectively that any even temporary impairment of the circulation to the brain is the exceptional result of postural change.

The effect of gravity upon the blood pressure in the veins of the arm has been experienced by everyone. Let the arm hang down at the side, and the veins of the hand distend with blood under pressure. Raise them above the heart, and the veins become nearly collapsed. Injuries involving the tearing of vessels of the hand illustrate the same thing. Because of the gravitational effect, a cut finger bleeds more profusely when the hand is held down at the side than when elevated.

Brain surgeons often utilize the influence of gravity by performing their operations with the patient in the sitting posture. Gravity then assists in diminishing hemorrhage by its action upon the pressure in the brain vessels.

VI. SUMMARY OF THE EFFECTS OF EXERCISE

The response of the circulatory system to muscular exercise furnishes an excellent example of the manner in which blood flow to an organ is adjusted to the rate of activity of that organ. The mechanisms involved have been presented in the text above, but it might be useful to recapitulate at this point. We may thereby gain a better insight into the course of events and a better appreciation of the automatically operating machinery involved. We know that muscular exercise causes an increased blood flow through the muscles. This increased flow is effected by all those adjustments which we might enumerate a priori from a consideration of the mechanics of the artificial circulation diagram in Figure 53 (p. 177): an accelerated heart rate, an increase in the volume of blood ejected at each stroke, a vasodilatation in the skeletal muscles, and a vasoconstriction in the internal organs (plus, in some animals, an increase in the circulating blood volume).

Our problem now is to see what makes these reactions occur. Just what components of muscular exercise effect these changes? Let us first consider exactly what are the immediate effects of exercise. There are chemical reactions which involve the liberation of heat and of carbon dioxide and other waste products. There is an increased kneading action of muscles upon the veins. The spleen may also contract. There is usually a certain amount of excitement. So we must look to the heat, carbon dioxide (and other metabolites), the kneading action of the muscles upon the veins, possibly splenic contraction, excitement, and probably increased epinephrine secretion for an explanation of the results. Even these primary effects of exercise are interrelated, for carbon dioxide apparently stimulates epinephrine secretion, which, in turn, effects splenic contraction. Let us now examine the end results one by one and see how they might be caused by these concomitants of muscle contraction.

The accelerated heart rate is effected through nervous channels by the excitement and by reflexes and perhaps through stimulation of the pacemaker by the heat and epinephrine.

The increased stoke volume is a consequence of the greater degree of filling of the heart in diastole (Starling's law), which is caused by the increased kneading action of the muscles and by the increased action of the respiratory movements in aspirating venous blood to the heart. We shall see later that the accelerated breathing is, in turn, dependent upon the carbon dioxide produced in the exercise. Carbon dioxide itself seems also to increase the strength of the heartbeat.

The vasodilatation in the muscles is caused mainly by a local action of carbon dioxide and, perhaps also, other metabolites such as lactic acid.

The vasoconstriction in the internal organs we ascribe to the action of carbon dioxide upon the vasoconstrictor center and possibly also to epinephrine acting upon the vessels themselves.

Finally, the increased circulating blood volume is explainable on the basis of splenic contraction, at least in some animals.

We see that, by these somewhat complicated and interrelated devices, muscle action sets machinery in motion which automatically increases blood flow, making possible a continuation of that action. This is a far cry from saying that the blood supply to muscle is increased in activity because the muscle needs more blood.

VII. COMMON DISORDERS OF THE CIRCULATION

In the chapter on the heart, derangements of the circulation affecting that organ were discussed. Disorders of the blood vessels are also fairly common, and their effects are to be understood on the basis of the derangements of blood-vessel physiology which they entail. Some blood-vessel defects have already been referred to in connection with intravascular clotting of blood. Plugging of an infected blood vessel with a thrombus might serve the useful end of decreasing the likelihood of bursting of the damaged vessel, but it may also produce serious damage or death. Plugging of vessels to parts of the brain or to the heart may be suddenly fatal.

The effects of occlusion depend, first, on the importance of the organ whose vessel is occluded and, second, on whether or not other vessels also carry blood to that organ, as is generally the case.

Allied in their effects to those of complete obliteration of vessels are conditions which abnormally narrow the caliber of vessels. This may be in the nature of a more or less localized blood-vessel spasm, or a "hardening"—actually a loss of elasticity—and thickening of the walls of the arterioles and arteries. Again, the effects depend upon what blood vessels are involved. Hardening (and narrowing) of the arteries of the brain may so interfere with proper nutrition of that organ as to cause serious mental derangements. Similar affection of the heart blood vessels may be rapidly fatal. Involvement of the kidney vessels may so damage those organs as to make impossible the proper elimination of wastes.

Hardening of the arteries is essentially a disease of the old. It seems in part to be a specific manifestation of the unexplained but rather general loss of elasticity of many tissues in old age. This effect upon the skin of the aged is known to all, but many aspects of hardening of the arteries are puzzling. The narrowed caliber of the vessels and the change in structure of the walls have not as yet been satisfactorily accounted for.

Usually associated with hardening of the arteries and narrowing of the internal caliber of the vessels is a chronic elevation of the arterial blood pressure.[14] The systolic pressure may rise to 250 or 300 mm. Hg. By some, this is looked upon as a compensatory adjustment, by which blood is forced through the narrowed vessels, and a more or less adequate circulation is maintained. The seriousness of high blood pressure (*hypertension*) lies partly in the danger that some blood vessel may rupture. Rarely does this occur in large vessels, with dangerous extensive hemorrhage. More often it involves vessels of such size that the loss of blood per se is not important. But if this bleeding occurs in a vital, and particularly a friable, structure like the brain, serious damage might result. Many brain cells may be torn and damaged by the blood escaping under high pressure. Everyone knows of individuals who have suffered such a "cerebral accident," which often causes paralysis, and is commonly known as a "stroke."

The rationale of having people with high blood pressure lead as quiet a life as possible is obvious when we recall that muscular exercise and excitement elevate the arterial blood pressure. This, of course, would increase the danger of rupturing a blood vessel. It should be realized that to lead a quiet life involves more than a reduction of physical exercise. Emotional tensions may be major factors in producing the vasoconstriction of hypertension, and emotional equanimity—aided by psychotherapy—may be as difficult and as important to achieve as is avoidance of excessive physical exercise.

Hypertension may also be fatal because of heart failure. Unable to eject sufficient blood against the high peripheral resistance, the overloaded, overtaxed heart may gradually dilate more and more and eventually fail entirely.

The less common defect of chronic low blood pressure (*hypotension*) becomes incapacitating when the reduced arterial pressure fails to drive the blood through the tissue and organ capillaries at an adequate rate. It is encountered in some endocrine disorders, in the latter stages of heart failure, in certain nervous disorders, and in debilitating diseases such as advanced tuberculosis and cancer.

14. See also p. 366 for relationship of kidneys to chronic high blood pressure.

Traumatic shock is a condition following injury or blood loss, resulting from surgery, wounds, burns, bomb damage, and a variety of traumatic insults to the body. The major component of this complex is a low arterial blood pressure and a consequent inadequate circulation or even a complete circulatory collapse, too commonly fatal.

Probably the most important cause of this low blood pressure is the loss of *circulating* blood, which may be very obvious, as in external bleeding and loss of blood volume from an injury, or it may be highly disguised. An example of the latter is the loss of blood, or at least plasma, into a traumatized part of the body in which extensive capillary damage provides microscopic or larger openings for blood or plasma to escape from the vessels into the tissues of the injured part. Similar losses of blood liquid also occur in extensive burns, in which great volumes of plasma liquids ooze through the injured area.

Under the conditions described the circulating blood volume is reduced, and the cause of the low blood pressure and means for its correction are clear. But sometimes shock occurs when there is too little obvious loss of blood volume to explain the low blood pressure. The cause of shock under these conditions remains an unsolved problem.

Any defect of the vessel wall is a serious threat to life, especially if the vessel is large or located in a vital organ. One of the many structures which is attacked by the organism of syphilis is the aorta. There is often damage to the semilunar valves in this infection. The organisms may also so weaken the aortic wall that a large thin-walled outpocketing of the vessel develops. A sudden elevation of blood pressure may burst the vessel at this point with a rapidly fatal hemorrhage.

Large swellings called *varicosities*, particularly in the legs, are common. These have no relationship whatsoever to syphilis. They seem rather to be dependent upon an inadequate return flow of venous blood to the heart. Probably a loss of elasticity of the veins with age is a factor here also. Certainly, gravity plays an important part, for varicosities are almost always limited to the legs, and seem to occur more frequently in people who must stand a great deal. This deprives the circulation of the action of alternate contractions and relaxations of muscles, so important in pumping venous blood against gravity.

VIII. THE LYMPHATIC CIRCULATION

A. Lymph and Lymph Vessels

The lymphatic system is a circulatory system having rather intimate anatomical and physiological interrelationships with the blood circulatory system, similar to it in certain respects and quite different in others. Lymph vessels, like blood vessels, are distributed to nearly all parts of the body. We may liken the system anatomically to the capillaries-plus-veins portion of the blood circulatory system. The lymphatic system possesses no counterpart of the arteries and, consequently, does not have a true continuous closed circulation. The

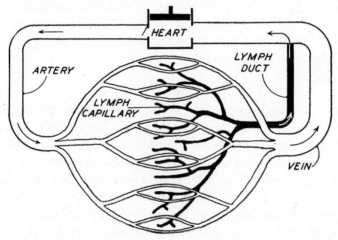

Fig. 66.—Relationship of the blood circulatory and lymphatic systems (*latter shown in black*). Arrows indicate direction of blood or lymph flow.

liquid of the vessels, called *lymph*, enters the system in the lymph capillaries, which resemble other capillaries except that they appear to be closed at their terminal ends. The lymph flows from the capillaries of all parts of the body into larger and larger vessels resembling veins in that they possess valves and are thin walled. In fact, the lymph vessels have even thinner walls than the veins. Larger and larger vessels finally converge to the left-shoulder region in the thorax, where the lymph empties into a large vein of the blood circulatory system, near the heart.

But where does the lymph come from? From where does it pass into the closed lymph system? Lymph, in general, originates as tissue fluid—the fluid which surrounds all cells. This, in turn, has reached the cells from the capillaries of the blood circulatory system.

Thus liquid may reach the cells by only one route—the arteries, arterioles, and finally capillaries, through whose walls it passes by diffusion and filtration. But there are two possible return routes. Either the liquid may re-enter the capillaries and be carried onward in the venous stream or it may enter the lymph capillaries. The exact mechanism of the latter is not known. At any rate, once in a lymph capillary, the fluid slowly moves on in a devious course through larger and larger vessels, empties into the large vein mentioned, and so eventually returns to the heart.

From this it is apparent that in composition the lymph must resemble blood. It consists of those constituents of blood which are able to penetrate the capillary wall, plus elements that may be added to it by the tissues. It contains only occasional red corpuscles and has much less protein than blood plasma. Otherwise it closely resembles blood plasma in composition.

B. MOVEMENT OF LYMPH

What makes the lymph move in the vessels, especially in the legs, where the movement is against gravity? Some vertebrates have lymph hearts, which are merely pulsating enlargements of the vessels. Mammals and man have no lymph hearts. Yet the "circulation" described continues. The mechanisms involved are exactly those which, in the case of the veins, assist the action of the heart; namely, muscle contractions and breathing movements. Alternate compression and release of the lymph vessels by skeletal muscles effectively move the lymph along because the vessels are well equipped with valves. Breathing movements, just as they aspirate air into the lungs and venous blood into the right atrium of the heart, also aspirate lymph toward the termination of the lymphatic system in the upper-left chest region. These devices serve effectively, despite the absence of a lymph heart, although the flow of lymph is considerably slower than that of blood in the veins.

The flow of lymph from the intestine is aided by another factor. The lymphatic capillaries here originate mainly in the millions of finger-like projections of the mucosa into the cavity of the intestine—the *villi* (see p. 268 and Fig. 94, p. 269). Microscopic examination of the exposed lining of the intestine in an anesthetized animal shows the villi rhythmically contracting and relaxing, shortening and lengthening. Apparently each villus acts like a tiny heart, squeezing the lymph onward.

C. Functions of the Lymphatic System

In one other important respect the lymphatic system differs from the blood circulatory system. At rather frequent intervals along the course of a lymph vessel are structures called *lymph nodes*. These are made up essentially of a network of connective tissue, in the meshwork of which are located two special kinds of cells. There are (*a*) cells which mature into one kind of white blood cell—the lymphocytes—and (*b*) phagocytic cells, which possess the same capacity as amebae for engulfing particles. As the lymph flows through the vessels, its course is interrupted by the nodes, through which it must pass, trickling through between the packed lymphoid cells and phagocytes. The lymph nodes play a large part in determining the peculiar functions of the lymphatic system (see Figs. 67 and 68).

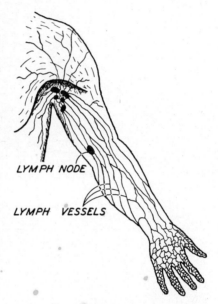

FIG. 67.—Lymph vessels and lymph nodes of the arm.

What are these functions? In the first place, the lymph system helps return tissue fluids to the blood circulation. In these fluids are some of the waste products of metabolism on the way to excretion. But why cannot this return be carried out adequately by the capillaries and veins of the blood circulatory system? Why have the two routes of return? With our present information we cannot answer these questions entirely satisfactorily. It seems, however, that solid particles are able to get into the lymphatic capillaries much more easily than into the blood capillaries. Then, as the fluid trickles through the lymph nodes, some of these solid particles are filtered out and thereby prevented from entering the blood stream. In the lymph nodes near the lungs of city dwellers, for example, so many particles of dust and soot are filtered out that in the course of a normal lifetime the nodes become very dark or even black in appearance.

Bacteria which may have gained entrance into the tissues also enter the lymphatics and may be filtered out and phagocytized in the lymph nodes (see chap. 13).

The lymph vessels which drain the intestinal walls serve another function of a special kind. Sugars and amino acids, as we shall see, are absorbed chiefly into the blood capillaries directly. But the fats that we ingest, digest, and absorb pass mainly into the lymphatics. Of course, even the fat soon gets into the blood stream via the lymphatics, but it does so by this indirect route. The significance of

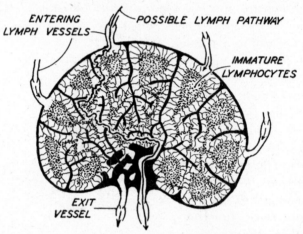

Fig. 68.—Diagram of a lymph node. Lymph flows slowly through the node by labyrinthine pathways. One such pathway for lymph is indicated. In the nodes some of the bacteria or solid particles are filtered from the lymph and engulfed by phagocytes (*not shown*). Certain of the white blood cells (lymphocytes) are formed in the nodes and added to the lymph as it flows slowly through the node. Arrows indicate direction of lymph flow.

this peculiarity is not entirely clear, although there is evidence that the digestion products of fat are injurious to red blood cells, which are virtually absent from lymph. These products are diluted before lymph enters the blood stream.

Another function of the lymphatics is the manufacture of lymphocytes. Young lymphocytes can be seen in active mitotic reproduction in the nodes, and examination reveals that mature lymphocytes are more abundant in the vessel leaving a node than in that entering it. They course along with the lymph and, with it, enter the blood stream.

The lymphatic vessels probably also provide an important mode of entry into the blood stream for proteins and for other useful products synthesized in various organs of the body.

CHAPTER SIX

MECHANISMS AND FUNCTIONS OF BREATHING

I. The Breathing Apparatus
   A. Evolutionary development
   B. Structure of the lungs in man

II. Mechanics of Breathing
   A. Inspiration and expiration
   B. Mechanical model of the breathing movements
   C. Effectiveness of the machinery
   D. Pressure changes in the chest
      1. Intrapulmonic pressure   2. Intrathoracic pressure

III. The Respired Air
   A. Quantity of air exchanged
      1. At rest and in exercise   2. Tidal and reserve air   3. Dead space
   B. Composition of the respired air

IV. The Nervous Control of Breathing
   A. The motor nerves
   B. The breathing center
   C. Respiratory reflexes
      1. Protective reflexes   2. Rhythm of the center: function of the vagi
      3. Afferents stimulated chemically

V. Chemical Control of Breathing
   A. Stimulating effect of carbon dioxide
   B. Adaptation to varied rates of body activity
   C. Indispensability of carbon dioxide
   D. Oxygen debt

VI. Artificial Respiration

VII. Nonrespiratory Functions of the Breathing Movements
   A. Voluntary modifications of breathing
      1. Vocalization   2. Straining
   B. Reflexes
      1. Panting   2. Coughing and sneezing
   C. Blood and lymph flow

VIII. Exchange of Gases between Alveoli and Blood in the Lungs: Diffusion

IX. Transport of Respiratory Gases in the Blood
A. Oxygen transport
1. Union of oxygen with hemoglobin   2. Factors controlling the reaction   3. Gas tensions and partial pressures   4. The oxyhemoglobin dissociation curve   5. Carbon monoxide poisoning
B. Transport of carbon dioxide
1. Bicarbonate formation   2. Acid-base balance
X. The Use of Oxygen by Cells
XI. Disorders of the Respiratory Tract
XII. Interrelations of Respiration and Circulation
A. Correlated adjustments in exercise
B. Effects of training

Energy is indispensable for the continued existence of the cell. At all times in all cells—plant and animal—energy liberation occurs uninterruptedly. In some forms of life, energy is liberated anaerobically, that is, in the absence of free oxygen; indeed, certain bacteria cannot exist in an environment containing abundant free oxygen. It is also possible for certain cells in the bodies of the highest animals, including man, to obtain energy independently of combustion, at least temporarily (see energy relationships in muscle, p. 379).

But, for the most part, the cells of the higher plants and animals obtain their energy from the burning of foods, in the process called *respiration*. In this, gaseous or dissolved oxygen molecules are utilized. Such combustion is indicated in the following formula:

$$C_6H_{12}O_6 + 6O_2 \rightarrow 6CO_2 + 6H_2O + \text{Free energy.}$$
(contains stored energy)

In this example a simple-sugar molecule is taken to illustrate foods in general. The oxidation of fats or proteins might likewise be represented. The end products of combustion are carbon dioxide, water, and energy. The energy yielded does not, of course, arise *de novo* in the reaction but is merely energy which has been released from the "bound" or potential form in which it existed in the sugar (or fat or protein) molecule. The ultimate origin of this intramolecular energy is, of course, the sunlight.[1] In the process of photosynthesis in green plants radiant energy is incorporated into the molecules of sugar (or starch) in the synthesis of these molecules from carbon dioxide and water.

1. The same statement might be made also for energy in general upon the earth. The energy of coal, fuel gases, wind, rivers, and waterfalls is traceable to the energy of sunlight. The energy of the tidal movements is an exception to this.

I. THE BREATHING APPARATUS

A. Evolutionary Development

This chemical reaction, though complicated enough when considered in all its details, especially as regards the role of the enzymes, is the essence of the story of respiration in all cells. But a whole new chapter on respiration must be written for those cells which constitute the tissues of complex, bulky, multicellular organisms. Large animals could not have evolved without special provision for the transportation of the respiratory gases (oxygen and carbon dioxide) between the exterior of the animal and the cells involved. Several schemes effecting this have been evolved in nature.

In the most highly specialized group of the annelid-arthropod series of animals—the insects—there is a series of openings along the sides of the body which communicate with a system of hollow tubes that ramify to all parts of the body. Air is circulated through these tubes by contraction-expansion movements of the abdomen.

Systems of this kind are limited to creatures which either are rather small or have a low rate of oxygen consumption, or both. For other animal forms it is inadequate. Early in the evolution of the circulatory system we find an intimate functional relationship between the transport of the respiratory gases and the circulation. In the annelid worms, of which the common earthworm is an example, oxygen diffuses into the vessels of the circulatory system lying near the external surface of the animals and is transported to all parts of the animal in the circulating blood stream. Carbon dioxide takes just the reverse course. This relatively simple scheme persists even up into the vertebrates, where in amphibians like the frog a considerable quantity of gaseous exchange occurs between exterior and the blood through the moist skin.

In the larger invertebrates and in the vertebrates we find that additional structures have been evolved which effectively bring oxygen from the exterior into intimate contact with the circulating blood. In fish there are the gills, where thousands of fine capillaries come into contact with the oxygen-containing water, which is moved continuously past the gills by a pumping or "breathing" action. Though some primitive fish did possess true functional lungs, these organs did not come into their own as the sole organs of breathing until the evolution of the reptiles and higher vertebrates.

In any case, the various devices for gaseous exchange between blood and exterior possess these features in common: there is a thin, moist, epithelial membrane of a large total surface area, having blood on one side and an oxygen-containing medium on the other. This applies alike to the skin of the earthworm and the frog, the gills of the fish, and the lungs of reptiles, birds, and mammals.

Thus in mammals and man a treatment of the cellular phenomena of respiration—called *internal respiration*—must be supplemented by study of the means whereby the respiratory gases get from the outside air to the cells and, again, from cells to exterior. This transportation we refer to as *external respiration*. There are two successive phases to this process: (*a*) breathing, which brings air and blood together in the lungs, and (*b*) the transportation of oxygen and carbon dioxide between lungs and cells. This latter is effected mainly by the circulatory system.

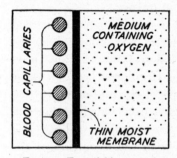

Fig. 69.—Essential features of all organs of respiration. The oxygen-containing medium might be air (e.g., man) or water (e.g., fish). The membrane might be the skin (e.g., frog), gills (e.g., fish), or air sacs (e.g., man). Dots represent oxygen molecules.

The primitive lungs of the amphibians and lungfish, as well as the swim bladder of other fishes, with which the lungs are apparently homologous, are fairly simple structures. They are paired hollow bags, communicating by tubes with the upper end of the alimentary canal; they are richly supplied with capillary blood vessels. The inner surface of the bag is thrown into folds and wrinkles, which increase the total surface area. In birds and mammals each of the two bags has become internally separated into innumerable tiny, intercommunicating compartments, the walls of which are very vascular. The total surface area of the respiratory epithelium of an adult man has been estimated to be nearly 100 sq. m., or about fifty times the skin area. The importance of the large area in effecting rapid gaseous exchange is obvious.

B. Structure of the Lungs in Man

In man the lungs develop embryologically, as they appeared in evolution, as an outpouching of the alimentary canal; in the adult, the respiratory tube or windpipe communicates with the tract in the region of the *pharynx*, just back of the mouth cavity. At this point lies the *larynx*, or voice box, in which are located the vocal cords.

Extending downward from this, lying in front of the gullet, is the *trachea*, or windpipe. This soon branches into two tubes, the *bronchi*, which terminate one in each lung. In the lung itself the bronchi branch profusely into smaller and smaller tubes, finally reaching a size just discernible to the naked eye. Each of these tubes terminates in a cluster of blind sacs, or *alveoli*. There is, then, a direct air channel from the exterior to the alveoli by way of the nasal or the oral cavity, the pharynx, larynx, trachea, bronchi, and the branches of the bronchi. These features are shown in Figures 70 and 71.

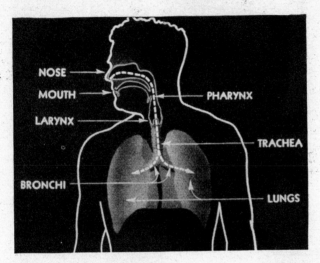

Fig. 70.—The breathing passages in man. Dotted line indicates air pathway. (From the sound film, *Mechanisms of Breathing*.)

Microscopic examination reveals the presence of several tissues in the external respiratory system. The larger tubes and cavities, such as larynx, trachea, and bronchi, are made up of an inner layer of epithelium and an outer layer of connective tissue. Between the two is an intermediate layer containing a small amount of smooth muscle, elastic and nonelastic connective tissue, and cartilage, or gristle. The cartilage is arranged in incomplete rings (split at one point) which nearly encircle the tube and give it a certain rigidity. The tubes are kept open by this tissue, allowing a free passage of air and preventing collapse except by some abnormally great, externally applied pressure. The innermost lining of epithelium consists of cells which secrete a moist, sticky lubricant called *mucus* and of *ciliated cells*. These cilia exert a continuous beating action so directed that,

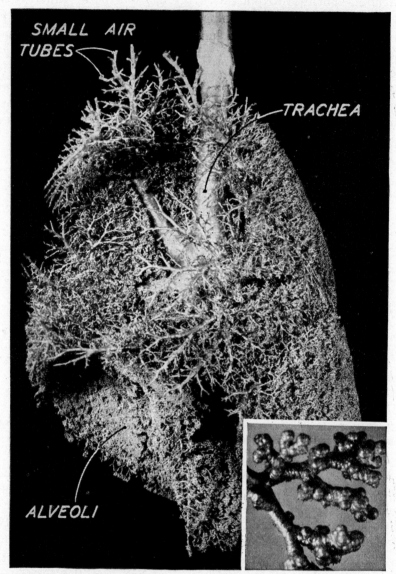

SMALL AIR TUBES

TRACHEA

ALVEOLI

Fig. 71.—Metal cast of air spaces and passages of the lungs of a dog. Where the metal filled the alveoli well, the metal forms an almost solid mass. Where the metal failed to fill the alveoli, the small branching air tubes can be seen. The insert (*lower right*) shows a cast of clusters of alveoli at the terminations of tiny air tubes. The magnification of the insert is about eleven times the actual size.

if solid particles like dust lodge upon the moist surface, they will be swept slowly upward to the pharynx. This constitutes a most useful though not perfect mechanism of defense against infectious disease, since harmful bacteria may be lodged upon solid particles which are inhaled (see also pp. 585–86).

Just as in the case of the blood vessels, the walls of the respiratory passages become thinner and thinner as we progress to tubes of smaller and smaller caliber. The cartilaginous layer is soon entirely lost, the other layers become thinner, and ciliated cells are replaced by thin, flat epithelial cells. The alveolar walls consist entirely of these thin, flat cells, which in shape and proportions resemble the cells which constitute the capillary walls. Immediately outside the alveolar walls are numerous capillary blood vessels of the pulmonary circulation.

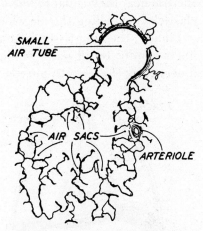

FIG. 72.—Microscopic appearance of lung tissue. Capillaries of the alveolar (air sac) walls are too small to be seen in a magnification of this scale. Those air sacs which seem entirely closed actually open into air passages which are not shown in this section.

The thinness of the alveolar walls has the same significance as the thinness of the capillary walls. At this point only, in the system, is there any appreciable passage of materials (gases) through the walls of the respiratory "tree." Although some inconsiderable gaseous exchange may occur through the linings of the larger tubes, by far most of it takes place in the alveoli, with their large combined total surface area. It is as true that "the lung passages all serve the alveoli" as it is that "the circulatory system serves the capillaries." For oxygen to pass from an alveolus into a capillary, it needs diffuse at most through two thin-cell thicknesses—that of the alveolus and that of the capillary wall.[2] Figure 72 is a diagram of a thin section of lung tissue, showing alveolar structure and the relationship of the alveoli to the respiratory tubes.

Generally distributed throughout the lung tissue, between adjacent alveoli and respiratory tubes, is a quantity of elastic connective tissue. This gives the lungs a remarkable degree of elasticity, which

2. In fact, the alveolar epithelium may be entirely lacking in some regions, so that the air in the alveoli is actually exposed to the walls of the blood capillaries.

is of considerable importance in the mechanics of breathing and in
the circulation of blood through the lungs. Lungs freshly removed
from an animal may be inflated through the trachea. Upon release of
the trachea, they collapse like a rubber balloon because of their own
elasticity, expelling the air which had been blown in.

Finally, the lung tissue is abundantly supplied with nerves whose
fibers ramify to all parts of the organ. Some of these are efferent,
innervating the smooth muscle of the respiratory tubes. They are of

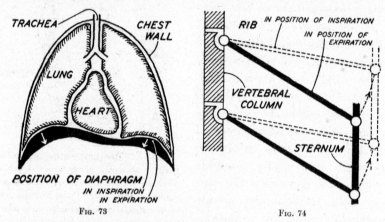

Fig. 73   Fig. 74

Fig. 73.—Contraction of the diaphragmatic muscle in inspiration causes the diaphragm
to descend (*indicated by white arrows*). The resultant increase in chest volume is indicated in
black. A corresponding volume of air is drawn into the lungs. The center of the diaphragm
and the heart also descend somewhat in inspiration.

Fig. 74.—Diagram of rib movements in breathing. Elevation of the front ends of the ribs,
or of the sternum to which the ribs are attached, causes an increase in the front-back diameter
of the chest.

no known significance in the normal breathing of mammals. In addition, there are numerous afferent fibers whose function is fairly well
understood.

II. MECHANICS OF BREATHING

In mammals the lungs lie completely inclosed within the chest
cavity, or *thorax*. One of the organs lies on either side of the heart.
The sides and the dome of the thorax are made up of the chest wall,
which is rendered rather rigid by the ribs. Below, it is bounded by
the *diaphragm*. This structure divides the general body cavity into
the upper *thoracic cavity* and the lower *abdominal cavity*. It is roughly
dome-shaped, attached to the body wall at its margins, and arching
upward at its center, as depicted in Figure 73. The diaphragm consists mainly of skeletal muscle.

A. INSPIRATION AND EXPIRATION

The structure of the chest cavity is such that its total volume can be increased or decreased. An increase in chest size is effected by a contraction of the diaphragm and by a number of muscles which move the ribs. The muscles which act upon the ribs[3] lift the ventral ends of the ribs and increase (chiefly) the dorsoventral dimensions of the chest cavity (see Fig. 74). Contraction of the diaphragm diminishes the convexity of its upward arching, increasing the vertical dimension of the chest.

Since the chest is a closed cavity, with only one opening from the outside—namely, the trachea—it follows that the increase in size will aspirate a quantity of air into the lungs by way of the trachea. In this manner the *inspiratory phase* of respiration is effected.

Expulsion of air from the lungs in the *expiratory phase* is done entirely by the elastic contraction of the lungs. All during inspiration the lungs are distended more and more by the increasing chest size. They may also be maintained for a time in the distended state by a sustained contraction of the muscles of inspiration.[4] As soon as these muscles relax, several factors contribute to a restoration of the thoracic cavity to its previous smaller size and shape. The elasticity of the thoracic walls and, in the case of man, the action of gravity lowering the front end of the ribs tend to reduce the size of the chest cavity. The lungs contract elastically as rapidly as the diminishing chest volume permits and, in contracting, expel a quantity of air.

During muscular exercise, when breathing movements increase in rate and amplitude, reduction of the size of the thorax in each expiration is hastened by the action of certain muscles of expiration. Contraction of some of these muscles lowers the front end of the ribs; contraction of others produces an upward pressure upon the diaphragm. The latter action is brought about by the muscles of the abdominal wall. They compress the abdominal contents, which push upward against the relaxed diaphragm. Expiration of this kind is known as "forced" or *active expiration,* as contrasted with the *passive expiration* in the breathing movements of a resting individual.

In either case, there is no actual compression of the lungs. The chest walls do not squeeze the lungs and forcibly expel the air. In

---

3. The *external intercostal* muscles, the "elevators of the ribs," etc.

4. Or by a closure of the *glottis.* The glottis is the slitlike orifice between the vocal cords in the larynx. Contraction of certain laryngeal muscles brings the vocal cords together in the midline and closes off the respiratory passage (see Fig. 87, p. 247).

expiration the diminishing size of the thorax simply *makes possible* the elastic contraction of the lungs themselves. In passive expiration the volume of the chest is diminished slowly; in active expiration, more quickly and completely, permitting a faster and more nearly complete emptying of the elastic lungs.

Thus the active agent in drawing air into the lungs is the contraction of the muscles of inspiration. The active agent in expelling air from the lungs in expiration is the elastic recoil of the lungs themselves. It is plain that the air passages themselves are entirely passive in the whole process. Even the upper air passages contribute nothing actively to the process.[5] This can readily be demonstrated by connecting the trachea with an artificial opening in the neck. The breathing movements continue quite normally. Sometimes it is even necessary for human beings to be subjected to such an operation, when the larynx must be removed for cancer of the throat.

B. Mechanical Model of the Breathing Movements

All the essential features of inspiration and expiration can be duplicated in a purely physical system. Through an opening at the top of a glass bell jar is inserted a short glass tube which is open to the air at its upper end and has a rubber balloon (or the trachea and lungs of a freshly killed animal) tied securely to its lower end inside the jar. The bottom of the jar is made of a sheet of rubber stretched across the opening and secured to the sides of the jar. The rubber sheet constitutes an artificial diaphragm, which increases the volume of the artificial chest when its center is pulled downward (see Fig. 75). With each downward pull upon the center of the diaphragm, air is sucked into the rubber balloon (or the lungs). With each release of the diaphragm, the stretched bag contracts and expels some of its air.

C. Effectiveness of the Machinery

The method of getting air into the lungs by this aspiration process is not the only device evolved in nature. In the frog the lungs are filled by the swallowing of air. Air is thereby pushed into the lungs under pressure. As far as filling the lungs is concerned, this scheme is as effective as the aspiration principle, in which air is drawn into the lungs by suction. Yet the latter is a more efficient device for aerating the blood, for, in breathing, air and blood must be brought

---

5. Except in the larynx, where the vocal cords are widely separated in each inspiration.

together. If air is pushed into the lungs under pressure, as in the frog, there is a simultaneous partial compression of the lung capillaries, whose blood is to be aerated. But in mammals, in which air is drawn into the lungs, there is simultaneously an aspirating effect upon the capillary blood vessels of the lungs. Blood as well as air is drawn to the alveoli, and a more effective air-blood intimacy is produced.

D. PRESSURE CHANGES IN THE CHEST

1. *Intrapulmonic pressure.*—It should be plain that the pressure within the lung cavities changes with each breathing movement.

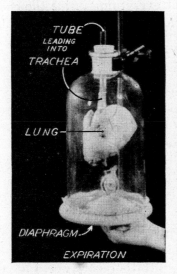
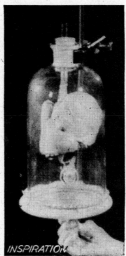

FIG. 75.—Mechanical duplication of breathing movements. Manipulation of the rubber diaphragm of the glass bell jar draws air into the lungs (*right*) or permits the lungs to collapse (*left*). (From the sound film, *Mechanisms of Breathing.*)

During the brief pause between successive breathing movements the pressure within the lung cavities—the intrapulmonic pressure—is the same as atmospheric. With the glottis open, there is a free communication from the interior of the lungs to the outside air. If the intrapulmonic pressure were anything but atmospheric, air would be leaving or entering the lungs, depending upon whether the intrapulmonic pressure were higher or lower than atmospheric. As the inspiratory movement commences, there is a rarefaction of the air in the lungs; that is, the intrapulmonic pressure falls below atmospheric (by about 1–2 mm. Hg.). It is this pressure drop which is responsible for the rush of air into the lungs. With a reduction in the rate of enlargement of the chest as inspiration comes to a close, the intrapulmonic pressure gradually rises and at the end of inspiration is again at atmospheric pressure, provided the glottis is open. As the expiratory movement commences, the recoil of the elastic lungs compresses the air inclosed within the lungs; that is, the intrapulmonic pressure exceeds atmospheric pressure (by about 2–3 mm. Hg.), and, consequently, air passes out of the lungs. Figure 76 schematically represents these pressure relationships in the lungs.

2. *Intrathoracic pressure.*—Less obvious but nonetheless real and important changes occur simultaneously in what is called the *intrathoracic space.* By this is meant everything inside the chest cavity but outside the lungs. The term is unfortunate because it suggests

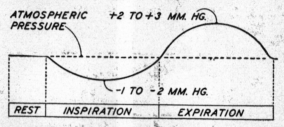

ATMOSPHERIC   +2 TO +3 MM. HG.
PRESSURE

–1 TO –2 MM. HG.

| REST | INSPIRATION | EXPIRATION |

Fig. 76.—Pressure changes in the lungs. The pressure is below atmospheric during normal inspiration and above atmospheric during expiration. Note that the pressure is atmospheric at the *end* of inspiration. At this time the intrathoracic pressure reaches its lowest point (see Fig. 79).

the existence of an air space between the lungs and the chest wall, like the air space in the bell-jar model of Figure 75. But there is *not a bubble of gaseous air* in the intrathoracic space. Indeed, if air is injected into the "space" between the lungs and the chest wall, it slowly dissolves in the fluids of the chest, and after a time it disappears entirely. Over the lateral and dorsal aspects of the lungs there is merely a thin film of watery liquid separating the lungs from the chest wall. At all times, in the course of every breathing cycle, the lung surfaces are in contact (through the thin film of liquid) with the inner surface of the chest wall. Included also in the intrathoracic space are the heart, the large vessels entering and leaving the heart, the esophagus or gullet, and the large lymph vessels carrying lymph from all parts of the body to the (*subclavian*) vein in the upper-left chest region. All these structures are subjected to the cyclic respiratory changes in the intrathoracic pressure.

Just what are these changes? One can measure them by inserting through the chest wall of an anesthetized animal a tube connected with a water manometer. The writing point of the float of the manometer can be made to inscribe a continuous record of the pressure changes in the chest in a manner similar to a mercury manometer record of blood pressure. An experiment similar in principle can be performed upon an intact animal like man in this manner: A small rubber balloon is attached to the end of a rubber tube. The balloon is passed part way down the gullet (i.e., swallowed part way) so that it comes to lie in that part of the gullet which is within the thoracic cavity, with the rubber tube protruding from the mouth. The balloon is inflated with a small quantity of air, and the tube is attached to a

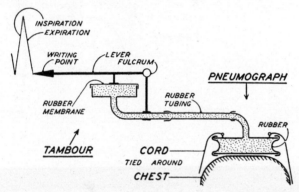

Fig. 77.—Apparatus for recording breathing movements. Expansion of the chest in inspiration compresses the pneumograph; this causes the rubber membrane of the tambour to bulge upward and the writing point to inscribe an upstroke on the paper of a kymograph. Pneumograph, tambour, and connecting tubes are filled with air (*shown dotted*).

water manometer. Again, a record of the intrathoracic pressure changes can be made. When the intrathoracic pressure rises, the balloon is compressed and the water manometer float rises. When the intrathoracic pressure falls, the balloon expands and the float falls (see Fig. 78).

Records made in either of these ways show that the intrathoracic pressure continues to fall throughout the entire inspiratory period and is lowest (in absolute terms) at the height (end) of inspiration. Recall that, at this point in the cycle, the intrapulmonic pressure has returned to atmospheric. During expiration the intrathoracic pressure rises, approximating atmospheric pressure at the end of expiration. Figure 79 illustrates these cyclic changes. Between successive respirations the intrathoracic pressure is about 2–4 mm. Hg. below atmospheric pressure. At the height of a normal inspiration

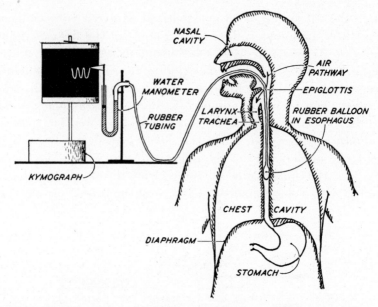

Fig. 78.—Method used to record the intrathoracic pressure changes of breathing in man. (See discussion in text.)

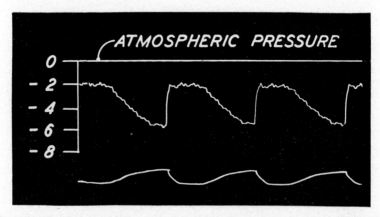

Fig. 79.—Kymograph record of intrathoracic pressure changes in an anesthetized dog. The lower curve is a record of the chest movements inscribed by the apparatus shown in Fig. 77. Upstroke is inspiration. The upper curve shows the intrathoracic pressure changes, recorded by means of a tambour connected with the intrathoracic "space." Figures to the left indicate millimeters of mercury *below* atmospheric pressure. Note that the intrathoracic pressure continues to fall throughout inspiration and reaches its lowest point at the *end* of inspiration. At this time the intrapulmonic pressure is atmospheric (see Fig. 76).

it is about 6–8 mm. Hg. below atmospheric pressure. These changes are entirely due to the mechanics of the elastic, stretched lungs and the changes in chest volume.

A mechanical model like that of Figure 80, *B*, helps to explain the pressure changes. In this, a movable piston fits into an airtight

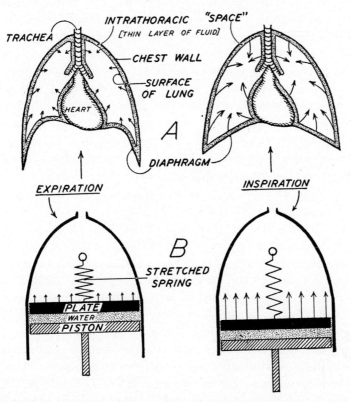

FIG. 80.—Mechanics of the intrathoracic pressure changes of breathing. In *A* the arrows represent the tendency of the lung surfaces to pull away from the chest wall. The lengths of the arrows represent the degree of that pull. The layer of fluid in the intrathoracic space is actually much thinner than shown here, being a small fraction of a millimeter thick. In *B* the water corresponds to the intrathoracic "space," and the arrows represent the tendency for the sliding plate to pull away from the piston. (See discussion in text.)

cylinder. A second movable piston, or sliding plate, is separated from the first by only a thin layer of water. Any movement of the piston effects a like movement of the plate. But this plate is so attached to a spring that, when the piston is pulled downward (in the diagram), the stretched spring *tends* to pull the plate upward, and the pressure of the water between plate and piston is thereby diminished. Note that, at any position of the piston which puts the spring on a stretch, the pressure of the water will be less than atmospheric and that, the greater the stretch of the spring, the lower will be that pressure.

This is an exact duplicate of the mechanics involved in the intrathoracic pressure changes in the mammal. The piston corresponds to the diaphragm (or chest wall, in general), and the plate to the lower (or other) surfaces of the lungs. In inspiration the diaphragm descends (and the sides of the chest wall expand laterally). The pressure of the layer of liquid between lungs and chest wall diminishes, since the lung surfaces tend to pull away from the diaphragm and the thoracic walls because of the elasticity of the stretched lungs.

In the infant the lungs just fit inside the chest cavity, completely filling it without being stretched in the position of respiratory rest between breaths. Therefore, at the end of expiration or the beginning of inspiration, the intrathoracic pressure is the same as atmospheric pressure. With each inspiration the lung surfaces remain apposed to the enlarging chest wall, but now the lungs are placed on a stretch, and the intrathoracic pressure falls. In growth the chest gradually becomes larger at a rate which exceeds the rate of growth of the lungs. In this process the lung surfaces and chest wall always remain in contact. Consequently, the lungs are now being stretched throughout the entire respiratory cycle, and the intrathoracic pressure is always below atmospheric, but more so at the end of inspiration than at the end of expiration. Even in the position of expiration the lungs are under tension; their surfaces tend to pull away from the chest wall, causing a pressure less than atmospheric.[6] During inspiration this pull is accentuated and reaches its height at the point of greatest stretch of the lungs, that is, at the height of inspiration.

The mechanics of this situation also make clear the part played by the chest wall and the lung elasticity in the expulsion of air from the lungs in expiration. In the mechanical model (Fig. 80) simple release of the pull on the piston is sufficient for the restoration of the resting condition. The piston does not *push* on the plate; it is the recoil of the spring which restores the plate to its former position of "expiration." Even if the piston is actually pushed upward,

6. Therefore, if a hole is made in the chest wall, accidentally or experimentally, exposing the external surfaces of the lungs to atmospheric pressure (*pneumothorax*), the lungs at once collapse. Also, inspiratory movements, instead of drawing air into the lungs, will now draw it through the abnormal opening into the chest cavity outside the lungs. Air at this point is of practically no use in the aeration of blood.

simulating forced expiration, the movement of the plate is still due to the recoil of the spring, provided the cavity opens freely to the exterior. Thus also in forced expiration, if the glottis is open, the sudden diminution in chest size in expiration does not compress the lungs from the outside. This act simply makes possible a rapid elastic recoil of the lungs. At all times during this process the intrathoracic pressure is less than atmospheric.

We have intimated that the respiratory movements are important in the flow of venous blood into the heart. This organ lies within the intrathoracic space and is subjected to the intrathoracic pressure changes. They exert less effect upon thick-walled ventricles or aorta than upon the thin-walled large veins and the atria. The maintained lower-than-atmospheric pressure in the thorax also causes the venous pressure near the heart and the intra-atrial pressure in diastole to be less than atmospheric; that is, the decreasing gradient of pressure from the veins to the heart is made steeper. This effect is still further accentuated during inspiration, when the intra-atrial pressure falls still lower. With the accelerated and amplified breathing movements in muscular exercise this effect becomes significant in accelerating blood circulation by hastening the return of venous blood to the heart. Similarly, the intrathoracic pressure changes are important in facilitating the flow of lymph.

III. THE RESPIRED AIR

A. QUANTITY OF AIR EXCHANGED

1. *At rest and in exercise.*—In man at rest about sixteen breathing movements are made per minute, and with each breath about a half-liter (about a pint) of air is inspired and expired. But it must not be concluded that at the end of a normal, quiet inspiration there is only 500 cc. of air in the lungs. This can be proved by forcibly expelling from the lungs all the air possible. If this is done, a volume of about 1.5 liters can be expelled in addition to the half-liter taken in during the preceding inspiration. Even after this, the lungs still contain about a liter of air which cannot be forcibly expelled. There is, therefore, a reserve supply of about 2.5 liters of air in the lungs, with which the half-liter taken in at each inspiration mixes.

It is also common experience that a good deal more than a half-liter of air can be inspired at will. After a normal inspiration, it is possible to inspire an additional volume of about 1.5 liters. In all, by making the deepest possible inspiration and the greatest possible expiration, it is possible to increase the quantity of air inspired and expired at each breath from the usual half-liter to as much as 3.5 liters. Thus in muscular exercise there may be a seven-fold increase in volume exchanged per breath. Ordinarily, this mechanism is not used fully, even in severe exercise, but is supplemented by an increase in the breathing rate.

2. *Tidal and reserve air.*—The volume of air inspired and expired in each breathing movement is called the *tidal volume*. At rest, this is about 500 cc. in man. Remaining in the lungs after a normal expiration is the *expiratory reserve volume* (or *supplemental air*), which can be forcibly expelled, and, in addition, the *residual volume*, which cannot be expelled by even the most forceful expiration. If the lungs are removed from the thorax, almost all the residual air is expelled, but a small quantity of *minimal air* still remains in the lungs. This minimal air gives the lung tissues a specific gravity less than water, and for this reason lungs are often referred to as "lights" by butchers. This property of lungs is of some medicolegal importance in certain cases of fatalities in newly born infants. If the infant were born dead, there would be no minimal air in the lungs; and a bit of lung tissue placed in water would not float. If even a single breath had been taken before death, there would have been minimal air in the removed lungs, and they would float on water. This simple test may be of importance in cases in which courts must decide if infanticide has been committed.

The volume of air which can be inspired in addition to the normal tidal air is called the *inspiratory reserve volume* (or *complemental* air). The tidal volume, plus the inspiratory and expiratory reserve volumes, constitute the *vital capacity*, or the greatest volume exchange possible with each breath. The relationships of these various volumes are shown in Figure 81.

3. *Dead space.*—Not all the air taken in with each breath contributes oxygen to the blood or receives carbon dioxide from it in significant quantities. Of the total air inhaled, in the human adult, approximately the last 150 cc. does not reach the alveoli. It remains in the larger air passages, where practically no exchange of gases between blood and air occurs. This volume of air is, of course, immediately forced out of the system at the beginning of the next expiration. On the other hand, the very last 150 cc. of air which is forced from the alveoli in expiration remains in these same tubes. Although laden with carbon dioxide, this air is at once drawn into

the alveoli with the beginning of the next inspiration. Such air is called *dead air*; the tubes which contain it, the *dead space*. There is, then, a dead space of about 150 cc. in the nasal cavity, larynx, trachea, bronchi, and larger branches of the bronchi. Only 350 cc. of the tidal air actually reaches the alveolar regions, to mix there with the expiratory reserve and residual air. If the dead space is artificially increased, by breathing through a long tube, there will be a gradual depletion of oxygen and accumulation of carbon dioxide in

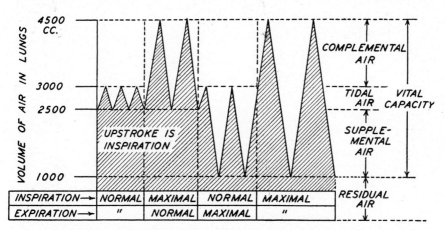

Fig. 81.—Diagrammatic tracings of breathing movements showing volume of air in the lungs (*shaded*) and the volume of air inspired and expired under various conditions. By "normal" is meant the quiet breathing movements of an individual at rest. The term "inspiratory reserve volume" is now widely employed instead of complementary air and the term "expiratory reserve volume" for supplementary air.

the alveoli and blood sufficient to stimulate augmented and accelerated breathing movements. A large increase in dead space may be fatal.

B. Composition of the Respired Air

It is obvious that the composition of the inspired (atmospheric) air differs from the expired air, since oxygen leaves the air in the lungs, and carbon dioxide is added to it. But it must not be thought that all the oxygen of the inspired air is removed at each breath. The expired air still contains three-fourths of the oxygen it had when it was inspired (see Table 3).

Thus air which has been breathed once is still suitable for rebreathing. It contains plenty of oxygen and not a dangerously high concentration of carbon dioxide. But, if air is rebreathed again and again, its oxygen content will become so low and its carbon dioxide concentration so high as to be incompatible with life. Such a state is rarely approximated even in a poorly ventilated room. The discomfort in such a room is chiefly due to the gradually increasing water-vapor content and the increasing temperature of the rebreathed air.

Nitrogen, which comprises the great bulk of air, is entirely inert so far as any utilization of it by the animal organism is concerned.

TABLE 3

DIFFERENCES IN COMPOSITION OF INSPIRED AND EXPIRED AIR

	Inspired Air	Expired Air*	Difference
Nitrogen...........	79%	79%	0
Oxygen............	20	16	−4
Carbon dioxide......	0.04	4	+4
Rare gases.........	Traces	Traces	....
Water vapor........	Variable	Almost saturated	....
Temperature........	Variable	Nearly body temperature	....

* The expired air may also contain certain volatile substances absorbed from the alimentary canal, such as alcohol, or certain volatile abnormal metabolic products, such as the *acetone* of diabetes.

Some of it becomes dissolved in the plasma of the blood, but it is not used in any known manner. As much of it is expired as is inspired.

IV. THE NERVOUS CONTROL OF BREATHING

A. THE MOTOR NERVES

The automatic rhythmic contractions of the muscles of breathing might be compared with the automaticity of the heart. But, in contrast with the heart, it is plain that the respiratory muscles themselves possess no intrinsic rhythmicity. If their nerves are cut, they cease contracting and are as completely paralyzed as is any skeletal muscle so denervated. There are several of these efferent or motor nerves, chief among which are the *phrenic nerves*, which innervate the diaphragm, and those nerves (the *intercostal nerves*) which innervate the (intercostal) muscles whose contractions elevate the ribs. Figure 82 shows the anatomy of these nerves. The phrenics arise from the neck (*cervical*) region of the spinal cord and pass downward through the thorax, one on each side of the heart, to the diaphragmatic muscle. The other (intercostal) nerves arise at various levels of the thoracic region of the spinal cord.

The fact that all the muscles of breathing contract in a harmonious manner indicates that the efferent nerves of these muscles are controlled and synchronized in their actions by some co-ordinating device.

B. The Breathing Center

Just what and where this co-ordinating device is we can determine by experiment. Observing the diaphragm as an example of a muscle of breathing, we note that its rhythmic contractions cease if the

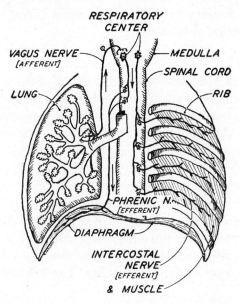

Fig. 82.—Some of the nerves of breathing. All structures shown are present on both sides. The intercostal muscles move the ribs as shown in Fig. 74 (p. 222). Arrows indicate direction of transmission of nerve impulses.

spinal cord is cut across at any point between the medulla and the origin of the phrenic nerves from the spinal cord. This suggests that the nervous discharges controlling this muscle originate in the medulla and pass downward in the spinal cord (see Fig. 82). As we should expect, the continuation or the rate of the movements is not affected by cutting the spinal cord below the origin of the phrenic nerves or above the medulla. The rhythmic contractions of the diaphragm continue.

If now we proceed with the method employed in the localization of the vasoconstrictor center and make successive thin slices of the medulla, starting from above that structure, we find that eventually a cut will be made which will immediately and permanently stop the diaphragmatic movements. This cut has removed the breathing or respiratory center.

Thus we form the picture of a paired cluster of nerve cells in the medulla oblongata which discharge volleys of nerve impulses rhythmically down nerve pathways of the spinal cord, which, in turn, connect functionally with the various efferent nerves of breathing. Unlike the situation in the case of the vasomotor or cardioregulatory centers, we find here no accessory centers in the spinal cord. The paired cluster of nerve cells in the medulla is the chief and only governor of external respiration.

C. RESPIRATORY REFLEXES

The problem of how the rate and the depth of the breathing movements are adjusted to varying physiological conditions must therefore be solved in terms of the activity of the respiratory center. What are the factors which influence the rhythmic discharge of volleys of nerve impulses from center to inspiratory muscles? A number of reflexes figures here—numerous afferent nerve systems play upon the respiratory center and stimulate or inhibit it. For example, stimulation of sensory nerves which would cause pain in the unanesthetized animal produces a reflex acceleration of respiration. Almost any afferent nerve in the body has some nervous connection with the respiratory center, and artificial or physiological stimulation may therefore modify the rate and depth of breathing. The following will serve as examples.

1. *Protective reflexes.*—Certain afferent nerves which furnish the sensory innervation of the lining of the larynx[7] and of the pharynx[8] have a protective function. Artificial stimulation of the sensory fibers (i.e., the central end) of either of these nerves produces a brief inhibition of respiration. They are activated physiologically by the presence of irritating agents in the larynx or by solid objects like a bolus of food passing through the pharynx.

Anyone who has worked in a chemistry laboratory has experienced the sudden "catching of the breath" when fumes of an irritating gas like nitric acid are inhaled. The gas, reaching the larynx and trachea, chemically stimulates the sensory endings here, setting up afferent impulses which inhibit the action of the respiratory center. The

7. The *superior laryngeal nerve*, a branch of the vagus.
8. A branch of the *glossopharyngeal* (the ninth cranial) *nerve*.

breath is "held" quite involuntarily. By this reflex the irritating gas is prevented from reaching and injuring the more delicate structures deeper in the lungs. Also, solid particles in the larynx mechanically stimulate the sensory nerves, evoking a similar reflex inhibition of breathing. There may even be a forcible, sudden, reflex expiration —a cough.

The afferent nerves of the pharynx are especially significant in the act of swallowing. Notice, in Figure 106 (p. 293), that the pathway for food—mouth to pharynx to gullet—crosses the pathway for air—nose to pharynx to larynx. Now, it does not matter particularly if air gets into the gullet, but for a particle of food to enter the larynx is serious or even fatal. We are not surprised to find, therefore, that in evolution mechanisms for the prevention of this accident have developed. One of these consists of an involuntary inhibition of breathing at the commencement of the swallowing act, while food or liquid is passing the opening into the larynx. It is virtually impossible to take a breath and swallow at the same time. The mechanism here is a reflex inhibition of the respiratory center. The bolus of food or the liquid swallowed mechanically stimulates the sensory fibers in the pharynx, which fibers carry impulses to the respiratory

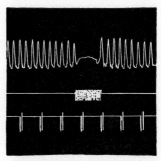

Fig. 83.—Reflex inhibition of breathing. Kymographic record of breathing movements of an anesthetized dog, showing inhibition during stimulation of the central end (afferent nerve fibers) of the sensory nerve of the larynx (superior laryngeal). Respiratory tracing made with apparatus shown in Fig. 77, so that the upstroke represents inspiration. Time in 5-second intervals.

center, inhibiting it. By means of this and other safeguarding mechanisms the aspiration of food into the air passages is normally prevented.

2. *Rhythm of the center: function of the vagi.*—What influences the cells of the respiratory center to discharge nerve impulses in their characteristic rhythmic, intermittent fashion? Some light is shed upon the problem by studies upon the vagus nerves, which furnish a sensory innervation of the lung tissue. If the central end of these vagus fibers is stimulated electrically, the respiratory center is immediately inhibited, and the breathing movements temporarily cease. The same fibers can be stimulated by artificially inflating the lungs. If this is done, the contractions of the breathing muscles again

cease temporarily. Now, physiologically, the lungs are distended in each inspiration. We might expect, therefore, that toward the end of an inspiration the resultant inflation and stretch of the lungs would stimulate these vagus nerve fibers. Nerve impulses would pass on up the vagus nerves to the respiratory center, whose action would be inhibited. It would therefore cease its discharge of impulses through the efferent system, the muscles of inspiration would relax, and passive expiration set in. But, with the collapse of the lungs, the inhibitory afferents in the vagi are no longer stimulated, the center is no longer inhibited, and a new inspiratory discharge should emanate from it.

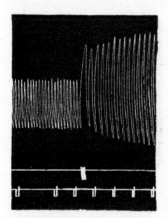

Fig. 84.—Respiratory tracing of an anesthetized dog (taken as in Fig. 77), showing the effect of cutting both vagus nerves (*indicated by signal*). Breathing movements are deeper and slower because inspiratory movements are no longer cut short by the normal reflex. Time in 5-second intervals.

Two experiments demonstrate that this actually occurs physiologically. If electrical instruments suitable for the detection of action potentials are attached to the intact vagi, it is found that toward the end of each inspiration a series or volley of nerve impulses passes up each vagus nerve. It is these nerve impulses which stop every normal inspiratory act. This is further shown by cutting the vagi in an anesthetized animal. The respiratory center discharges as before. The lungs inflate, and, when a certain degree of distention has been reached, the sei ,ory endings of the vagi in the lungs are stimulated. The nerve impulses so initiated, however, are unable to cross the cut in the nerve. Consequently, the respiratory center continues to discharge through the efferent system, and inspiration goes on for a longer period than normal. But note that, if this vagus mechanism were the only factor affecting the rhythm of the respiratory center, we should expect that, when these nerves were cut, there would be a continuous discharge of the center and a maintained spasm of the inspiratory muscles. However, when the vagi are cut, the inspiratory period is merely prolonged, and the depth of each inspiration is increased. But the center ultimately ceases discharging in each inspiration, even in the absence of the vagi, and expiration begins. Thus, while normally the vagi are responsible for the cessation of each inspiratory movement, permitting passive expiration to set in, thereby partly controlling the rhythm of the center, nevertheless there is also some other factor which functions in the absence of the vagi. What this may be is uncertain. Perhaps it is something in the center itself which contributes to the rhythmicity of the inspiratory discharges.

In fact, experiments indicate that, even in the absence of all afferent influences, the center still discharges intermittently and that the characteristic rhythm of the breathing movements is still maintained. The cells of the respiratory center apparently resemble the tissue of the pacemaker of the heart in the possession of an intrinsic rhythmicity, which is not dependent directly upon nervous influences.

3. *Afferents stimulated chemically.*—Evidence has been accumulated indicating that certain sensory nerves in various parts of the body affecting respiration are stimulated by chemical changes in the blood. If the oxygen content of the blood is diminished, or its carbon dioxide content increased, nerve endings near the carotid sinus and in the aorta are chemically stimulated, effecting a reflex acceleration of respiration. The carotid sinus and aortic arch areas (see pp. 197–98) contain two types of receptors: those responding to stretch, lying in the walls of the carotid sinus and aortic arch, and those responding to chemical changes (oxygen lack or carbon dioxide excess), lying in the adjacent carotid body and aortic body, which are small glandlike structures. It is the latter receptors which are involved in the reflex respiratory acceleration described. There is also some evidence that certain sensory nerves in muscles are similarly stimulated and produce the same reflex respiratory effect. Such nerves would contribute to the acceleration of external respiration which occurs in muscular exercise when oxygen is being used rapidly and carbon dioxide is produced in excess by the active skeletal muscles.

Yet there seems to be little question but that the chief effect of changes in the chemical composition of the blood is produced directly upon the cells of the respiratory center itself.

V. CHEMICAL CONTROL OF BREATHING

It is a common experience that during muscular exercise the rate and the depth of breathing are increased. Associated with the muscular exercise are a number of initial chemical changes, first, in the muscles and, second, in the blood. There is a reduction of the oxygen content of the blood because this gas is now being used by the active muscles at an increased rate; there is an increase in the products of muscular activity, chiefly carbon dioxide and lactic acid. Thus we should expect that one or more of these changes might exert a direct chemical stimulating effect upon the respiratory center.

It would be quite reasonable to expect, for example, that lack of oxygen in the blood might stimulate the respiratory center, just as we found that it stimulates the cells of the red bone marrow. In the latter case the mechanism admirably adapts the rate of red-cell production to the requirements of the body for oxygen. In breathing, the blood oxygen might conceivably function equally effectively to adapt the rate and depth of breathing to the varying requirements of the tissues for oxygen. But experiment reveals that, while oxygen lack does play a minor role, the prepotent stimulus to the respiratory center is carbon dioxide. To prove this, we must experimentally produce the changes in the blood characteristic of muscular exercise, one at a time.

A. Stimulating Effect of Carbon Dioxide

If air from a small closed chamber is breathed and rebreathed, and care is taken to remove all the expired carbon dioxide, the oxygen of the chamber will gradually be used up. The concentration of oxygen in the blood gradually diminishes, with no appreciable change in the blood carbon dioxide concentration. In such an experiment breathing is accelerated relatively little, even though the experiment is carried to the point where the oxygen content of the blood is considerably reduced.

However, if the same experiment is repeated, except that the expired carbon dioxide is not removed from the system but allowed to accumulate to be rebreathed again and again, a very marked acceleration of respiration, as well as extreme discomfort ("air hunger"), will result. In this experiment oxygen is being depleted from the blood as before, but, also, carbon dioxide is accumulating. Having already demonstrated that oxygen lack is not a very potent stimulus, we may conclude that carbon dioxide accumulation is the responsible agent.[9]

9. We must admit the possibility of the primary stimulus being "something intimately associated with carbon dioxide," because it is difficult to determine whether the stimulus is carbon dioxide itself or acid. Carbon dioxide in water solution forms the acid H_2CO_3, so that an excess of carbon dioxide always means also an excess of acid. On the other hand, the pres-

Finally, if an individual breathes air containing the normal, or even more than the normal, percentage of oxygen, but containing only a slight excess of carbon dioxide, respiration will again be accelerated. Here the oxygen content of the blood has been maintained practically unchanged, and the carbon dioxide content is increased. Unquestionably, then, the chief chemical stimulus to the respiratory center is carbon dioxide, or something intimately associated with carbon dioxide, such as an increased acidity or, rather, a decreased alkalinity.

The chemical control of the center is usually much more potent than any afferent nervous influence. While it is possible to inhibit the action of the center by stimulation of certain afferent nerves, such inhibition is of brief duration. Despite prolonged stimulation of these inhibitory nerves, respiration will usually start again. The carbon dioxide stimulation of the center is more effective than any nervous inhibition. Likewise, no "voluntary" holding of the breath can overcome the stimulating effect of carbon dioxide for long. When carbon dioxide reaches a great enough concentration, the most determined effort of the "will" cannot nullify its effect on the center. One could not commit suicide by voluntarily holding the breath.

B. Adaptation to Varied Rates of Body Activity

The manner in which this chemical mechanism operates in physical exercise is plain. The very waste product of muscle contraction (or other tissue activity) is itself the stimulus to mechanisms which hasten the removal of that waste from the body. Carbon dioxide enters the blood from the active muscles (and other tissues), and, as this blood circulates through the respiratory center, the rate of breathing is automatically accelerated and the excess carbon dioxide more rapidly eliminated. If the muscular activity has been particularly severe, an excess of carbon dioxide (and lactic acid) persists in the blood for a time, and external respiration remains correspondingly rapid even after the exercise has stopped.

---

ence of an excess of acid in the blood drives carbon dioxide from the blood into the tissues, including the respiratory center. Thus, while an excess of acid in the blood and tissue fluids does accelerate respiration, it is difficult to say whether this is an effect of the acid itself or of the carbon dioxide which has been driven from the blood into the respiratory center.

It is found that heat also stimulates the respiratory center, just as it accelerates the heart rate.

C. Indispensability of Carbon Dioxide

But the importance of carbon dioxide to breathing does not end here. Not only does this chemical adjust the respiratory rate to the varying rates of activity of the organism but apparently it is indispensable for the normal function of the center even during rest. Experimentally, it is possible to diminish the carbon dioxide content of the blood to a low point by voluntarily breathing rapidly and deeply, or overventilating an experimental animal by vigorous and rapid application of artificial respiration. After this has been done for a minute or so, automatic breathing ceases entirely, until the carbon dioxide concentration of the blood again approaches normal.

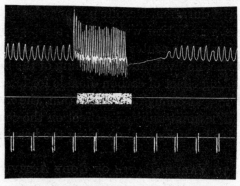

Fig. 85.—Effect of carbon dioxide diminution upon the breathing of an anesthetized dog. The breathing movements (*upper curve, recorded as in Fig. 77*) temporarily cease after administration of rapid and deep artificial respiration for 15 seconds (*indicated by signal*). A temporary diminution in the carbon dioxide of the blood is responsible for the cessation of spontaneous breathing. Time in 5-second intervals is shown on the lowest line of the tracing.

Apparently, the normal continuous action of the center is primarily dependent upon the normal level of carbon dioxide in the blood. Without such carbon dioxide stimulation, breathing would fail, oxygenation of the blood would cease, and death would result. In this respect the breathing machinery is apparently somewhat defective. Sufficient lack of oxygen or excess of carbon dioxide are both fatal, yet oxygen lack itself is not a sufficiently strong stimulus to the center to provide for the oxygen requirements of the body even at rest.

Thus we add here to the already impressive list of mechanisms in which carbon dioxide plays a major role. Through its effects on the heart, the blood vessels, and the respiratory center, it occupies the position of one of the most important chemical regulators of the body. It started its career in animal evolution as a waste product. In mammals and man it is still a waste product, but it has also become most useful, and even indispensable, to the organism.

D. OXYGEN DEBT

In a later discussion of muscle metabolism (see p. 378) it will be pointed out that, even with the maximal rate and depth of respiration, and the maximal rate of blood flow through skeletal muscles, there is an insufficient supply of oxygen to furnish, by the combustion of foods, the energy necessary for hard muscular work. The muscles obtain their energy from other reactions. Among these is the breakdown of glycogen to lactic acid. However, some of this lactic acid, in turn, is burned to carbon dioxide and water. For this latter reaction free oxygen is needed. But, since this oxidative reaction releases no energy required directly for muscle contraction, it need not be completely carried out during the period of muscular activity. Such oxidation does go on during the actual muscle activity, but much of it is necessarily deferred by the limitations of the oxygen supply until the exercise and the lactic acid production have ceased. During the period of muscle activity, while the lactic acid is accumulating, the body is said to be in *oxygen debt*. It has borrowed against the time when the exercise will cease, and, with a continuation of the rapid respiratory rate, the debt can be repaid, and the lactic acid disposed of. Most of the lactic acid produced is reconverted into glycogen, for which reaction a supply of energy is needed, which is furnished by the oxidation of a small part of the lactic acid.

For example, during a period of exercise lasting 2 minutes, the anaerobic energy-liberating processes may yield such a quantity of lactic acid that 10 liters of oxygen are required for its combustion or (indirectly) its reconversion to glycogen. If this volume of oxygen would be taken into the blood in the lungs and delivered to the muscles during the exercise period, the lactic acid could be dealt with as rapidly as it is produced, and there would be no oxygen debt. But actually only about 5 extra liters of oxygen might be delivered to the active muscles in such an exercise period. A 5-liter debt of oxygen is therefore incurred. This is paid off by continuation of the rapid breathing, at a gradually decreasing rate, for perhaps 5–10 minutes after the exercise is stopped.

The borrowing capacity of the body is limited. In extreme cases a total debt of about 20 liters of oxygen may be accumulated, but usually before this limit is reached fatigue prevents further muscle contractions.

Prolonged work or exercise of a less severe nature is often associated with so-called "second wind." In this condition the augmented breathing and blood flow deliver oxygen to the muscles rapidly enough so that the rate of disposition of lactic acid tends to keep pace with the rate of its production.

VI. ARTIFICIAL RESPIRATION

When the breathing machinery fails, it is sometimes possible to reinstitute breathing by the prompt administration of artificial respiration. This procedure is of especial value in resuscitating a near-drowned person who is still alive but whose breathing has stopped. There are mechanical devices which rather faithfully reproduce the normal intermittent pressure changes in the chest. The main trouble with most of these devices is that they are rarely available at the beach, where the need for immediate institution of artificial respiration is most likely to occur. Artificial respiration is possible with no apparatus at all by several methods. The method which best combines effectiveness and relative ease of administration is the "back-pressure–arm-lift" method, in which pressure on the back forces air from the lungs and release of that pressure plus the "arm lift" draw air into the lungs.

In this method the subject is placed face down, elbows bent, one hand upon the other, with the face, turned slightly aside, placed on the hands. Debris or froth should be swept from the opened mouth and the tongue brought forward. The operator places his left knee close to the subject's right arm just beside his forehead, and his right foot near the opposite elbow. Or the operator may kneel on both knees, one on each side of the subject's head, as comfort may dictate. These foot and knee positions may be shifted at will throughout the procedure as the operator desires. He places his hands just below the subject's shoulder blades, fingers spread, thumbs pointed toward the spine, their tips 2 or 3 inches apart.

All this (positions illustrated in Fig. 86*A*) must be effected as speedily as possible. The actual artificial respiration now commences.

To expel air (and water) from the lungs, the operator now rocks forward (Fig. 86*B*), keeping his arms straight and allowing the upper part of his body to exert a steady pressure downward upon the subject's back.

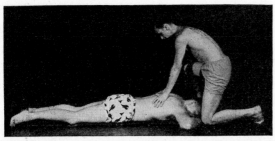

FIG. 86A.—The "back-pressure–arm-lift" method of artificial respiration. The initial positions of subject and operator (described in text) are shown. Speed is important in achieving this orientation.

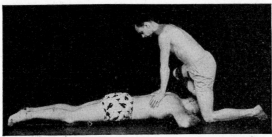

FIG. 86B.—Back pressure, or artificial expiration. The weight of the upper part of the operator's body, arms kept straight, presses upon the subject's back, expelling air (and water) from the lungs.

FIG. 86C.—Release of back pressure, or beginning artificial inspiration. Some air enters the lungs because of release of pressure and elasticity of the chest wall. The kneeling position of the operator is optional with him and may be alternated with the one-knee–one-foot position.

FIG. 86D.—The arm lift, or completion of artificial inspiration. This is effected by the operator pulling the subject's arms toward the operator's shoulders, effecting further aspiration of air into the lungs. The operator now reverts to the position shown in Fig. 86A. All motions should be in a steady, even rhythm. (Photographs courtesy *Armed Forces Medical Journal* and David B. Dill, Scientific Director, Chemical Corps Medical Laboratories, Army Medical Center, Maryland.)

The operator then releases that pressure quickly, avoiding an extra push with the release. This release of pressure draws some air into the lungs, as the elastic structures of the chest wall recoil. As the operator releases pressure, he rocks backward, allowing his hands to slide to a position just above the subject's elbows (Fig. 86C).

Continuing the backward motion, the operator, his elbows unbent, pulls the subject's arms upward and toward himself (Fig. 86D). The lift should be continued until a resistance and tension are felt at the subject's shoulders. This maneuver draws more air into the lungs by arching the subject's back and lifting the weight of his own body from his chest.

This four-stage cycle of (1) back pressure, (2) release, (3) arm lift, and (4) release should be continued regularly at twelve cycles per minute, or 5 seconds per cycle, with each of the four stages in the cycle occupying an equal time. These manipulations should be continued for at least a half-hour, unless spontaneous breathing supervenes.

This procedure is simpler than its description, is often very effective in saving life, and should be familiar to every adult.

This method is not practicable when artificial respiration must be administered over relatively long periods of time. In *poliomyelitis*, or *"infantile paralysis,"* the nerves of the breathing muscles may be temporarily paralyzed, and artificial respiration is necessary for days, weeks, or, in rare instances, years. Here the "iron lung" is employed. The entire body, except the head, is inclosed in a rigid, airtight tank. Intermittently, air is withdrawn from and then forced back into the tank by a pump. Withdrawal of air expands the chest, drawing air into the lungs much as does a normal inspiration. Forcing air back into the tank, raising the pressure against the chest, produces expiration.

VII. NONRESPIRATORY FUNCTIONS OF THE BREATHING MOVEMENTS

A. VOLUNTARY MODIFICATIONS OF BREATHING

1. *Vocalization.*—The breathing movements serve a number of useful functions besides the chief one of providing an exchange of the respiratory gases (oxygen and carbon dioxide) between lungs and outside air. Vocalization of any kind—talking, singing, laughing—is a complex of muscular movements involving not only the lips,

tongue, jaws, and mouth but the breathing machinery as well. Sound production is a modification of breathing in which currents of air set the vocal cords into vibration. The pitch of the sound is regulated by the tautness of the vocal cords, which can be modified voluntarily. The volume of sound, its intensity, can also be varied at will by changing the force of the air current. Speech as we know it depends upon a degree of voluntary control of the breathing machinery— upon a limited control of the respiratory center by the higher conscious centers of the brain. Not only does such control make possible all degrees of modulation of the voice, or emphasis of this or that word or phrase, but it also permits the characteristic uninterrupted

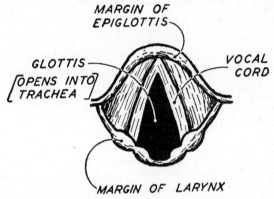

FIG. 87.—View of the larynx from above (i.e., from the pharynx), showing the slitlike opening into the trachea between the vocal cords (see also Fig. 106, p. 293).

flow of speech. Vocalization is limited ordinarily to the expiratory phase of respiration. If there were no possibility of voluntary control of the breathing movements and the duration of the phases of the respiratory cycle, speech would necessarily be intermittent, seizing its opportunity during expiration, and waiting during inspiration. Vocal music as we know it would not be possible under such circumstances.

2. *Straining.*—Straining is an act involving a modified respiratory movement, which aids in the normal evacuation of the bladder or rectum. In straining there is a sharp contraction of the skeletal muscles of the abdominal wall which serves to elevate the pressure within the abdomen. If at such a time the muscles of breathing, especially the diaphragm, are relaxed, and the glottis is open, much of the compressing effect would be lost. The abdominal muscles compress the abdominal contents and would simply push them up-

ward against the relaxed diaphragm, which would, in turn, be forced upward. Only a forced expiration would be effected. But, if the breath is held by maintaining the inspiratory muscles in a state of contraction, or if the glottis is closed after an inspiration, then the abdominal organs compressed by the contraction of the abdominal wall cannot be forced upward. They encounter a rigid diaphragm which does not yield to pressure. Consequently, the full compressing effect is exerted upon the contents of the bladder or of the rectum, as the case may be.

Of course, these various physiological acts involve changes in the specific organ involved also. In urination or in evacuation of the rectum the bladder or rectum contracts, and there is a correlated relaxation of the smooth muscle bands which encircle and guard the exits of these organs. Thus the modified respiratory movement contributes only a part, though an important one, toward the total effect.

Straining furnishes another example of the voluntary control of breathing movements—the deep inspiration at the beginning of the act and the holding of the breath by closing the glottis.

B. Reflexes

1. *Panting.*—The panting of a dog from heat is another instance of control of the respiratory center by centers lying higher in the brain. Panting, of course, is an involuntary reaction not depending upon the activity of the centers of consciousness in the brain. It is effected reflexly through the so-called *temperature-regulating center* (see p. 335) which lies in the brain above the medulla. The reflexly accelerated—though shallow—respiration in panting owes its cooling effects to the copious secretion of saliva, whose evaporation from the protruded tongue is hastened by the rapidly moving air currents.

2. *Coughing and sneezing.*—Coughing and sneezing result from irritation of the linings of the respiratory passages. Essentially, these are modified respiratory acts, in which at first a strong forced expiratory movement is started with the vocal cords tightly apposed (i.e., with the glottis closed). This greatly elevates the pressure within the lungs, so that, when suddenly the glottis is opened, a blast of air is abruptly forced from the lungs through the mouth in a cough, or through the nose in a sneeze, tending to expel the irritating object from the breathing passages.

Yawning, sighing, snoring, and hiccoughing are all modifications of the respiratory movements whose mechanisms are imperfectly understood and whose adaptive significance is even less clear.

C. Blood and Lymph Flow

We have had an occasion earlier to refer to the importance of the breathing movements in the flow of lymph and of venous blood. These circulatory fluids are aspirated into the thin-walled vessels of the chest: the blood into the right atrium and the lymph into the large terminal lymph duct (the *thoracic duct*) which empties into a large vein near the heart. Valves in the veins and lymph vessels prevent any backward flow of blood or lymph toward the periphery in expiration.

The breathing movements are even more nicely adapted to this function than is indicated by the foregoing. Each descent of the diaphragm in inspiration mechanically compresses the abdominal contents. This can be demonstrated by the same rubber-balloon–water-manometer technique described for detection of changes in pressure in the thorax in breathing (see Fig. 78). To do this, the balloon needs merely be passed down into the abdominal cavity instead of into the chest cavity. That is, it must be swallowed into the stomach instead of into the gullet. Under such conditions the manometer float rises with each inspiration instead of falling.

This mechanical effect is admirably adapted in man to force blood and lymph upward against gravity. At the time that an aspirating effect is being exerted upon the right atrium, large veins, and (thoracic) lymph duct, there is a simultaneous compression of the veins and lymph vessels of the abdomen. These fluids are then pushed up from below, as well as sucked from above, in each inspiration.

VIII. EXCHANGE OF GASES BETWEEN ALVEOLI AND BLOOD IN THE LUNGS: DIFFUSION

Thus far we have observed the mechanisms which operate in transporting oxygen to the alveoli and in expelling carbon dioxide from the lungs to the exterior of the body. Since lung tissue is made of living cells, these organs also use some of this oxygen and contribute their quota of wastes. But most of the actual utilization of the oxygen inhaled and the production of carbon dioxide exhaled occur in organs other than the lungs in the combustion of foods. So we must consider further the machinery for transporting oxygen and carbon dioxide between alveoli and the cells throughout the body. It is clear that the distribution of oxygen and the collection of carbon dioxide are carried out by the circulating blood, although the way in which this is done and the chemical combinations in which oxygen and carbon dioxide are transported require further consideration. But, first, we must inquire into the entrance and exit of the gases into and out of the blood stream. How and why does oxygen enter the capillaries in the lungs and carbon dioxide leave the capillaries here? How and why do exactly the reverse movements of these gases occur through the walls of the capillaries of other organs of the body?

One might suspect that absorption of oxygen from the alveoli is similar to the absorption of foods from the intestines. There are indeed, certain details in common in these two processes, but there are also certain fundamental differences. The cells lining the intestinal walls are capable of taking even very dilute materials from the interior of the intestine and passing them into the blood, where the concentration of those materials may be even greater. These cells, in other words, can absorb solutes against a diffusion gradient, or water against osmosis. They can move materials in a direction opposite to that in which materials would move if only known physical forces operated.

But this does not hold for the cells forming the walls of the alveoli or capillaries. Accurate measurements disclose the fact that there is normally a greater concentration[10] of free oxygen in the alveoli than in the blood of the lung capillaries. That is, under normal conditions there is a diffusion gradient from alveoli to blood, and we find a continuous movement of oxygen in that direction, apparently by the physical process of diffusion alone. Neither the alveolar cells nor the capillary cells seem capable of secreting oxygen into the blood stream. They are quite passive in this transport, simply offering no obstacle to the diffusion of the gas. These findings accord perfectly with the structure of the cells, which are flat and extremely thin and quite different in appearance from those cells which participate actively in the movement of materials, such as the cells lining the intestine or those which make up the digestive glands or the kidney tubules.

Similarly, in the passage of carbon dioxide into the alveolar cavity from the capillaries, diffusion is the driving force. The lung cells cannot actively secrete carbon dioxide into the alveoli, as evidenced by the fact that the concentration of carbon dioxide in the capillary blood normally exceeds that in the alveoli. Indeed, if air containing an abnormally high percentage of carbon dioxide is breathed, directly elevating the carbon dioxide concentration in the alveoli, there is marked interference with the passage of carbon dioxide

10. See discussion of gas tensions and partial pressures, pp. 253–55.

into the alveoli. If the alveolar concentration is sufficiently elevated, there will actually be a movement of the gas into the blood. The cells through which the gas passes seem quite incapable of secretory activity. They can neither concentrate carbon dioxide in the alveoli nor move it against a diffusion gradient.

At the capillaries of the tissues throughout the body, diffusion occurs in the reverse direction. Oxygen passes out of the blood to the cells, which use it up just about as rapidly as it is diffused. Carbon dioxide is most concentrated in the cells which produce it. It diffuses from those cells, soon finding its way into the capillaries.

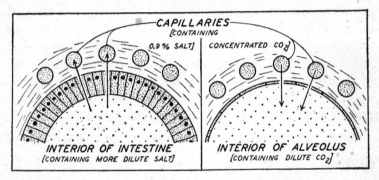

Fig. 88.—The cells lining the intestine can move materials (e.g., salt, *indicated by dots*) against a diffusion gradient. The cells lining the alveoli are unable to do this. Materials (e.g., carbon dioxide, *indicated by dots*) move through these cells in the direction of the diffusion gradient.

Thus, everywhere in the system, between lungs and cells, the gases move in the direction of decreasing diffusion gradients. The oxygen gradient decreases from alveoli to blood to tissue cells. The decreasing diffusion gradient of carbon dioxide is in the reverse direction— tissues to blood to alveoli.

IX. TRANSPORT OF RESPIRATORY GASES IN THE BLOOD

The mechanisms for the transport of the respiratory gases in the blood are somewhat similar in principle; that is, each gas is carried mainly in chemical combination with special blood constituents. Far more of these gases is carried in the blood than could be present in simple solution in the plasma. At atmospheric pressure, only about 0.2 cc. of oxygen and about 0.3 cc. of carbon dioxide could be carried by 100 cc. of blood plasma in simple solution. Yet, actually, we find that in 100 cc. of blood there are 12–20 cc. of oxygen and 30–50 cc.

of carbon dioxide. Obviously, these gases are present in the blood otherwise than in simple solution for the most part.

Oxygen is transported practically entirely in loose combination with hemoglobin within the red cells, and carbon dioxide is carried largely in the form of bicarbonate ions (HCO_3^-), both in the plasma and in the red cells.

A. Oxygen Transport

1. *Union of oxygen with hemoglobin.*—The function of hemoglobin in oxygen transport can readily be ascertained. One can compare in a test tube the quantity of oxygen which will enter into combination with either whole blood or plasma alone. Similarly, one can contrast the oxygen-combining capacity of a colloidal suspension of hemoglobin (or hemolyzed blood) with water or plasma. The results indicate that the hemoglobin is the essential oxygen-carrying agent and that it can function in this capacity just about as well in a colloidal suspension[11] as it does when present within the red cells. Chemically, we represent the reaction which occurs as follows:

$$\text{Hb } (Reduced\ hemoglobin) + O_2 \rightarrow HbO_2\ (Oxyhemoglobin).^{[12]}$$

The relationships between oxygen and hemoglobin can be further studied outside the body in a test tube. We find not only that hemoglobin readily unites with oxygen but that the combination formed is relatively unstable. The oxyhemoglobin is readily disintegrated or dissociated into its two constituents. In other words, the interaction of these two molecules is a reversible one, which can be expressed:

$$\text{Hb} + O_2 \rightleftarrows HbO_2.$$

These test-tube reactions are easy to follow, partly because reduced hemoglobin and oxyhemoglobin are of different colors: Hb is purplish, and HbO_2 is a bright crimson.

We may expect, then, that these reactions take place physiologically. The reaction proceeds to the right in the lungs, giving the oxygenated blood the crimson color of arterial blood, and to the left in the capillaries of the body tissues, converting the blood into the purplish hue characteristic of venous blood.

2. *Factors controlling the reaction.*—But *why* does the reversible reaction go as it does *when* it does? Why are there a combination in

11. Hemoglobin in this form in the blood is rapidly excreted in the urine, however, and therefore lost from the body.

12. Actually, each molecule of hemoglobin combines with four molecules of oxygen.

the lungs and a dissociation in the tissue capillaries? The principles involved here are clear. In any given sample of blood, with all the conditions kept constant, the volume of oxygen with which the hemoglobin is in combination is determined within certain limits by the concentration of free oxygen (dissolved or gaseous) in the medium to which the hemoglobin or blood is exposed. If we expose containers of blood to air containing 20, 10, and 5 per cent oxygen, and allow time for equilibrium to become established, we find that the first sample will contain most oxygen in combination with hemoglobin; the second, an intermediate amount; and the third, least.

Or, if we alternately expose a single sample of whole blood to air containing 20 per cent oxygen and air containing almost no oxygen, we find that (if time for establishment of equilibrium is allowed) our blood sample will contain more oxygen in the oxygen-rich environment than in the oxygen-poor environment. That is, the reaction $Hb + O_2 \rightleftarrows HbO_2$ is driven to the right by the presence of an abundance of oxygen, and to the left, freeing the oxygen, when the concentration of oxygen in the environment is low. Thus the factor which determines the direction in which the reaction goes is chiefly the concentration of oxygen to which the blood is exposed. In the lungs this concentration is relatively high, and the oxygen which has diffused into the blood combines with the hemoglobin. In the tissues elsewhere it is relatively low (almost zero), and oxyhemoglobin dissociates, freeing oxygen into solution in the plasma, whence it diffuses out of the capillaries to the oxygen-poor tissue cells.

3. *Gas tensions and partial pressures.*—In this analysis we have laid down the broad general principles of the chemistry and physics of oxygen transport. There remain certain additional physical and chemical principles which must be understood for a full appreciation of this physiological mechanism. A special aspect of the question of diffusion gradients is involved here. When the respiratory gases are in simple solution, the gradient of diffusion is entirely dependent upon the relative concentrations of solute in various parts of the system. But in oxygen transport we deal with oxygen not only in simple solution in the plasma and tissue fluids but also in the gaseous form in the alveoli and in combination with hemoglobin in the red cells. A consideration of merely the actual percentage concentration of total oxygen in these various states is inadequate to explain the direction of diffusion in all cases. This becomes apparent when we consider that, if plasma or water is allowed to come into equilibrium

with normal air, each 100 cc. of the liquid will contain only about 0.2 cc. of oxygen. Though the air in this case has a concentration of oxygen one hundred times that of the liquid, there is no effective diffusion gradient from the gas to the liquid at equilibrium point.

The concept of *partial pressure* of gases in a gaseous mixture and of *tensions* of gases in solution helps to clarify the situation. The partial pressure of a gas is simply the total pressure of a mixture of gases multiplied by the volume percentage of the gas in question. The total pressure of the mixture of gases in the atmosphere is about 760 mm. Hg. Oxygen, making up about one-fifth of the total mixture, has then a partial pressure of about 152 mm. Hg. In a mixture of gases of a total pressure of 760 mm. Hg., containing one-tenth oxygen by volume, there would be a partial pressure of oxygen of 76 mm. Hg.

The tension of a gas refers to gas in solution or in combination in a solution. We express it numerically in terms of the partial pressure of the overlying gas *with which it is in equilibrium*. If we expose any sort of liquid—be it water, plasma, blood, or a colloidal suspension of hemoglobin—to air, equilibrium will be reached when the rate at which molecules entering the liquid equals the rate of departure from it. The absolute quantity or concentration of oxygen will vary considerably with different liquids. The tension, however, which is an expression of the tendency for the dissolved or combined gas to diffuse out, is exactly the same in each case by definition.

By way of example, suppose we place in a large airtight container filled with air an open beaker of plasma, one of whole blood, one of hemoglobin in suspension in water (15 gm. per 100 cc.), and another of a more dilute suspension of hemoglobin in water (7.5 gm. per 100 cc.). Eventually, all the liquids will come into equilibrium with the air. Since the partial pressure of the oxygen of the air is 152 mm. Hg., and since equilibrium has been reached, then, by definition, the tension of oxygen in each of the liquids is also 152 mm. Hg. Yet, at this equilibrium point, the actual concentration of oxygen is quite different in the different liquids. In every 100 cc. there will be 0.2 cc. of oxygen in the plasma, 20 cc. in the whole blood, 20 cc. in the first hemoglobin suspension, and 10 cc. in the second (see Fig. 89). Essentially, that oxygen which is combined with hemoglobin is temporarily not diffusible.

When we examine the partial pressures and tensions of oxygen in the significant gases or liquids of the body, we find the relationships which are shown in Figure 90.

Note that the oxygen decreases in partial pressure, or in tension, from alveoli to blood to tissues, though the actual concentration of oxygen in arterial blood exceeds its concentration in the alveoli. The partial pressures and tensions, and not the concentration, determine the direction of the diffusion. The blood passes through the capillaries too quickly for equilibrium to be reached between interior and exterior of capillaries, both in the lungs and in the tissues of the body. Arterial blood thus has a tension below the partial pressure of alveolar air, and venous blood has a tension higher than that of the

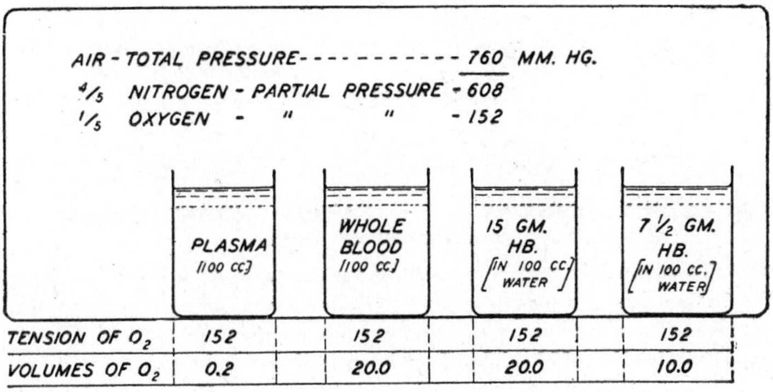

FIG. 89.—Gas tensions in liquids. The tension of gas (e.g., oxygen) in any liquid is equal to the partial pressure of that gas in the gas mixture (e.g., air) with which the liquid is in equilibrium. At equilibrium the rate of entrance of oxygen molecules into the liquid equals the rate of exit of oxygen molecules from the liquid. Note that, though the tension of oxygen is the same in each case, the actual concentrations (i.e., volumes) of oxygen in the liquids differ markedly. "Hb." is hemoglobin.

tissues. Venous blood, in other words, still contains most of the oxygen, though in a lesser quantity than arterial blood.

4. *The oxyhemoglobin dissociation curve.*—The relationship between the volume percentage of oxygen held in union with hemoglobin and the tension of oxygen to which blood is exposed is expressed in the so-called "oxyhemoglobin dissociation curve" (Fig. 91). In the figure two curves are shown—one for venous blood and the other for arterial blood—because of the finding that, at the same oxygen tension, venous blood combines with less oxygen than arterial blood. The reason for this is that venous blood is slightly less alkaline than arterial blood. The reaction of the blood definitely affects its oxygen-carrying capacity, which capacity is reduced by lesser alkalinity and increased by greater alkalinity. In fact, the method for determining the oxygen content of blood depends partly upon this relationship.

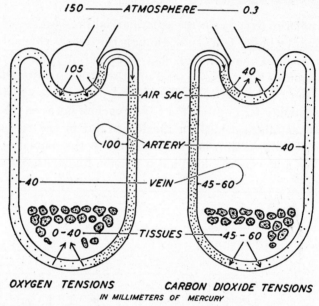

FIG. 90.—Tensions of the respiratory gases in various parts of the body, showing a diffusion gradient of oxygen from air sacs to arteries to tissues and a diffusion gradient of carbon dioxide from tissues to veins to air sacs. Straight arrows represent direction of diffusion. Curved arrows represent direction of blood flow from lungs to tissues to lungs.

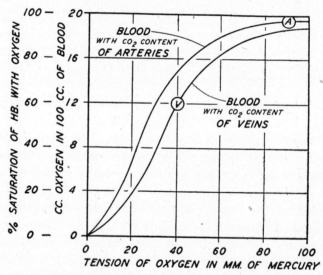

FIG. 91.—Curves of dissociation of oxyhemoglobin, showing the quantity of oxygen which can be held by hemoglobin, as affected by these two physiological variables: (1) the oxygen tension to which blood is exposed (high in lung capillaries, low in tissue capillaries) and (2) the carbon dioxide content of the blood (high in veins, lower in arteries). Note that each 100 cc. of normal arterial blood (point *A*) has 7 cc. more oxygen than normal venous blood (point *V*).

The addition of acid to blood "drives off" oxygen, whose volume may then be measured. In the animal organism this reaction occurs also at the capillaries of the various organs, for at this point carbon dioxide, in the form of carbonic acid, enters the blood. Oxyhemoglobin dissociates physiologically, then, both because of the low oxygen tension in the tissues and because it is driven out of combination by the entering carbonic and other acids.

The curve also shows that, when the oxygen tension is 100 (i.e., that of arterial blood), the blood is 95 per cent saturated; that is, 100 cc. of blood will hold 18–19 cc. of oxygen in combination. At an oxygen tension of 40 (i.e., that of venous blood), the hemoglobin is about 60 per cent saturated, containing about 12–13 cc. of oxygen per 100 cc. of blood. Each 100 cc. of blood, therefore, in passing through the capillaries of an organ, loses about 6–7 cc. of oxygen on the average.

5. *Carbon monoxide poisoning.*—Carbon monoxide (CO), a component of some illuminating gas and of automobile exhaust gas, owes its injurious or even fatal effects upon the organism to the fact that it combines with hemoglobin, forming *carbonmonoxi-hemoglobin* (CO Hb), even more readily than oxygen combines with hemoglobin. In the form of CO Hb, the pigment is unable to combine with oxygen and is therefore useless for oxygen transportation. The body cells will be furnished inadequate supplies of oxygen to an extent which may be fatal to the organism. If the air breathed contains only 0.5 per cent of CO, approximately one-half of the hemoglobin of the blood is found to be in CO Hb form. In this condition an individual would be as badly off as if he had suddenly lost half the red cells of the body.

The reaction CO + Hb → CO Hb is also reversible, so that, in an atmosphere free from CO, the CO Hb combination breaks up, freeing hemoglobin for use in normal oxygen transport. Accidental CO poisoning, therefore, should be dealt with by removing the victim to the fresh air and by at once instituting artificial respiration if breathing has stopped.

B. Transport of Carbon Dioxide

1. *Bicarbonate formation.*—Very little of the carbon dioxide produced in metabolism remains as dissolved molecules. It is soon converted into carbonic acid:

$$CO_2 + H_2O \rightarrow H_2CO_3 .$$

This acid, in turn, reacts with salts of the plasma proteins and the salt of hemoglobin within the red cell in the following manner:

$$H_2CO_3 + Na\ Proteinate \rightarrow NaHCO_3 + H\ Proteinate$$
$$H_2CO_3 + K\ Hb \qquad\qquad \rightarrow KHCO_3 + H\ Hb .$$

Since $NaHCO_3$ and $KHCO_3$ are practically completely ionized, the carbon dioxide is present mainly as HCO_3^-. In addition, there is a direct combination of some of the carbon dioxide with hemoglobin, forming a loose union of the two.

These reactions proceed to the right as shown when carbon dioxide is added to the blood in the tissue capillaries. The reactions are reversed in the lungs, and carbon dioxide is freed from bicarbonate and hemoglobin and diffuses into the alveoli. Again, the concentration of carbon dioxide to which the blood is exposed determines the direction in which these reactions will go. The equilibrium shifts in the direction of bicarbonate formation where an abundance of carbon dioxide enters the capillaries in the body organs. It shifts in the direction of carbon dioxide liberation in the capillaries of the lungs, where carbon dioxide is diffusing out of the blood stream.

2. *Acid-base balance.*—These reactions are of considerable importance quite apart from carbon dioxide transport as such. They prevent the carbonic acid from significantly changing the reaction of the blood in the acid direction. Hydrogen proteinate and hydrogen hemoglobinate are much weaker acids than hydrogen bicarbonate (i.e., H_2CO_3).[13] Consequently, the alkalinity of the blood changes but little despite the large quantities of carbon dioxide which enter the blood in the capillaries.

Furthermore, we find that oxyhemoglobin ($HHbO_2$) is a stronger acid than reduced hemoglobin (HHb). Thus, when carbon dioxide enters the blood, the tendency to increase the blood acidity is partially offset by the oxyhemoglobin changing to the less acid reduced form at the same time.

This is of importance because the tissues, in general, are very sensitive to changes in the reaction of the blood. In fact, life is possible only within a very narrow range just to the alkaline side of neutrality, so that, no matter how effective a carbon dioxide carrying device might have been evolved, it would be of no use to the organism unless it could simultaneously prevent the development

13. That is, the ionization of these acids is less than that of H_2CO_3. The sodium and potassium salts in the reactions given above are practically 100 per cent ionized.

of an acid reaction. Life as we know it would be quite impossible in a medium as acid as blood would be if the carbon dioxide were present entirely as carbonic acid.

This mechanism is especially efficient because the neutralization and excretion of the acid entail no loss of base from the body. The acid-forming carbon dioxide is eliminated in the lungs without excretion of sodium or potassium.

The maintenance of the proper reaction in the blood and tissue fluids—in the internal environment—depends upon the neutralization or *buffering* of other acid products of metabolism as well as the H_2CO_3. The lactic acid of muscle metabolism, for example, reacts as follows:

$$\text{H Lactate} + \text{Na Proteinate} \rightarrow \text{Na Lactate} + \text{H Proteinate}.$$

Again, the acid form of the protein is less ionized, and therefore less acid, than is lactic acid. Consequently, the reaction of the blood is not markedly changed in the acid direction, despite the addition of lactic acid; the conditions suitable for life, so far as acid-base balance is concerned, are maintained in the internal environment.

When metabolic acids (other than H_2CO_3) are produced in excess, there may be a considerable loss of sodium from the body. In the foregoing reaction, for example, the Na lactate is excreted by the kidneys, and, unless the lost sodium is replaced by the ingestion of new sodium, there will be a depletion of the *alkaline reserve*—of the bases (chiefly sodium) of the blood. The term *acidosis* is applied to such depletion of alkali and does not indicate the actual development of acidity in the tissues, as the name implies.[14]

X. THE USE OF OXYGEN BY CELLS

All the mechanisms so far discussed deal with getting oxygen from the outside atmosphere to the cells, and carbon dioxide from the cells to the outside air. Respiration proper, the actual utilization of oxygen by the cells in the combustion of foods, is only now ready to commence. The problem of tissue oxidations is beyond the scope of this book. Suffice it to say that in tissue oxidations there is the same energy liberation as would occur in the oxidation of the same quantity of the same materials outside the body.

A striking difference is seen in the fact that these oxidations occur very rapidly at the relatively low temperatures found in the body,

14. See further discussion of acid-base balance on p. 315.

while in the test tube they occur only very slowly or only at temperatures incompatible with life. The difference is dependent upon the presence in cells of a number of specific respiratory enzymes. These comprise an extremely complicated system, in which several enzymes are required at successive stages in the oxidative breakdown of any one of the several kinds of foods.

XI. DISORDERS OF THE RESPIRATORY TRACT

The respiratory passages, from mouth or nose to alveoli, are in a position vulnerable to infections. After birth the mouth and nasal cavities always contain microörganisms, which may attack the system either in the upper passages or deep down in the lungs. Bacteria-laden particles or droplets of water may also be aspirated at any time. Many respiratory infectious diseases derive their names from the site of the infection. One may suffer from an infection and inflammation of the nasal cavity (*rhinitis*), the pharynx (*pharyngitis*), or the larynx (*laryngitis*). All these may be variations of the common cold, which is thought to be caused by infestation by a kind of submicroscopic infectious agent called *virus*. Various kinds of infections in these cavities may spread secondarily into the middle-ear cavity (*otitis media*) (see p. 484) or into the nasal sinuses (*sinusitis*), which open directly into the nasal cavity.

Tissue like that of the tonsils may also become badly infected and serve as a focus of infection in which bacteria grow and multiply, liberating their injurious metabolic products into the blood, secondarily migrating to remote parts of the body, there to set up new sites of infection.

If an infection in the upper respiratory passages extends downward, there may be involvement of the tracheal linings (*tracheitis*) or the bronchi (*bronchitis*) or the alveoli themselves (*pneumonia*).

A number of special organisms may grow in the respiratory system and effect injury which may be fatal, either by mechanical interference with breathing or blood circulation or by the liberation of metabolic products which injure even remotely situated organs. Included among these are the organisms of scarlet fever and diphtheria, localized in the throat, and tuberculosis, which most often localizes and extends through the deeper tissues of the lungs.

Several mechanisms have been evolved for the defense of the organism against such infections. The hair which lines the nasal cavity and the thick mucous secretion on the linings of the tract entrap some of the bacteria or bacteria-laden particles. Arrested in their progress, they may then be expelled by a reflex cough or a sneeze. Ciliary motion also helps to move the infected particles up from the deeper regions of the lungs (see also pp. 585–86).

These defense mechanisms tend to prevent the entrance of bacteria into the system. They keep the deeper passages relatively free from bacteria and prevent the bacteria from lodging in locations favorable for growth, where they might multiply and produce varying degrees of damage or even death. Another associated scheme of defense is found in that category of responses of the organism which we call *immunity*, which will be dealt with at length in chapter 13.

There are several noninfectious abnormalities of the respiratory system which cause considerable impairment of health, especially during certain seasons of the year. These include "hay fever," certain types of *asthma*, and related conditions, which are special manifestations of "sensitivity" of people to specific (usually protein) constituents of foods and of dusts and pollens in the inspired air. These reactions are closely allied biologically to the general phenomena of immunity, and their underlying principles will be considered at a later time. At present, we may say that people respond in different ways to proteins to which they are sensitive. The reaction may be limited to the nasal linings and consist of an engorgement of the blood vessels and a copious secretion of mucus, giving the sniffling, sneezing condition frequently observed (or experienced). In other instances the reaction may take the form of a constriction of the finer air passages, effected by a spasm of the smooth muscle fibers which encircle those tubes. The result is a mechanical interference with the movement of air, causing the labored, struggling breathing characteristic of asthma.

XII. INTERRELATIONS OF RESPIRATION AND CIRCULATION

A. Correlated Adjustments in Exercise

On more than one count it should have become clear that breathing and circulation are intimately related physiologically. A bare consideration of the functions of each should lead us to this conclusion. In the exchange of gases between cells and external air, the breathing movements and the circulation of the blood are successive stages in the same process. We have found many instances, also, where the factors controlling blood flow simultaneously influence breathing. In general, these two physiological processes are attuned to the rate of tissue and organ activity, and the rate of breathing and blood flow vary in parallel fashion.

When the circulatory adjustments discussed in an earlier section produce a greater blood flow to active muscle, breathing is also accelerated by carbon dioxide stimulation of the respiratory center. Thus not only does the increased blood flow deliver oxygen to the active muscles at a faster rate but the carbon dioxide which that blood carries more rapidly to the lungs is eliminated more swiftly. On the other hand, the increased breathing caused directly by the muscular exercise aids in the rapid flow of blood by facilitating the return of venous blood to the heart.

A simplified schema of the causal relationships and interrelationships in the adjustments to exercise is as shown in the accompanying tabulation.

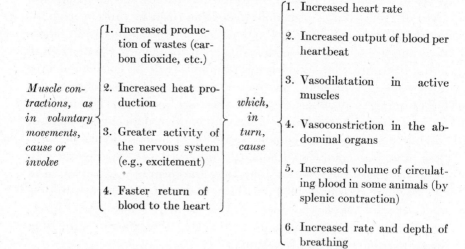

The net result is a richer oxygen supply to the active muscles and a faster elimination of carbon dioxide from them.

B. Effects of Training

Besides the immediate responses of the circulatory and respiratory systems described, there are rather definite chronic or persisting effects of repeated exercise. Physical training produces certain long-lasting effects. Primary among these is that upon the muscles themselves. They grow to a larger size by virtue of an increased size of the individual fibers comprising the muscle.[15] This makes the muscle stronger and more capable of doing its work.

Even more important than this are changes in the central nervous system associated with learning. Repeated practice of any muscular act leads to such changes in the manner of performing the act that it is accomplished with less energy expenditure. Better co-ordination of muscle groups and nicer adjustment of rate and degree of contraction to the requirements for smooth performance of the act are effected.

In addition, training modifies the circulatory and respiratory responses. The heart increases somewhat in size and in the trained subject beats more slowly both during rest and during activity. A given amount of muscular action in the trained subject produces less cardiac acceleration than in the untrained. However, both at rest and during work, the cardiac stroke volume is greater. During exercise the total cardiac output is increased to a larger extent by more forcible beats than by more rapid beats. This, apparently, is a more economical way of increasing the cardiac output, so far as energy expenditure by the heart is concerned.

Changes in the breathing system are somewhat similar. The trained subject breathes more slowly but more deeply than the untrained, and the increase in respiratory exchange during exercise is accomplished to a greater extent by an increased depth than an increased rate of breathing, which is again probably more efficient.

The net result of this is to enable the trained subject to bring oxygen to the active muscle cells and to eliminate wastes from those muscles more effectively and rapidly. Lactic acid accumulation is less because it is burned or reconverted to glycogen to a much larger extent during the actual exercise period. Therefore, accelerated respiration persists for a shorter time at the close of the exercise period; that is, the oxygen debt is less for the same quantity of work in the trained subject. Reduction in lactic acid accumulation is probably a factor also in delaying the onset of fatigue in the physically trained individual.

15. There seems to be no increase in the *number* of fibers in a "trained" muscle.

CHAPTER SEVEN

THE WORK OF
THE ALIMENTARY CANAL

I. Physiological Anatomy of the Alimentary Canal
  A. Evolutionary history
  B. Organs and glands of the digestive tract
  C. Microscopic structure
    1. Mucous membrane—glands 2. Smooth muscle 3. Connective tissue 4. Nerve cells and fibers 5. Blood and lymph vessels

II. The Nature of Secretion

III. Chemical Factors in Digestion
  A. Saliva
    1. Functions of saliva 2. The splitting of starches
  B. Digestion in the stomach
    1. Action of pepsin 2. The acid of the stomach
  C. Bile and pancreatic juice
    1. Emulsification of fats 2. The enzymes of the pancreatic juice
  D. Digestion in the intestines
  E. Bacterial action in the colon

IV. Control of the Digestive Glands
  A. Reflex control of salivary secretion
  B. Chemical control of the pancreas
    1. Pancreatic secretin 2. Secretin control of bile flow
  C. Gastric secretion
    1. Experimental collection of gastric juice 2. Phases of gastric secretion
  D. The intestinal juice

V. Movements of the Alimentary Tract
  A. Observations of gastrointestinal movements
  B. Swallowing
  C. Stomach movements
    1. Peristalsis 2. Emptying of the stomach
  D. Intestinal movements
    1. The small intestine 2. The colon

VI. Control of the Digestive Movements
  A. Nervous control of swallowing
    1. Reflex nature 2. Directing the food into the gullet
    3. Esophageal peristalsis

B. Function of the nerves of stomach and intestines

C. Control of the pyloric sphincter

D. Mechanical stimulation of gastrointestinal movements

    1. The function of bulk   2. Diarrhea and constipation   3. The action of cathartics

E. Evacuation of the rectum

VII. HUNGER, APPETITE, AND THIRST

A. Hunger contractions

B. Cause of the contractions

C. The mechanism of thirst

VIII. ABSORPTION OF FOOD AND WATER

A. Physical factors

B. Cellular work

C. Semipermeability of the intestinal wall

D. Entrance of foods into the blood

    1. Amino acids and sugars   2. Indirect route of the fats

IX. DISORDERS OF THE ALIMENTARY CANAL

A. "Indigestion" and "acid stomach"

B. Ulcers and cancer

C. Appendicitis—infections

D. Gallstones

The digestive tract modifies the foods which are eaten and absorbs them into the blood stream for distribution throughout the body. The modification of the foods is essentially a chemical one, in which the digestive enzymes convert the large molecules of proteins, fats, and carbohydrates into simpler, smaller particles, dissolved or finely suspended in water. Various mechanical factors normally assist in this process, preparing the food for rapid chemical digestion by breaking the relatively large masses into smaller pieces. Thus we recognize not only chemical but also mechanical factors in the digestive process. The mechanical movements of the alimentary canal, in addition, move the intestinal contents along, to be acted upon successively by a series of enzymes, and also mix the mechanically broken-up food with the digestive juices. And, when the foods are in a state in which they can be absorbed, the churning movements expose all parts of the entire semiliquid mass to the absorbing surface of the intestine. Finally, the intestine excretes certain wastes, which are, in turn, moved along the canal, eventually being eliminated from the body in the fecal matter, or *feces*.

I. PHYSIOLOGICAL ANATOMY OF THE ALIMENTARY CANAL

A. EVOLUTIONARY HISTORY

In the simpler multicellular animals the alimentary system not only digests food but also distributes it throughout the body: the gastrovascular cavity combines the activities of digestion and circulation. The gastrovascular cavity of *Hydra* (see Fig. 12, p. 64) extends through all parts of the relatively simple, two-cell-layered body, sending branches into each tentacle. With the evolution of the mesoderm, exemplified by the condition found in the flatworms, the problem of adequate distribution of food to all parts of the relatively thick mesodermal layer is met by the profuse branching of the gastrovascular cavity. No cell of the body is very far removed from the terminal ramification of some branch of the alimentary cavity. The movements of the whole animal serve to move the digested materials through this cavity. With still further specialization and increase in the bulk and complexity of the mesodermal structures, as in segmented worms like the earthworm, even this scheme becomes inadequate. Here we find the digestive system simplified into a tube running from mouth to *anus*, and an entirely new device, the blood circulatory system, serving to distribute the digested, absorbed materials, bridging the anatomical gap between the digestive cavity and the cells of the body which ultimately make use of the food.

B. ORGANS AND GLANDS OF THE DIGESTIVE TRACT

Although no longer required to distribute the food throughout the body, the alimentary canal of the higher animals, including the vertebrates, is, nevertheless, a rather complex system, comprising a series of distinct organs and structures: the mouth, or *oral cavity;* the *pharynx;* the gullet, or *esophagus;* the stomach; the small intestine, of which the first portion is called the *duodenum;* the large intestine, or *colon;* and the *rectum*, which latter communicates with the exterior through the anal orifice, or anus. In addition to these portions of the digestive tube proper, certain glands, such as the *salivary* glands, *liver*, and *pancreas*, develop from the primitive alimentary canal embryologically and remain in communication with it through their ducts.

The digestive canal of mammals lies mainly within the abdominal cavity; the esophagus lies in the thorax, traversing that cavity in the midline behind the heart and trachea, connecting the mouth cavity with the stomach after penetrating the diaphragm. From stomach to rectum, a fold of tissue (the *mesentry*) attaches the otherwise freely lying tube with the dorsal wall of the abdominal cavity. Through this fold of tissue run the blood and lymph vessels and the extrinsic nerves which supply the digestive organs. Lining the entire abdominal cavity and covering the abdominal organs is a moist layer of

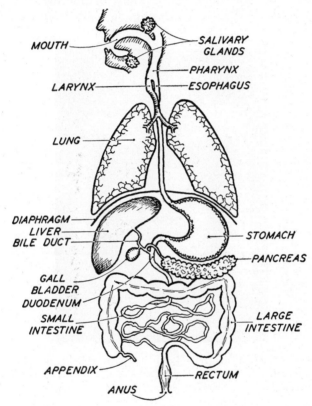

Fig. 92.—The organs and glands of the digestive tract. Part of the respiratory system is also shown.

thin, flat, semitransparent epithelial cells, known as the *peritoneal membrane*. From this membrane is derived the term "peritoneal cavity," as synonymous with the abdominal cavity.

C. Microscopic Structure

From esophagus to rectum there is a similarity in the general structural plan of the wall of the digestive tube. From the *lumen*[1]

1. The cavity of the tube, or of any vessel or duct or canal, is called the lumen.

outward are encountered the following layers: the *mucous membrane*, or *mucosa*, the muscular layers, and the connective-tissue layer. Nerve cells are also abundant in the wall.

1. *Mucous membrane—glands.*—The mucous membrane, especially in the stomach and intestines, is thrown up into numerous folds which markedly increase the secreting and absorbing surface of the tube. In addition, there are millions of microscopic finger-like projections of the mucosa into the lumen, called *villi* (see Fig. 94).

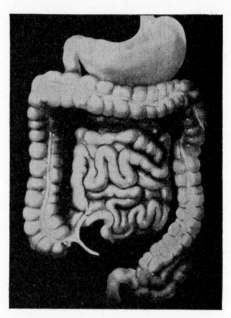

Fig. 93.—The gastrointestinal organs of man. The stomach lies above, the small intestine is coiled, and the large intestine outlines three sides of a square. The large intestine terminates in the rectum. (From the sound film, *Body Defenses against Disease.*)

It has been estimated that, because of the villi, the intestinal surface is increased as much as fifteen times in some areas. Its inner lining is composed of columnar epithelial cells of several varieties. Some of them secrete *mucus*, a viscid fluid which is a good lubricant; others are concerned with the absorption of digested foods; others secrete digestive enzymes.

In the main, the enzyme-secreting cells are arranged into tiny glands, with a saclike secreting portion, and a tiny duct leading directly into the lumen of the canal. Millions of such minute glands are found in the walls of the stomach and small intestine. The digestive glands, then, are of two main types anatomically: those of microscopic size, distributed profusely in the walls of the stomach and intestine, and the glands which lie as separate organs outside the digestive tract proper, communicating with it by large macroscopic ducts, through which their secretions are poured into the lumen of the tract. These are the three pairs of salivary glands, the pancreas, and the liver, which is the largest gland in the body.

2. *Smooth muscle.*—The muscular layer consists of smooth muscle. In most digestive organs it is arranged in two layers—circularly around the tube and longitudinally along the long axis. In the stomach the muscle is arranged in several layers. It is by virtue of

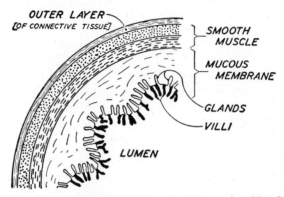

FIG. 94.—Structure of the intestinal wall shown in cross-section. Note how the mucous membrane is thrown into folds. Its surface area is still further increased by the many finger-like villi (*shown in black*). The muscles are arranged in outer longitudinal and inner circular layers.

co-ordinated contractions of these muscle layers that the movements of the digestive tube are possible.

At several points along the alimentary canal the circular muscle layer is thickened into bands of fibers which, upon contraction, can occlude the lumen of the tube at that point. Such bands of muscle are called *sphincters.* By relaxing or contracting, they permit the food to be moved onward or prevent or delay this movement of food. Such sphincters are located at the junction of the esophagus and stomach, the junction of stomach and small intestine—the *pylorus*—the junction of the small intestine with the large, and, finally, at the anus. Contraction of the sphincter at the junction of the esophagus with the stomach tends to prevent the stomach contents from being pushed into the esophagus by the digestive movements of the stomach. The pyloric sphincter between stomach and intestine is of considerably less importance in controlling the rate at which the stomach contents are emptied into the small intestine than was formerly thought to be the case.

3. *Connective tissue.*—Connective tissue not only pervades the mucous and muscular layers, binding them together, but also makes up the tough, elastic, flexible outer layer of the tube.

4. *Nerve cells and fibers.*—In the walls of the stomach, intestines, and esophagus are numerous nerves. Some of these are the terminal fibers of the extrinsic nerves of these organs, connecting them anatomically and functionally with the brain and spinal cord. But, in addition, complete neurones are to be seen especially just under the mucosa and between the circular and longitudinal muscle layers. These are nerve-cell bodies with short dendrites and axones, entirely contained within the wall of the tract. They comprise a diffuse network of nervous tissue which, in a sense, resembles the nerve net of some lower invertebrates (see p. 418). It is a kind of decentralized or peripheral brain, by which a degree of nervous control of the intestinal movements is possible independently of the spinal cord or brain. The full significance of this network, however, is not entirely clear.

5. *Blood and lymph vessels.*—The walls of the tract are richly supplied with blood and lymph vessels, which carry nourishment and oxygen to the cells of the digestive organs and drain away the wastes. In addition, they transport digested and absorbed foods from the alimentary canal to places of storage or utilization.

II. THE NATURE OF SECRETION

Secretion involves the movement of material across cell membranes. In the secretion of sweat, certain constituents of the blood pass through the cells of the sweat glands into the ducts which transport the sweat to the body surface. In salivary secretion, likewise, materials enter the gland cells from the blood and migrate through them into the salivary ducts.

Secretion involves more than osmosis, filtration, or diffusion, although these processes may contribute to it. First of all, secretory cells have a more or less characteristic structure. Whereas the cells of the capillary walls through which materials move largely by the purely physical processes mentioned are thin, flat structures, the gland cells generally are not. They are cuboidal or columnar in shape and usually possess clearly defined cytoplasmic inclusions. During the process of secretion the internal structure of the gland cells changes its appearance, certain of the cytoplasmic inclusions[2] becoming smaller in size and, more especially, fewer in number. In periods of rest these granules increase in number. This suggests that in secretion some specific activities of the living cells intervene in the process. The cells are more than simply passive channels through which materials are moved by forces external to the cell.

Additional information fully substantiates this conclusion. Some glands will continue to secrete even though physical forces operate to oppose the action. Experimentally, the pressure within the ducts of the salivary gland can be elevated considerably above the blood pressure. This, of itself, would tend to force fluid from the ducts, through the cells, and on into the blood by filtration. Yet in such an

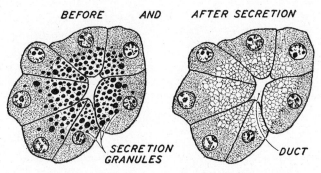

BEFORE   AND   AFTER SECRETION

SECRETION GRANULES

DUCT

Fig. 95.—Visible changes in secretory cells. Cell products seen as granules collect in the protoplasm during rest and are liberated (in solution) into the duct during secretion.

experiment the gland continues to secrete, continues to take up materials from the blood, and continues to force them into the ducts against this filtration pressure.

Secretion usually involves also the formation of specific new materials not found in the blood, or, at least, an increase in the concentration of certain compounds taken from the blood. Some dissolved components of urine are more concentrated than in the blood stream. Activities of the cells of the kidneys bring this about. In saliva we find compounds not present as such in the blood stream— products which have been manufactured in the cells from simpler ingredients of the blood or body fluids. Gland cells are factories of chemical synthesis, which involves protoplasmic activity.

Finally, it has been shown that during secretory activity the oxygen consumption and the carbon dioxide production of the gland

2. Called *secretion granules*.

increase. The manufacture of the secreted compounds, then, and the liberation of them from the cell require the expenditure of energy derived from oxidations and also the performance of work by the cell.

Ordinarily, we think of secretion as a polarized process in a cell in which it takes up materials from the blood and tissue fluids at one end and passes materials into ducts or cavities at the free end of the cell. A secretion-like cellular activity is also known in which materials are moved by cells in the reverse direction back into the blood. The cells lining the intestine, for example, take up materials from the cavity of the intestine, in absorption, and "secrete" them into the tissue spaces, whence they enter the blood stream. Likewise, certain cells in the kidneys absorb some of the constituents of the urine of the kidney tubules and pass them back into the blood. These activities are not usually referred to as "secretion," but apparently a similar type of cell activity is involved. It is, in a sense, the reverse of secretion, so far as direction of movement of fluid and solutes are concerned.

III. CHEMICAL FACTORS IN DIGESTION

Perhaps the first demonstration of the dissolving action of the juice of the stomach was made by the French scientist, René A. Réaumur, in 1752. He collected *gastric juice* in a picturesque manner from his pet bird—a kite. Since these birds eject indigestible material from the stomach, Réaumur had the bird swallow small sponges placed within short lengths of perforated tubing. Later, when the kite obligingly delivered back the tubes, he squeezed out the gastric juice soaked up by the sponges and tested its dissolving effect upon meat.

Chemically, digestion involves the splitting of the large molecules of the three major fuel and protoplasm-building foods into smaller, simpler, soluble, absorbable molecules. Proteins are broken down to their constituent amino acids, carbohydrates into simple sugars, and fats into fatty acids and glycerol. Many kinds of digestive enzymes are concerned. Not only are there specific enzymes[3] for each of these three kinds of foods but, in some instances, there are specific enzymes for successive stages in the breakdown of one kind of food.

3. We shall consider only the more important of the digestive enzymes. A review of the chemistry of carbohydrates, fats, proteins, and enzymes (see chap. 2) will be useful at this point.

A. SALIVA

1. *Functions of saliva.*—In all mammals the secretions of the salivary glands carry on certain functions unrelated to chemical digestion. Saliva dissolves certain of the constituents of the food and is therefore important in initiating the sense of taste, which can be aroused only by materials in solution. Saliva also protects the lining of the mouth against drying and facilitates expectoration of injurious or distasteful objects. It makes speech easier in man and also lubricates his food, thus aiding its passage down the esophagus.

In some animals, including the dog, the functions of the saliva seem to be entirely of a nondigestive nature. In a few mammals, including man, saliva contains a digestive enzyme called salivary *amylase*[4] (or *ptyalin*), which is manufactured by the salivary gland cells.

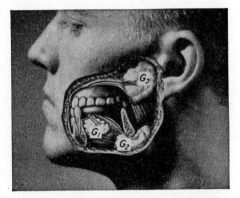

2. *The splitting of starches.*— Salivary amylase is an enzyme which splits (cooked) starches. It breaks down the large complex molecule to simpler components. Although this action is effected rather quickly, salivary digestion is usually in-

FIG. 96.—The salivary glands in man. Superimposed on the face of a man is a drawing showing the location of the three salivary glands (G_1, G_2, G_3) on the left side of the face. Note that from each gland a duct leads into the mouth cavity. (From the sound film, *Digestion of Foods.*)

complete by the time food is swallowed, even though the chewing process has been conscientiously prolonged. Most of the enzyme action occurs within the saliva-saturated masses of food during the early part of the sojourn of the food in the stomach, before the gastric juice has been thoroughly admixed with the food.

Remarkably enough, salivary amylase cannot split starches into simple sugars, and double sugars taken by mouth as such—for example, ordinary cane sugar—are unaffected by this enzyme. This striking example of enzyme specificity is puzzling. Why should salivary amylase be unable to split the bond between the two simple sugars in a double-sugar pair when it can easily split the apparently

4. The suffix -*ase* designates an enzyme. Amylase is an enzyme acting upon *amylon* (from the Greek for "starch").

similar bonds between successive pairs of molecules? We know no reason for this but accept it as indirect evidence for the existence of different kinds of bonds between the successive molecules of simple sugars in the starch molecule.

FIG. 97.—Digestion of starch by saliva. Into two beakers (*A* and *B*) are placed opaque suspensions of starch in water. Some saliva (which contains the enzyme salivary amylase) is mixed into beaker *A*. Later, the saliva has digested the starch, converting the large molecules into smaller, invisible molecules of double sugar, rendering the solution transparent (*A'*). The beaker lacking saliva remains unchanged (*B'*). (From the sound film, *Digestion of Foods*.)

Double sugars cannot be absorbed as such. Further enzyme action is necessary in their breakdown into absorbable simple sugars. This occurs later in the small intestine.

B. DIGESTION IN THE STOMACH

1. *Action of pepsin.*—The chief enzyme secreted by the numerous glands in the walls of the stomach is *pepsin*, a protein-splitting enzyme. But the large protein molecules are not completely digested to the amino acid stage in the stomach. Pepsin can effect only a partial breakdown. The products of peptic digestion,[5] although soluble, still cannot be absorbed in any appreciable quantities. Further digestive action is effected by pancreatic and intestinal juices.

2. *The acid of the stomach.*—Along with pepsin, the stomach secretes hydrochloric acid (HCl), which is indispensable for peptic digestion. In a neutral or alkaline solution pepsin acts upon proteins very slowly, if at all. Any acid hastens peptic action, but, of all the acids investigated, HCl is most effective in this. Therefore, though HCl is not itself an enzyme and is itself unable to split the proteins to any significant degree at body temperature, nevertheless it is an important digestive component of the gastric juice.[6]

5. Called *proteoses* and *peptones*. In time, pepsin can split these substances still further, even to the amino acid stage, but only extremely slowly.

6. An insignificant amount of fat digestion sometimes occurs in the stomach, which may contain some fat-splitting enzyme derived from the intestine. The acid nature of the juice, also, causes a slight amount of carbohydrate breakdown.

C. Bile and Pancreatic Juice

1. *Emulsification of fats.*—The bile contains waste material—the decomposition products of hemoglobin, or bile pigments. These are of no significance in digestion and are eventually eliminated in the feces, but bile also contains certain salts (the *bile salts*) which act upon fats. They break up the droplets, converting the fat into a fine emulsion[7] like the finely divided fat droplets of cream. There is no actual chemical breakdown of the fat molecules by bile but simply a fine physical subdivision of the fat particles. This greatly increases

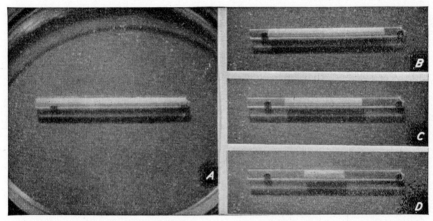

Fig. 98.—Digestion of protein by gastric juice. Into a flat dish containing gastric juice (*A*) is placed a glass tube with open ends, previously filled with boiled egg white. The gastric juice (which contains the enzyme pepsin) acts at the open ends of the tube, converting the solid protein into smaller soluble molecules (proteoses and peptones), which are invisible in solution. In *B, C,* and *D* are shown three successive stages in the progressive digestion of the egg white. (From the sound film, *Digestion of Foods.*)

the surface of fat which is exposed to the pancreatic fat-splitting enzymes, thereby hastening the chemical digestion of fat.

Perhaps even more important than this function, which can be effected by other substances normally present in the small intestine, is the role of the bile salts in the absorption of the digested fats. The bile salts appear to enhance the absorption of digested fat through the intestinal wall, perhaps by forming an easily absorbed molecule of combined fat and bile salt.

2. *The enzymes of the pancreatic juice.*—The pancreas secretes enzymes capable of splitting all three of the major food substances—carbohydrates, fats, and proteins. *Trypsin* (or pancreatic *pro-*

7. An *emulsion* is a colloidal suspension of one liquid (in this case, fat) in another (water).

teinase)[8] converts proteins into small custers of a few amino acids, and pancreatic amylase (or *amylopsin*) converts starches into a double sugar. The fat-splitting enzyme *steapsin* (or *lipase*) breaks

TABLE 4

SUMMARY OF THE MAIN FEATURES OF THE CHEMICAL FACTORS IN DIGESTION

Gland	Activated Mainly	Secretes	Digestive Juice Acts Upon	Digestion Products
Salivary	Reflexly	Salivary amylase (ptyalin)	Cooked starch	Double sugar
Stomach (gastric glands in stomach walls)	Reflexly and chemically (gastrin)	HCL* and pepsin (proteinase)	Proteins	Intermediate stages between proteins and amino acids
Liver	Chemically (secretin)	Bile salts*	Large fat droplets	Emulsified fat (small droplets)
Pancreas	Chemically (secretin)	Pancreatic amylase (amylopsin)	Intact or partially digested starches	Double sugar
		Trypsin† (proteinase)	Intact or partially digested proteins	Small amino acid groups (*peptids*)
		Steapsin (lipase)	Fats	Fatty acids‡ and glycerol‡
Small intestine (glands in intestinal wall)	?	Erepsin (*peptidase*)	Split products of gastric and pancreatic digestion of proteins	Amino acids‡
		(Enterokinase)	(Inactive trypsinogen)	(Active trypsin)
		Several carbohydrate-splitting enzymes	Double sugars	Simple sugar‡

* Not enzymes.
† Secreted as inactive trypsinogen, which is converted to active trypsin by enterokinase.
‡ In a state which can be absorbed.

up the fat molecule into fatty acids and glycerol. The enzymes of this important gland are thus capable of duplicating the digestive action on starches produced by the saliva and on proteins produced by pepsin; and, unlike salivary amylase, they are able to attack either cooked or uncooked starches. The pancreas is the

8. Secreted in an inactive form, called *trypsinogen* (see Table 4). The conversion of inactive trypsinogen to active trypsin by enterokinase is reminiscent of the conversion of inactive prothrombin into active thrombin by thrombokinase.

only gland, also, which secretes appreciable quantities of fat-digesting enzyme, so that one should expect a profound derangement of digestion if this gland is destroyed or removed.[9] The greatest disturbance of digestion resulting from removal of the pancreas is in fat digestion. The fats pass through the alimentary canal almost unchanged chemically and are lost to the body in the feces. The effects upon protein and carbohydrate digestion are less marked, because the salivary and gastric juices can digest these foods about as well as can the pancreatic juice.

The digestive juices secreted into the intestine—bile and pancreatic and intestinal juices—are alkaline in reaction. The pancreatic enzymes are found to act best in such a medium. Thus, before any appreciable digestion occurs in the intestines, the acid contents of the stomach are neutralized by the juices of the intestines.

D. Digestion in the Intestines

The term "intestinal juice" (*succus entericus*) is reserved for the digestive liquid secreted by the tiny glands lying in the walls of the small intestine, although it should be noted that bile and pancreatic juice are also secreted via the pancreatic and bile ducts into the intestine, where their digestive activities take place. Likewise, the contents of the stomach, including the gastric juice, also eventually enter the intestine, contributing to the intestinal fluids.

Several enzymes are present in the intestinal secretion. One of them, *erepsin* (or *peptidase*), although unable to affect intact protein molecules, can break down the products of gastric and pancreatic protein digestion to the amino acid stage. Other enzymes attack double sugars, derived either from salivary or pancreatic digestion (*maltose*) or directly from the food. Milk sugar (*lactose*) and cane sugar (*saccharose*) are examples of the latter. A number of double sugars find their way into the intestine, and, as we should expect from the general principle of the specificity of enzymes, each kind of double sugar apparently requires a specific enzyme for its conversion into simple sugars.[10]

9. Indeed, it was the desire to learn more about its effects upon digestion that caused investigators to remove the pancreas in experimental animals. This led to the discovery of the immediate cause of diabetes mellitus in man. Diabetes is not a condition of deranged digestion but is due to disturbance in function of that part of the pancreas which is not concerned with the elaboration of the digestive juices (see p. 509).

10. *Maltase* acts upon malt sugar, *lactase* upon milk sugar, and *invertase* upon cane sugar. The enzyme *enterokinase*, which converts pancreatic trypsinogen into trypsin, should also be included in a list of intestinal enzymes.

E. BACTERIAL ACTION IN THE COLON

There is an abundant growth of bacteria in the intestine, especially the lower small intestine and the large intestine. This is a perfectly normal condition, existing from soon after birth until death. If they remain in the intestinal tract, the bacteria are usually harmless, although they may produce disease and death if they invade the blood stream or escape into the abdominal cavity in large numbers.

Just what the significance of this extensive bacterial growth is, in health and disease, has not been clearly demonstrated, at least in carnivores. The bacteria themselves obtain energy mostly from the undigested foods which find their way into the colon. These energy-liberating processes are largely fermentations, in which several gases are produced.

Herbivorous animals are dependent to a considerable extent upon the action of the intestinal bacteria in the digestive process. The cellulose of plant cells effectively imprisons the stored food of the cells, unless it is destroyed by cooking or by bacterial action. The ordinary enzymes of the digestive tract are ineffective in attacking cellulose, though the intestinal bacteria can do so, at least in herbivores. Most herbivorous animals are equipped with a large out-pocketing of the colon, in which extensive bacterial fermentation occurs, liberating plant-cell products which can then be digested and absorbed. The cecum of man, from which the appendix is an outgrowth, is homologous with this colonic sac of herbivores.

It is undoubtedly true that some of the products of the metabolism of these bacteria (such as phenol and the amines) would be injurious to the body if they were absorbed into the blood stream in considerable quantities. It seems questionable, however, whether injurious quantities are ever absorbed, even in chronic constipation.

The deleterious effects of absorbed "poisons" of this origin are probably very much exaggerated, especially by unscrupulous promoters of remedies for "poisoning of the system."

IV. CONTROL OF THE DIGESTIVE GLANDS

Our interest in the control of the secretion of the digestive glands is as great as in the action of the enzymes secreted. What is it which determines when the digestive juices will be secreted? By what machinery is the rate of secretion correlated with the presence of food to be digested by those juices? What determines the copious flow of salivary juice into the mouth when food is present in the mouth or even when food is smelled or seen or imagined? What causes the gastric juice to be formed in great quantities when food enters the stomach, or even before the food arrives in that organ, while food is still being chewed? And what is the mechanism which insures marked pancreatic activity just at the time when food passes through the pylorus into the intestine?

To answer these questions requires experimental study of the glands in action. The rate of secretion of certain glands can be studied by collecting the fluid poured into the duct. A fine glass tube (or cannula) can be inserted into the duct of a salivary gland or the pancreas. The rate at which the juice emerges from the tube (drops or cubic centimeters per minute) will be a measure of secretory activity. Such procedures are usually limited in their application to anesthetized animals. Occasionally in man injuries may expose gland ducts and permit similar observations.

It is also possible to observe the rate of flow in an animal previously so operated upon that the duct of the gland in question leads to the surface of the body. The juice formed by the gland, instead of emptying into the alimentary canal, is carried to an opening in the skin, where it can be collected and its rate of formation determined. The latter method can be used in the unanesthetized animal, provided the procedures used to stimulate the gland can be employed without inflicting pain on the animal. The mere ingestion of food, for example, will cause a great increase of salivary flow.

A. REFLEX CONTROL OF SALIVARY SECRETION

If the nerves of the salivary glands are stimulated electrically, there is an immediate pronounced increase in rate of secretion of saliva. Nerve impulses in the efferent fibers are capable of initiating activity of the gland cells just as a motor nerve activates a muscle. But are such efferent impulses initiated physiologically, and, if so, by what mechanisms? That they do occur when food is chewed is indicated by the fact that, if the nerves are severed beforehand, the chewing of food produces no acceleration of salivary flow. We deal here with a reflex. Food in the mouth stimulates neither the salivary glands themselves nor the efferent nerves, but it does stimulate the endings of sensory or afferent nerves. Sensitive organs called *taste buds*, located chiefly in the tongue, are activated chemically by the food in the mouth. Other nerve endings in the mouth are stimulated mechanically in chewing. The impulses set up in these nerves are transmitted to a region of the brain and are "reflected" back along the efferent secretory nerves to activate the salivary gland cells. This occurs independently of sensations of taste. Severing afferent nerves[11] will therefore also interfere with, or prevent, salivary secretion when food is being chewed.

B. Chemical Control of the Pancreas

The mechanism controlling pancreatic secretion is of quite a different nature, though the adaptive result is the same—that is, a copious secretion of the juice at the time food is present to be acted upon. Stimulation of the nerves of the pancreas has but little effect upon secretion, and severing of all the nerves of the gland fails to prevent the augmented flow which occurs normally very soon after the acid gastric contents start to enter the intestines. Experiment has shown that the same increased flow occurs if HCl is artificially introduced into the upper intestine. Physiologically, of course, HCl enters the duodenum along with other gastric contents. It is apparently this acid, then, which is involved. But how? How can acid in the duodenum cause an increased activity of the pancreas, lying at some distance away? We have already shown that nervous connection between the two is not a factor.

1. *Pancreatic secretin.*—These problems began to be solved when it was discovered that, upon placing a dilute solution of HCl into a washed loop of intestine previously removed from a recently killed animal, and allowing it to remain for a while, a substance is formed by the walls of the isolated intestinal loop which, when injected into the blood stream, has a stimulating action upon the pancreas. The injection of HCl into the blood has no effect upon pancreatic secretion. But the injection into the blood of only a few cubic centimeters of the fluid from the intestinal loop into which HCl has been placed has a marked stimulating action. The nature of the substance formed

11. The sensory nerves of the mouth are not the only ones whose stimulation, physiologically, will initiate reflex salivary flow. Increased salivation at the sight or smell of food indicates that the *optic* and *olfactory* nerves may also be involved. And, as all have experienced, even the thought of food may make the mouth "water." These reactions are of a special nature, being examples of *conditioned reflexes* (see p. 457). As contrasted with the inherited or unlearned reflex salivary response to stimulation of the taste buds by food, these are learned responses, dependent upon the experience of the individual in associating the taste of food with its odor and appearance. Abnormal irritation of nerve endings in the stomach and intestines, associated with the feeling of nausea, may also increase salivary flow.

by the cells of the intestinal wall under the influence of acid[12] has been studied in detail; and, though it has been prepared in fairly pure form, its exact formula is unknown. It has been given the name *secretin*, and, because it acts upon the pancreas specifically, it is sometimes referred to as pancreatic secretin.

From the foregoing and other evidence we may picture the mechanism as follows: As material passes through the pylorus into the in-

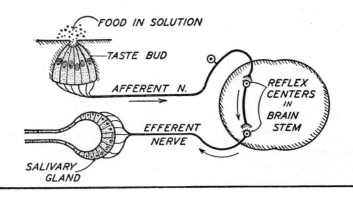

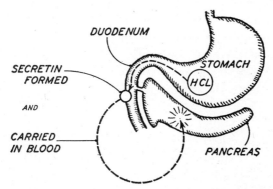

Fig. 99.—Nervous control of salivary secretion (*above*) and chemical control of pancreatic secretion (*below*). By means of these distinctly different mechanisms, saliva flows when food is present in the mouth, and pancreatic juice is secreted when food (and HCl) enters the small intestine. Arrows (*above*) indicate direction of nerve impulses. Dotted arrows (*below*) indicate course taken by chemicals. (See discussion in text.)

testine, the HCl in it acts upon some substance (*prosecretin*) found in the cells of the intestinal wall, converting it into secretin. Some of the secretin may escape into the intestine, where it is probably of no physiological significance. But some of it is liberated directly into the blood stream. Being carried to the pancreas, it exerts a

12. Acids other than HCl have the same effect.

chemically stimulating effect upon the secretory cells of that gland, increasing the secretory rate (see also Fig. 205, p. 568).

Apparently there are two components of the crude "secretin" described: one stimulates the pancreas to secrete water and salts; the other excites the secretion of enzymes. The two substances are chemically related but can be separated and their two different effects demonstrated.

2. *Secretin control of bile flow.*—Secretin also stimulates the liver cells to increase production of bile. This accounts in part for the fact that bile flow, as well as pancreatic secretion, is augmented by the emptying of the stomach contents into the intestine.

Further evidence indicates that a substance is also formed by the intestinal walls under the stimulus of fats especially and liberated into the blood stream. This substance has been given the name *cholecystokinin*. Its action seems to be specific upon the gall bladder, which lies against the liver and is an outpouching of the liver duct leading from the liver to the duodenum. Its walls contain smooth muscle, and in its hollow interior is stored some of the bile formed by the liver between meals. Cholecystokinin causes the gall bladder to contract and to expel into the intestine the stored bile it contains.

Thus the entrance into the intestine of the gastric contents (containing HCl and fats) not only stimulates bile production by the liver but also causes previously formed bile to be expelled from the gall bladder into the duodenum.

C. Gastric Secretion

1. *Experimental collection of gastric juice.*—The secretory activity of the stomach is somewhat more difficult to study than that of the pancreas or salivary glands because here we find no single duct collecting all the juice, which duct can be transplanted conveniently to the surface of the body. The juice must be collected from the organ itself. This can be done fairly easily upon man or animal by inserting a rubber tube down the esophagus into the stomach and aspirating the gastric contents. Or an operation can be performed upon an experimental animal, making an artificial opening from the stomach through the abdominal wall. Important studies have also been made upon human beings with such openings, produced by accident or in surgical treatment for the alleviation of closure of the esophageal tube by disease; semiliquid food must be pushed directly into the stomach by such people through the artificial opening.

A serious objection to all these methods of collecting gastric juice and determining what factors affect the secretion is that the juice so obtained is contaminated by food. To be sure, pure juice can be collected when there is no food being eaten or present in the stomach. But the most desirable time to study secretion is during digestion.

The great Russian physiologist, Ivan Pavlov, overcame this difficulty by the surgical construction of a pouch made of part of the stomach. A Pavlov pouch is so constructed that its cavity is completely separated from the lumen of the main stomach, and there is no interference with digestion of ingested food or its normal passage through the main stomach. Care is taken not to disturb the innerva-

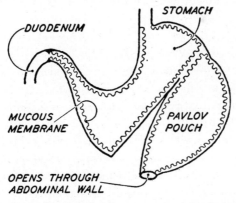

FIG. 100.—The Pavlov pouch. The pouch is so constructed that its cavity is separated from the cavity of the stomach proper; it opens to the outside, so that its juice can be collected; its nerves and blood vessels remain intact.

tion of the pouch. The perfectly normal lining of the pouch secretes at a rate parallel with that of the main stomach. The pouch opens through the abdominal wall, and the juice it secretes can easily be collected (see Fig. 100). The effect of food ingestion on secretion can be observed without contamination of the juice with food.

2. Phases of gastric secretion.—Gastric secretion is accelerated with the ingestion of food, just as is the secretion of salivary juices. In the stomach, however, there is a combination of nervous and chemical stimuli. Both in man and in an animal with a Pavlov pouch there is an increased rate of secretion even before food is swallowed. This early secretion is completely abolished in animals by cutting the vagus nerves,[13] which carry efferent fibers to the

13. Or by the injection of *atropine*, which paralyzes the vagi. Such "chemical" cutting of the vagi can be done on intact animals, including man.

stomach. Here again is a reflex effect, in which the afferent fibers are chiefly from the mouth. Presence of food in the mouth of the healthy and hungry animal initiates reflex gastric as well as salivary secretion, the afferent nerves in each case being stimulated by food in the mouth. Again, as with saliva, the gastric juice can also be produced by conditioned reflexes developed as a result of learning to associate the sight and smell of food with its taste: the sight or smell of food initiates the reflex.

The adaptive significance or utility of this mechanism is quite apparent. Considerable quantities of gastric juice are poured into the stomach even before the arrival of the food it is to digest.

Although section of the vagus nerves abolishes this so-called "nervous" or *cephalic phase* of gastric secretion, it does not abolish all gastric secretion. Even with all the nerves of the stomach cut, the organ still secretes copiously when food is present inside its cavity. This is, in some measure, due to a stimulating effect of distention of the stomach by the food. In part, the secretion is controlled chemically by a device closely resembling the pancreatic secretin mechanism. The partially digested products of proteins cause the liberation by the stomach wall of a material or materials into the blood stream. Such material, called *gastrin*, chemically stimulates the stomach glands to increased secretory activity. If a piece is cut out of the stomach and transplanted with blood vessels intact to some new site, such as the skin, it is found that the transplanted mucosa will secrete when certain food substances, especially protein decomposition products, are placed into the stomach. Since all nerves of the transplanted tissue have been cut, the effect must be chemical, reaching the tissue through the blood stream. That the chemical stimulating agent is not some absorbed product of digestion is indicated by the fact that such products injected intravenously fail to elicit a significant gastric response. On the other hand, decoctions of gastric mucosa, upon which protein digestion products have acted, do have this stimulating action upon injection. This second stage of gastric secretion is known as the *gastric phase*.

D. The Intestinal Juice

The mechanisms controlling the secretion of intestinal juice have not been clearly worked out. The roles of nervous, chemical, and mechanical stimuli are not apparent from the data available at present. That there are correlating devices we know from the fact that intestinal juice also, like the other juices studied, is secreted in great amounts at the time food is present in the intestine to be digested.

V. MOVEMENTS OF THE ALIMENTARY TRACT

The most common type of movement displayed in the alimentary canal is called *peristalsis*. This occurs in the esophagus, stomach, small and large intestines, and rectum. Peristalsis is a progressing band of constriction which slowly passes along the digestive tube for some distance. It is effected, of course, by a series of co-ordinated contractions of the circular smooth-muscle fibers of the various

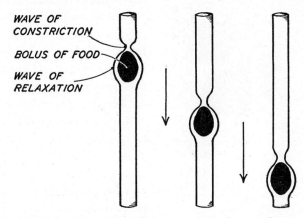

WAVE OF CONSTRICTION

BOLUS OF FOOD

WAVE OF RELAXATION

FIG. 101.—Peristalsis in the esophagus. Three stages in the downward movement of a solid *bolus* of food are shown.

organs. Generally, the contraction wave is preceded by a wave of relaxation. Figure 101 shows the nature of peristalsis in diagrammatic form.

A. Observations of Gastrointestinal Movements

To a limited extent peristalsis and the other movements of the alimentary canal can be studied by direct visual inspection of the exposed tract in the anesthetized animal. For the most part, however, anesthetics abolish or at least markedly diminish these movements, and the tract lies quite motionless even when filled with food. It is necessary, therefore, to make indirect observations on the intact animal. The methods commonly employed are the X-ray and the balloon-manometer techniques.

If an unanesthetized man or animal is fed a meal containing some material opaque to X-rays, such as an innocuous salt of barium, the organs filled with the material will cast a shadow upon an X-ray photographic plate, or upon a fluoroscopic screen, which can be observed visually. The dark shadow of the organ stands out in contrast with the surrounding tissues, through which X-rays easily penetrate. Contractions of the organ, or progressive waves of contraction passing over an organ, or the onward movements of the contents of the alimentary canal, can be fairly well observed.

The rubber-balloon–water-manometer method for determining the motility of hollow organs has already been outlined in the discussion of changes in intrathoracic pressure during breathing. With such a rubber balloon in the esophagus, moderately inflated, it is found that at each swallowing movement the writing point of the manometer float rises, indicating that the balloon has been compressed by a wave of contraction which has passed down the gullet. The balloon may likewise be swallowed into the stomach, in which case the contractions or relaxations of that organ also will produce movements of the writing point of the manometer. In any case, permanent records of the contractions may be inscribed by means of the kymograph (Fig. 103).

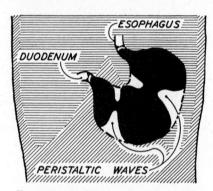

Fig. 102.—Peristalsis in the stomach as it appears by X-ray. The stomach, viewed from in front, has been filled with a liquid containing barium salts, which are opaque to X-rays. Two peristaltic waves are shown. The stomach is shown closed off from esophagus and duodenum by contraction of the *sphincters* (see also Fig. 104).

It is difficult to introduce such a balloon with the attached rubber tube into the intestine beyond the stomach. Studies upon the movements of the intestines by this technique are almost completely limited to animals in which openings from intestine through the abdominal wall have been produced surgically with aseptic technique.

B. Swallowing

Both the X-ray and the balloon techniques demonstrate peristaltic waves in the esophagus. In this case direct inspection of the exposed esophagus in anesthetized animals can also be employed. Each wave sweeps downward from the upper end of the tube to the stomach in a continuous, progressing movement, pushing semisolid pieces of masticated material down to the stomach. In general, there is one esophageal peristaltic wave for each swallowing movement initiated from the mouth region. Water or semiliquid foods immediately pass rapidly down to the lower end of the esophagus partly by the action of gravity. The peristaltic wave coming down a few seconds later merely sweeps the fluid past the sphincter guarding the entrance into the stomach, which relaxes in advance of the peristaltic wave.

C. Stomach Movements

1. *Peristalsis.*—Very soon after the food has reached the stomach, peristaltic waves are initiated in the organ. In fact, swallowing may start them, although these waves are not simply an onward continuation of the esophageal peristalses. They arise in the full stomach even though the esophagus is quiescent, and, in any case, early in the digestive process they do not involve the entire organ. At first the waves arise about midway between the esophageal orifice and the pylorus and pass from left to right toward the pylorus. At this stage the main body of the stomach, containing the bulk of the swallowed food, is nearly quiescent. As digestion proceeds, the peristaltic waves originate closer and closer to the esophageal end of the stomach, gradually encroaching upon the main mass of the food. Finally, two or three hours after a meal is ingested, the waves originate almost at the esophageal orifice and sweep over the entire organ. In one to four hours after the in-

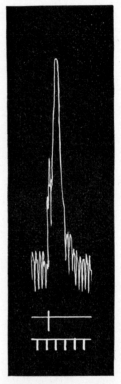

Fig. 103.—Kymographic record of a swallowing movement, record-ed with the apparatus shown in Fig. 78. The tracing shows respiratory deflections (see p. 227), and soon after the subject voluntarily swal-lows (*indicated by signal*) there is a marked rise in the curve caused by a peristaltic wave compressing the balloon in the esophagus. Time in 5-second intervals.

gestion of food, depending upon the kind and quantity, the stomach has emptied practically all its contents into the duodenum.

2. *Emptying of the stomach.*—For a long time it was thought that intermittent opening and closing of the pyloric sphincter determined

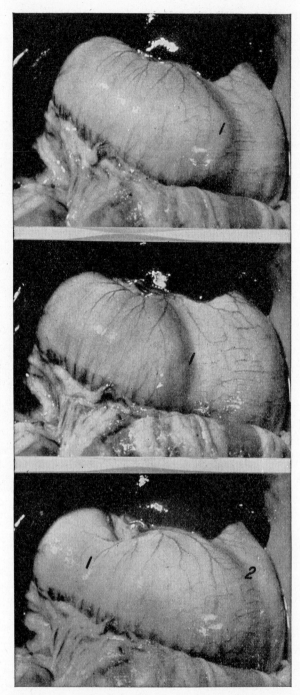

Fig. 104.—Peristalsis of the stomach. The organ is viewed from in front and below, so that the left margin of the stomach is on the right side of the pictures. A peristaltic wave (*1*) passes over the stomach from left to right (*top picture to bottom*).By the time this wave has reached the right (pyloric) end of the stomach (*bottom picture*), a new peristaltic wave (*2*) has started from the left. (From the sound film, *Digestion of Foods*.)

the rate at which the stomach contents are moved into the duodenum. Apparently such sphincter action has little, if any, such control. The major factors seem to be the general tone of the stomach and the force of the gastric peristalses. As gastric digestion proceeds, and the stomach as a whole gradually contracts, the increasingly stronger peristaltic waves push the more liquid contents of the stomach into the duodenum.

The presence of certain food substances or digestion products in the duodenum inhibits gastric tone and peristalses. This effect is particularly marked in the case of fats. As a consequence meals with a high fat content remain in the stomach a relatively long time. It is probable that this effect is produced primarily reflexly, since cutting the vagus innervation of the stomach and duodenum greatly reduces or may even abolish the phenomenon.

D. INTESTINAL MOVEMENTS

1. *The small intestine.*—Throughout the length of the small intestine the movements are of two main kinds. Peristaltic waves occur, slowly moving the digesting materials onward. These waves ordinarily do not progress for any considerable distance. In human beings each wave will move along for several inches and then "die out." In addition to the peristalsis, there are also movements of a churning nature called *rhythmical segmentation.* These are alternating contractions and relaxations of successive segments of intestine, which do not progress onward in a co-ordinated manner, as do the peristaltic waves.

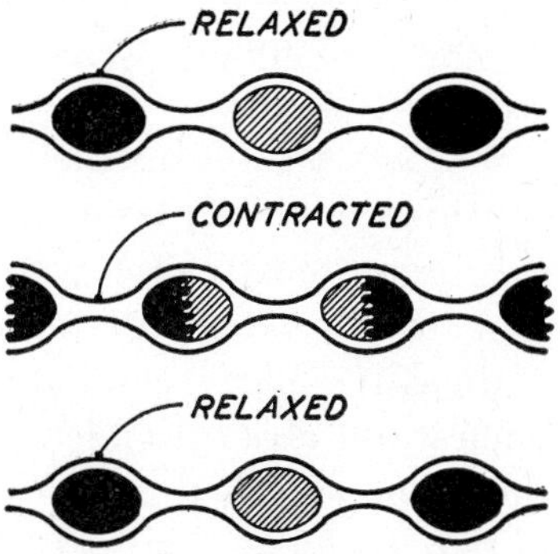

We might compare peristaltic waves to "stripping" a fluid-filled rubber tube by a gentle progressing compression between the fingers, which are moved slowly along the tube. The

FIG. 105.—Churning movements (rhythmical segmentation) in the small intestine. Three successive stages are shown in a small segment of bowel. Digesting masses of food are shown crosshatched and in black. The masses of food become thoroughly admixed with one another, and with the digestive juices by means of these movements.

churning movements would resemble rather an alternating compression and release of the tube at numerous points, with no onward movement of the compressing fingers.

This kind of movement mechanically breaks up the intestinal contents, which are now pretty well macerated, and thoroughly mixes them with the juices of the pancreas, liver, and intestines. In addition, all parts of the liquid mass are brought into contact with the absorbing surfaces of the intestinal walls, through which the digested food enters the blood stream. This is especially important in the lower small intestine, where most of the absorption of digested foods takes place.

Thus, in any segment of the intestine, these churning movements will continue for a time, and then a peristaltic wave will move the materials onward a short distance into the next segment of intestine, where the process will be repeated. And so, slowly and interruptedly, the materials are moved onward, ultimately into the large intestine. Approximately three to six hours are required for the passage of material from the pylorus to the end of the small intestine, a distance of approximately 22 feet in the human adult.

By this time digestion has gone to completion, and nearly all the digested food particles have been absorbed. The materials which pass into the large intestine are of semiliquid consistency, containing considerable quantities of water derived either from ingested liquids or from the digestive juices.

2. *The colon.*—In the colon there occur both peristalsis and churning movements somewhat like those of the small intestine. In this organ most of the water is absorbed. This consists of not only some of the water ingested with the food but also some of the water of digestive juices secreted higher up in the alimentary canal as well. Such water absorption leaves the contents of the terminal portion of the colon—reached after a stay of about twenty-four to thirty hours in that organ[14]—semisolid, having the consistency normally of butter in the summertime. This fecal material owes its normal color mainly to the bile pigments and the decomposition products of the pigments. It is made up of a residuum of indigestible materials such as cellulose, excretions of the large intestine,[15] some unabsorbed secretions of the small intestine, and, finally, large masses of bacteria, which form approximately half of the dry material of feces.

14. The total time required for materials to pass from mouth to rectum is normally thirty to forty hours.

15. In addition to absorbing water, the colon is an important organ of excretion. Certain constituents of the blood, notably excess calcium and iron, are excreted as salts into the lumen of the colon.

VI. CONTROL OF THE DIGESTIVE MOVEMENTS

A. NERVOUS CONTROL OF SWALLOWING

The factors responsible for initiating and controlling the gastro-intestinal movements are nervous and mechanical in the main. The swallowing act, including the peristalsis of the esophagus, is entirely dependent upon nerves; and, if these are injured or destroyed experimentally or by disease, the individual is unable to swallow. Swallowing is reflex in nature. We think of the act ordinarily as being "voluntary"—that we can swallow, or refrain from swallowing, as we wish. This is true only of the initial act of the complex mechanism. The first part of swallowing consists of contractions of a series of muscles in the mouth region which push the oral contents back into the pharynx; from this point on, the series of reflexes is entirely beyond the control of the will. We are quite unable to stop the act of swallowing once materials in the mouth have reached the pharynx. Correspondingly, we are not consciously aware of most of what goes on from this time forward.

1. *Reflex nature.*—There is a sharp contraction of the muscles comprising the walls of the pharynx, which moves the food downward. In this reflex the afferent side of the arc is made up of certain of the cranial nerves (*trigeminal, glossopharyngeal,* and the superior laryngeal branch of the vagus) which innervate the mucosa of the back of the mouth, the pharynx, and the laryngeal region. Artificial stimulation of the central end of these nerve trunks even in an anesthetized animal will initiate the swallowing complex quite as readily as will mechanical stimulation of the nerve endings in the pharynx of a normal individual by a bolus of food or a quantity of water or saliva.

Besides the afferent nerves mentioned, sensory nerve fibers in the vagi, innervating the mucosa of the esophagus itself, are also capable of initiating reflex esophageal peristalsis. Such reflexes are set off when a bolus of relatively dry food becomes lodged in the esophagus, mechanically stimulating it. The resultant peristalses serve usually to move the obstructing bolus downward.

2. *Directing the food into the gullet.*—In directing the food into the gullet, the contraction of the pharyngeal muscles is only a small part of the complex. It should be obvious that such contraction can do no more than squeeze upon the pharyngeal contents. If this alone occurred, the material might be forced in one of several directions

besides into the gullet: back into the mouth proper, into the nasal cavity, or into the trachea. But, in swallowing, all these possible pathways for food except the normal one are closed off by muscular contractions. The tongue muscles, pressing that organ against the roof of the mouth and fitting against the cheeks on both sides, close off the anterior portion of the mouth cavity. Other muscles elevate the roof of the mouth (the *soft palate*) and prevent passage of food into the nasal cavity. In those infants in whom the palate remains open in the midline (*cleft palate*) due to faulty embryonic development, there is a regurgitation of food into the nose and out of the nostrils, with each swallow. This may be serious enough to cause death of the infant from undernourishment unless the defect is soon corrected surgically.

The most complicated of the activities directing the course of the food into the proper channel is associated with preventing it from entering the breathing passages. Such a contingency is serious even if the bit of food is not large enough to cause mechanical obstruction, because food is contaminated with bacteria, and the entrance of even a small particle into the lungs might start a serious infection.

Considering the fact that breathing goes on at all times, it would seem that it would be very easy to aspirate particles into the lungs if an inspiration should occur just at the time a swallowing act is initiated. But the swallowing machinery is so geared with the breathing mechanisms as to make this virtually impossible. Not only is inspiration involuntarily inhibited during swallowing (see chap. 6, p. 237), but it is well-nigh impossible by an act of the will to swallow and to inhale at the same time. This is because those same sensory nerves whose stimulation reflexly causes the swallowing movements at the same time transmit impulses to the respiratory center and inhibit it. This is easy to demonstrate experimentally. Electrical stimulation of the central end of such a nerve (e.g., the superior laryngeal) not only causes swallowing but also inhibits respiration. The machinery is so adjusted that the respiratory inhibition is only momentary—just long enough for the material being swallowed to get past the opening of the larynx.

A still further guaranty of correct passage of the material is found in the contractions of the muscles which elevate the larynx. By palpation one can verify the fact that with each swallowing movement the larynx is lifted upward. Study of the anatomical relationships in Figure 106 shows why this is effective. Elevation of the larynx pulls it up under the base of the tongue, which presses the over-hanging flap (the *epiglottis*) against the opening into the larynx, effectively closing it off.

Thus, in swallowing, the mechanical stimulation of afferent nerves by the bolus of food causes a reflex discharge through a number of motor nerves to a series of muscles whose contractions close off the nasal, oral, and laryngeal cavities, besides pushing the food down into the gullet.

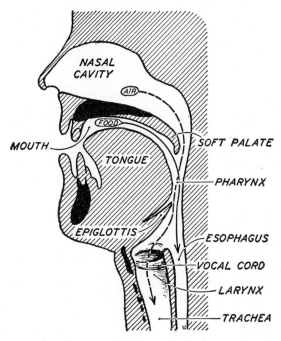

Fig. 106.—Relationships of the passageways for food and air. Numerous devices (discussed in text) prevent swallowed food from entering the larynx or trachea.

3. *Esophageal peristalsis.*—Part of the efferent discharge in the swallowing reflex is over the motor nerves (branches of the vagi) to the esophageal musculature, which initiates a peristaltic wave in that tube. The wave of peristalsis passes slowly down from the pharynx to the opening into the stomach. Such a wave is quite capable of pushing food or water along ahead of it into the stomach.

As pointed out earlier, however, ordinarily well-chewed food is finely divided and semiliquid, so that it needs no peristaltic wave to transmit it through the esophagus. The impetus given it in the initial phase of the swallowing act is sufficient, aided often by gravity, to project the mass at once down the gullet. It comes to rest at the lower end of the gullet, prevented from immediately entering the stomach by the sphincter located at that point. By listening with a stethoscope just below the sternum, one can hear a mouthful of swallowed water splash lightly as it encounters the resistance of the closed sphincter.

Nevertheless, a reflex peristaltic wave is set up in the tube and slowly moves downward. As is generally true of peristalses everywhere, the wave of constriction is preceded by a wave of relaxation. When this latter wave reaches the sphincter, it opens the contracted band, and the waiting material is propelled into the stomach. The sound from this—a gurgling—can also be heard with a stethoscope about five seconds after the initiation of the swallowing act. The five seconds is the period of time required for the peristaltic wave to traverse the gullet.

These peristaltic movements can be seen in an experimental animal whose esophagus is exposed under anesthesia. Even in intact animals or in man they can be studied by the balloon technique. Each time a peristaltic wave passes down the esophagus it compresses the balloon and causes the writing point to be elevated. A typical record of such an experiment has been shown in Figure 103.

B. Function of the Nerves of Stomach and Intestines

When we pass to the gastrointestinal tract beyond the gullet, we find extrinsic nerves playing a much less prominent role. It can easily be shown that stimulation of one set of efferent nerves (branches of the vagi) augments the movements of the stomach and intestines, and stimulation of another set (branches of the *splanchnic nerves*) tends to inhibit these movements. However, if all the extrinsic nerves of the stomach and intestines are cut, the digestive movements —though somewhat modified—will still continue. In fact, even if a strip of intestine is removed from a recently sacrificed animal and placed in a suitable warm salt-water medium, the movements continue. In this respect the movements of the stomach and intestine are quite unlike those of the esophagus, which is completely paralyzed (except at its lower end) when deprived of its nerves. But cutting the extrinsic nerves of the stomach or intestines or removing a strip from the body does not remove these organs from all nervous control. Recall that there is a network of nerves within the walls of the digestive tube made up of complete nerve cells with their short processes, constituting the same sort of diffuse nervous system encountered in some of the lower invertebrates. Just what role these neurones play in controlling the gastrointestinal movements is not clear at present. Determining their function by removing them seems impossible, since such removal would entail extensive damage to the muscles.

In this we are reminded of the heart, whose spontaneous movements are not stopped by denervation or even by isolating the organ from the body. The heart also is equipped with a similar intrinsic network of nerve cells and fibers. Here, also, the full significance of the peripherally located neurones is still a problem.

C. Control of the Pyloric Sphincter

We have seen that the stomach is a reservoir for food and that it empties itself gradually, forcing only small quantities of semiliquid material into the intestine at a time. The pyloric sphincter is closed part of the time during digestion and is relaxed much of the time during the digestive activities of the stomach.

Just what factors control the activity of this sphincter are not entirely known. The consistency of the stomach contents is apparently a factor. Water which is drunk (e.g., a pint or so) is passed out of the stomach through the sphincter in about ten minutes or less, while solid foods may not be completely emptied for four or five hours. The strength of the peristaltic contraction of the stomach also seems to be involved. In general, a powerful peristaltic contraction sweeping over the stomach is more likely to open the sphincter in advance of the wave of constriction. Weaker contractions are less likely to do this.

Until recently it was thought that the reaction (acidity or alkalinity) of the gastrointestinal contents in the neighborhood of the sphincter, especially on the intestinal side, was the main factor controlling the sphincter. It was said that, when acid is introduced into the duodenum just beyond the sphincter, the muscle band tightens, closing the opening. When alkalies are introduced, it opens. A most satisfying, but probably inaccurate, theory resulted: The gastric juice is acid in reaction, and, as soon as some of it is moved into the intestine by gastric peristalsis, the sphincter is closed. Gradually the acid is neutralized by the alkaline bile and pancreatic juice, and the sphincter relaxes, so that a peristaltic wave of the stomach will force through a little more material. Again the sphincter closes at once. Repetitions of the process lead to a gradual emptying of the stomach.

The persistence of this theory is to be ascribed to its attractiveness, for it seems to explain so well what happens. Here, indeed, we may well heed Francis Bacon's golden counsel, "Whatever the mind seizes upon with peculiar satisfaction is to be held in suspicion." Suspicious scientists repeated the earlier experiments and found them inconclusive. Not only acid, but even strong alkalies, or almost any stimulation of the intestinal mucous membrane beyond the pylorus, cause closure of the sphincter. Furthermore, many apparently normal people secrete no acid whatsoever in the gastric juice, and yet the action of the pyloric sphincter and the rate of emptying of the stomach seem to be practically normal.

It is probable that the passage of food from the stomach depends far more upon the gradual contraction of the whole stomach, the strength of the gastric peristalses, and the consistency of the food than upon any acid control of the pyloric sphincter.

D. Mechanical Stimulation of Gastrointestinal Movements

1. *The function of bulk.*—One of the most important factors in stimulating the gastrointestinal digestive movements is the presence of food within the interior of the organs. Mere bulk within the canal seems to exert some sort of mechanical stimulation of the motor machinery of the surrounding wall. Perhaps the adequate stimulus here is stretch of the muscles and nerves; perhaps local reflexes, limited to the nerve plexuses within the gastrointestinal walls, also play a role.

In the stomach and intestines the food itself plus the digestive juices seem adequate to promote peristalsis. But, when digestion is complete, and the digestion products have been absorbed—that is, in the colon—the peristaltic movement is likely to be sluggish unless the food eaten contains some *roughage*. Roughage refers to indigestible components of the food which reach the colon practically unchanged and, by their bulk, promote and stimulate normal colonic movements.

2. *Diarrhea and constipation.*—If the contents of the colon are moved rapidly through that organ, there is diminished opportunity for water absorption, and the material reaches the rectum in a semifluid condition. The frequent, loose bowel movements of *diarrhea* are the consequence. If materials are moved through the colon slowly, there is excessive water absorption. The fecal matter becomes hard and dry, bowel movements are infrequent, and the rectum is evacuated with difficulty. This is called *constipation*.

Between these extremes is the normal condition, in which the stools passed are of approximately the size, shape, and consistency of an overripe banana. Although normally there is about one bowel movement per day, the number of movements is a less adequate criterion of normal colon motility than the consistency of the stools.

The maintenance of normal colon motility and the avoiding of constipation depend largely upon the inclusion of some roughage in the diet. Yet too much roughage may be harmful and cause "bowel distress," by mechanical irritation of the mucosa of the colon.

Proper bowel function is important, but not all the troubles and failures of man are due to constipation, even though a multitude of advertisers continually urge buying one or another product to induce "proper elimination" and to dissipate that afternoon drowsiness which bars the way to business and financial conquest. It would seem that some cathartics are almost indispensable for social success, not to mention capturing a wife or a husband. Millions of dollars are spent annually by advertisers to frighten, cajole, and encourage us to resort to cathartics to "help nature." Needless to say, these millions come from the pockets of those who are "sold." There is a real place in medicine for cathartics, but in most cases of self-diagnosis and self-medication cathartics probably do more harm than good.

3. *The action of cathartics.*—Certain of the natural foods appear to have a mild stimulating action upon the colon movements—a laxative effect—independently of bulk. This seems to be the case with orange juice and prunes. How the laxative effect is produced is not known, but it is suggested that there may be some chemical stimulating action. Cathartics like "Epsom salts" ($MgSO_4$) owe their action to the fact that the intestinal wall is relatively impermeable to this salt. It remains in the intestinal lumen and by osmosis draws water into the canal or diminishes normal absorption of water. The result is that the intestine is filled with a watery bulk, which is about as effective as any other bulk in promoting peristalsis. Other cathartics, such as the various mineral oils, owe their effectiveness partly to a softening of the relatively dry colonic contents.

It is generally agreed that habitual and excessive use of any cathartic is harmful, tending to aggravate the constipation. Physicians are familiar with this condition, known paradoxically as *cathartic constipation.*

E. Evacuation of the Rectum

The emptying of the rectum—*defecation*—is a rather complicated process. It involves a strong contraction of the rectal walls and a relaxation of the anal sphincter. This of itself may suffice to empty the rectum, especially if the rectal contents are semifluid, as in diarrhea. But usually other mechanisms aid in the expulsion of the fecal matter. There is a strong contraction of the muscles of the abdominal wall, which compresses the abdominal contents. The compression is made more effective by a maintained contraction of the diaphragm and a closure of the glottis. In defecation, then, first a deep breath is taken, and the diaphragm pushes downward upon the abdominal organs. The breath is held by closure of the glottis. This fixes the diaphragm, so that the compressing action of the contracting abdominal muscles is not dissipated in pushing the diaphragm upward. This whole complex is known as *straining* and is employed not only in aiding the expulsion of feces from the rectum but also in urination, in vomiting, and again in the expulsion of the fetus through the birth canal at childbirth.

VII. HUNGER, APPETITE, AND THIRST

The importance of hunger and appetite in evolution is indicated by the fact that our scientific information about quantitative and qualitative food requirements is so recent as to have had virtually no effect upon man or domestic animals. The important guides in the selection and ingestion of foods have been the inborn urges of hunger and appetite.

What is hunger? Like all sensations or urges, it is difficult to describe, but there are some things about it which seem clear. Two components of the feeling of hunger can be recognized. One of these is a sort of generalized weakness and restlessness, referred to no special part of the body. Little is known about the cause of these states, but they are probably related to sensory nerve impulses from the alimentary canal and, perhaps in some cases, to a diminution of the sugar content of the blood.

Considerably more is known about the second component of the hunger complex. This is a more definite sensation, usually localized clearly in the upper part of the abdomen, over the stomach—in the "pit of the stomach." These sensations consist of intermittent pangs, as of tension or pressure. They may even be cramplike. A hunger pang lasts only a few seconds, then disappears, only to recur again in a minute or two.

Hunger is not to be confused with appetite, which is a desire for food. Hunger is usually accompanied by appetite, but one may have appetite without hunger. A tempting dessert may arouse the desire to eat even after all hunger has been appeased by a large meal. Appetite depends in large measure upon experience, upon acquiring a liking for certain foods, and upon the memory of pleasant experiences with food. On the other hand, the hunger mechanism appears to be inherited and is not essentially modified by the experience of the individual.

A. Hunger Contractions

It has been definitely established that hunger pangs are caused by contractions of the empty stomach.[16] Experiments upon human subjects, by the balloon technique, have given this information. The balloon is swallowed into the stomach and partially inflated. Every time the stomach contracts, it squeezes upon the balloon within it, causing the manometer writing point to be elevated. The subject, who does not see the record being made by his stomach, presses the key of a signal when he experiences a hunger pang. It is found that these sensations generally occur just at the time the record shows that the stomach is contracting (see Fig. 107).

Hunger is not a continuous process; the contractions do not follow one another without interruption. Characteristically, the contractions occur in series, between which the stomach is nearly quiescent and no hunger pangs are felt. The contractions follow one another for a period of about 30 minutes. Then for a period of time ($\frac{1}{2}$–2 hr.) all contractions stop, after which a new period of activity again sets in. The active periods are called *hunger periods.*

In any one hunger period the gastric contractions are weak at first, becoming stronger and stronger as the period progresses. Correlated with this is a gradually increasing intensity in the sensations felt. But then at the end of the period all hunger pangs cease, and, even though no food is taken, a quiet period will follow.

---

16. Contractions of the lower end of the esophagus and of the duodenum occurring at the same time probably also contribute to the hunger pains.

Ordinarily, of course, we eat a meal after one or two hunger periods have occurred, but, if fasting is prolonged, the periodic activity continues. Strangely enough, the hunger periods are not essentially different, or longer in duration, after a ten-day fast than after an all-night fast broken by the morning meal. In this sense, hunger does not increase in intensity as fasting is prolonged. The pangs are no sharper or no more frequent after four days than after four hours. It is true, however, that the feeling of generalized weakness is progressive. Aside from this there is no cumulative discomfort. In this regard, water deprivation is far worse, for thirst becomes extremely uncomfortable, even very painful, in a few days. Furthermore, man can live without food for five weeks, while water deprivation is fatal in five to ten days.

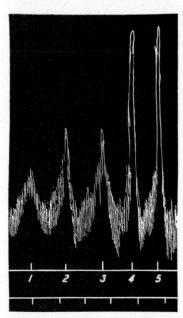

Fig. 107.—Hunger contractions in man, recorded with the apparatus shown in Fig. 78 (p. 228), except that the rubber balloon is in the stomach. Contractions of the empty stomach compress the balloon and cause the (five) rises shown in the curve. The signal indicates the points at which the subject experienced subjective hunger pangs. At *1* mild hunger was felt; at *4* and *5* there were sharp pangs; sensations at *2* and *3* were intermediate in intensity. Deflections caused by breathing also appear in the record. Time in minutes.

B. Cause of the Contractions

Thus, in the mechanism of hunger, the stomach really constitutes a sensory end organ, whose contractions stimulate sensory nerves which transmit impulses to the brain, where awareness of hunger arises into consciousness. But what controls the activity of the end organ? What is responsible for this characteristic activity of the empty stomach? The answer to these questions is not entirely clear. The contractions do not depend upon the extrinsic motor nerves, for they occur in a stomach whose nerves are severed. One factor may be the sugar content of the blood. At any rate, when the blood-sugar concentration is considerably reduced experimentally, the hunger machinery is stimulated. Just how important this is in normal physiology is not clear, since the fluctuations in blood-sugar concentration in the normal mealtime-fasting cycle are slight. It must be remembered, too, that hunger may set in even before the materials eaten at a previous meal are entirely absorbed.

A further puzzle is presented by the fact that all our observations indicate that the contractions of the stomach during hunger are in no way different from the peristaltic contractions occurring toward the end of gastric digestion. Yet we do not ordinarily experience these latter as pangs; in fact, we do not feel them at all. What accounts for this difference is quite unknown.

C. The Mechanism of Thirst

We have even less knowledge of the underlying mechanisms in thirst. This is a far more uncomfortable sensation and may even become exceedingly painful. It is not characterized by the intermittency of hunger and becomes progressively worse until death.

The sensation of moderate thirst seems localized to the back of the throat and is in the nature of an unpleasant feeling of tension, bordering on a sensation of tickling and mild burning. It is difficult to be objective in a description of thirst and to divorce the sensation itself from the desire for water. There is a tendency to think of thirst as a dryness in the throat, and an entire theory of the mechanism of thirst has been formulated on this basis: When there is a diminution of the water content of the blood and the tissues generally, the secretions of the body, including saliva, are diminished in volume. Because less saliva is secreted, the mouth and throat become dry. It is this sensation of dryness which has been called *thirst*.

Parts of the theory stand the test of experiment, but one of the most crucial links in the chain does not. Dryness of the surface of the mouth and throat does not by itself produce true thirst. It may cause a desire to moisten the parts; but, unless this surface dryness is also accompanied by a water deficiency in the deeper tissues as well, true thirst is not experienced. A dog deprived of all its salivary glands drinks no more water and is therefore presumably no thirstier than a normal animal whose mouth and throat are kept moist by a continuous flow of saliva.

It seems more likely that the significant water lack in initiating thirst sensations is in the deeper tissues of the mouth and pharynx. In an animal deprived of water all the body tissues become dehydrated—deficient in water. Apparently, this deficiency is effective in stimulating sensory nerves mainly in the throat region, which thus appears to be a sense organ for thirst, just as the stomach is for hunger.

VIII. ABSORPTION OF FOOD AND WATER

Materials within the alimentary canal are, in a sense, still outside the body. Only after absorption into the circulating fluids do the digestion products and the salts and water and vitamins become part of the internal environment and actually available for use by the tissue cells. Although absorption takes place to some extent throughout the length of the gastrointestinal tract, much less absorption occurs in the upper portion than in the lower. Little of anything is absorbed from the stomach. Alcohol is a notable exception. The fact that this occurs explains the rapidity with which effects are noticeable after ingestion of alcohol. Since the mucosal cells are apparently impermeable to intact starch, fat, or protein molecules, these substances are not absorbed until digestion is complete—that is, in the lower reaches of the small intestine.

The upper portion of the gastrointestinal tract is engaged chiefly in secreting rather than in absorbing, while the reverse is true lower down. However, water drunk when the stomach and small intestine are free from food is absorbed in the small intestine, and, if water is placed into an empty loop of small intestine of an animal under conditions preventing peristalses from moving it into the colon, the water is absorbed. But, when water is ingested with food, even though some of the water is absorbed in the small intestine, this is largely replaced by water of the digestive juices secreted into the intestine. Under such conditions considerable quantities of water remain unabsorbed at the time the intestinal contents reach the colon, and the watery consistency of the material which enters the large intestine is approximately the same as that which leaves the stomach. Such water is absorbed from the large intestine.

A. PHYSICAL FACTORS

Absorption is a complex phenomenon involving purely physical factors as well as activity of the living cells of the intestinal mucosa. This can be well illustrated by an experiment in which salt solutions of various concentrations are placed successively into a loop of intestine tied off from the rest of the bowel and the quantities of water and salt which disappear from the loop in a given period of time measured. Table 5 gives typical results of such an experiment. If only the physical factors of diffusion and osmosis operated, we should expect, first, that nothing would be absorbed from solution No. 2, because its salt concentration is about equal to that of blood; second, that solution No. 1, being less concentrated than blood, would lose water to the blood but might gain in salt content by diffusion of salt from the blood; and, third, that solution No. 3 might lose salt to the blood by diffusion and draw water from it by osmosis. But the experiment shows that from all three solutions both water and salts are absorbed.

Osmosis and diffusion, however, are seen to modify the absorption process, because the least water is absorbed from the solution which has the highest osmotic pressure (No. 3). From this same solution, in

TABLE 5

EFFECT OF CONCENTRATION OF SALT ON THE ABSORPTION
OF SALT AND WATER FROM THE INTESTINE

SOLUTION	PERCENTAGE CONCENTRATION (OF NACL)	PUT INTO INTESTINE		ABSORBED	
		H_2O (Cc.)	NaCl (Gm.)	H_2O (Cc.)	NaCl (Gm.)
1...............	0.5	100	0.5	75	0.3
2...............	1.0	100	1.0	50	0.6
3...............	1.5	100	1.5	25	1.0

which the diffusion gradient from intestine to blood is greatest, most salt is absorbed in a given period of time.

B. CELLULAR WORK

Thus we must conclude that, while these physical forces play a role in absorption, some other phenomena, probably involving specific cellular action, are also involved. A further experiment brings this out perhaps even more strikingly. If blood is withdrawn from an animal and the serum from it placed into an intestinal loop of the same animal, it will all be absorbed in a short time. This occurs despite the fact that the fluid within the intestine (the animal's own serum) is in perfect osmotic and diffusion equilibrium with the blood. Apparently, the cells lining the intestine are capable of actively taking up materials from the lumen side and passing them into the blood. This reminds us of the process of secretion, in which cells actively participate in the moving of materials. It depends upon the activity of living cells and involves cellular work with the expenditure of energy. The very shape of the tall, columnar cells lining the intestine, so different from the thin, flat cells of the lung alveoli or the capillary walls—which are apparently incapable of secretion and through which materials are moved by physical forces—suggests that cellular activity is involved in the processes of absorption. The mucosal cells look like gland cells.

C. Semipermeability of the Intestinal Wall

The layer of cells lining the wall of the intestine also acts as a semipermeable membrane, which permits certain materials to pass through and rejects others. The common salt, NaCl, for example, is quickly absorbed, while another salt, $MgSO_4$, is not. One might suspect that the size of the molecule is a factor here, for the $MgSO_4$ molecules are larger than NaCl molecules. However, molecules much larger than $MgSO_4$, such as glucose or amino acids, are readily absorbed. Reference has previously been made to the fact that semipermeability, like absorption, cannot at present be accounted for upon purely physical grounds (see pp. 54–57) and that some unknown factors dependent upon the integrity and activity of the cell are also concerned.

D. Entrance of Foods into the Blood

1. *Amino acids and sugars.*—The amino acids from digested proteins and the simple sugars are absorbed directly into the blood stream. They pass into the capillaries of the intestinal vessels and are carried away by the circulating blood. It will be recalled that these vessels are branches of the portal vein (see Fig. 30, p. 128) and that, therefore, these absorbed molecules are transported at once to the liver and are carried through the liver capillaries before passing on into the general circulation. We relate this to the fact that the liver plays an important part in the modification of these foods (see chap. 8).

2. *Indirect route of the fats.*—The fats, on the other hand, get into the blood mainly by an indirect route. A number of peculiarities are encountered in fat absorption. The fat molecule is broken down to fatty acid and glycerol in the intestine, but in the very process of absorption, even as the molecules are passing through the absorbing cells, this chemical process is reversed, and at least some of the fatty acid and glycerol is reconverted into fat.

The tiny droplets enter mainly the lymphatic vessels instead of the blood capillaries, so that during the absorption of a fatty meal the lymphatic vessels of the intestine (the *lacteals*) look white; if a lacteal is cut open, there escapes a lymph which looks milky from its rich fat content. But the fat eventually gets into the blood, because the intestinal lymph vessels, like lymph vessels from all other parts of the body, empty finally into a large vein at the apex of the left-chest cavity near the heart.

What is the significance of this indirect way in which fat gets into the blood? There is evidence that this absorption mechanism is useful because certain digestion products of fat (especially fatty acids) are injurious to red blood cells, which are not present in lymph. By the time these substances enter the blood stream, they have been greatly diluted by lymph entering the thoracic duct from areas of the body other than the intestine.

IX. DISORDERS OF THE ALIMENTARY CANAL

A. "INDIGESTION" AND "ACID STOMACH"

The commonest disorder of the alimentary canal is what is vaguely termed "indigestion." Occasional minor gastrointestinal upsets occur in all people, disappear spontaneously in a short time, and require no special treatment. At times these are associated with acute inflammation of the stomach lining from some irritant like alcohol, or with dietary indiscretions such as gorging. Frequently, the discomfort of "indigestion" is really a large-bowel phenomenon, associated with faulty colon motility or spasticity of that organ. Needless to say, any chronic state of "indigestion" should be investigated by the physician and not treated by the patient, even with the willing and proffered aid of drug advertisers. What goes by the name of "chronic indigestion" may indeed be relatively harmless, or it may prove to be a very serious ailment requiring expert medical care.

Emotional states or chronic "nervous tension" may markedly affect the gastrointestinal system. Worry, fear, and anger may so interfere with secretion and motility that digestion is considerably impaired. Such conditions are not uncommon in our modern high-speed "civilization," especially in tense, high-strung people. The causal factor to be dealt with here, of course, is not the indigestion itself but the mental and emotional tension of the individual.

Much is made of "acid stomach" by people who wish to (and do) profit by popular ignorance. We have seen that gastric juice is normally acid, and should be acid, and that the acid is an aid to peptic digestion, not to mention its function in stimulating pancreatic secretion by secretin formation. "Acid stomach" should no more be treated with pills than any other normal state.

Indeed, we have seen that a lack of acid in the stomach, though of itself not usually serious, is associated in some cases with the very serious abnormality—pernicious anemia.

It is true that there is a condition in which it is desirable to neutralize the acidity of the stomach with alkalies taken by mouth. This constitutes one of the best-known treatments for ulcers of the stomach and duodenum. Acid apparently retards or may even prevent healing of the ulcer. But this neutralization of acid can effectively be carried out only under the supervision of a competent physician.

B. Ulcers and Cancer

"Stomach ulcers" usually do not occur in the stomach but much more frequently in the duodenum just beyond the stomach. Round areas of the duodenal (or sometimes gastric) lining break down and present an eroded appearance. The cause of ulcer formation is not clear, but it seems that the normal gastric acidity not only retards the healing of ulcers but, according to the best evidence, is partly responsible for the pain associated with them.

Emotional factors are undoubtedly involved in the formation and chronicity of some duodenal ulcers. Ulcers are more common in emotionally tense individuals and during periods of exceptional emotional stress in less tense people. Examples of emotional states causing body changes are multiple: embarrassment causes the vasodilatation of blushing; fear may induce a quickening of the heart, sweating, etc.; and the anticipation of food initiates salivary flow. The machinery operating from brain to effector organ in these examples is fairly clear.

What possible connecting links are discernible to provide a channel for the emotions to produce or affect ulcers? Acid from the stomach retards ulcer healing, once an ulcer is formed. Secretion of acid by the stomach is increased by stimulating the vagus nerves, which innervate the muscles and glands of the stomach. The vagus nerves arise in the brain. Thus there exists a series of mechanisms for the emotions to influence ulcers which is no more complex than the mechanisms involving the dilation of the face blood vessels in blushing.

Besides the discomfort or pain of ulcers, they also are dangerous. A blood-vessel wall may be eroded, and a fatal hemorrhage occur. Or repeated minor hemorrhages may render the individual anemic and dangerously lower his resistance to ever threatening infection. Occasionally an ulcer erodes through the entire stomach wall. This precipitates a grave crisis, to be treated surgically at once. The danger in any perforation of the gastrointestinal tract lies in the escape

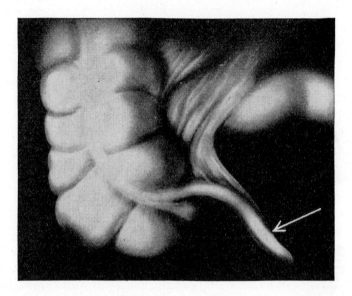

FIG. 108.—The vermiform ("worm-shaped") appendix of man (*indicated by arrow*) lying at the junction of the small and large intestines (see also Fig. 92, p. 267). (From the sound film, *Body Defenses against Disease.*)

of bacteria into the general peritoneal cavity, and death from a generalized peritoneal infection, or *peritonitis.*

There is also evidence which indicates that chronic ulcers may develop into cancer, though most cancers of the alimentary tract are not caused in this way. Cancer is a condition in which a group of cells, for some entirely unknown reason, commences to grow and multiply at a tremendous rate, invading and destroying surrounding tissues, spreading through the lymphatic system or the blood stream to distant parts of the body, where the destructive growth is repeated. Almost any part of the alimentary canal can be affected, though cancer is commonest in the stomach and rectum. Surgical excision of the growth is the only known remedy, at least of cancer of the alimentary canal. If the cancer is discovered early enough, before spread to other organs has occurred, and is completely removed, a permanent cure may be effected.

C. Appendicitis—Infections

Appendicitis is a condition of inflammation of the appendix, associated with infection and more or less destruction of the wall of that organ. The genesis of the condition is obscure, but it is effectively treated by surgical removal of the offending organ. If this is not done, there is the danger of early rupture of the appendix, setting free into the peritoneal cavity myriads of organisms from the bacteria-laden intestines, causing generalized peritonitis and death.

Other parts of the alimentary canal are subject to infection by a number of specific organisms, such as the typhoid, dysentery, and tubercle bacilli, as well as by a species of ameba which causes amebic dysentery. These and other alimentary infections can only be mentioned here.

D. Gallstones

Bile, secreted continuously by the liver, is stored in the gall bladder between meals. During its stay here the bile is concentrated by absorption of water from it. Certain of the constituents of bile—especially *cholesterol*—are only slightly soluble, and, in the course of this concentration, they may precipitate, forming the nucleus of a gallstone. The gallstone grows larger and larger as more and more precipitation occurs.

In other instances, infections of the gall bladder seem to predispose to stone formation. Clumps of bacteria have been described forming the nucleus of a stone.

If the gallstones remain in the gall bladder, there may be no serious ill-effects. But, if one of them should be forced into the main duct leading from the liver to the duodenum, the drainage of bile from the liver is obstructed, sometimes with serious consequences.

Once formed, stones of these or other origins are probably rarely, if ever, redissolved. Nor is there any known solvent of gallstones which can be taken by mouth or injected. Surgical removal is the only method of dealing with them.

CHAPTER EIGHT

THE HISTORY OF FOODS IN THE BODY

I. UTILIZATION OF FOODS
   A. Carbohydrates
      1. Storage in the liver   2. Use by cells
   B. Fats
      1. Function of stored fats   2. Relationship of the fats to carbohydrate combustion   3. Structural uses of fat
   C. Proteins
      1. Combustion and urea formation   2. Significance of the portal circulation   3. Storage of proteins
   D. Interconversion of foods

II. ANIMAL HEAT
   A. Source of animal heat
   B. Determination of heat production
      1. Direct calorimetry   2. Heat value of foods   3. The respiratory quotient   4. Ratio of heat produced to oxygen used   5. Indirect calorimetry
   C. Factors modifying total heat production

III. BASAL HEAT PRODUCTION
   A. Relationship to size of animal
   B. Influence of sex and age
   C. Abnormal basal metabolic rates
   D. Body weight

IV. REGULATION OF BODY TEMPERATURE
   A. Evolution of the thermostat
   B. Balance of heat production and heat loss
   C. Modifications of heat production
      1. The thyroid hormone   2. Skeletal muscle activity
   D. Regulation of heat loss
      1. Skin blood vessels   2. Evaporation   3. Accessory factors
   E. Controlling mechanisms
      1. Reflex regulation   2. The temperature-regulating center
   F. Fluctuations in body temperature
      1. The normal range   2. Abnormal temperatures—fever

V. The Fuel: Diet and Nutrition
  A. Energy requirements: carbohydrates and fat
  B. Protein
     1. Nitrogen balance   2. Quantitative protein requirement   3. Biological value of proteins: adequate and inadequate proteins   4. Animal and vegetable proteins
  C. Water
  D. Inorganic salts
  E. Vitamins
     1. Early experiences with deficiency diseases   2. Early experiments on vitamins   3. Action of vitamins   4. General distribution in foods   5. The "sunshine" vitamin   6. Complexity of vitamins
  F. Roughage
  G. Variety in the diet
  H. Unknown factors

VI. The Use of Alcohol and Tobacco
  A. Alcohol as a food
  B. Alcohol as a drug
  C. Alcohol as a poison
  D. Tobacco

VII. The Fate of Metabolic Wastes
  A. Elimination of metabolic end products
  B. Elimination and excretion
  C. The work of the kidneys
     1. Excretion of wastes   2. Regulation of blood composition
  D. Structure of the kidneys
  E. Mechanisms of urine formation
     1. Filtration through Bowman's capsule   2. Reabsorption of water   3. Reabsorption of threshold substances   4. Secretion of special substances   5. Summary
  F. Factors controlling urine volume
     1. Relationship to blood volume   2. Influence of the solids excreted   3. Mechanism of sugar diuresis
  G. Derangements of kidney function

I. UTILIZATION OF FOODS

The foods we eat are used in the body as fuel, as building stones for the construction of protoplasm and tissues, and as factors contributing to the composition of the internal environment. The salts, water, and vitamins contribute practically no energy, and their significance will be dwelt upon in some detail in the section on nutrition. At present we shall consider chiefly the use made by the body of the energy-containing foods, which constitute the bulk of the diet; that is, the carbohydrates, fats, and proteins.

A. Carbohydrates

The carbohydrates enter the blood stream as simple sugars like glucose. After a starchy meal there is always an elevation of the sugar concentration of the blood. From the usual concentration of a little less than 0.1 per cent (i.e., 0.1 gm. in 100 cc. of blood) it may rise to 0.12 or 0.14 per cent. We shall see later that, when the glucose content of the blood rises above 0.14 per cent, some of it is excreted by the kidneys and is lost to the body in the urine. But usually before this level (the *kidney threshold*) is reached, the glucose is removed from the blood stream more economically. The excess accumulates, principally in the liver, but also in the skeletal muscles, in the form of glycogen. Chemically, glycogen is much like starch, consisting of many simple-sugar molecules which are condensed into a large compact molecule of "animal starch."

1. *Storage in the liver.*—The discovery of sugar storage in the liver was made by that pioneer French physiologist of the nineteenth century, Claude Bernard. His conclusions were based upon measurements and comparisons of the glucose concentration in the blood of the portal vein and the liver (*hepatic*) vein, which latter drains the blood from the liver. During absorption of a meal there is a higher concentration in the portal vein carrying blood from the intestines to the liver than in the blood leaving the liver in the liver vein. The obvious conclusion is that some of the glucose is removed or changed by the liver. Analysis of the liver substance shows that at this same time new animal starch makes its appearance, obviously being formed in the liver from the glucose which has disappeared from the blood as such. We may express this chemically as follows:

$$\text{Glucose} \rightarrow \text{Glycogen} + \text{Water} .$$

The storage of glucose as glycogen does more than simply prevent loss of the temporary excess of sugar in the urine. It makes available a reserve supply of sugar which may be drawn upon by the organism in the interval between meals. Bernard found that, after a brief fast (two or three hours) when no more sugar was being absorbed, the quantitative relationships in blood-sugar concentration described above were reversed. That is, there was now more glucose in the blood leaving the liver (hepatic vein) than in the blood entering the liver (portal vein). This represents a reversal of the reaction indicated above, which now may be written as a reversible reaction:

$$\text{Glucose} \rightleftarrows \text{Glycogen} + \text{Water} .$$

The apparent simplicity of this formula may be deceptive. There are intermediate stages in the reaction, and enzymes and even hormones figure in the events. The reaction to the right seems to require the presence of insulin, the internal secretion of the pancreas, while

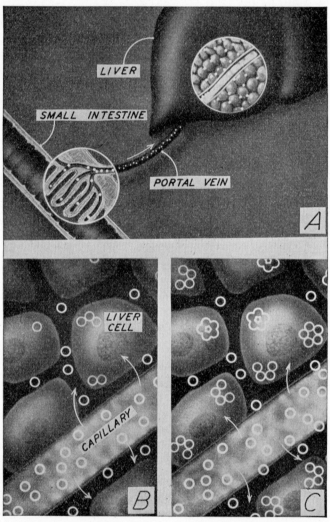

Fig. 109.—Storage of carbohydrate in the liver. In *A* absorbed glucose molecules (*white dots*) are seen (*enlarged circular insert*) entering the capillaries of the intestinal wall. They flow through the portal vein (*arrow indicates direction of blood flow*) which enters the liver. In the liver (in *B* and *C*) glucose (*white circles*) diffuses out from the capillaries (*arrows indicate direction of diffusion*), entering liver cells. Here the glucose molecules cluster together to form large glycogen molecules. Actually, one glycogen molecule contains many more than the five glucose molecules shown in the diagrams. (From the sound film, *Foods and Nutrition.*)

epinephrine, the internal secretion of the medulla of the adrenal gland, facilitates the reaction to the left. Both require the presence of at least one catalytic enzyme.

We may consider this an equilibrium reaction, which proceeds to the right, under normal conditions, whenever there is an excess of available glucose in the blood, and to the left as soon as the blood-sugar concentration begins to fall. We have here an important cog in the machinery which maintains the blood-sugar level at a relatively constant value, independently of food intake or fasting. Its importance is clear when we recall that a lowering of the blood-sugar level to about half the normal value (to 0.04 per cent) is incompatible with normal body function. The animal becomes increasingly irritable, and muscle twitchings and convulsions set in, followed by coma and even death of the individual. It is in this sense that we ascribe to this fuel food, glucose, a role in the composition of the internal medium, which must be kept fairly constant for normal cellular life and activity.

2. *Use by cells.*—The burning of glucose furnishes energy for the cells. All cellular activity requires energy as surely as activity of an automobile motor requires energy. And in both cases this energy comes largely or entirely from combustion: either of glucose or other fuel or of gasoline. The formula for the combustion of glucose,

$$C_6H_{12}O_6 + 6O_2 \rightarrow 6CO_2 + 6H_2O + \text{Energy},$$

again oversimplifies the case. Intermediate reactions occur which would require several pages to indicate in full. And then there are the special cases, such as the relationship of glucose and glycogen to the energy transformations in muscle, to be considered later (see pp. 378–83).

Combustion of glucose by cells is like the burning of that sugar "in the test tube" so far as concerns the quantitative relationships of the compounds involved and the quantity of energy liberated per gram of sugar burned. A striking difference is that in the body the combustion takes place at low temperatures, while outside the body the burning can occur only at temperatures too high to be compatible with the life of a cell. The difference is made possible by the ubiquitous enzymes. There are several oxidative enzymes in cells which catalyze the oxidative processes not only of glucose but also of

other foods as well, making possible an even liberation of energy at
body temperature.[1]

We have seen that glucose constitutes a part of protoplasm and
that combinations of glucose with other compounds enter into the
architecture of protoplasm. To a certain extent, therefore, the carbohydrates of the diet are used in growth of cells and as building stones
to replace broken-down, worn-out protoplasm.

B. FATS

The absorbed fats enter the blood as microscopic droplets largely
via the lymphatic vessels. After a fatty meal, blood serum which
separates from clotted blood, or blood plasma from centrifuged
blood, presents a milky appearance owing to this abundance of fat.

1. *Function of stored fats.*—There is a storage of this fuel in the
specialized fat cells, described earlier. We have seen that these fat
depots are rather generally distributed throughout the interior of the
body, as well as just under the skin, where they aid in the conservation of body heat, fat being a poor conductor of heat.

But insulaton is only one function of stored fat, usually of minor
importance. Like glycogen, it constitutes a food reserve for the organism. In times of fasting or inadequate food intake, these energy
stores may be tapped, and the fat removed from the fat cells and
burned by cells for heat and energy. We have pointed out that the
fat in storage fluctuates in quantity much less rapidly than the quantity of stored glycogen. Glycogen stores are drawn upon between
meals and replenished with every meal. Fat stores are drawn upon
to a significant degree only after days or weeks of inadequate food
intake or fasting and are relatively slowly built up over days or weeks
or months by daily ingestion of more food than is burned.

Like carbohydrate, the fats are burned to carbon dioxide and
water. The reaction indicating the combustion of a typical fat
(tripalmitin) is

$$2C_{51}H_{98}O_6 + 145O_2 \rightarrow 102CO_2 + 98H_2O + \text{Energy} .$$

As in all cellular combustion, oxidative enzymes are required as
catalysts here also.

2. *Relationship of the fats to carbohydrate combustion.*—There is
evidence that fats are not completely oxidized to carbon dioxide and

1. For the functions of insulin in sugar combustion see chap. 12.

water unless considerable glucose is also being properly metabolized. Such incomplete fat oxidation leaves an excess of acid products which are injurious or even fatal (see p. 514).

3. *Structural uses of fat.*—Fats and fatlike molecules are built into the structure of protoplasm. They are especially important in the composition of cell membranes, where to a considerable extent they determine the phenomena of cellular semipermeability, as discussed in chapter 2.

C. Proteins

To a greater extent than either the fats or the carbohydrates, proteins are used to build new protoplasm. Not merely in growth but also in the replacement of cellular protoplasm which breaks down under the wear and tear of living, a constant supply of those amino acids used for the synthesis of cellular proteins is essential. Though this is apparently their chief function in the body, the proteins are also burned for energy.

1. *Combustion and urea formation.*—The first stage in the breakdown of the amino acid molecule is *deaminization;* that is, the NH_2 (or amino) group is split off from the amino acid molecule. In this chemical reaction the amino group is converted into ammonia, NH_3, most of which, in turn, is immediately converted into urea. Some of the NH_3 formed in the kidneys by deaminization unites with water and becomes NH_4OH ($NH_3 + H_2O \rightarrow NH_4OH$), which is an alkali. In emergencies, when excessive acid is produced in metabolism, this ammonia is apparently used to neutralize the acid:

$$NH_4OH + H \text{ Acetoacetate} \rightarrow NH_4 \text{ Acetoacetate} + H_2O.$$
$$\text{(acid formed in}$$
$$\text{diabetes)}$$

The neutralized acid, in the form of the ammonium salt, is now eliminated by the kidneys. This is done without loss of the plasma alkalies, such as sodium, which would occur if these acids were neutralized by the sodium salts of the plasma:

$$\text{Na Proteinate} + H \text{ Acetoacetate} \rightarrow \text{Na Acetoacetate} + H \text{ Proteinate}.$$
| (relatively strong | (excreted by the | (relatively weak |
| acid) | kidneys) | acid) |

Ammonia itself is a very toxic substance. Most of the ammonia of deaminization is immediately converted into urea, which is relatively innocuous:

$$2NH_3 + CO_2 \rightarrow H_2N-\underset{\underset{O}{\|}}{C}-NH_2 + H_2O \ .$$

The urea, in turn, is excreted by the kidneys. It becomes possible, then, to determine approximately how much protein is being broken down in the body in any given period of time simply by measuring the quantity of urea in the urine.

The deaminization process as well as the urea formation are known to take place mainly in the liver.[2] It has been found that an excised liver, perfused with ammonium salts, converts them into urea. On the other hand, removal of the liver stops deaminization and urea formation.

2. *Significance of the portal circulation.*—From the fact that the liver stores much of the absorbed glucose and deaminizes the amino acids, we see the utility of the blood from the intestines, containing these newly absorbed products, being carried first to the liver in the portal circulation. It is true that these substances, or anything else introduced into the blood stream at any point, would be carried to the liver anyway within a minute or so. However, experiment shows that when the blood from the intestines, containing absorbed amino acids, is diverted from direct flow to the liver via the portal vein,[3] urea formation is markedly impaired.

3. *Storage of proteins.*—There is very little true storage of proteins in the body. As a result, when the carbohydrate and fat stores near depletion in fasting, the proteins of the tissues themselves are used for energy. The energy requirements, present so long as the machinery remains alive, are supplied by combustion of part of the machine. The body literally burns itself. Needless to state, this cannot continue for long without permanent damage or death.

D. INTERCONVERSION OF FOODS

The fats, proteins, and carbohydrates are distinct chemical entities, with characteristic molecular structure. They are also much

2. Some deaminization occurs in the kidneys, giving rise to the ammonia used to neutralize excess acid produced in metabolism.

3. This experiment is performed by joining the portal vein to the inferior vena cava, in what is known as an *Eck fistula*, after the Russian physiologist, Nikolai Eck.

alike, however, especially in that the framework of the molecule consists of a chain of carbon atoms.[4] Certain transformations of one of these substances into another are possible: In plants the simple carbohydrates manufactured in photosynthesis can be converted into fats or, with the addition of soil salts, into amino acids.

In the animal organism it is apparently extremely uncommon for amino acids and proteins to be manufactured from starch plus salts or ammonia; almost entirely, they must be synthesized from amino acids obtained directly from plant food or indirectly by eating meat, whose proteins, in turn, were derived from plants.

But all other possible transformations of the three major foods can and do take place. Evidence for this comes in part from quantitative determinations of glycogen formation in the liver of animals fed carefully controlled diets of known composition. Any glucose which is fed, of course, increases the glycogen content of the liver. But so also does the feeding of protein or fat. These experiments demonstrate that nearly 60 per cent of the protein molecule and about 10 per cent of the fat can be transformed into glucose.

The ability of the animal body to convert starch into fat is well known, and for centuries this capacity of the animal organism has been exploited in animal husbandry. The farmer feeds his hogs mainly starch in the form of corn, and the animated chemical laboratory

4. Examination of the complete structural formula of a typical carbohydrate, a typical fatty acid, and a typical amino acid reveals this similarity. Molecular structural features shared by all three are indicated in boldface type:

Glucose   Fatty acid   Amino acid

Note that the only difference between the fatty acid and the amino acid given here is that there is an NH_2 group in the amino acid instead of an H in the fatty acid.

converts it into the economically more valuable fat. In terms of animal economy this phenomenon is a conversion of an excess of quickly available energy reserves into a more permanent storage form, which may again be converted back into glucose in a time of need. The food interconversions of the animal organism may be schematically represented in this way:

$$\text{Proteins} \rightarrow \text{Carbohydrates} \rightleftharpoons \text{Fats} \,.$$

II. ANIMAL HEAT

The origin of animal heat has interested biologists for centuries. In the "warm-blooded" animals especially the phenomenon is striking because enough heat is produced even in extremely cold environments to keep the body temperature above that of the environment.[5] There have been many fanciful notions as to the source of the heat. It was thought to be derived chiefly from the friction of the moving blood against the vessel walls, or to be produced by some unknown process in one or another of the internal organs and distributed by the blood. In general, it was felt that animal heat was fundamentally different in origin from heat generated in nonliving systems and that animal heat depended in some mysterious manner upon activity of living cells and organs, of a sort quite different from the burning of fuel outside the body.

A. Source of Animal Heat

To the French chemist and physiologist, Antoine Lavoisier, we owe the concept that animal heat is essentially the same in origin as heat produced in any combustion process—that animal heat is derived from oxidations. Not content with merely propounding this hypothesis, Lavoisier did a great deal more. He devised an ingenious though crude experiment to test the truth of his hypothesis.

Lavoisier constructed two chambers completely surrounded with ice. Into one chamber he placed a burning candle; into the other, a living guinea pig. He determined the heat generated in each chamber over a period of time by measuring the quantity of ice which was melted. In addition, he measured the quantity of carbon dioxide produced in each chamber. Calculating for each chamber the ratio of the quantity of heat produced to the quantity of carbon which had been oxidized to carbon dioxide, he found a striking correlation,

5. In very warm environments the body temperature of birds and mammals, maintained always at approximately a constant, may be less than that of the environment.

such that he felt warranted in concluding that the same chemical processes were involved in both chambers:

$$\text{Carbon compounds} + O_2 \rightarrow CO_2 + H_2O + \text{Energy}.$$

(wax in the candle)   (heat)
(food in the animal)

A corollary of this conclusion was that the same total quantity of heat is liberated from the combustion of a given weight of any specific fuel food, whether that combustion takes place in the animal organism or in the test tube outside the body. This would seem a reasonable-enough conclusion when we consider that into a gram of glucose, for example, there has been incorporated a very definite total quantity of energy. Therefore, when the glucose molecules are shattered and the energy released (as heat), it should not matter whether the process occurs inside or outside the animal organism, provided that there is an equal breakdown of molecules in both cases and that the molecule fragments are alike. In either case, all the energy that can be obtained from the glucose molecule is that which was built into it from the sunlight. Reasonable though this may seem to us, we must remember that when Lavoisier formulated his conclusions the law of the conservation of energy had not yet been recognized.

It can scarcely be considered a discredit to Lavoisier, though it deserves mention, that he thought that the oxidations took place entirely in the lungs. Still others, who accepted his results, believed that the reaction took place in the blood. It was not until some years later that it was proved that these oxidations occur in all living cells.

B. DETERMINATION OF HEAT PRODUCTION

1. *Direct calorimetry.*—Since Lavoisier's crude experiments, refinements in the technique of measuring animal heat—*calorimetry*—have served to confirm and to extend his fundamental observations. Figure 110 shows the construction of a metabolism chamber large enough to accommodate a man for several days. It is equipped with a folding cot, a chair, a desk, and a bicycle *ergometer*, upon which measured quantities of work can be done. The walls are specially constructed, with successive layers of insulating materials separated by layers of air, so that no heat can escape.

A stream of water is led into the chamber and circulated through pipes within it and then led out. All the heat generated by the individual within is carried off in the water and can be measured. The unit of measurement is the (large) calorie, which is the quantity of heat required to elevate the temperature of a liter of water $1^\circ 0$ C. Measured quantities of oxygen are led into the chamber, and the carbon dioxide and water produced by the man are quantitatively determined. Through a special opening food can be passed, samples of which are previously analyzed chemically. All excreta are passed out the same opening for analysis.

2. *Heat value of foods.*—By the use of such a chamber complete "balance" studies can be made, accurately determining chemical and

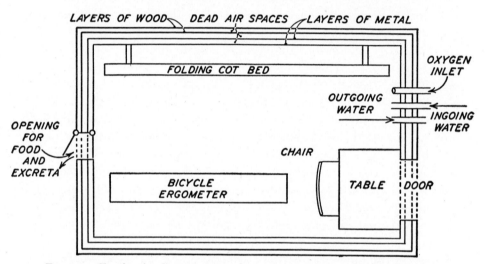

Fig. 110.—Chamber for direct determination of heat production. The chamber can accommodate a man for several days. The walls are well insulated, and all the heat generated in the chamber is carried off by water which circulates through radiators in the chamber. Measured quantities of work can be done on the ergometer. Food and excreta are analyzed. The oxygen used and carbon dioxide produced are measured.

physical quantitative relationships between food ingested, excreta eliminated, heat generated, oxygen used, and carbon dioxide produced. The heat produced by each of the three kinds of fuel food, even when burned simultaneously, has been determined and compared with the quantity of heat generated by burning these foods outside the body. The results are as shown in Table 6. The discrepancy in the case of protein is due to the fact that combustion is not so complete in the body as it is outside. Some of the energy of the original protein molecule is still held within the urea molecule; combustion of the urea itself liberates the additional calorie of heat. Carbohydrates and fats, on the other hand, are usually burned to the same end stage (carbon dioxide and water) both inside and outside the body.

A considerable difficulty is encountered in such studies. Regardless of the composition of the foods recently eaten, proteins, fats, and carbohydrates are being burned at all times in the normal organism. How can we determine just how many grams of each of the fuel foods are being burned in a given period of time? Obviously, it would be necessary to determine this in order to assign definite caloric values to the various foods. This involves a complicated procedure whose main outlines only can be given here. For example, we can determine how much protein is being burned by measuring the nitrogen in the excreta (chiefly in the urea of the urine), since there is a constant proportion of nitrogen (16 per cent) in the protein molecule. The

TABLE 6

HEAT GENERATED BY THE BURNING OF FOODS

MATERIAL BURNED	APPROXIMATE QUANTITY OF HEAT GENERATED (CALORIES PER GRAM)	
	Inside Body	Outside Body
Carbohydrate..........	4	4
Fat....................	9	9
Protein...............	4	5

principle involved is essentially the same as that used in the determination of the rate of red-cell and hemoglobin destruction by quantitatively measuring the bile pigment in the excreta.

3. *The respiratory quotient.*—In addition, we must determine the respiratory quotient, or R.Q., which is the ratio of the *volume* of carbon dioxide produced in the organism to the *volume* of oxygen used (CO_2/O_2). What information can be derived from these figures is apparent from the balanced formulas for the oxidation of glucose and of a typical fat:

$$C_6H_{12}O_6 + 6O_2 \rightarrow 6H_2O + 6CO_2$$

$$\text{R.Q.} = \frac{6 \text{ (volumes } CO_2)}{6 \text{ (volumes } O_2)} = 1.0 \text{ for carbohydrate;}$$

$$2C_{51}H_{98}O_6 + 145O_2 \rightarrow 102CO_2 + 98H_2O$$

$$\text{R.Q.} = \frac{102 \text{ (volumes } CO_2)}{145 \text{ (volumes } O_2)} = 0.7 \text{ for fats .}$$

The R.Q. for proteins, of a variable formula, is about 0.8.

Thus, if the R.Q. of an individual at any given time is 1.0, the presumption is that only carbohydrates are being oxidized. If it is 0.7, only fats are being burned. If it is intermediate between these figures (the normal condition), there must be combustion of a mixture of foods. Just what this mixture is can be learned by determination of the quantity of nitrogen excreted and by subtracting from the measured carbon dioxide and oxygen volumes the volumes of these gases which would be involved in combustion of the quantity of protein corresponding to the measured nitrogen excretion. The ratio of the remaining carbon dioxide to the remaining oxygen gives us the *nonprotein respiratory quotient*. Should this be 1.0, we know that, besides the protein, only sugar is being burned and can compute how much. If it is 0.85, halfway between 1.0 and the R.Q. for fats—0.7—we should know that, besides the protein, equal quantities of carbohydrate and fat were being burned, etc.[6]

4. *Ratio of heat produced to oxygen used.*—One quantitative relationship discovered by direct calorimetry has been of great practical importance. This is the finding that there is not only a definite quantity of heat liberated by the burning of a gram of carbohydrate, for example, but also that this combustion always involves the consumption of the same volume of oxygen. Practically, this means that heat liberation can be measured indirectly, simply by measuring the volume of oxygen used. It was found that, when the normal mixture of foods is being burned, nearly 5.0 calories[7] of heat are always produced for every liter of oxygen used up. *Indirect calorimetry*, then, by measurement of oxygen consumption, may be substituted for direct heat measurement.

5. *Indirect calorimetry.*—Indirect calorimetry is technically much easier to carry out than direct calorimetry. The direct calorimeter described is very expensive to build, keep in working order, and operate. Indirect calorimetry can be carried out in a few minutes by a single operator, using inexpensive, convenient, portable, more nearly foolproof apparatus. A vertical section through such an instrument is shown in Figure 111. It consists essentially of a closed system containing a supply of oxygen, from which the subject in-

6. These calculations and conclusions are based upon the assumption that no appreciable interconversion of one food into another is occurring during the experiment. Any such interconversion renders the calculations incorrect.

7. This figure (more exactly, 4.825 cal.) varies somewhat with variations in the proportions of fats, carbohydrates, and proteins being burned.

spires, and into which he expires. The carbon dioxide is removed from the expired air by passing it through a soda-lime tank. It is very simple to measure the quantity of oxygen removed from the tank by the subject in a measured period of time, and from this to

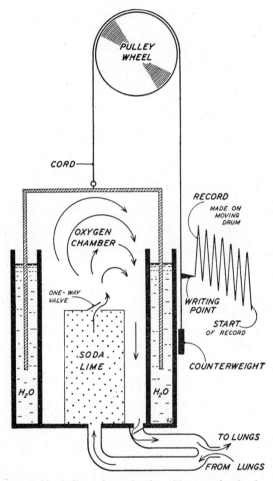

Fig. 111.—Apparatus for indirect determination of heat production by measuring oxygen consumption. The subject breathes into and from the oxygen chamber through the tubes at the bottom. The slope of the curve recorded by the movements of the upper cylinder is a measure of the rate at which oxygen is used by the subject. Arrows indicate direction of oxygen movement as the subject breathes.

calculate the rate of oxygen consumption, as well as the heat produced per hour or per day.

This simplification has made possible the measurements of the rate of heat production upon many more individuals than would have been possible by direct calorimetry and the establishment of significant relationships between the rate of heat production, or the metabolic rate, and various other factors in health and disease.

C. Factors Modifying Total Heat Production

Lavoisier made another discovery which laid the foundation for our concept of "basal" heat production, or *basal metabolism*. He found that in the normal individual essentially three variables modify the rate of total heat production—physical exercise, exposure to cold, and ingestion of foods. Figure 112 diagrams these effects. In

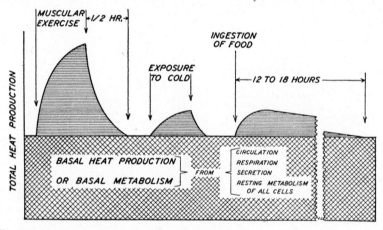

Fig. 112.—Factors modifying heat production in warm-blooded animals. There is the basal heat production, plus heat derived from muscular exercise (part of which heat is liberated after exercise is stopped), from exposure to cold, and from the ingestion of food.

muscular exercise the rate of heat production increases and continues at a higher level after the exercise has ceased, for a time determined by the severity and duration of the exercise. Everyone has experienced this heat-producing effect of exercise and has utilized it in combating external cold. In sleep, on the other hand, where muscular relaxation is more nearly complete, the metabolic rate has been found to be about 10 per cent lower than in the resting, waking state.

We shall see later that exposure to cold leads to a type of involuntary exercise, characterized by increased muscular tension and even shivering. Lastly, eating food has a warming effect, especially if the meal contains proteins. The metabolism-stimulating effect of foods, called *specific dynamic action*, is especially marked in the case of proteins and lasts for some twelve to eighteen hours after the protein ingestion. It must not be mistakenly thought that this effect is simply due to added combustion because the meal has made more fuel available. It is rather that certain products of the metabolism of foods, especially of amino acid metabolism, exert an actual chemical stimulation upon cellular heat-liberating processes especially in the liver, but perhaps elsewhere as well.

III. BASAL HEAT PRODUCTION

These three variables can easily be controlled. A human subject can fast for eighteen hours, refrain from exercise, and remain physically at rest for one-half hour and still keep comfortably warm. Under these *basal conditions* all the heat generated by the individual is due to the activities of the internal organs (heart, blood vessels, alimentary canal, glands, etc.) and to the "resting" metabolism of the cells in general. This is the basal heat production or the basal metabolism. We find the *basal metabolic rate*, or B.M.R., to be one of the physiological constants, depending upon the height, weight, age, and sex of the normal individual. Race is apparently also a factor, since some Orientals tend to have a lower rate than Occidentals.

A. RELATIONSHIP TO SIZE OF ANIMAL

We should expect to find some sort of relationship between the size or weight of an animal and the basal heat production: that the greater the mass of protoplasm involved, the greater would be the heat production. And it is true that a large mammal or bird produces more heat per day than a small one. But, when the heat output is expressed in terms of body weight, we find that small animals produce much more heat per gram of tissue than do large animals. However, when the heat output is expressed in terms of total surface area of the animal, we find a significant correlation. The caloric output per square meter of body surface is roughly the same in animals ranging in size from a mouse, weighing a fraction of a kilogram, to a pig, weighing many kilograms. In this respect, loss of heat by a mammal or bird follows the laws of cooling inanimate objects. Any object warmer than its environment will give off heat not in proportion to its mass but in proportion to its surface area. In birds and mammals, likewise, we see that the rate of the heat loss (normally equaled by the rate of heat production) is proportional to the body-surface area (see Table 7).

This relationship holds not only for different species of birds and mammals but also for individuals of different sizes within a given species, including man. In man, measurements of body-surface area are very laborious, particularly in the region of the face, hands, and feet, with their many convexities and concavities. Many sufficiently direct measuréments have been made to learn that there exists a fairly constant relationship between the surface area and the body weight and height. It is now possible, because of these empirical findings, to calculate the surface area of an individual by applica-

TABLE 7

RELATIONSHIP OF SIZE OF ANIMALS TO RATE OF
HEAT PRODUCTION

ANIMAL	WEIGHT IN KILOGRAMS	CALORIES PRODUCED PER DAY	
		Per Kilogram	Per Square Meter of Body Surface
Pig.................	128	19	1,078
Man................	64	32	1,042
Dog................	15	52	1,039
Mouse..............	0.02	212	1,188

tion of a formula which expresses the results of these findings.[8] Most adults have a surface area of from 1.5 to 2.0 sq. m.

B. INFLUENCE OF SEX AND AGE

The B.M.R. is affected by both sex and age. Females, in general, have a metabolic rate about 5 per cent less than males, and the rate is higher in the young than in the old, expressed in calories per square meter. So far, the reasons for this are not at all clear. It may be recalled that the red-cell count is also lower in females than in males. Neither of these differences seems unquestionably to be true sex differences, although we do find them in some other animals. The female dog has the same basal rate as the male. The human sex factor, as well as the age factor in B.M.R., may possibly be related in some way to the rate of skeletal muscle activity. Men are commonly more active physically than women, and the young of both sexes are more active than the aged. In addition, the growth process in the young requires not only building materials but also the expenditure of extra energy and, therefore, production of extra heat also (see Fig. 113).

Literally thousands of basal metabolism tests had to be made to establish these facts. This large number was possible only by the

8. Weight $^{0.425}$ (in kg.) $\times$ Height $^{0.725}$ (in cm.) $\times$ 71.84 $=$ Surface area (in sq. cm.).

indirect method, using the basal metabolism apparatus. From the figures obtained, complete tables have been constructed, so that, in the case of any individual, we need simply look under the proper age and sex and see what his metabolic rate per square meter of body surface should be if he is normal. Knowing also the individual's weight and height, we can calculate his surface area. The product of these (area times calories per square meter) gives the "normal" rate for this individual. We then actually measure his metabolic rate by the

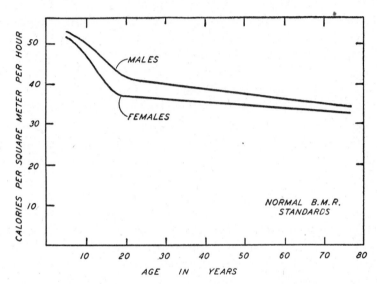

FIG. 113.—Basal metabolic rates of normal male and female human beings from early childhood to old age.

indirect method and can know at once whether his actual rate is normal or lower or higher than normal and by what percentage.

A typical instance gives an idea of the actual numerical values involved. A male of twenty-five years has a normal rate of 40 calories per square meter of body surface per hour, or 960 calories per day. If his height is six feet and his weight 175 pounds, his surface area would be 2.0 square meters. The product of these two equals 1,920 calories per day for the individual. Of course, in addition to this basal heat output, an individual liberates an extra quantity of heat, varying from 1,000 to 2,000 or more calories, depending upon the amount of muscular work done in the course of the day. This figure for a student, leading a relatively sedentary life, is about 1,000 calories.

It is common to express the basal metabolic rate as a plus or minus percentage of the normal. A rate of plus 20 per cent means that the B.M.R. is 20 per cent higher than is usually found for a person of the age, sex, height, and weight of the subject. It is not until a B.M.R. is more than 10–15 per cent plus or minus that it is considered as probably abnormal.

C. Abnormal Basal Metabolic Rates

Of the factors which produce abnormal basal metabolic rates, we need only mention fever and abnormal thyroid-gland function. In fever the B.M.R. is always high, increasing by about 5–6 per cent for each degree (F.) rise in body temperature. This effect might have been predicted from our knowledge of the fact that, within limits, elevation of temperature of any tissue or organ increases its activity. We found this to be true of the heart rate. Fundamentally, this is because heat accelerates the chemical reactions involved in these processes just as it hastens chemical reactions generally. Since metabolism consists of a complex of chemical reactions, it is not surprising that the B.M.R. is also influenced by the temperature of the body.

Chapter 12 deals at some length with the role of the thyroid glands in the regulation of basal metabolism. The secretion of these glands, the thyroid hormone, apparently controls, in part, the rate of the oxidations in all cells of the body. Definite modifications in rate occur, therefore, as a result of thyroid malfunction: basal heat production is increased when the thyroids are overactive, secreting an excess of the hormone; it is low when the thyroid glands are underactive. In the field of thyroid derangement a knowledge of the metabolic rate is indispensable to the physician. No physician would be content to diagnose thyroid overactivity or underactivity and institute measures to control the abnormality without repeated metabolic rate determinations upon his patient. The B.M.R. apparatus, then, is of practical as well as scientific importance: it is a useful tool of the physician, as well as a device which has enabled physiologists to learn a great deal about the metabolism of the normal individual.

D. Body Weight

The effect of certain drugs upon the B.M.R. is of importance in cases where it is desirable to reduce the weight of an individual. A number of factors operate to determine the body weight of the human adult, such as normal and abnormal functioning of some of the glands of internal secretion, especially the thyroids and the hypophysis. Hereditary factors are also involved; there is a tendency for overweight or slimness to "run in families." But, in large measure, body weight depends upon the balance between the caloric value of the foods eaten and the caloric value of the foods burned. An imbalance in the one direction tends to increase body weight; in the other, to decrease it. Imbalances may be corrected, or set up, when it is

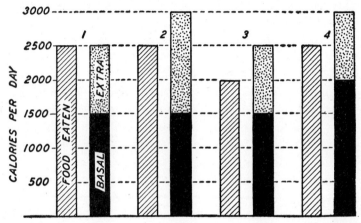

FIG. 114.—Relationship of caloric intake and output to body weight. Each crosshatched column represents the caloric value of the food eaten by an individual. In the second column of each pair the black area represents basal heat output; the dotted area represents additional heat produced by muscular exercise. If all the food eaten is digested, absorbed, and utilized, condition *1* should lead to a constant body weight. Conditions *2* (increased exercise) or *3* (decreased food intake) should produce loss in weight. Condition *4*, in which the basal metabolism is elevated, cannot be attained without danger to health.

desirable, by modifying either the caloric intake (the amount of daily food ingested) or the caloric output (the amount of daily work).

In addition, of course, an imbalance theoretically could be corrected or set up, as the case may be, and body weight gained or lost, by modifying the B.M.R. (see Fig. 114). Eating thyroid-gland substance, for example, will elevate the metabolic rate and tend to cause weight loss. Because it is much easier for people to take a capsule of thyroid substance every day than to exercise hard or reduce the food intake, it has been found profitable by unscrupulous drug manufacturers to include thyroid substance in weight-reducing preparations for fat people who like to eat and dislike physical work. Such preparations will reduce the body weight in spite of a large intake of food and limited exercise, but into the bargain it will also induce an artificial state of hyperthyroidism, with all its distressing and even dangerous sequelae.

The problem of control of body weight is not entirely one of aesthetics. There are conditions (e.g., cardiac disease or the arthritis or "rheumatism" of advanced age) in which it is highly desirable that overweight be corrected. Needless to say, the use of thyroid hormone for this purpose is safe only under skilled supervision. It may be categorically stated that no drug which will reduce body weight by increasing the metabolic rate is safe enough for the layman to administer to himself.

IV. REGULATION OF BODY TEMPERATURE

A. EVOLUTION OF THE THERMOSTAT

The evolution of an automatic thermostat in animals has done much to diminish the dependence of the organism upon the vicissitudes of the environment. This, together with the ability to manipulate and control the environment, is the essence of evolutionary progress in the last analysis.

Lower animals, including among the vertebrates the fish, amphibians, and reptiles, are more or less at the mercy of the environmental temperature. In the cold, metabolic processes and activity of the individual become sluggish. A return to a more active state must await a rise in external temperature. The mammal or bird, on the other hand, provides its internal organs with a fairly constant environmental temperature, making possible a more even tempo of life-activities. The metabolism and general activities of an arctic polar bear or an antarctic penguin or whale go on as effectively as those of a tropical bird or mammal. The thermal environment of the tissues of man is kept at about the same temperature (98°–99° F.) whether the external temperature is 120° F. above zero or 60° below.

This was apparently a gradual evolutionary development, with intermediate forms displaying an imperfect temperature control. In the most primitive of mammals, the egg-laying type, which represents a transition form between reptiles and mammals, we find fluctuations in body temperature considerably greater than in the more specialized higher mammals.

But a price is paid by the higher vertebrates for this adaptation, for through the long ages of later mammalian and avian evolution the body tissues have become so adjusted to this fairly constant temperature that relatively minor fluctuations in temperature, under abnormal circumstances, are now injurious or fatal to cells, tissues, or the organism as a whole. Cold-blooded vertebrates tolerate temperatures well below 80° F. indefinitely, but mammalian tissues can-

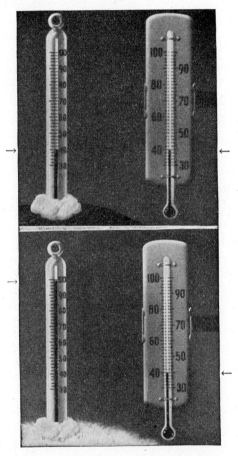

Fig. 115.—Body-temperature fluctuations. The thermometers on the left record the body temperature of a turtle (*above*) and of a rabbit (*below*). The thermometers on the right record the air temperature of the closed chamber in which the animals are placed. The temperature of the chamber can be changed at will. The temperature of the turtle's body fluctuates with that of the chamber. The temperature of the rabbit's body is maintained at a high level despite a low environmental temperature. (From the sound film, *Control of Body Temperature*.)

not long survive if the body temperature should temporarily fall to 80°F. or rise to 110° F. Needless to state, such temperatures would be reached only if serious defects in the temperature-regulating machinery should develop.

The terms "warm-blooded" and "cold-blooded" are misleading because, in environmental temperatures above that of the mammalian body, a "cold-blooded" animal is actually warmer than a "warm-blooded" animal. More accurately descriptive and to the point would be terms which signify "constant body temperature" (*homothermal*) or "variable body temperature" (*poikilothermal*).

B. BALANCE OF HEAT PRODUCTION AND HEAT LOSS

The temperature of any animal is the resultant of the rate of heat production and the rate of heat loss. The constancy of temperature in mammals and birds depends upon mechanisms which modify these two factors independently and so adjust their relative rates as to keep the resulting temperature at a nearly constant level.

In the last analysis, rate of heat production depends upon the rate of tissue oxidations. Heat is lost from the body by convection, radiation, and conduction from the surface, evaporation of water from the skin, warming of cold air taken into the lungs, and warming cold food or drink taken into the stomach.[9]

A certain amount of voluntary control over body temperature can be exercised. Heat production can be accelerated in the cold by voluntary exercise and by increased food consumption or diminished in summer by muscular relaxation[10] or by eating a minimum of proteins, with their marked specific dynamic action. Heat loss can be modified by wearing fewer or more clothes or by taking advantage of the cooling effect of water evaporation by wetting the body surface by bathing.

C. MODIFICATIONS OF HEAT PRODUCTION

We are interested here in the devices which operate altogether automatically and adequately, independently of voluntary effort. Practically the only means available to the body for modifying the rate of heat production is to alter the degree of muscular activity.

1. *The thyroid hormone.*—It would seem reasonable to guess that

9. The percentage total loss of heat by the body of man is distributed approximately as follows:
    a) From the skin, by radiation and convextion, and by evaporation of sweat.. 85
    b) Evaporation of water in respiratory passages............ 10
    c) Warming of food and air taken in.................... 5

10. "Voluntary" changes in muscle activity in the heat or cold often depend less upon an intelligent decision to relax or to exercise than a marked disinclination to muscular work in the heat and the invigorating effect of cold.

modifications of rate of thyroid-hormone production might also figure here. There is some evidence that animals exposed to cold for many days may develop an increased metabolic rate, which does not occur in animals whose thyroid glands have been removed. But the evidence is not clear. In any case, if such adjustments to cold do occur, they take some time. The evidence is that some twenty-four hours elapse between the appearance of an excess thyroid hormone in the body (e.g., by injection) and a stimulating effect on metabolism. The thyroid machinery is simply not adapted for the rapid readjustments required when a mammal goes abruptly from a warm to a cold environment or vice versa.

2. *Skeletal muscle activity.*—There is almost no lag between skeletal muscle activity and added heat production, and, in the cold, muscle activity increases. An increased muscle tension, or tone, develops, which may continue to the point of shivering. Shivering is more or less involuntary but not entirely so. It is possible to inhibit it to a certain extent. If this is done with any success under a cold shower, the body temperature will drop rapidly. In warm environments muscle tension diminishes. Thus in the cold there is an automatic stoking of the furnaces; in the warm, the fires are banked.

D. Regulation of Heat Loss

The first adjustments made by the body are on the side of heat loss. When a room becomes cool on an autumn day, the first thought of the occupant is to close any open windows rather than to open the furnace draft. In this regard the "wisdom of the body" is equal to that of the intelligent householder.

1. *Skin blood vessels.*—In mammals heat loss is modified in several ways. In the cold the skin blood vessels constrict, so that less of the warm blood from the internal organs circulates through the surface vessels. Less heat is therefore lost by radiation; heat is conserved. A vasodilatation of skin vessels, on the other hand, bringing more warm blood to the surface of the body and facilitating heat loss, normally occurs in a warm environment.

2. *Evaporation.*—Evaporation of any liquid from any surface absorbs heat from, and cools, that surface. Evaporation of sweat from the skin of man, or saliva from the protruded tongue of a panting dog, or water from the respiratory passages, cools the surface and the blood flowing through it, and any tendency toward an elevation of body temperature is offset. Any such tendency constitutes an

effective stimulus to sweating in man or to copious salivary secretion and panting in the dog, which lacks sweat glands.

The efficacy of the sweating mechanism in dissipating heat can be strikingly demonstrated by experiments exposing human beings to high temperatures in closed chambers. If the air in the chamber is maintained perfectly dry, permitting free evaporation of sweat, man can tolerate and adjust himself for a short time to external temperatures as high as the boiling point of water. But, if the chamber is saturated with moisture, preventing evaporation of sweat, the body temperature of the subject rises rapidly. Collapse may occur by the time the temperature in the chamber reaches 90° F.

3. *Accessory factors.*—Accessory factors involved in special cases are the fluffing of the feathers of a bird or the erection of the hairs of furred animals in the cold by contraction of smooth muscles at the hair roots. Both these effects serve to entrap about the body a thicker layer of stagnant air, which is a good heat insulator. In man, who has lost most of his body hair, this factor is negligible. The few hairs man has are equipped with the same tiny smooth muscle of the furred animals; these muscles actually contract in the cold. The tiny skin puckerings known as "goose flesh" apparently represent the evolutionary survival of a response once useful but now ridiculously futile.

E. Controlling Mechanisms

These are the adjustments, then, responsible for correlating heat production and heat loss. But what further mechanisms are involved? What causes muscles to become tense in the cold and sweat to flow when it is warm? So far we have only considered effector structures—the furnace, the windows, the cooling system. But what is the thermostatic device which automatically stokes the furnace or closes the windows or starts the cooler? It might be supposed that temperature changes have a direct action upon the organs and tissues involved in the final adjustment—that heat directly stimulates the sweat glands, and a lowered temperature, the skeletal muscles.

1. *Reflex regulation.*—Experimentally, the above can be shown not to be so, for, if the efferent nerves of these structures are cut, the reactions to temperature changes fail to occur. Sweat fails to be secreted, vasomotor responses do not take place, changes in muscle tension are absent, and fur is not fluffed if the glands, blood vessels, and skeletal and hair muscles are denervated, no matter what the temperature of the environment. Thus, to a large extent, the responses

are reflex in nature, resulting from stimulation of specialized nerve endings in the skin: the "cold" receptors, which are stimulated by a temperature fall, and the "heat" receptors, which are stimulated by a rise (see Table 8).

2. *The temperature-regulating center.*—We may inquire further into the location in the central nervous system of the center or centers through which these reflexes are mediated. We can learn the answer only by experiment, applying the same principle adopted for location of the respiratory center—making transverse cuts across the spinal cord and its upward extension, the brain stem. It becomes

TABLE 8

REFLEX ADJUSTMENTS TO EXTERNAL HEAT OR COLD

Skin Receptors Stimulated	Reflex Changes Produced	Effects of Reflexes
Sensory nerve endings for cold	1. Vasoconstriction of skin vessels (2. Erection of hairs) (3. Fluffing of feathers)	Decreased heat loss
	Increased muscle tension	Increased heat production
Sensory nerve endings for heat	1. Vasodilatation of skin vessels 2. Increased sweat secretion (3. Increased salivation and panting)	Increased heat loss
	Muscular relaxation	Decreased heat production

clear at once that the primary temperature-regulating center (unlike the respiratory center) is not located in the medulla, for, whether a cut is made either in the upper part of the spinal cord or above the medulla, the animal loses its ability to regulate the body temperature. In fact, any cut across other levels of the brain stem, even well above the medulla, produces this same effect. It is only when the cut is made anterior to that forward portion of the brain stem called the *thalamus* that the animal retains its temperature-regulating machinery intact despite the cut.[11] The temperature-regulating center, then, lies in the thalamus.

11. Such a cut would mean cutting between the anterior end of the brain stem (the thalamus) and the two lateral outgrowths of the brain stem, the *cerebral hemispheres*. A more detailed description of the anatomy of this region will be found in chap. 10.

We saw earlier that, although respiration is controlled in part reflexly through the respiratory center, conditions in the region of the center itself, namely, the carbon dioxide concentration, also determined the activity of the center and therefore of the efferent systems leading from it to the breathing muscles. Similarly, in the case of the temperature-regulating center, nervous discharges through the appropriate efferent systems can occur reflexly as a result of afferent influences from the skin playing upon the center, as we have seen. But they can also occur as a result of changes locally in the center itself. In this latter case the critical factor is probably not a chemical change but a physical one—temperature. When the center is warmed, efferent discharges from it to sweat glands and skin vasodilators lead to increased heat loss; when the center is cooled, the heat-conserving mechanisms are activated, and muscle tension is increased by nervous discharges through the efferent nerves of the muscles. In part, the evidence for this is direct, derived from actual warming or cooling of the temperature-regulating center itself. A more common experiment is to warm or to cool the blood in the carotid artery sufficiently to change the temperature of the brain, including the heat center, without directly changing the temperature of the rest of the animal.

Thus we see that the thermostat of warm-blooded animals is a complex inherited mechanism, governed mainly by the "involuntary" part of the nervous system, operating about as well as normally in the complete absence of the cerebral hemispheres and their centers of consciousness.

F. FLUCTUATIONS IN BODY TEMPERATURE

1. *The normal range.*—Like all physiological constants, the "constancy" of the body temperature is relative rather than absolute. In man there are normally diurnal fluctuations of approximately 2° F., the body temperature being highest in the afternoon or early evening and lowest in the early morning hours (2:00–5:00 A.M.) in individuals who sleep at night. Workmen who regularly sleep in the daytime may show a reversal of the times at which the high and low points occur, after a period of weeks or months.

2. *Abnormal temperatures—fever.*—No part of the animal machine is exempt from breakdown: the temperature-regulating machinery may go awry under abnormal conditions. The most commonly observed abnormality is fever—an elevation of the temperature above

the normal level of 87°–99° F. In fever the body temperature may rise to 110° F. before death occurs.

Fever is a common consequence of infection. Bacterial products liberated into the system by infectious organisms so regularly produce fever that the existence of fever is taken as presumptive evidence of infection. The bacteria produce this action in some way chemically, for, if dead bacteria or bacterial products or even nonbacterial "foreign proteins" are injected, fever likewise results.

The temperature-regulating machinery also breaks down in sunstroke, suffered during hard physical labor under a hot sun, or in the heatstroke of mill and foundry workers.

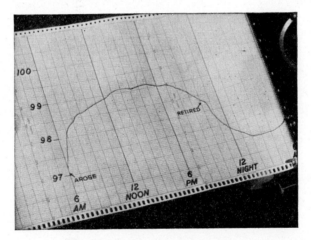

Fig. 116.—Body-temperature fluctuations. A continuous recording of the body temperature of a normal man over a 24-hour period, showing a maximum of about 99° F. in the afternoon and a minimum of about 97° F. in the early-morning hours. (From the sound film, *Control of Body Temperature.*)

The mechanisms of fever production are not entirely clear. The initial defect is more often one of diminished heat loss rather than increased heat production.

Interference with heat loss from the skin blood vessels is often striking. Despite the elevated body temperature, the skin vessels may be constricted and the skin pale. This may actually reduce the skin temperature so that the "cold" nerve endings are stimulated. The individual experiences a "chill," and there may be reflex shivering just as in exposure to external cold. Such shivering increases heat production and, in the absence of adequate heat dissipation, accentuates the fever.

V. THE FUEL: DIET AND NUTRITION

One might question whether the subject of diet and nutrition should not logically be presented earlier instead of after the topics of digestion, absorption, and metabolism. However, the requirements of an adequate diet can perhaps be best understood and the significance of the various constituents appreciated fully only after we know the use to which the foods are put in the body. What are the requirements of an adequate diet?

A. ENERGY REQUIREMENTS: CARBOHYDRATES AND FAT

There should be enough of the fuel foods to furnish calories sufficient for the basal metabolism plus the extra metabolism associated with physical exercise, exposure to external cold, and growth in the young. What this extra metabolism will be depends chiefly upon the amount of muscular work performed and the temperature in which that work is done. It may range from 1,000 calories in an individual leading a relatively sedentary life to as much as 7,000 calories in a lumberman working in the North Woods in winter. These calories must come from carbohydrates (4 cal. per gm.), fats (9 cal. per gm.), and proteins (4 cal. per gm.). Since protein is biologically and economically expensive, it is desirable to obtain most of this energy from the cheaper carbohydrates and fats.[12]

B. PROTEIN

1. *Nitrogen balance.*—Because protein is being continually torn down in the tissues and lost from the body, constant replacement must take place from proteins of the food, even in the adult, where growth has ceased and there is no increase in total bulk of protoplasm. It is convenient to determine the protein content of a food sample, or of the excreta, by determining its content of nitrogen, which regularly comprises 16 per cent by weight of the protein molecule. We may say, then, that the daily nitrogen intake must equal the daily average excretion. The individual, that is, must be maintained in a state of nitrogen equilibrium. Sometimes a positive balance is required—the nitrogen intake exceeding the output—during active growth in the young, during gestation in the pregnant female, and during convalescence from a debilitating illness, in all of

---

12. See following section for the role of the carbohydrates and fats as "protein sparers." Although fat is a concentrated form of energy, it cannot serve as the main source of energy, partly because large quantities of fat in the diet depress the digestive processes.

which conditions more protoplasm is manufactured than is daily destroyed. A condition of continued negative balance is abnormal and indicates that protein is being lost and that the organism is literally burning its own substance, as in starvation or debilitating diseases.

2. *Quantitative protein requirement.*—How much protein is required to maintain nitrogen equilibrium depends in part upon how much other fuel food is eaten. Let us attempt to learn experimentally

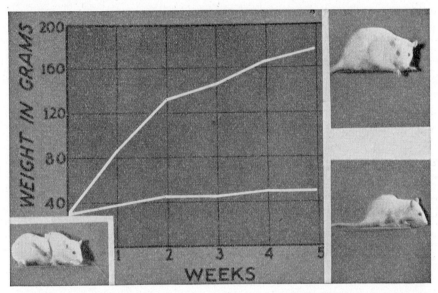

Fig. 117.—Effect of inadequate protein intake. A pair of litter-mate rats (*lower left*) weighed nearly 40 gm. each at the outset of the experiment. Both rats received identical diets, except that one received too little protein. At the end of five weeks this rat (*lower right*) had gained very little weight (*lower growth curve*). The control rat (*upper right*), receiving an adequate supply of protein, grew to nearly 180 gm. (*upper growth curve*). (From the sound film, *Foods and Nutrition.*)

what this requirement is with protein as the sole source of energy. Typical results of such an experiment are given in Table 9. If a dog loses 25 gm. of protein on a day when none is eaten, we might guess his daily requirement to be 25 gm. But, when this quantity is fed, still more is metabolized; and some, though less than before, is still lost. By increasing the intake each day to the amount used by the body the day before, we gradually diminish the negative balance until, with an intake of 60 gm. or more, equilibrium is established.

But, if we include carbohydrates and fats in the diet, we can

manage to establish protein or nitrogen equilibrium at a much lower
level of protein intake. As little as 25–30 gm. of an adequate protein have sufficed in some human experiments.[13] It is for this reason
that carbohydrates and fats are referred to as *protein sparers;* their
presence in the diet reduces the quantity of protein required to maintain the animal in nitrogen equilibrium.

TABLE 9

PROTEIN REQUIREMENT OF A DOG WITH PROTEIN AS THE
SOLE SOURCE OF ENERGY

Day of Experiment	Protein Fed (Gm.)	Protein Used Up (Deter-mined by N Excretion) (Gm.)	Protein Lost (Gm.)
1	None	25	25
2	25	35	10
3	35	40	5
4	45	48	3
5	60	60	None
6	Over 60	Over 60	None

So far as energy requirements are concerned, the following quantities of these three foods constitute a well-balanced diet for a person
requiring about 2,700 calories daily:

	Gm.	Cal.
Proteins	100	400 (15)%
Carbohydrates	400	1,600 (60%)
Fats	75	675 (25%)

3. *Biological value of proteins: adequate and inadequate proteins.*—
It has been pointed out that proteins differ from one another on the
basis of the number and kinds of amino acids of which they are made.
Obviously, then, when muscle protein is being synthesized, all the
specific amino acids which go into the structure of that specific kind
of protein must be available to the manufacturing muscle cells. If
any of these amino acids is lacking entirely or is available only in
inadequate quantities, the muscle protein cannot be made. Apparently animals, including man, can synthesize small quantities of
amino acids from simpler molecules, but this occurs to an extremely
limited extent. Fortunately, however, the animal body can fairly
readily transform some kinds of amino acids to other kinds. But

13. That this small quantity of protein will maintain an individual in nitrogen balance does
not indicate that this is the optimum intake (see further discussion below).

there are certain amino acids which cannot be supplied the tissues in this way. They cannot be manufactured from other related amino acids in animals. It goes without saying that these specific amino acids must therefore be present as such in the diet. Being essential dietary constituents, they are referred to as the *essential amino acids*. Any protein which is lacking or quantitatively deficient in essential amino acids is referred to as an *inadequate protein*, biologically inferior as a food to the so-called *adequate proteins*.

Zein, the principal protein of corn, is such an inadequate protein. If this is the sole source of protein, the animal loses nitrogen no matter how much of the protein plus carbohydrates and fats is fed. Body weight decreases, and death is inevitable. If one of the essential amino acids (*tryptophane*) absent from zein is added to the diet, nitrogen balance is restored, although in the young growth fails to take place. If still another of the missing essential amino acids (*lysin*) is now added, normal growth takes place. Gelatin is another inadequate protein, lacking three of the essential amino acids. Experiments of this sort have shown that, of the approximately twenty different kinds of amino acids which have been obtained by breaking down protein molecules, ten are of the essential variety.[14] It has been possible to raise apparently normal rats by feeding no native proteins but substituting for them appropriate mixtures of these essential amino acids.

4. *Animal and vegetable proteins.*—How can one be sure that his diet is adequate as regards proteins without actually knowing by chemical analysis of the dietary proteins their amino acid composition? By observing this principle: In general, eat more of those proteins which are most nearly like the proteins of the body. For man this means animal proteins—meat, sweetbreads, liver—as many kinds of animal tissues as are palatable.

There should be no misconception about animal versus plant proteins. While man can be more sure of obtaining all the required amino acids from a minimum of proteins by eating animal products, he can also obtain them from plant sources entirely. Some animals, the herbivores, eat only plant foods regularly. This possibility follows necessarily from the fact that animals cannot manufacture significant quantities of amino acids. The amino acids are traceable ultimately to plant synthesis. However, it is generally true that more

14. There is evidence indicating similarly that there are also certain of the fatty acids which must be present in the diet as such.

protein, and a greater variety, must be eaten where plants are the sole source. And in general, whether the protein is of animal or plant origin, it is wise to provide a comfortable margin of safety of protein intake. Though it may be possible for man to maintain health on 25 gm. per day, as some experiments have indicated, it is probably wiser to ingest 50–100 gm. daily.

C. WATER

Of the dietary constituents furnishing no energy to the body, water deserves first mention. We need not repeat what has already been emphasized of the importance of water to tissues. It is the medium indispensable for cell life and protoplasmic activity. Drying is usually equivalent to death. Water deprivation is much more rapidly fatal than deprivation of any or all other dietary constituents.

D. INORGANIC SALTS

There is a continuous turnover of salts in the body, quantities being lost daily in the excreta. On an inadequate salt intake the body

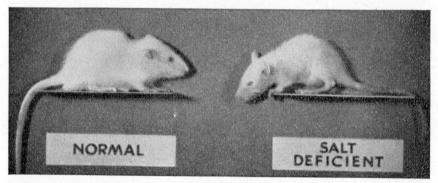

FIG. 118.—Effect of salt deficiency in the diet. Litter-mate rats of the same weight were fed the same diet, except that one rat received insufficient inorganic salts. This rat (*right*) grew poorly and was less healthy than the control rat (*left*). (From the sound film, *Foods and Nutrition*.)

salts are partly conserved. The urine contains less salts. Yet salt-free diets are rapidly fatal. Death occurs even more rapidly on a salt-free diet than in total starvation because, when salt-free food is eaten, excretion of the wastes of its metabolism entails the simultaneous loss of salts from the body in greater amounts than during total starvation.

Much information as to the needs of the body for certain of the metallic atoms has been accumulated by experiments. Only infinitesimal quantities of some of these minerals are required, but they are quite indispensable for life or health nonetheless.[15] Such common foods as meat, liver, eggs, cheese, milk, and vegetables are rich sources of the required food salts.

In part, the salts are built into body tissues and cells and cell products: Calcium and phosphorus are found especially in bone and teeth, iodine in the hormone of the thyroid gland, and iron in hemoglobin. The importance of the salts in maintenance of osmotic equilibria and, in general, of an internal environment suitable for normal cell activity has been stressed repeatedly in this volume.

E. VITAMINS

1. *Early experiences with deficiency diseases.*—Vitamin-deficiency diseases were apparently unknown as long as man and animals lived upon natural foodstuffs of a sufficient variety. Man's acquaintance with these deficiencies and the discovery of the vitamins were consequences of refining and preserving foods—the milling of flour, the polishing of rice, and the canning and drying of meats and vegetables. Experiences with the consequences of vitamin deficiencies and the almost miraculous curative powers of vitamin ingestion antedate by centuries man's knowledge of the vitamins themselves or of the real cause of these deficiency diseases. Sailors were acquainted with the bleeding gums and painful, swollen joints of *scurvy*, which was sure to strike mysteriously some members of the crew on long sea voyages. They also learned that hope of a cure was vain until land was reached and fresh vegetables and fruits were available. Otherwise, death was inevitable. The British sailors capitalized upon the empirical discovery that even fruit juices, especially of limes, were efficacious in preventing scurvy, and they carried the juice with them on long sea voyages. This has earned for the British sailor the term "limey." No conception of the nature of scurvy or the mechanism of its cure or prevention went with this practical knowledge. Here was a disease, there was the preventive or cure, and that completed the information. Lime juice might well have belonged to the same category as cinchona bark, which cured malaria, or foxglove, which improved dropsy. We know now that quinine (from cinchona) kills or inhibits the growth of an infectious agent in the blood, digitalis (from

15. Daily requirements of some of these elements have been estimated as follows: chlorine 1.2 gm., sodium 0.8 gm., calcium 0.7 gm., phosphorus 0.7 gm., iron 0.015 gm., copper 0.001 gm., manganese 0.0003 gm., and iodine 0.0002 gm.

foxglove) improves the circulation in some types of heart failure, and lime juice corrects a dietary deficiency.

2. *Early experiments on vitamins.*—The beginnings of a real insight into the role of vitamins came with the production of a vitamin deficiency in chickens. Christiaan Eijkman, a young Dutch physi-

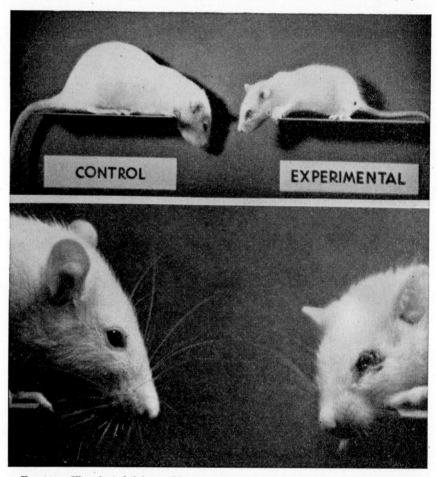

Fɪɢ. 119.—Vitamin A deficiency. Litter-mate rats of equal weight were fed identical diets, except that one rat received no vitamin A. This rat (*upper right*) grew poorly, as compared with the control animal. Also, the experimental animal developed eye infections (*lower right*), which is characteristic of avitaminosis A. (From the sound film, *Foods and Nutrition.*)

cian stationed in Java, was impressed by the ravages of a disease of the Orient called *beriberi*. It affects chiefly the nervous system, causing inflammation of nerves, paralyses, deranged sensations, weakness, and, if untreated, death. In Eijkman's time (latter part of nineteenth century) discoveries of infectious organisms responsible for various diseases were being made almost daily, and many people believed beriberi also to be caused by microbes.

Eijkman, stationed at a prison camp, discovered that chickens fed upon the rice left over from the prisoners' meals developed beriberi just as had the prisoners. He first thought that the chickens became ill and died because they ate too much starch. But later it was learned from experiments that the sick chickens could be cured by adding rice husks to the diet. They could now eat as much rice as before but did not become ill. The illness depended not upon anything eaten but upon a lack of something in the diet.

Many investigators have contributed to our present knowledge that beriberi is the result of a dietary deficiency which can be prevented by incorporation of the lacking substance in the diet. The substance, found in nature in rice husks as well as elsewhere, was called a vitamin.

3. *Action of vitamins.*—In the following decades several other vitamins were discovered. They came to be designated by letters, the antiberiberi vitamin being referred to as vitamin B, and the antiscorbutic vitamin as C. Although the vitamins are indispensable in the diet, only extremely small quantities are adequate to preserve health; therefore, vitamins can be of no significant caloric value. Some, and perhaps all of them, act as catalysts in vital physiological reactions. Chemically, our information about the vitamins is accumulating rapidly. Several of them have been prepared in apparently pure form and their chemical formulas determined.

4. *General distribution in foods.*—The subject of the vitamins lends itself to dramatization, and we are being made very vitamin-conscious indeed. Vitamins are added to many foods and even to beer, chewing gum, and candy. Vitamin-enriched white bread is a good food from which vitamins have been removed (by milling the flour) and to which vitamins are subsequently added. Whole-wheat bread, not enriched, but containing its natural vitamins, is more nutritious.

Like many clever advertising campaigns, this also rests upon an undeniable truth, upon which foundation is built an appealing (and profitable) superstructure of exaggeration. The truth is that vitamins are indispensable in the diet. But they are so generally distributed in natural foods—in fruits, vegetables, animal and plant fats, and grains—that if we eat an ordinary varied diet we are practically sure to obtain the required kinds and quantities of vitamins. A possible exception to this is during the early growing period of the human infant when the diet is necessarily limited. In very young children it has been found beneficial to take special precautions to add vitamins C and D to the diet.

5. *The "sunshine" vitamin.*—Perhaps the most exploited, and the most deserving of further mention, is the famed "sunshine" vitamin—D. This vitamin, especially concentrated in fish oils such as

Fig. 120.—Experimental rickets. The dogs are females of the same litter. The dog on the left received no vitamin D. The dog on the right was fed the same diet as the other, with vitamin D in addition. (From Steenbock, courtesy of *Journal of the American Medical Association*.)

that of the cod liver, is indispensable for growth and normal maintenance of bone and teeth. Lack of it is not fatal, though it may produce bowed legs and other bone abnormalities more or less disfiguring or crippling.

A peculiar feature of this vitamin is that it can be formed by the action of sunlight (ultraviolet light) upon certain fatlike compounds (*sterols*, especially *ergosterol*). These compounds are normally present in the skin, so that this vitamin, unlike all the others, can actually be formed in the body by exposure of the skin to sunlight.

In conclusion, it might be stated that some of the vitamins are soluble only in fats and, therefore, are found in nature only in fats. These include A, D, E, and K. Here, then, is another function which the natural fats of the diet serve—as a vehicle for the fat-soluble vitamins.

6. *Complexity of vitamins.*—Vitamin B, once thought to be a single substance, has now been shown to consist of a complex of at least twelve distinct chemical components, each with its peculiar chemical formula and properties and actions on the animal organism. Each may be considered a separate vitamin. Examples of the various components or "fractions" of the vitamin B complex are *thiamine, nicotinic acid,* and *riboflavin,* the chemical formulas of which are well known. A deficiency of thiamine in the diet causes the condition described earlier as beriberi, first studied in detail by Eijkman. Nicotinic acid deficiency causes pellagra, characterized by digestive, nervous, and mental disturbances and skin changes. Riboflavin is involved in the enzyme catalysis of oxidative processes within the cells of the body. The B_{12} fraction of the vitamin B complex is essential for the prevention of per-

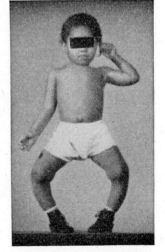

nicious anemia (see pp. 103–4). Each of the other components of the vitamin B complex, when present in adequate quantities in the diet, prevents its own peculiar deficiency in growth, metabolism, or general health. In some instances, two or more of the components may be essential for some special physiological process.

The original mistake of assuming that vitamin B was a single substance was natural enough, since the various fractions are often closely associated in plant and animal tissues, that is, in man's foods, and because a dietary deficiency of one fraction produces certain defects which are similar to some defects caused by an inadequate intake of another fraction.

There is evidence that still other vitamins may consist of more than one substance.

Partly for the sake of completeness, partly to emphasize the wide distribution of the vitamins in nature, a simplified table

Fig. 121.—Rickets in the child. The marked degree of bone deformity in this child can probably be only partially corrected by vitamin D administration. It could have been prevented. (From the sound film, *Foods and Nutrition.*)

is given (Table 10) of the better-known vitamins, their natural sources, and the deficiencies caused by their lack.

TABLE 10

THE MOST IMPORTANT VITAMINS, THEIR MAIN SOURCES, AND THE CHIEF
VITAMIN-DEFICIENCY CONDITIONS

Vitamin	Some Excellent Natural Sources	Chief Results of Deficiency in the Diet
A............	Yellow vegetables Cod-liver oil Butter Cheese Cream Egg yolk	Infections, especially about the eyes Retarded growth and development Night blindness Dryness of skin
B complex (thiamine)*....	Pork Whole cereals Egg yolk Yeast	Beriberi: paralysis of nerves; muscle incoordination; digestive upsets; loss of appetite; faulty growth
B complex (nicotinic acid)*	Pork Liver Yeast Eggs Milk Fresh vegetables	Pellagra: alimentary, nervous, and mental disorders; pigmentation and thickening of the skin; may be fatal
B complex (riboflavin)*...	Liver Milk Meats	Impaired cell oxidations; retarded growth; nerve degeneration; cataract
B complex (B$_{12}$)*......	Liver	Pernicious anemia
C*...........	Grapefruit Green vegetables Tomatoes Orange juice Lime juice	Scurvy: hemorrhages, bleeding gums, painful joints, fragile bones
D...........	Sunlight† Egg yolk Fish oils (especially cod liver) Salmon	Rickets: faulty development of bones and teeth; bones become soft, fragile, and deformed Decay of teeth
E‡..........	Lettuce Whole wheat	Sterility in the male Death of unborn young in the pregnant female
K...........	Liver Cabbage Spinach Kale Tomatoes	Diminished prothrombin content of blood and delayed coagulation

* Necessary for the life of the individual.   ‡ Necessary for the life of the race.
† By exposure of the skin to sunlight.

F. Roughage

We have seen the importance of bulk in the intestine in stimulating the gastrointestinal movements (see p. 296). Foods should therefore contain some residue of indigestible materials. This generally is in the form of cellulose and is especially abundant in fruits, vegetables, and whole grains.

G. Variety in the Diet

A varied diet is the best insurance that there will be an adequate intake of certain required dietary constituents, such as essential amino acids, vitamins, and salts. Variety constitutes a margin of safety in nutrition. The constantly increasing list of specific dietary substances found to be essential to health or even life emphasizes the extreme importance of dietary variety, the folly of food fads which restrict variety, and the importance of medical supervision over diets restricted for such specific purposes as weight reduction.

In addition, variety is an important stimulus to the appetite. Even though all the known elements may be present in a standard diet of rats, for example, they often do not eat as much or grow as well or live as long as they do when the diet is varied.

H. Unknown Factors

Undoubtedly, more remains to be learned about the dietary needs of the organism. If we knew all the kinds and quantities of substances necessary, it should be possible to make up a diet of purified compounds entirely adequate to support all the normal life-processes of growth, metabolism, and reproduction over the entire life-span. Thus far this has never been done, either for the entire organism or for isolated groups of cells. It is always necessary, for instance, in culturing cells and tissues outside the body to furnish nutrient juices derived from embryonic tissues. Needless to say, such crude extracts contain many chemical compounds of an unknown nature.

Despite the large number of known and unknown dietary essentials, their distribution is fairly widespread, so that a diet of natural foods adequate in calories, protein, calcium, vitamin A, riboflavin, thiamine, nicotinic acid, and vitamin C is likely also to contain all the other essential constituents.

VI. THE USE OF ALCOHOL AND TOBACCO

Any scientific discussion of the effects of alcohol[16] on man is fraught
with difficulty, because of the emotions and prejudices which may be
aroused and of the moral and social implications of the subject, not
to mention the incompleteness of our knowledge, especially about the
effects of repeated small doses of alcohol. But let us attempt an im-
partial survey of the available data. Is alcohol a food, a drug, or a
poison? Or is it all three? The latter possibility must be admitted
at the outset, for these three categories of substances are not mutual-
ly exclusive. An excessive intake of an indispensable dietary ingredient—table salt, or even water, for example—can be decidedly injurious or poisonous.

A. ALCOHOL AS A FOOD

To answer the question, "Is alcohol a food?" we must first agree
upon a definition of "food." A food is any substance which upon ingestion can furnish the body with useful constituents of the cell environment, building stones for body parts, or energy. Now, alcohol,
in the concentrations reached by drinking the substance, is clearly
not a necessary or useful component of the internal environment,
and apparently it cannot be used to manufacture body parts. But it
is a food, since it can be and is burned, yielding energy which the
body can use. Numerous experiments on man and animals have amply demonstrated this. Animals given moderate quantities of alcohol
in addition to an adequate weight-maintaining diet will store fat and
gain weight. This is not because alcohol is converted into fat but
because the combustion of alcohol spares the body the necessity for
burning all the carbohydrate and fat ingested, and some is stored.
Furthermore, alcohol can act as a protein sparer (see p. 340). The
evidence indicates that over 90 per cent of the alcohol which is ingested is burned. None seems to be stored or changed, and very little
(1–10 per cent of that taken) is excreted through the lungs and kidneys. Some of it may remain in the blood and tissues for hours, depending upon the quantity taken; it gradually disappears as it is
oxidized.

But, while we recognize all this, we hasten to add that alcohol is
not commonly used because it is a food. Certainly as a source of en-

16. Chemically, there are many alcohols, with widely differing properties. Here we are concerned only with ethyl alcohol (C_2H_5OH), the most important ingredient of intoxicating
alcoholic beverages.

ergy it has no advantages over glucose. In fact, it is inferior to glucose, since it cannot be stored in the body.

B. ALCOHOL AS A DRUG

Alcohol is used mainly because of certain effects considered pleasurable by some people, quite independently of any food value. These we may call *drug effects*. A drug has been defined as a substance which, when introduced into the body, modifies the activity of the body organs otherwise than by increasing the supply of available energy.

Alcohol affects mainly the central nervous system, especially the brain. The drug is sometimes thought to be a stimulant, because an individual who has taken a moderate amount may be unusually active, apparently alert, talking freely. Actually, alcohol depresses at least some parts of the nervous system. By depressing first the highest critical faculties, it may promote uninhibited and courageous behavior in even the most timid and self-conscious. This dulling of self-criticism and disregard of minor unpleasant sensations provides a dubious means of escape from the cares of life and produces the feeling of careless well-being that accounts for the wide use of the drug.

Many experiments have shown, however, that mental and motor performance of almost any kind is less efficient and accurate in even a moderately inebriated individual. At the time, the drinker *thinks* he is doing very well—better than normal. This has lent credence to the oft-repeated myth that some people work better when slightly intoxicated. But objective measurements show clearly that accuracy of performance and judgment are definitely impaired. Tests reveal that even a mildly intoxicated man drives a car less well than when he is sober. He tends to react more slowly, drive faster, and take more chances, and he is less likely to use his brakes than to swerve aside to avoid objects—all of which increase the likelihood of accidents.

Alcohol possesses no specific curative powers in any disorder or infection. Its medical use has now been greatly curtailed, and its medical value is limited almost entirely to its effect in dulling the senses, so that it may sometimes be employed for minor surgery, when other anesthetics cannot be used. It is also used extensively as a solvent for drugs.

C. ALCOHOL AS A POISON

The line of demarcation between a drug and a poison is not sharp. Practically every drug, no matter how useful it may be medically, is injurious (i.e., a poison) when used in excess. We may say that alcohol becomes a poison when an individual takes so much that normal body functions are impaired. The depressing action of the drug may affect those parts of the central nervous system which control speech, which becomes thick and inco-ordinate, and bodily movements, which become poorly integrated. The depression may continue, to produce a sleeplike state, or coma, commonly known as "passing out." Still larger quantities may even kill.[17]

It has been extremely difficult to find an answer to the question, "Does alcohol produce chronic poisoning?" Is the habitual use of alcohol injurious to the body, and, if so, how? Numerous mistakes have been made in this field. For example, the relatively high incidence of a certain brain disease (*general paresis*) in drinkers led to the erroneous conclusion that alcohol was directly responsible, by destroying brain cells. Now we know that general paresis is caused by syphilitic infection. The reason for the high incidence in drinkers seems to be that alcohol may diminish restraints in sex matters. Hence promiscuity and exposure to infection are likely to be more common. Again, the causal relationship between chronic alcoholism and insanity is deceptive. In many instances the alcoholism is probably a symptom of the disorder rather than a cause.

Studies upon man and laboratory animals ingesting moderately large quantities of alcohol over long periods of time fail to show conclusively that there are seriously deleterious effects on body organs, longevity, or reproduction. But the often-repeated ingestion of larger quantities seems definitely to lead to a diminution of mental function, longevity, resistance to infection, and general health. Rabbits which have been rendered very resistant to pneumonia, for example, lose virtually all their artificially induced immunity to this infection immediately following a stupor-producing dose of alcohol.

Perhaps the most important problem in this field is that presented by the excessive use of alcohol by some people who are unable to control the amount they drink and become alcohol addicts, or "alcoholics." Why are some people able to use alcohol in moderate

17. There are objective indications of intoxication when the level of alcohol in the blood reaches 0.1 per cent. Profound stupor, sometimes terminating fatally, may occur when the level reaches 0.6 per cent. A fatal blood alcohol level may be reached by the ingestion of 6 cc. of absolute alcohol per kilogram body weight. One quart of 100 proof (50 cer cent alcohol) whiskey contains about 6 cc. of alcohol per kilogram for an adult of average weight.

amounts or discontinue its use at will, while others become alcohol addicts, driven to frequently repeated excesses? Certainly the person who consistently seeks refuge from the trials of life in the euphoria of alcoholic intoxication creates a vicious cycle difficult to break. Life becomes increasingly exacting, because drunkenness lowers his efficiency and therefore multiplies the troubles from which he must seek escape.

Education and an understanding of the effects of alcohol seem to be relatively ineffective in preventing the habitual intemperate use of alcohol by an important percentage of the population. It appears likely that nervous instability and emotional conflicts, involving a sense of dependence or basic insecurity, play a vitally important role in the development of alcohol addiction. Until we know more about these emotional difficulties and how to measure, control, and prevent them, the problem will remain with us.

Contrary to common opinion, addiction to alcohol is by no means limited to the ignorant, the unsuccessful, or the poor—the traditional "drunkard" who reels on the streets and sprawls in the gutter. It occurs as commonly in the well-to-do, the intelligent, and the successful, although there may be such a deterioration of ability and behavior that the apparent success in life is short-lived. It is often difficult to differentiate between the true alcoholic and the so-called "social drinker" who may also consume considerable quantities of alcohol. In general, the alcoholic is usually unable to control the extent of his drinking and carries it to great excess. Danger signs indicating that social drinking may be passing into addiction to alcohol are a dependence upon alcohol to bolster courage or to allay anxiety or drinking to such an extent that there results an interference with the quality or quantity of work done, an impairment of previously good marital, family, or social relationships, or an increasing personal unhappiness. Anyone in whom such signals of developing chronic alcoholism appear should seek the aid of a physician and psychiatrist. "Alcoholics Anonymous," an organization of former alcohol addicts who assist one another in maintaining total abstinence, has contributed materially to the rehabilitation of alcoholics.

D. Tobacco

Though it is not a dietary constituent in the ordinary sense, tobacco may be appropriately considered at this time. It is highly desirable to know the physiological effects of tobacco, harmful or otherwise, since tobacco is used even more extensively than alcohol. There seems to be little dispute about the occurrence of irritation of the throat and upper respiratory tract, sometimes producing chronic inflammation, coughing, and burning. Some studies seem to indicate tendencies toward subnormal body weight and susceptibility to colds in smokers. Nausea, indigestion, diminished appetite, vasomotor disturbances, palpitation of the heart, and shortness of breath have sometimes been observed.

A very specific effect upon human subjects studied under controlled conditions is that of vasoconstriction. This affects the skin blood vessels, so that the skin tends to become pale and cool, and the arterioles in the interior of the body, so that the arterial blood pressure is elevated. Therefore, it seems reasonable for physicians to curtail or forbid smoking in people with a tendency toward high blood pressure or in those who have had heart attacks caused by constriction of the arterioles of the heart muscle. Smoking also increases the heart rate.

Statistics have been accumulated which seem to indicate that cigarette smokers are more likely to develop cancer of the lung than are nonsmokers, presumably because of irritation of the lung tissues by the inhaled smoke.

On the whole, it seems safe to say that, whenever definite effects are produced by excessive smoking, these effects are deleterious.

The common notion that nicotine is responsible for all the effects of tobacco is probably unfounded. The smoke contains a number of other irritating or deleterious substances;[18] all of these, including nicotine, are present in very small concentration.

The reason for the almost universal appeal of tobacco is not clearly apparent. Smokers list as pleasures derived from the habit sociability, fragrance, relaxation, stimulation, steadying of nerves, quieting of hunger, sight of the smoke, and feel in the lips. Some of these are trivial or hard to measure; some undoubtedly reflect the unfounded claims of the high-power advertising of the cigarette-makers.

The pronounced drug effect of alcohol seems entirely lacking in tobacco-smoking. The use of tobacco probably serves smokers mostly by providing a motor outlet for tensions and nervousness through the necessary manual, oral, and respiratory manipulations involved.

18. These include carbon monoxide, ammonia, aldehydes, furfural, and acids. The heat of the smoke is also irritating, especially in pipe-smoking.

VII. THE FATE OF METABOLIC WASTES

A. Elimination of Metabolic End Products

The constant flow of materials through the metabolizing organism, in the course of which matter yields up its energy or is subjected to more or less complex chemical transformations, is completed by elimination of useless or injurious end products from the organism. We have had occasion to examine the mechanisms for the eliminations of some of the wastes in foregoing sections and chapters. The lungs eliminate waste carbon dioxide, and the liver excretes the end products of hemoglobin breakdown—the bile pigments.

The sweat glands are of first importance in body-temperature regulation. They are of practically no importance in the excretion of wastes and are mentioned here mainly because of a popular misconception to the contrary. Perspiration somewhat resembles in composition very dilute urine, having about one-eighth the concentration of total solids, most of which is NaCl. On an ordinary, comfortable day about 600 cc. of water is lost from the skin.[19] On a very hot day sweat secretion may amount to 2 or 3 liters. But, in any case, the wastes of the sweat are so diluted by water, being more dilute when more sweat is secreted, that the total waste elimination by this route is not significant. With maximal perspiration, no more than 2–2.5 gm. of urea per day are lost by this route, an amount equal to that excreted in about 100 cc. of urine, which is less than a tenth of the normal daily output. When the kidneys are seriously damaged by disease, the sweat becomes more concentrated, especially in urea. This we recognize as a compensatory adjustment, but the compensation is not nearly of a degree sufficient to have any significant effect upon the serious outcome of kidney damage.

The excretion of the waste heavy metals, such as calcium, by the colon, has also been discussed.

B. Elimination and Excretion

A distinction brought to mind by mention of the excretory function of the colon should be made at this point—the distinction between "excretion" and "elimination." These terms are often loosely used interchangeably. This practice is quite acceptable and is good usage, provided the fundamental difference between the two distinct processes involved is clear.

19. Part of this may be by evaporation of water from the epidermis directly.

Strictly speaking, by *elimination* is meant the evacuation of those hollow organs in which waste materials have been accumulated. *Hydra* eliminates wastes which have collected in its gastrovascular cavity. Specialized cells—the flame cells—of the flatworm help to eliminate materials which have accumulated in the many-branched excretory tracts. In mammalian breathing waste carbon dioxide is eliminated from the lungs, and in defecation and urination the rectum and bladder are emptied of their accumulated waste products. Elimination depends essentially upon mechanical changes (e.g., pressure) which force wastes into the outside world. For example, the intermittent evacuation of the urinary bladder depends partly upon contraction of the muscular wall of that organ, a reflex phenomenon dependent upon intact sensory and motor innervation of the bladder. In a sense, the interior of the bladder or the rectum or the lungs is not truly inside the body; that is, substances in these organs are not part of the internal environment of the cells and tissues of the body generally.

Excretion, on the other hand, deals with the extracting of wastes from the internal environment, from the circulating body fluids, and the passing of them into temporary waste depositories—into the lungs, bladder, and gastrointestinal tract. The cells of the colon *excrete* waste calcium and iron from the blood into the lumen. These atoms are *eliminated* from the body in defecation.

We have seen that some excretion takes place by the operation of physical forces alone, apparently, such as the diffusion of carbon dioxide from the blood into the alveoli of the lungs. But, in large part, excretion is carried out by cellular activity, involving cellular work and the expenditure of energy by the cell. This holds for the excretion of calcium into the colon and bile pigments into the bile ducts of the liver. This cellular activity may be modified by physical forces such as osmosis, filtration, and diffusion but cannot be accounted for entirely by them. Activity of living cells contributes largely to the phenomena.

Excretion, therefore, closely resembles secretion, so far as the involvement of the activity of living cells is concerned. Thinking of the process itself rather than of its role in the bodily economy, we frequently use the term "secretion" instead of "excretion." It is common to speak of secretion of urine by the kidney, for example, and to designate the kidneys as glands.

C. The Work of the Kidneys

1. *Excretion of wastes.*—The paired kidneys lie in the back of the abdominal cavity, one on each side of the body. These organs are as indispensable for life as the lungs or liver. They function as excretory glands in the formation of urine, which flows steadily from the kidneys to the bladder (see Fig. 123), about 1,500 cc. being excreted daily by a human adult. Chemical analysis of the urine reveals the nature of the products of kidney excretion. Urine normally contains salts, urea (from protein metabolism), *uric acid* (from metabolism of

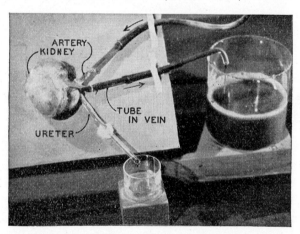

Fig. 122.—Functioning isolated kidney. A kidney removed from an anesthetized animal continues for a time to secrete urine through a glass tube inserted into the ureter (note the drop collecting at the end of the tube), provided blood is circulated through the organ. Arrows indicate direction of artificial blood flow through tubes inserted into the kidney artery and vein. The large flask collects the blood after flow through the kidney. The small flask collects the urine, which is normal in composition. (From the sound film, *The Work of the Kidneys.*)

a modified protein called *nucleoprotein*, abundant in cell nuclei), *creatinine* (from protein and muscle metabolism), and a number of other materials.[20]

In addition to the wastes of normal metabolism, the kidneys also excrete abnormal products of deranged metabolism, such as the acid bodies produced in diabetes.

2. *Regulation of blood composition.*—More than mere excretors of waste, the kidneys are important in the regulation of the composition

20. In 1,500 cc. of urine—the average daily excretion of an adult man—the following substances are dissolved: urea, 30.0 gm.; sodium chloride, 15.0 gm.; sulfuric acid, 2.5 gm.; phosphoric acid, 2.5 gm.; potassium, 3.3 gm.; creatinine, 1.0 gm.; uric acid, 0.7 gm.; ammonia, 0.7 gm.; hippuric acid, 0.7 gm.; magnesium, 0.5 gm.; calcium, 0.3 gm.; and other substances, 2.8 gm. The dissolved substances total about 60 gm.

of the blood and in the preservation in the internal environment of the conditions necessary for life. When even some of the normal constituents of the blood, such as sugar, rise above a certain concentration, they are excreted by the kidneys. When excess acid is produced in metabolism, the urine becomes more acid than usual; when bases are abundantly present, the urine becomes alkaline in reaction. Even the total blood volume is controlled in part by the kidneys. When the blood volume tends to increase, as after ingestion of large vol-

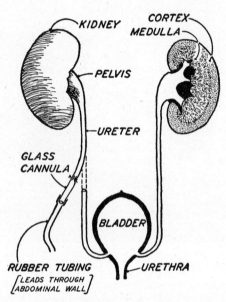

Fig. 123.—The urinary system. The kidney on the right has been cut open to reveal its internal structure. The ureter on the left has been cut and cannulated for the collection of urine as it flows from the kidney.

umes of water, the urine is copious and dilute; when the blood volume is low, the urine volume diminishes. This latter adjustment occurs after a hemorrhage, resulting in a conservation of the remaining water of the organism. Such a beneficial adjustment is not without its dangers, because, if urine secretion is scanty, the resultant accumulation of wastes in the blood may become a more serious threat to life than the diminished blood volume.

D. STRUCTURE OF THE KIDNEYS

Any understanding of the machinery by which the kidneys carry out their work entails first some knowledge of the anatomy of these glands. Gross examination (see Fig. 123) displays the relationships of the kidneys, *ureters*, bladder, and *uretha*. The kidney itself, upon being dissected, reveals a division into an outer layer (the *cortex*) and an inner portion (the *medulla*), which latter connects with the expanded beginning of the ureter. Microscopic examination shows the kidney to be composed of hundreds of thousands of tiny tubules, coiled in some places, straight in others, in a characteristic complex arrangement. One such secreting unit is diagrammed in Figure 124. The tubules join one another in the deeper regions of the

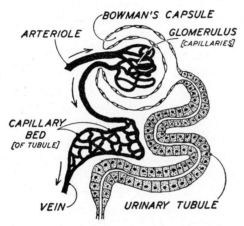

Fig. 124.—A secreting unit of the kidney. Arrows indicate direction of blood flow. Filtration takes place from glomerular capillaries into Bowman's capsule. As the urine passes through the tubule, its composition is changed, as described in text. Each human kidney contains nearly a million such units.

kidney (in the medulla), forming larger and larger tubes, which finally empty into the central cavity, or *pelvis*, of the kidney, which is continuous with the ureter. Each tubule begins as a blind sac known as a *Bowman's capsule*,[21] lying in the outer (cortical) layer.

Urine is derived from blood. Everything which is present in the urine has first been present in the blood,[22] from which the kidneys "separate" it. We should therefore expect to find an intimate association between the kidney blood vessels and the urinary tubules, of a nature different from that of blood vessels and tissue cells elsewhere. A terminal ramification of the kidney artery passes to each Bow-

21. After Sir William Bowman, English anatomist.

22. An exception is the *hippuric acid* of the urine. This is a relatively nontoxic material synthesized partly by the kidney cells from *glycocoll* (an amino acid) and *benzoic acid*, a toxic product present in some foods and also formed by bacterial action in the intestine, from which it is absorbed into the blood. The kidneys can also convert urea into ammonia.

man's capsule, in the cuplike depression of which it breaks up into capillaries. This capillary tuft is called a *glomerulus*.

Blood leaves the glomerulus through tiny vessels of capillary size, which are then profusely distributed to the tubular epithelium.

E. Mechanisms of Urine Formation

1. *Filtration through Bowman's capsule.*—Our first clue as to how the kidney works comes from anatomy—from a careful consideration of the structure of Bowman's capsule and its relationship to the glomerular capillary tuft. Here we have capillaries in close apposition to a membrane made up of thin flat cells, quite unlike secretory cells in structure. The walls of Bowman's capsule closely resemble the thin walls of the capillaries themselves and also of the lung alveoli. We have seen that these latter cells apparently play little or no part in moving materials—that molecules are moved across the cell membranes by purely physical forces like diffusion, osmosis, and filtration. So it would seem reasonable to form the hypothesis that physical forces are also responsible for movement of materials from the blood into the urinary tubules—that the cells of Bowman's capsule are not actively secretory in function. "Observation," said Claude Bernard, "starts an hypothesis, and experimentation tests whether the hypothesis be true."

What deductions can be made from this hypothesis which can be tested experimentally? If filtration were responsible for the entrance of urine into the capsule, then the capsular urine should be of essentially the same composition as blood, minus those blood components, such as the formed elements and the plasma proteins, which normally filter through capillary walls to a negligible extent. Obviously, collection of capsular urine would be extremely difficult technically, and at best only very minute quantities could be made available for chemical analysis. Yet this difficult feat has been accomplished. A finely drawn-out quartz tube is inserted into the capsule under the microscope, and capsular urine is collected in quantities sufficient for chemical analysis. Analysis shows that the urine here has essentially the same percentage composition as blood plasma minus its proteins. In this first step in urine formation, filtration is quite adequate to explain the findings. The capsular cells may conceivably play an active role, but there is no clear evidence for this concept.

2. *Reabsorption of water.*—Urine as it is eliminated from the body

is quite different from blood plasma. There is no change in the urine
in its passage through the ureters, bladder, or urethra. These structures add nothing to, or remove nothing from, the urine. It follows,
therefore, that certain changes take place in the composition of the
urine in the kidneys themselves between the time the urine enters the
capsule and emerges from the organ into the pelvis and ureter. These
changes in the urine occur, that is, in the tubules.

What are these changes? In the first place, voided urine (or bladder
or ureteral urine) is more concentrated than capsular urine. Therefore, water must have been reabsorbed from the urine by the tubules
and returned by them to the blood. In this conclusion we have the
support not only of the foregoing logic but, more dependably, of
direct evidence. Certain dyes injected into the blood filter into the
capsule with the other urinary constituents. Watching this colored
material under the microscope trickling slowly down the tubules,
pushed onward by formation of additional urine in the capsule, one
can observe a deepening of its color. Water cannot actually be seen
to be reabsorbed from the urine, but the gradually increasing concentration of the dye in the tubular urine indicates that this is taking
place.

This alone would lead finally to a urine whose composition is
simply that of very concentrated blood plasma minus proteins.
Analysis shows that such is not the case. Certain constituents are
more concentrated than others. Urea is seventy times as concentrated in urine as in blood, while NaCl is about two times as concentrated. Something else must be taking place in the tubules to
account for this.

3. *Reabsorption of threshold substances.*—Bladder urine, we know,
contains no glucose; capsular urine does. This substance filters
through the capsule walls. Sugar also, as well as water, must therefore be reabsorbed by the tubules. The cells comprising the walls of
the tubules, resembling in appearance the cells of any gland, are apparently endowed with the capacity for taking up glucose from the
tubular-lumen end of the cell and passing it into the capillary end—
secreting glucose, that is, from tubule into blood. Reabsorption of a
portion of some of the other urinary solids probably also takes place,
accounting in part for the differential concentration of the urinary
constituents.

The term "kidney threshold" has been used as a graphic expression
of a well-known phenomenon. Certain substances present in the

blood normally do not make their appearance in the bladder urine until the concentration of that substance in the blood exceeds a certain critical level. This level is called the *threshold level* for that substance. Sugar is such a threshold substance; its threshold is about 0.14 per cent. It is as though elevation of the sugar level in the blood above this point causes it to spill into the urine over a threshold in the kidney. Actually no such simple state of affairs exists. Kidney thresholds really are determined, as far as present evidence goes, by the tubular cells and their reabsorption of urinary constituents. The reason glucose is found in bladder urine when its threshold is exceeded is apparently that at or above this blood-sugar level the renal tubular epithelium is no longer able to remove all the glucose from the tubular urine and secrete it back into the blood.

We should be under no delusions as to the adequacy of this explanation. Granted that the reabsorption process largely determines the phenomenon of kidney threshold, we still do not know how these cells are enabled to maintain definite thresholds for different substances. Why is reabsorption of a compound affected by the concentration of that compound in the blood? We shall probably be unable to solve these problems until we know more about the cellular processes of secretion in general.

4. *Secretion of special substances.*—There is some evidence that there may also be an actual outward secretion of substances from blood into the urine of the tubules. After the injection of certain dyes into the blood of a number of animals, if individuals are then sacrificed at intervals and the kidneys examined microscopically, the following is observed. Soon after injection, dye particles can be distinguished in the basal portions of the tubule cells (i.e., in the blood end of the cells). Later, dye particles appear in the midportion of the cell and, still later, at the lumen end and actually in the lumen itself, as though the cells actively secrete the dye from the blood into the tubule. Dyes, it will be said, are abnormal materials. Does anything like this occur normally? We cannot be sure of this. It is found that uric acid crystals often form in the tubular epithelium of birds, and it is supposed that these are derived from uric acid being secreted by the tubular cells. Although the evidence is not conclusive, we must recognize the probability that there occurs an active secretion by these cells of some of the urinary constituents, perhaps including urea or uric acid.

5. *Summary.*—We have formulated this conception of urine for-

mation, which is admittedly tentative, subject to correction when further data are available: The urine is apparently filtered into the capsule from the blood and, in its passage through the tubules, is made more concentrated by reabsorption of water, has the *relative* concentration of its constituents changed by reabsorption of some of these constituents, and receives more waste solids by active tubular secretion.[23]

We see in this that part of the energy for urine secretion comes actually from the heart, for the heart pumps blood into the glomeruli under pressure, and the pressure is responsible for filtration into Bowman's capsule. In addition, work is also done by the kidney cells, especially those of the tubules. They use oxygen and produce carbon dioxide, and the consumption of oxygen and the production of carbon dioxide by these cells vary with the rate of urine formation.

F. Factors Controlling Urine Volume

1. *Relationship to blood volume.*—This picture of urine formation helps us to understand, partly, various factors influencing the volume of urine secreted.[24] Why is the urine volume diminished after a hemorrhage? This follows directly from what has just been said and from what we learned about blood volume and blood pressure. Diminution in blood volume lowers the blood pressure. Blood pressure determines filtration through capillary walls; in fact, capillary blood pressure *is* the capillary filtration pressure. Thus, when the blood pressure is reduced, the glomerular filtration pressure is likewise less, filtration of urine into the capsule is diminished, and the urine volume is scanty. The body fluids are thereby conserved.

Similarly, any tendency for the blood volume to increase, as after the ingestion of great quantities of fluids, is offset in part by an increased filtration of urine into the capsule. The kidneys excrete more water, and the result is a restoration of the blood volume to its normal level.

There is thus a direct relationship between the blood pressure, or, essentially, glomerular capillary filtration pressure, and urine vol-

23. The kidneys of certain species of lower vertebrates contain no glomeruli, being made up of secretory units which are the counterpart of the kidney tubules of mammals. Urine formation in the aglomerular kidney is essentially a matter of secretary activity of the tubule cells, apparently, without the occurrence of true filtration.

24. As a matter of fact, the experimental chronological sequence has been the reverse of this to a large extent. Study of the factors we shall mention has helped us to fill out the details of the picture of the work of the kidneys we have given above.

ume and, similarly, between the rate of blood flow through the kidneys and the volume of urine formed.

2. *Influence of the solids excreted.*—It is a common observation, easily demonstrable experimentally, that, when there are more dissolved solids in the blood plasma for the kidney to excrete, the urine volume is increased also. Stated purely descriptively, for each molecule of dissolved solid excreted, a certain quantity of water must also pass out of the kidney. Let glucose serve as an example. When its concentration in the blood is above the threshold level, as in diabetes, not only is there glucose in the voided urine but the urine volume is greatly increased. In fact, the greatly increased urinary output in diabetes gives the condition its name[25] and was noticed long before the abnormal presence of sugar in the urine was discovered.

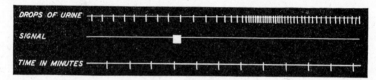

Fig. 125.—Flow of urine from the kidney of an anesthetized dog, collected as shown in Fig. 123. Note the continuous normal flow of urine and the accelerated rate of excretion soon after the intravenous injection (*indicated by signal*) of 15 gm. of glucose dissolved in salt solution. The urine excreted at the accelerated rate contained glucose.

Similarly, an elevation of the urea concentration of the blood increases urine flow. By stimulating urine flow, the urea of the blood thus acts to hasten urea excretion. In this we are reminded of the analogous action of carbon dioxide, another body waste. By its effect on the respiratory center, waste carbon dioxide serves as the stimulus for its own removal.

The same effect can be shown experimentally by inserting a glass tube into one ureter (see Fig. 123) of an anesthetized dog and by leading off the urine secreted by the corresponding kidney through a rubber tube. We note a constant flow of urine, indicating a continuous secretion by the kidney, in contrast with the normally intermittent emptying of the bladder. Glucose injected into the blood is detectable in the urine in a few minutes. At the same time there is a marked increase in the rate of flow of urine from the tube.

3. *Mechanism of sugar diuresis.*—Why does this occur? We get some notion of the factors involved from what we have learned about

25. The word *diabetes* (from the Greek *diabainein*) means "to pass through": water rapidly goes through the body.

reabsorption in the tubules. In the diabetic, and in our experiment, the tubules fail to reabsorb all the sugar from the tubular urine. This sugar, now unable to get through the tubular epithelium, acts as does any molecule at a semipermeable membrane; it exerts an osmotic effect, tending to pull water into the tubular lumen from the tissue fluids and blood. But, inasmuch as reabsorption of water by the tubular cells takes place normally, the osmotic action here tends to counteract and diminish the rate of that absorption. This results in an augmented volume of urine formation, or diuresis.

G. Derangements of Kidney Function

Any agent which causes destruction of kidney tissue may be serious or fatal because of accumulation of harmful waste products in the blood. Apparently, under many abnormal conditions, the cells of these organs possess but little capacity for regeneration. When glomeruli or tubules are destroyed by disease, they are apparently not replaced to any significant extent.

Infectious agents are often responsible for kidney damage. Occasionally, after a sort throat, kidney disease sets in. It seems that the bacteria in the throat sometimes gain entrance into the blood stream and may lodge in the kidney tissues. Here the bacteria either may produce an acute kidney insufficiency or, growing very slowly, may cause a gradually progressing destruction of the glands.

Sometimes the kidneys may be damaged by toxins of bacteria which themselves never lodge in the kidneys. Toxins liberated into the blood by diphtheria bacilli growing in the throat may produce this effect.

In disease of the blood vessels generally, there may also be changes in the kidney vessels of such a nature as to diminish the blood supply to those organs and produce injury or death of kidney cells from lack of oxygen and nourishment.

Stones forming in the urinary tract passages by precipitation of one or another of the urinary solids may block the flow of urine from the kidney. If the obstruction is marked enough and of long standing, the involved kidney degenerates. Early surgical removal of the stone cures the patient—until and unless another stone forms and causes obstruction.

The association of chronic kidney disease with high blood pressure (in "Bright's disease") has long excited the interest and concern of physicians and physiologists, but so far the relationships between these phenomena are not clear. Apparently, the same vascular changes inducing an elevated blood pressure may also cause kidney damage from an inadequate blood supply. But there is also evidence that damaged kidneys may lead secondarily to high blood pressure, for if the kidney arteries of dogs are partially compressed, so that the blood flow through the kidneys is diminished, a chronic hypertension ensues. Just how such poorly nourished kidneys produce the hypertension is not yet entirely clear. There is no apparent impairment of the excretion of the known wastes ordinarily removed from the blood by normal kidneys. Experimentation indicates that the damaged kidneys manufacture abnormal blood-pressure-elevating substances which then circulate through the blood stream and produce a generalized vasoconstriction.

We cannot yet be sure whether the high blood pressure of human Bright's disease is secondary to the kidney damage and similar in origin to this experimental hypertension. In man there is at present no direct evidence of the existence of any blood-pressure-elevating substance in the blood.

One kind of kidney derangement in which much of the plasma protein is lost in the urine, causing the collection of excessive water in the tissues, has already been described. The mechanism of this water imbalance is well understood (see pp. 67–91).

Fortunately, the normal body is equipped with more kidney tissue than it needs. At any one time only a fraction of the renal glomeruli and tubules is active. Removal of one whole kidney, plus sometimes even as much as half of the other, seems in no wise to impair urine secretion in an experimental animal. There is here, as with most organs, a considerable margin of safety, and destruction of a great deal of kidney tissue by disease is compatible with adequate renal function. Likewise, when one kidney is badly infected by some organism such as the tubercle bacillus, which threatens to kill the individual by spreading to the body in general, this margin of safety makes possible the surgical removal of the infected organ, often effecting a complete cure.

CHAPTER NINE

THE ACTION OF MUSCLE AND NERVE

I. Movements of the Skeleton

II. Physiological Properties of Muscle
  A. Mechanical changes
  B. Irritability independent of nerves
  C. The refractory state
  D. The all-or-none principle
      1. Application of the law to motor units   2. Effect of repetitive stimuli: staircase   3. Effect of tension and load   4. Fatigue
  E. Action potentials

III. Kinds of Contraction
  A. The single twitch
  B. Tetanus
      1. The curve of tetanic contraction   2. Summation
  C. Tonus

IV. Transformations of Matter and Energy in Muscle Contraction
  A. The source of energy
  B. The role of oxidations
  C. Oxygen debt
  D. Heat production
  E. The site of fatigue

V. Chronic Effects of Exercise on Muscle

VI. The Smooth Muscle of the Viscera

VII. The Stimulation of Nerve
  A. Nerve conduction
      1. Chemical and thermal changes   2. Electrical changes   3. The nerve impulse   4. Independence of intensity of stimulus and size of impulse   5. Recovery: the refractory state   6. Impulses in different nerves alike   7. Activation only at the end of the fiber
  B. Threshold stimuli: the law of Bois-Reymond
  C. Duration of the stimulus

VIII. The Nature of the Nerve Impulse
  A. The membrane theory
  B. Current of injury: the demarcation potential
  C. Current of action
  D. Electrotonus

        E. Polar stimulation
        F. Summation of stimuli
        G. The iron-wire model of nerve
        H. Excitation in other tissues
   IX. Functions of the Neurone Sheaths
        A. The myelin sheath
        B. The neurilemma sheath: degeneration and regeneration of nerve
    X. The Law of Bell and Magendie
   XI. The Autonomic Nervous System
        A. The sympathetic, or thoracolumbar, system
        B. The parasympathetic, or craniosacral, system
        C. Experiments with drugs
  XII. Inhibition: Neurohumeralism

In the skeletal muscles we encounter a system concerned primarily with adjustments of the organism to the external environment. In previous chapters we have devoted ourselves mainly to the many complex nervous and chemical internal adjustments of part with part, organ and system with organ and system, and to the many interrelated processes involved in the preservation of the conditions of life in the internal environment. In this chapter, and the following two chapters of the book, our inquiry will deal mainly with the manner in which the organism as a whole adjusts itself to the ever changing external environment.

1. MOVEMENTS OF THE SKELETON

The skeletal muscle system contributes to such adjustments by moving the body and its parts. The skeletal muscles derive their name from their intimate relationship, anatomically and physiologically, to the skeleton. The main outlines of the anatomy of the skeleton can best be obtained from a study of the skeleton itself or of a picture of it (see Fig. 126). Built about the main axis—the bony spinal column—are the shoulder and hip girdles, to which fore and hind limbs are articulated; the skull, or *cranium*; and the ribs, which provide a certain rigidity to the chest wall.

The separate bones of the skeleton are articulated to one another by means of *ligaments*, consisting of tough fibrous connective tissue. Some of the joints are immovable (those between the various skull bones); others are only slightly movable (the joints between the vertebrae); others are freely movable.

The freely movable joints are of several types: the hinge joint, exemplified in the fingers; the pivot joint, illustrated by that component of the elbow joint which permits the rotation of the lower arm on the upper arm, as in using a screw-driver; and the ball-and-socket joint, like that of the shoulder and hip. Muscles are able to produce movements depending upon the nature of the joint and the points of attachment (*origin* and *insertion*) of the muscles (see Fig. 127). An understanding of these movements is best obtained by manipulating

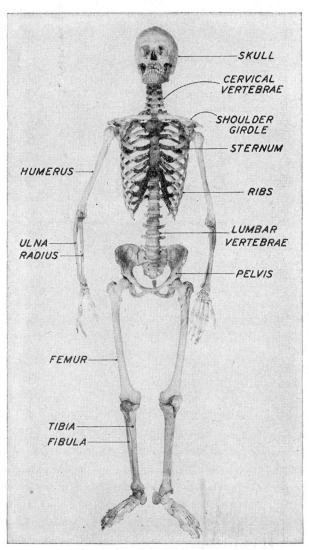

FIG. 126.—The human skeleton

the parts of a skeleton, duplicating typical movements of one's own skeletal parts, and observing the kinds of movements which occur at each of the joints involved. Neither description nor diagrams can convey a clear picture.

A few of the muscles are attached not from bone to bone but from bone to skin and even from one part of the skin to another. In man such muscles are well developed mainly in the face, where they are employed in speech and in emotional expression.

The anatomist studies single muscles, their attachments, and the isolated movements they could theoretically produce. Such ana-

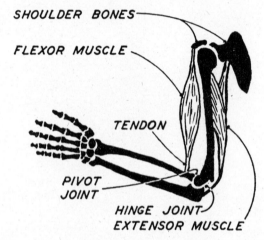

FIG. 127.—Muscles of the arm. Contraction of the flexor muscle (*biceps*) bends the arm at the elbow. Contraction of the extensor muscle (*triceps*) straightens the arm. The pivot joint at the elbow is also shown. Rotation of the lower arm on its long axis involves movement at this joint.

lytical treatment must not lose sight of the fact that physiologically muscles never act singly but in groups. We cannot control individual muscles but can only produce movements. We cannot contract the biceps muscle; we can only flex the arm—a complex involving a number of muscles besides the biceps, acting together as *synergists*. Usually, such movement involves also the inhibition of an *antagonistic* group of muscles. Ordinarily, when a *flexor* group of muscles contracts in a bodily movement, the antagonistic group of *extensors* relaxes. These co-ordinated muscle changes are effected by the correlated action of the nerves of these muscles, as will be discussed later (see pp. 428–30). In the maintenance of a fixed position of a body part, or of the posture of the body as a whole, such reciprocal muscle action may not occur, and antagonistic muscle groups may be maintained in a continuous steady state of contraction.

II. PHYSIOLOGICAL PROPERTIES OF MUSCLE

The property of shortening upon being activated, together with the development of action potentials and the liberation of heat, are common to all types of muscle—skeletal, smooth or nonstriated, and cardiac.

A. MECHANICAL CHANGES

The essential mechanical feature of muscle contraction is a rapid change in shape of the muscle cells or muscle fibers. Each activated fiber becomes shorter and thicker, with essentially no change in its total volume. The whole muscle, also, composed of thousands of fibers, does not change its volume in contraction. This was experimentally demonstrated years ago by stimulating an isolated muscle in a fluid-filled closed chamber from which a narrow fluid-filled tube extended vertically. During contraction no change occurred in the level of the fluid in the upright tube.

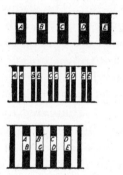

Fig. 128.—Changes in a skeletal muscle fiber during contraction. As each muscle fiber shortens and thickens in contraction, the dark bands (*A*, *B*, etc.) split into halves. Each half moves away from its partner (*A–A*, *B–B*, etc.) and at the height of contraction joins the next adjacent half-band (*AB*, *BC*, etc.)

The structural changes which occur in a muscle fiber in contraction have been studied at great length, and changes in the appearance of the cross-striations described in detail. While these are of interest, and are undoubtedly intimately associated with the shortening process, yet that process itself remains obscure. Just what change occurs in the organization of the protoplasm to make each fiber shorter and thicker? We cannot say at present. Ameboid movement, ciliary motion, and muscle contraction are related phenomena in that each involves a change in shape of a cell, or a protoplasmic part of a cell. The fundamental nature of none of these is well understood.

B. Irritability Independent of Nerves

It might be supposed that the irritability of muscle could be demonstrated simply by application of a stimulus directly to an excised muscle. All skeletal muscles are richly supplied with motor nerves, however, so that we could not be sure that we were not, in this case, stimulating the muscle indirectly through the nerves imbedded in it. But, when degeneration of all nerves of a muscle is produced experimentally, the muscle still retains a certain irritability for a time. Furthermore, if the limb of an animal is treated with the drug *curare*, impulses in nerve fibers are no longer able to pass across the nerve-muscle junction and activate the muscle, although curare does not interfere with conduction of nerve impulses in the nerve trunk itself. In such a preparation, application of a stimulus to the muscle itself is effective in eliciting a contraction; stimulation of the nerve is not. The stimulus must have affected muscle fibers independently of nerves, indicating an independent irritability of muscle.

It must not be supposed from this, however, that any voluntary or reflex control of muscle contraction whatsoever is possible except through the medium of nerves. If the nerve of a muscle is cut, the muscle retains its independent irritability for some weeks and can be artificially stimulated by application of an electric current through the skin. But a physiological activation of the muscle is impossible. The muscle is paralyzed and cannot be made to contract either by voluntary effort or by any reflex activation. Smooth muscle and heart muscle are different in this respect, for they still continue a more or less normal activity when their extrinsic nerves are cut. The heart and intestines or stomach persist in their contractions when all their nerves are severed, although certain fine adjustments of these contractions to the varying levels of body activity are no longer possible. This peculiarity of cardiac and smooth muscle may be due in part to the nerve cells present in such muscle.

C. The Refractory State

Like cardiac muscle or nerve, the skeletal muscles also display the phenomenon of refractory state, though this is of much shorter duration in skeletal muscle than in the heart. For only about 0.005 second after the application of one effective stimulus to a muscle is that muscle refractory, failing to respond to a second stimulus. During this period, presumably, chemical and physical processes are occurring which restore the muscle to its normal resting condition.

D. The All-or-None Principle

The all-or-none phenomenon, so easily demonstrable in heart muscle, is less clearly in evidence in skeletal muscle. For example, it is common experience that voluntary muscle contractions can be graded from a barely perceptible shortening to a maximal contraction. The arm flexors can bend the arm at the elbow to any desired angle by graded contractions. Also, if graded intensities of stimuli are applied to an isolated muscle, varied strengths of response result. There is no question, therefore, but that the all-or-none law does not apply to the whole muscle, as is also true of a nerve bundle composed of many fibers, any number of which may be active at a given time. But does this law apply to single muscle fibers as it does to single nerve fibers?

1. *Application of the law to motor units.*—The evidence on this point is conflicting. If stimuli to an isolated nerve-muscle preparation are gradually increased, the contractions become greater and greater. However, beyond a critical intensity of the stimulus, no further increase in contraction amplitude occurs, no matter how intense the stimulus. One interpretation of such findings is that, at this critical level of stimulus intensity, *all* the muscle fibers have been activated. Any further increase of stimulus strength is now ineffective because no additional fibers can be activated, and each of the activated fibers, in accord with the all-or-none mode of behavior, can contract only maximally if it contracts at all.

Most of the evidence indicates that single skeletal muscle fibers display the all-or-none mode of behavior just as does the muscle mass of the heart ventricles. Although the evidence has been questioned, it should be borne in mind that, in the intact animal, skeletal muscles are normally activated via motor nerves, and, since the nerve impulses are of an all-or-none character, the resulting contractions of the muscle will also be all-or-none. Each motor nerve fiber innervates about a hundred muscle fibers. Such a nerve-muscle complex is termed a *motor unit.* Thus, while the all-or-none law does not apply to a whole muscle, it apparently does hold for the motor unit.

Thus gradations in strength of reflex or voluntary contraction of muscles depend partly upon the number of motor units acting. This holds not only for physiological activation of muscle but apparently also for electrical stimulation of a muscle-nerve preparation. A weak stimulus activates fewer motor units; a strong stimulus activates more. In either case, the activated motor units respond maximally, the inactivated units not at all.

2. *Effect of repetitive stimuli: staircase.*—Changes in the muscle tissue itself can and do alter the strength of the contraction. A second stimulus to either skeletal or cardiac muscle, following shortly after a first stimulus of equal magnitude, will elicit a greater response (see Fig. 129). The muscle tissue is in some manner so altered by its response to the first stimulus—perhaps, chemically, by metabolites—that the second stimulus induces a greater contraction.

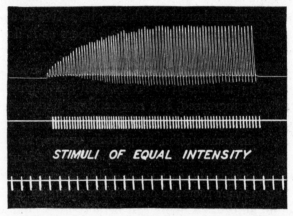

Fig. 129.—The staircase phenomenon. Application of electrical stimuli in rapid succession to a frog heart whose spontaneous beat is arrested induces contractions of gradually increasing strength. Changes occur in the heart muscle as a result of one stimulus, making it more contractile when stimulated a second time, unless too long an interval elapses between stimuli. Skeletal muscle also displays this phenomenon. Time in 5-second intervals.

A second stimulus applied while a skeletal muscle is still in the process of responding to the first stimulus also seems capable of initiating a further liberation of contractile energy from the muscle, so that the contraction resulting from the two successive stimuli is somewhat greater than can be induced by either of the stimuli alone.[1]

3. *Effect of tension and load.*—Tension placed upon a muscle also affects its contractile power, just as stretch increases the power of the heartbeat, within limits. It is partly for this reason that a muscle will do more work when it is made to lift a weight than when it is not so "loaded." If one stimulates an isolated muscle with a weight of 10 gm. suspended from one end, the contraction will lift the weight to a height of perhaps 5 mm., performing 5 gm. cm. of work If 20 gm. are suspended, a greater tension is placed upon the muscle,

1. Referred to as *summation of contractions*, or responses, or *superposition of contractions*.

and now a stimulus of the same strength will cause a stronger contraction. The 20-gm. weight will not be lifted as high as before—perhaps only to 3 mm. But the work done—6 gm. cm.—will be greater. This relationship applies only within limits, of course. If too heavy a weight is suspended, the muscle will be unable to move it, and all the energy freed in such a muscle, upon stimulation, is dissipated as heat.

4. *Fatigue.*—Finally, the state of fatigue of a muscle affects the contractility of the fibers in the direction of decrease. Fatigue can be carried to such a point that the contractions cease entirely. This is probably due to chemical changes in the muscle, associated with lactic acid accumulation (see Fig. 132, p. 383).

E. Action Potentials

Skeletal muscle, in common with other tissues, develops electrical changes when it is activated. These changes, in which the actively contracting portion of muscle or muscle fiber is electrically negative with respect to resting portions, have been discussed earlier (see p. 141). These action potentials are similar in nature and origin to those developed in nerve (see p. 386).

III. KINDS OF CONTRACTION

A. The Single Twitch

A single stimulus to a muscle, as from a single electric shock, produces a quick twitch of the muscle. An isolated muscle attached to a muscle lever with a writing point can be made to inscribe a record of a single twitch upon a very rapidly moving drum, which shows the time relations of the twitch. Figure 130 is such a tracing of a frog muscle twitch. The mechanical changes in the contraction process last about 0.1 second. Three distinct phases of the reaction are recognized—the *latent* period, the period of *shortening*, and the period of *relaxation*. The latent period, about 0.01 second in duration, is the interval elapsing between the application of the stimulus and the beginning of the mechanical shortening. Although no mechanical change occurs here, the tissue is not inactive. The excitation process is going on, and it is during this phase that the action potential is developed. Simultaneously, energy-liberating reactions are taking place with explosive rapidity. Tension is now developed, internal friction (viscosity) and the inertia of the system[2] are being overcome, and the shortening phase commences.

2. Since part of the latent period is employed in setting the recording system into motion, it follows that the latent period will be shorter when the moving parts of the recording system are lighter and have less inertia.

Shortening lasts about 0.04 second, and work may be done, such as the lifting of weights. In the succeeding period of relaxation, lasting about 0.05 second, there is a return to the previous relaxed state. The mechanical change in the fiber during relaxation is passive, caused by the elasticity of the muscle fibers and a pulling upon the muscle, stretching it back to its previous shape. The ends of the muscle are not actively pushed apart in relaxation.

But it is not to be thought that, when the twitch is over, recovery is complete. Heat production continues for a time, and chemical

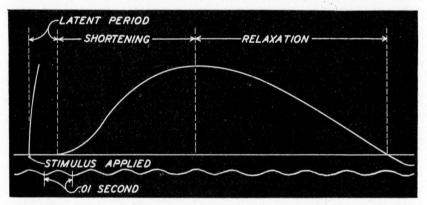

Fig. 130.—A single twitch of a frog muscle recorded on a rapidly moving smoked plate employing a lever somewhat like that shown in Fig. 36 (p. 137).

reactions associated with reformation of glycogen and organic phosphates (see pp. 380-81) are taking place.

B. TETANUS

1. *The curve of tetanic contraction.*—Isolated twitches do not ordinarily occur physiologically. Skeletal muscle action is usually a sustained contraction in response to the normal physiological stimulus, which is not a single stimulus but a volley of separate stimuli in rapid succession in the form of a series of impulses over the efferent nerve of the muscle. Such a muscle response is called *tetanus.*[3] Its nature can be analyzed experimentally by applying a series of electrical stimuli to an isolated muscle at varying frequencies. If a record is inscribed upon a moving kymograph drum, we obtain results typified by Figure 131. Inspection of such tracings shows that in tetanic

3. This should not be confused with the *tetany* of parathyroid gland deficiency, or the muscle spasms produced by the *tetanus* bacillus.

contractions (a) there is a sustained, smooth contraction and (b) the amplitude or degree of contraction is greater than that attained in a single twitch. How may we account for these findings? The smoothness of the curve is due to each stimulus in the series reactivating the muscle before it has had time to respond fully to the preceding stimulus. The resultant curve from two stimuli, or a series, separated by intervals of time a little greater than the refractory period will then be a fusion of two or a series of twitches. We saw that tetanus could not be produced in heart muscle, because the refrac-

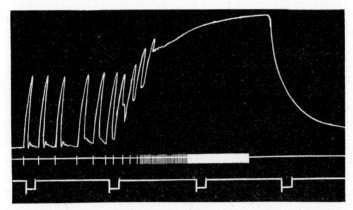

Fig. 131.—Tetanus in skeletal muscle. When stimuli are applied to a muscle at a gradually increasing rate of frequency (*indicated by signal*), the individual muscle twitches blend together so that, when the stimuli are applied in very rapid succession, a smooth, sustained contraction results. Ordinary muscle movements are of this tetanic nature. Note that the height of contraction is greater in tetanus than in a single twitch. Record inscribed by use of apparatus similar in principle to that shown in Fig. 36. Time (*bottom line*) in 5-second intervals.

tory period is very long, persisting not only throughout the latent period but for practically the entire duration of systole. And even after the absolute refractory period of the heart is over, in the *relatively refractory state* following, the latent period of a second stimulus is so long that, before the effect of this second stimulus is manifested, diastole has already proceeded part way.

2. *Summation.*—For an understanding of the fact that the amplitude of contraction in tetanus exceeds that of a single twitch, we refer again to the discussion above, where it was pointed out that the response of a muscle to one stimulus changes its state in such a way that a second stimulus applied just outside the refractory period induces a second response greater than the first. We also saw that the curve of contraction obtained from two stimuli in rapid succes-

sion is of greater amplitude than the contraction from either stimulus alone. This is referred to as *summation of responses* in the muscle. It is manifested even when each stimulus is strong enough to activate all the fibers of the muscle, indicating that it is not a consequence of activation of more fibers by the two stimuli together than were activated by either alone. In skeletal muscle tetanus this helps to account for the high amplitude of contraction.

C. Tonus

In tetanus, though the contraction is always smooth, the degree or amplitude of the contraction will depend in part upon the number of muscle fibers or nerve-muscle motor units activated. A mild contraction may involve perhaps only 5 per cent of the motor units, while a strong contraction may involve 20 or 30 or 50 per cent of the fibers. The term *tone*, or *tonus*, is applied to the state of sustained partial contraction which is present in all normal skeletal muscle so long as its efferent nerve and its reflex connections are intact. Tone appears to be a mild tetanus of the muscle, dependent upon a continuous series of efferent nerve impulses, involving a small fraction of the muscle fibers. It is modified reflexly in various ways, such as in adjustments to external temperature changes or in the maintenance of body posture.

In the tetanus of tonus, or even the stronger tetanus involving actual mechanical movement, there seems to be an alternation in the fibers which are active at a given time. Whether 5 or 30 per cent of the fibers are active, it is not always the same 5 or 30 per cent which are contracting at any one instant. The fibers work in relays, spelling one another and providing any one fiber with opportunity for rest and recovery.

IV. TRANSFORMATIONS OF MATTER AND ENERGY IN MUSCLE CONTRACTION

The energy-liberating reactions involved in muscle contraction are complex. The difficulties involved in attempting to investigate and understand them are great because of the rapidity with which the reactions occur and because the separate reactions are not really distinct from one another in time. While one reaction is taking place in a forward direction, reversal of another recovery reaction may be occurring. Yet we have gained a surprisingly comprehensive view of what happens.

A. The Source of Energy

The attack upon such a problem commences with a determination of the compounds which are used up in contraction and those which are formed. Analysis reveals that glycogen, oxygen, and organic phosphates diminish in quantity in active muscle and that lactic acid, carbon dioxide, inorganic phosphate, and other products are formed. That oxygen is used and carbon dioxide formed suggests at once that something is burned in the process. This oxidation, reflected in the more rapid breathing characteristic of muscular exercise, is not essential to the contraction, because a muscle deprived of all free oxygen will continue for some time to respond to stimuli by contracting. Even in moderate exercise, also, the accelerated oxidations continue for an appreciable time after actual muscle contractions have ceased, as evidenced by the fact that breathing continues at an accelerated rate for some time after the actual physical exertion is stopped. We might associate the oxidative processes, then, not with actual contraction but with recovery. Additional evidence for this view comes from the observation that a muscle repeatedly stimulated in the absence of free oxygen fatigues more quickly than when an abundance of oxygen is present.

The disappearance of glycogen and the appearance of lactic acid are related, the following reaction[4] taking place:

$$n(C_6H_{10}O_5) + nH_2O \rightarrow nC_6H_{12}O_6 \rightarrow 2nC_3H_6O_3 + \text{Energy} .$$
$$\text{(glycogen)} \qquad\qquad \text{(glucose)} \qquad \text{(lactic acid)}$$

Note that this reaction involves no molecular oxygen. It resembles the energy-yielding breakdown of sugar to alcohol and carbon dioxide in fermentation by yeast (see pp. 49–50). In the breakdown of glycogen, energy is liberated with explosive rapidity. It was once thought that this was the energy directly responsible for the actual mechanical shortening. We know now that this is not necessarily true because, when a muscle is treated with a certain compound (mono-iodoacetic acid), a *pharmacological* effect is produced which prevents all lactic

---

4. This reaction is very much simplified. There are several intermediate steps, including also a stage in which the glucose is in combination with phosphate, in a *hexosephosphate* complex.

acid formation, all glycogen breakdown. Yet a muscle poisoned with this drug will contract, when stimulated, in typical manner.

Having thus eliminated both oxidations and glycogen breakdown as the immediate source of energy for contraction, we are left with the following:

Organic phosphate[5] → Phosphate + Organic compounds + Energy .

In this reaction, also, energy is liberated with explosive rapidity. According to the present concept, it is from such reactions that the energy utilized in actual muscle contraction is derived, at least when more than mild exercise is taken. However, any physiologist would freely grant that new information might be unearthed at any time, necessitating a recasting of this hypothesis. This has happened more than once in past decades.

If glycogen breakdown and oxidations are not indispensable for the contraction itself, what is their role? The clue to the answer comes from a further study of the recovery process. After contraction has ceased, organic phosphates reappear as such in muscle, apparently being resynthesized from their breakdown products. Now, just as the breakdown *frees* energy, the synthesis *requires* energy from some outside source.[6] Here, then, may be the function of the glycogen breakdown. The energy set free in this reaction may be borrowed for the synthesis of the organic phosphates, which in this way can be used over and over again.

B. The Role of Oxidations

In recovery not only are the organic phosphates resynthesized but also glycogen is reformed from lactic acid. This, too, requires energy, which may well come from the oxidative processes known to be occurring at this time, whose function until now we have been unable to fit into the scheme. Careful quantitative analyses have indicated that about four-fifths of the lactic acid formed is reconverted into glycogen. The other one-fifth disappears, presumably

5. Included in this category are *adenosine triphosphate*, whose breakdown probably liberates the energy most directly employed in muscle contraction, and *creatine phosphate*, whose breakdown probably provides energy for the resynthesis of adenosine triphosphate.

6. This parallels the reversible respiration-photosynthesis reaction:

$$C_6H_{12}O_6 + 6O_2 \rightleftarrows 6CO_2 + 6H_2O \text{ Energy} .$$

In the reaction to the right, energy is liberated. For a reversal of this, energy from an external source (sunlight, in photosynthesis) must be put into the system.

being burned, furnishing by its combustion the energy for the resynthesis of the remaining four-fifths.

By way of summary, and as a tentative hypothesis, we may list
the following as the reversible reactions of muscle contraction:

Organic phosphates $\rightleftarrows$ Phosphate + Organic compounds + Energy ;

(used more directly in ac-
tual contraction process)

Glycogen $\rightleftarrows$ Intermediate stages $\rightleftarrows$ Lactic acid + Energy ;

(used mainly in resynthesis
of organic phosphates)

One-fifth of Lactic acid + $O_2 \rightarrow CO_2 + H_2O$ + Energy .

(used in resynthesis of remaining four-
fifths of lactic acid to glycogen)

The postulated role of lactic acid oxidation in the schema explains
certain objectively demonstrable facts, which is, of course, the only
justification for any theory. It makes clear the finding that, when
stimulated muscle is deprived of oxygen, there is no glycogen
resynthesis. This compound gradually disappears, and lactic acid
accumulates. Therefore, muscle contracting anaerobically (without
free oxygen) fatigues more quickly than when oxygen is freely available to dispose of the lactic acid.

The chemistry of muscle contraction described is based largely
upon classical work on isolated frog muscles. While probably applying, in the main, to intact muscles of the body, these concepts must
be modified somewhat when applied to muscles functioning in the
body, with the circulation intact. For example, under normal conditions, some of the lactic acid produced in contraction escapes into
the blood and may be burned or resynthesized to glycogen elsewhere
in the body.

What is the utility of such a complex, apparently roundabout
mechanism of energy liberation? The clue to the puzzle lies in a consideration of the oxygen available to muscle in the intact organism.
Almost no molecular oxygen is stored in or about the cells of muscle
or any other tissue. The oxygen it uses comes directly from the blood
and is used up about as rapidly as it reaches the tissues. There are
practically no reserves of free oxygen outside the blood.

As we have seen earlier, as soon as muscular activity commences,
the quantity of the blood flowing to the active muscles is considerably increased, thereby greatly increasing the quantity of available

oxygen. However, these circulatory adjustments require a certain
time for their consummation, and, even after the blood flow to the
muscle becomes maximal, the oxygen thus made available may be
quite inadequate to support oxidation at a rate fast enough to liberate the energy required for muscle contraction. If skeletal muscles
had to depend solely upon oxidations for their energy, they would not
be able to function with their characteristic rapidity and endurance.

C. Oxygen Debt

The foregoing hypothesis of energy relations in muscle contraction and the role of aerobic oxidations in recovery clarify the phenomenon of oxygen debt (see pp. 243–44). In the intact individual
muscle exercise can take place rapidly without waiting for the
occurrence of circulatory adjustments to carry more oxygen to the
muscle. The energy supporting the contractions is derived from the
anaerobic reactions described. However, if the source of energy in
the organic phosphates and glycogen is not to be rapidly exhausted,
energy must be repaid to the system by resynthesis of those compounds. This is possible, apparently, only by the oxidation of some
of the lactic acid. The energy debt, then, really becomes a debt of
oxygen. Usually this is repaid to a large extent after the muscular
exercise itself has been completed. The rapid breathing commenced
during the muscular work continues for a variable period after
muscular work is stopped. During this period the debt of energy—
the oxygen debt—is being repaid.

D. Heat Production

In a physical machine driven by combustion of fuel, only a part
of the energy liberated in combustion is converted into work done
—into moving the parts of the machine. Most of the energy is liberated as heat. This applies also to the machinery of muscle. Measurements indicate that only approximately 20–30 per cent of the
energy is converted into mechanical work, the rest being dissipated
as heat. In this respect the mechanical efficiency of muscle is about
that of the best physical machines. But, while the heat produced in
engines is largely wasted energy, the heat produced in muscle
contraction is put to good use, especially in mammals or birds exposed to the cold. The heat prevents a fall of body temperature,
which would entail a retarded rate of chemical and physiological
activity. We have seen that, in the cold, muscles are involuntarily

activated in an adaptation which utilizes the heat produced in contractions. The muscular movements of shivering are of no significance in themselves. It is the heat liberation accompanying the movement which is of utility to the organism.

From what we know of the chemistry of muscle contraction we should expect that heat liberation would continue even after the actual contractions had ceased, parallel with the oxidation of lactic acid occurring in the recovery phase. Experiment indicates that only about half of the total heat production occurs during contraction. The other half makes its appearance as *delayed heat* in the recovery period. Muscle contracting in the absence of oxygen fails to liberate this delayed heat, which suggests that its source is the oxidation of lactic acid.

E. THE SITE OF FATIGUE

A further word needs to be said about fatigue. Undoubtedly, a muscle in which considerable lactic acid has accumulated and from which glycogen and organic phosphates have been exhausted will lose all power of contraction. The term *fatigue* may be applied to

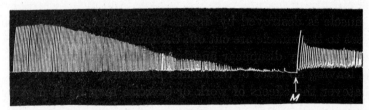

FIG. 132.—Fatigue. Contractions of a muscle (recorded by means of a muscle lever) induced by rapidly repeated stimuli to the nerve of the muscle. When the muscle no longer responded to nerve stimulation, stimuli applied directly to the muscle (beginning at *M*) caused definite contractions, indicating that the fatigue was not primarily in the muscle.

this state. In the intact animal, however, fatigue sets in long before this condition is reached in muscle. Actually the most significant and effective fatigue impairing muscular effort occurs in the spinal cord and brain (see pp. 424–25). In an isolated muscle-nerve preparation the point of easiest fatigability is not the muscle, nor is it the nerve itself. In nerve it is only with the most refined techniques that we can demonstrate anything like fatigue, in the sense of reduced reactivity as a result of previous activity. The point in the muscle-nerve system most susceptible to fatigue is the junction between nerve and muscle fibers—at the point where nerve impulses activate muscle. This can be demonstrated, since we know that nerve fibers display almost no fatigue, in the following experiment: Stimulate repeatedly the nerve of a nerve-muscle preparation until fatigue reduces the muscular contractions considerably, or even to zero. Then apply the electrodes to the muscle itself and stimulate with shocks of the same intensity as before. The muscle will now respond vigorously (see Fig. 132).

V. CHRONIC EFFECTS OF EXERCISE ON MUSCLE

The muscle enlargement with correspondingly greater strength, which is developed by work or training, is a commonly observed phenomenon. There is evidence that the muscle enlargement is not due to any increased number of muscle fibers, as might be suspected, but rather to an increase in size of each fiber. The exact nature of this effect, apparently a growth phenomenon, or the mechanism of its production is not yet known.

The effect seems in some way to be related to the influence of nerves and nerve impulses reaching the muscle by way of the efferents. At any rate, if the number of nerve impulses is reduced, as in disuse of the muscle, the muscle fibers become smaller, and the whole muscle shrinks in size and becomes weaker. And, if the efferent nerve of a muscle is destroyed by disease or accident, so that all nerve impulses to the muscle are cut off, the muscle shrinks in size greatly and may entirely disappear and be replaced by fibrous connective tissue, leaving no trace of the former structure.

Whatever the effects of work or exercise may be in making the muscle larger and stronger, there seems little doubt but that the most beneficial effects of work or training are in other directions, so far as increased muscular efficiency is concerned. In training, the increased effectiveness of the circulatory and respiratory adjustments to exercise and the better muscular co-ordination and decreased waste movements that result are probably of much greater significance than mere increase in muscle size and strength.

VI. THE SMOOTH MUSCLE OF THE VISCERA

The smooth muscle in the walls of such internal structures as the stomach, intestines, uterus, blood vessels, ureters, and ducts displays certain differences from skeletal muscle. The contraction caused by a single stimulation is very much slower, lasting some 20 seconds. All phases of the contraction are prolonged. The latent period, for example, may have a duration of as long as 3 seconds.

The tonus of smooth muscle is much more variable than that of striated muscle. The tone of the stomach may be of such a low degree as to permit the organ to reach into the lower regions of the abdominal cavity, or it may be so great as to convert the organ into a narrow, thick-walled tube. The metabolism of smooth muscle has not been worked out very satisfactorily, but it seems that, to a large extent, tonus can be maintained at least in some smooth muscles independently of extrinsic nerves and without considerable expenditure of energy, as if some internal change takes place in the organization of the protoplasm which maintains the shortened state structurally. Again, this peculiarity of smooth muscle may be related to the nerve cells in the visceral muscles.

VII. THE STIMULATION OF NERVE

In recent decades persistent attacks upon the problems of the physiology of the nervous system, using new and refined weapons, have been encouragingly successful. Eventually, man would like to understand fully that portion of his body which in a measure sets him somewhat above his animal cousins—the brain—especially in its performance of the so-called highest functions of memory, learning, intelligence, and reasoning. We still are a long way from this goal, but we are approaching it surely, no matter how great the distance at present.

One point of attack has been to concentrate on that part of the nervous system which is least complex—the peripheral nervous system—to study the nature of the nerve impulse and the phenomena associated with it.

A. NERVE CONDUCTION

Unlike ameboid, ciliary, or muscle action, nervous activity per se involves no changes which are easily detectable. Except by the employment of very special techniques, we cannot directly see anything happening in a nerve or a nerve fiber during the transmission of a nerve message, or, as it is technically termed, a *nerve impulse.* A useful index of nerve activity is the effect of nerve impulses upon some structure such as a muscle or gland with which the nerve is in physiological association, that is, which is innervated by the nerve. A muscle with its nerve can even be removed from the body, and upon stimulation of the nerve the muscle will twitch. Here the muscle is not directly activated by the artificial stimulation but is thrown into activity by some influence reaching it via the nerve. This influence is the nerve impulse, or a series of impulses.

1. *Chemical and thermal changes.*—Indirect evidence of the activity can be observed also in the nerve fiber itself. In active nerve the oxygen consumption and the carbon dioxide production increase, indicating that oxidations occur; these presumably liberate the energy concerned with conduction or with the recovery of the nerve after conduction. Heat is a by-product which has been quantitatively determined. Though they have been studied at length, the chemical reactions and energy relationships in nerve are not nearly so well understood as in muscle.

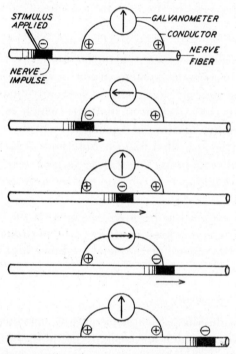

Fig. 133.—Electrical changes in active nerve. A nerve impulse traveling along a nerve fiber betrays its presence by its negativity, as compared with the positivity of resting, inactive regions of the nerve fiber.

2. *Electrical changes.*—With each nerve impulse there are accompanying electrical changes like those described for muscle, which are easily detectable with suitable sensitive instruments. These changes spread from the site of stimulation along the fibers to their termination and serve as a most useful index of impulse conduction in nerves (see Fig. 133).

3. *The nerve impulse.*—Yet the nerve impulse is not simply an electrical current like that of a charged wire, as might be expected from these electrical changes. If this were true, nerve fibers should not have to be alive to function. Dead nerves are still good conductors of electricity and yet quite incapable of conducting a nerve impulse. Also, if only a small segment of a nerve is crushed, nerve impulses set up in uninjured regions will travel only to the injured area and there will stop. They will be unable to traverse this section, even though simple electrical conductivity through the injured region is unaltered.

Early investigators were confused upon this question. A nerve, made up of bundles of parallel fibers, strongly suggested an electric cable made of many wires. Furthermore, investigators thought that nerve impulses traveled with the speed of electricity in a wire. The lapse of time between stimulation of a nerve and the response of a muscle seemed almost zero, indicating a very high rate of transmission of the impulse.

Therefore, because one could obtain only relatively short segments of nerve from animals to work with, it was asserted dogmatically that the rate of transmission of the nerve impulse could never be determined.[7] Only six years after this prediction was made, Helmholtz measured the speed of the nerve impulse in a segment of frog nerve only a few inches long. Instead of traveling at the rate of electricity—186,000 miles a second—the nerve impulse in frog nerve was found to travel at a rate of about 30 meters per second.[8]

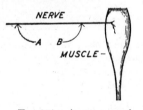

Fig. 134.—A nerve-muscle preparation. Stimulation of the nerve at *B* causes the muscle to contract sooner than from stimulation at *A* because, in the latter instance, the nerve impulse must travel a greater distance before activating the muscle.

The principle that Helmholtz employed was the application of two successive stimuli at two points along the nerve of an isolated nerve-muscle preparation, and measurement of the time elapsing between application of each stimulus and the beginning of muscular response, as diagrammed in Figure 134. Contraction of the muscle followed stimulation at *A* after a slightly longer pause than that following stimulation at *B*. The difference in these two periods indicated the time required for transmission of the impulse over the

7. By Johannes Peter Müller, a leading German physiologist of that time (1846).

8. The rate is about 100 meters per second in human nerve fibers.

measurable span of nerve from *A* to *B* and gave the rate of conduction.[9]

Thus the nerve impulse is not simply an electrical current flowing along a conductor. Its nature, in fact, is quite independent of the mode of stimulation. Whether it be initiated electrically, mechanically, thermally, chemically, osmotically, or in any other way, the impulse travels at the same rate of speed and is in all observable respects exactly the same.

Fig. 135.—An ignited powder fuse resembles somewhat a nerve which is conducting a nerve impulse. The energy involved in the transmission of the disturbance comes from the nerve fiber or the powder fuse and not from the stimulating agent. (From the sound film, *The Nervous System.*)

The impulse is a physicochemical disturbance in the fiber, which, once started, is self-propagating. As the impulse travels, each segment of the nerve, as it becomes active, is itself the stimulus to the adjacent segment of the fiber. In this respect the disturbance is like the combustion of a powder fuse, in which each active region stimulates the adjacent regions.

4. *Independence of intensity of stimulus and size of impulse: the all-or-none law.*—The analogy to a powder fuse may be extended. Here,

9. The actual technique employed by Helmholtz was the inscription on a rapidly moving sensitive plate of the curves of the two muscle twitches (see p. 375) made by stimulation at the two points, successively. The difference between the two *latent periods* gave the transmission time.

too, the nature of the transmitted activity is independent of the type of stimulus employed to set it off. Whether it be set off by a blow with a hammer, by heat, or in any other way, the rate of transmission and the chemical changes which occur will be identical. An important deduction can be made from this hypothesis. In the fuse the degree of activity—the magnitude of all the physical and chemical changes which occur—is independent not only of the kind of stimulus employed but also of the strength of the stimulus. If we strike the powder a weak blow, or a sharp blow of much greater strength, in either case the propagated disturbance in the fuse will be the same. The stronger stimulus induces no stronger or faster or more extensive propagated explosion. The only requirement is that the stimulus be strong enough to start the activity.

If what we have said of nerve is true, the same relationships should hold. Testing this deduction experimentally, we find it to conform with observable facts. Unless a sufficient stimulus is applied, the nerve fails to respond. When a sufficiently strong stimulus is used, the nerve responds.[10] But, if we increase the stimulus any amount above this threshold strength, we in no wise affect the impulse which is set up in any fiber.

In a word, the energy for the conduction of the impulse comes from the nerve itself and not from the stimulating agent. Therefore, though the impulse is independent of the strength of the stimulus, the nature and strength and rate of transmission of the impulse can obviously be affected by the condition of the nerve, just as the propagated disturbance in the fuse can be affected by the condition of the powder, such as dampness or dryness.

Thus we observe that nerve, also, behaves according to the all-or-none law. This states, summarizing the foregoing, that, if a nerve fiber responds at all to a stimulus, it responds maximally for the condition of the fiber at that particular time. The law can also be expressed thus: The strength of the response of a nerve fiber (i.e., the strength of the nerve impulse) is independent of the strength of the stimulus, provided that stimulus is of at least threshold strength.

This conclusion is based upon fairly direct evidence—upon measurements and records of action currents led off even from single nerve fibers. The experimenter teases apart the strands of nerve fibers in a small nerve, under the microscope, and cuts all the fibers

10. Such a stimulus, just strong enough to induce a response in any irritable tissue, is called a *threshold stimulus*.

but one. If stimuli are now applied on one side of the cut, and action currents led off from beyond the cut, the electrical changes will be those produced by impulses in the single intact fiber (see Fig. 136). The deflections of the galvanometer are found to be always of the same magnitude, no matter what the intensity of the stimulus, provided only the condition of the nerve fiber (such as its temperature) remains unchanged.[11]

5. *Recovery: the refractory state.*—In considering what happens after a nerve impulse has passed down the fiber, we must soon part company with the powder-fuse analogy. Once set off, the fuse can never be reactivated. It never recovers, never is restored to its original "irritable" condition. Now, immediately after the transmission of an impulse in a nerve, it also is incapable of further activity. For a time it cannot transmit a second impulse. During this period the fiber is said to be in a refractory state.

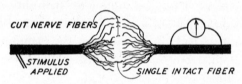

FIG. 136.—Preparation of nerve used to measure electrical changes in single nerve fibers

But this condition is of only brief duration, lasting about 0.001–0.005 second in nerve. It is thought that during this period chemical and physical changes are occurring in the nerve which are just the reverse of the changes during activity. The nerve restores itself to the resting state, regaining its former irritability and conductivity.

The all-or-none principle and the refractory state are by no means peculiar to nerve cells. They have also been observed in other cells which lend themselves to suitable observations.

Of the all-or-none phenomenon, this further might be said. It applies to individual cells and not always to structures composed of many cells. In a nerve trunk, composed of hundreds or thousands of fibers, increasing the strength of a single stimulus increases the activity in the total nerve trunk simply by activating more of the fibers. In a nerve, then, graded stimuli can produce varying degrees of activity in a nerve trunk either by setting up more impulses in any given fiber or by setting up impulses in more fibers. In terms of a

11. When stimuli producing a series, or volley, of impulses are used, it is found that graded intensities of stimuli lead to fewer or more impulses per volley. Stronger stimuli, that is, lead to more nerve impulses, each of which is maximal intensity.

common experience, a mild pinprick in the finger may stimulate only one or a few nerve fibers and cause a mild sensation; a sharp prick stimulates more fibers and sets up more impulses in each, producing a sharp pain.

6. *Impulses in different nerves alike.*—Not only are all impulses in one nerve fiber like all others in that fiber, provided the condition of the fiber remains unchanged, but, also, impulses in different nerves are alike, except for minor variations. The impulses in the nerves of glands or skeletal muscles, or, on the sensory side, of vision or hearing or touch or pain, are all essentially alike, so far as we now know. The effects of nerve stimulation, whether it be experimental or physiological, are dependent upon the destination of the nerve and the structures to which the impulses are carried and not upon any functional peculiarity of the nerve fibers or specificity of the nerve impulses. This fact finds formal expression in what is known as *Müller's law.* It states that specific sensory or motor or other effects produced by nerve stimulation are dependent upon the structure or region to which the impulses travel, and which are set into activity by the impulses.

We might draw an analogy between some of these principles of nerve behavior and a telegraph system in which only one message can be sent. This message is: "Act, go ahead—do your stuff." There is no possibility of altering the content of the message. If the sender desires emphasis, no amount of vigorous pounding of the key will alter the quality or the urgency of the delivered message. However, the message may be repeated as often or at as frequent intervals as desired. The effect produced by the message depends upon who receives it and what the characteristic activity of the receiver happens to be.

7. *Activation only at the end of the fiber.*—Normally, nerve fibers are not stimulated along their course. Activation occurs only at one end of the fiber, from which point the impulse travels to the other end.[12] Depending upon the direction of transmission of the impulse, nerve fibers are designated as *afferent* or *efferent.* Afferent fibers transmit impulses from various regions of the body to the brain or spinal cord. They include all the sensory fibers. Efferent fibers trans-

12. Exceptions to this may occur abnormally: the "funny bone" at the elbow, which is really a nerve trunk lying near the bone, is often stimulated mechanically; also there are the osmotic stimulation of pain fibers by salt in a wound and the chemical stimulation of pain fibers in a stomach ulcer by the acid of gastric juice.

mit impulses from the brain or spinal cord outward to various special body structures. They include all the secretory, motor, and inhibitory nerves.

In the central nervous system more or less complicated interconnections are effected, by intermediate nerve cells and fibers, so that stimulation of afferent fibers may lead to activation of efferents, in what is known as a *reflex*. In a reflex, then, the general pathway of a nerve impulse is from the point of stimulation at the termination of an afferent fiber to the spinal cord or brain in that afferent, through nerve cells and fibers in the brain or cord to efferent fibers, which, in turn, transmit the impulses to some such structure as a muscle or gland (see Fig. 151, p. 419).

B. Threshold Stimuli: The Law of Bois-Reymond

Much of our information about the activity of nerves has been derived from studying the conditions which lead to activation, the characteristics which our stimulating agent must possess. These laws have been worked out, in the main, for electrical stimulation because it is easier to express stimulus strength quantitatively than when mechanical, thermal, osmotic, or chemical energy changes are used as stimuli.

We may consider any environmental energy change to be a stimulus. Whether it will be an effective or an adequate stimulus depends essentially upon several things about the stimulus. First, it must be above a certain minimal intensity, which we refer to as the *threshold intensity*. Second, this intensity must be reached rapidly enough. A very slow change may not stimulate at all, while a rapid change will be quite effective, even though the total change is exactly the same in both cases.

This principle, known as the *Bois-Reymond law*,[13] is merely a more accurate statement of a phenomenon we have all observed in a crude way. We are especially aware in consciousness not of constant stimuli but of *changes* in strength of stimulus. The *cessation* of a continuous noise may first call our attention to the fact that there was a noise in the first place. Very gradual increases in light intensity or in temperature may be unnoticed, while the same total change occurring suddenly will be immediately and clearly perceptible.

13. After the German pioneer in experimental physiology, Emil Du Bois-Reymond.

C. Duration of the Stimulus

Third, this threshold intensity, reached in a certain minimum time, must persist for a definite temporal period if the stimulus is to be effective. Even if the stimulus is intense, that is, unless it *lasts* long enough (some thousandths of a second), no excitation occurs. Partly upon this depends the fact that alternating currents of extremely high frequency fail to stimulate nerve or produce damage to tissues even when tremendous voltages are used, although the heating effects of such currents may be considerable.

VIII. THE NATURE OF THE NERVE IMPULSE

The most generally accepted notion of the nature of the nerve impulse is known as the *membrane theory*. Let us first present the theory and then give the evidence in its support.

A. The Membrane Theory

The membrane of the nerve cell and its fiber is electrically polarized, with positively charged ions on the outside and negatively charged ions on the inside (see Fig. 137, *A*). The maintenance of this

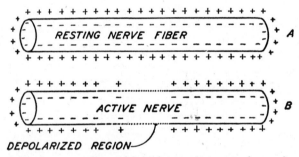

Fig. 137.—*A:* Resting nerve fiber, with its intact plasma membrane charged positively on the outside, negatively on the inside. *B:* Active nerve fiber, with an increased permeability of the membrane (*shown dotted*) and a depolarization at the site of a nerve impulse.

polarization depends in part upon the semipermeability of the fiber membrane. If the membrane were not impermeable to the charges (really to the ions carrying the charges), the plus and minus charges would be drawn together and would neutralize one another. The concept states further that not only does the impermeability to ions make the polarization possible but also that the polarization, in turn, helps in some way to maintain the membrane semipermeability. A breakdown of either one of these—the polarization or the semipermeability—would then cause a breakdown of the other.

Activation of nerve is thought to be associated with both factors. In the active region of the nerve, at the exact site of the moving impulse at a given instant of time, the membrane is thought to have lost both its polarization and its impermeability. Such a state is shown in Figure 137, *B*. If this constitutes the essence of the physical change at an activated point of nerve, it is easy to see how this state

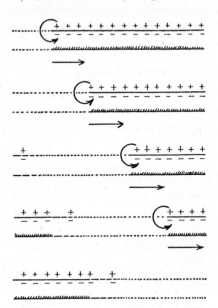

propagates itself along the fiber —how the nerve impulse is conducted—for, now, in the depolarized region, the membrane is permeable to the charges on the adjacent, not-yet-activated region. These charges, or ions, migrate through the permeable gap and neutralize one another. But, when this new area is thus depolarized, its impermeability is also lost, opening a route for the charges still farther down the nerve. In this manner the nerve impulse is believed to "roll" down the surface of the nerve fiber, as depicted in Figure 138. The figure also shows the recovery process—the restoration of the membrane and its impermeability to ions and the rebuilding of the polarization.[14]

Fig. 138.—Transmission of an impulse along a nerve fiber, shown in five stages (*top to bottom*). The increased membrane permeability at the active region (*shown dotted*) permits a movement of ions (*curved arrows*) resulting in depolarization of the next segment of the fiber. Straight arrows show direction of impulse transmission. The recovery process, involving repolarization and restored membrane impermeability, is also shown.

A number of demonstrable facts about nerve conduction accord very well with this view and, therefore, lend support to it. The refractory period would depend upon the persistence of the depolarized state for a time. Only when the recovery processes in the nerve have restored the membrane impermeability and polarization can stimulation again take place. The all-or-none phenomenon

14. It has been demonstrated that, in some nerve fibers, the active region of the nerve fiber develops an actual reversal of the resting polarization. That is, at the point along the nerve fiber where the nerve impulse is passing, the membrane is not merely depolarized but even develops an excess of positive charges *inside* the fiber and an excess of negative charges *outside* the fiber. The basic concepts presented are not appreciably altered by this finding.

would be explained by the resting nerve always having a positive charge of a given magnitude and the active portion of nerve a zero charge, so that, no matter what the strength of the stimulus, the depolarization can go only to zero, and the action current could be only of an intensity determined by the extent of the positive charge present on the surface of the resting region of nerve. We have the demonstrable fact, also, that a nerve trunk stimulated artificially along its course can and does conduct in both directions, as would be expected from the theory.

Further justification for the membrane theory of nerve conduction comes from a number of experimental sources. The evidence is in part fairly direct and in part indirect, but it all points in the same direction. Let us examine this evidence by asking certain questions which must be satisfactorily answered to prove our theory tenable.

B. Current of Injury: The Demarcation Potential

Is resting nerve polarized, as described? If we could place one of the electrodes of a galvanometer upon the outer surface of a nerve fiber and one inside the fiber, we should be able, according to the theory, to detect the polarization by recording a flow of current (positive stream) from outside to inside through the conducting system of the instrument. Precisely this experiment, although difficult, has been done. A more indirect experiment is done by crushing the end of a nerve trunk, mechanically destroying the membrane. An electrode placed upon the crushed end will now be leading off from the interiors of the fibers. Such an electrode attached to a galvanometer registers a negativity at this injured region as compared with a point on the uninjured surface of the nerve trunk. The phenomenon of injured regions of nerve (or muscle) being negative electrically to uninjured regions was discovered more than a century ago (1843) by Du Bois-Reymond and was given the name *potential of injury*, or *demarcation potential*.

C. Current of Action

Is the active region of nerve depolarized? The most direct evidence for this is the action current, or *action potential*. The region of nerve conducting an impulse is always electrically negative with respect to resting nerve. This objective finding is quite in accord with our theory, for in this case one electrode is placed on the outside of the positively charged membrane of resting nerve; the second electrode is placed on the active, depolarized region, in which the charges have neutralized one another through the temporarily permeable membrane. The latter would be at a lower electrical potential (i.e., more negative), and a current would flow through a conductor joining the two electrodes, just as in leading the current from the two poles of a battery through a conductor (see Fig. 139).

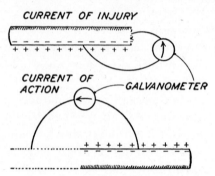

FIG. 139.—Current injury: When a galvanometer is connected with the normal surface and an injured end of a nerve, the direction of current flow indicates that the normal surface is positively charged.

Current of action: The direction of current flow through a galvanometer connecting active (*shown dotted*) and resting regions of nerve indicates that the active region is more negative than the resting region. Arrows indicate direction of flow of positive current through the galvanometer.

D. ELECTROTONUS

Is depolarization required for stimulation? If this were true, then it would follow that any increase in the polarization should decrease the sensitivity to stimulation and that any decrease in the polarization should render the nerve more sensitive. These conditions can be produced by passing through a nerve a constant current, from a battery, of an intensity itself *too weak to stimulate the nerve.* From all we know about the physics of electric currents, we can be sure that this produces a condition like that shown in Figure 140, namely, an accumulation of plus charges on the semipermeable membrane of the nerve fibers at the *anode* (i.e., positive pole of the battery), and the reverse change at the *cathode* (i.e., negative pole). The behavior of nerve with its polarization changed in this way is exactly as the theory predicts. In unaffected normally polarized regions of the nerve (N, Fig. 140) the threshold of stimulation is normal, but in the region of the anode, where polarization is increased (at A), a stimulus stronger than normal is required to excite the nerve. In the cathodal region, where polarization is decreased (C), a stimulus weaker than normal will be effective. These changes produced by a subthreshold constant current are called physiological polarization, or *electrotonus;* the cathodal effect, increasing the sensitivity of the fiber, is called *catelectrotonus;* the anodal effect of decreased irritability, *anelectrotonus.* Anelectrotonus can be built up in this way to such an extent that an impulse started in some distant point on the nerve will be unable to pass the region of increased polarization.

E. POLAR STIMULATION

The phenomena of *polar stimulation* also lend weight to the concept that, to be effective, a stimulus must depolarize the membrane. When a nerve is stimulated by a constant current of more than threshold strength, activation of the nerve usually takes place only at the instant when the circuit is made or broken. If the electrodes are placed some distance apart, it is found that the stimulus produced at make of the circuit (i.e., closure of the key) always occurs at the cathode (*C*, Fig. 141). The break stimulus always occurs at the anode (*A*). This can be proved by leading off action currents at some

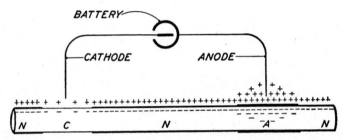

FIG. 140.—Electrotonus. Passing a weak constant current through a nerve decreases the polarization of the membrane at the *cathode* and increases it at the *anode*. Correspondingly, the irritability of the nerve at *C* is greater than at *N*. The irritability of *A* is much less than at *N*.

distant point (*E*). When the key is closed, it takes less time for the impulse to reach this point (*E*) than when the key is opened because, in the first case, the impulse had to travel a shorter distance (*C* to *E*) than in the second (*A* to *E*). This finding also is in accord with the theory. Presumably, *when the key is closed*, negative charges from the cathode at once depolarize the nerve at that region, while at the anode not only is there no activation of the nerve but the membrane polarization is actually increased during the passage of the stimulating current.

More puzzling is the fact that, at the opening of the key, activation of the nerve occurs at the anode. It is plain to see that, when the key is opened, the *excess* of charges at the anode (*A*) will disappear; the ions carrying them will diffuse away, being no longer held by the polarizing current. However, our theory postulates that this process goes on to complete depolarization, inasmuch as activation is effected at this time. It is difficult to see why the depolarization should go to completion when the key is opened and not come to a halt when the resting condition (as at point N) is attained, though plausible explanations for this have been advanced.

However, we should certainly expect that this "overshooting" of the depolarization process at the anode would be a less effective stimulus than the active depolarization induced at the cathode on closing the circuit. Experiment bears out this expectation. The make shock of a constant current is a more effective stimulus than the break shock. A stronger stimulus is needed to cause a break stimula-

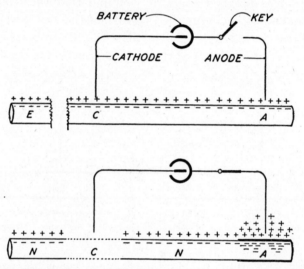

Fig. 141.—Application of a constant current of strength sufficient to stimulate a nerve. When the circuit is closed (*below*), the nerve is stimulated only at the cathode because the negative charges from the cathode depolarize the nerve membrane. Stimulation at the anode occurs when the circuit is broken. (See discussion in text.)

tion (at the anode) than to cause a make stimulation (at the cath-ode). When a stimulus is used of a strength just great enough to produce activation at the make, there will be no activation on break of the current. This latter effect is strictly in accord with the predictions of the polarized membrane theory.[15]

15. It must be remembered that these effects are obtained by stimulating with a *constant* current. When *induced* currents are used for stimulation, it is found that the response of the nerve at break of the current *in the primary circuit* is greater than at the make. This depends upon purely physical phenomena in the induction apparatus, or *inductorium*. The secondary coil of the inductorium actually delivers stronger shocks when the primary circuit is broken than when it is made, because of "self-induction." Setting up a current in the primary circuit by closing the key causes a current of reverse direction to be set up momentarily not only in the secondary coil but also in the coils of the primary itself. This, of course, opposes the

F. Summation of Stimuli

It can easily be shown that the application of two or three or a very few stimuli in rapid succession will activate a nerve even though each stimulus is of itself less (by a little) than the threshold strength. We should explain this, in terms of the membrane theory, as follows: The first stimulus effects only a partial depolarization. Immediately following this inadequate stimulus, polarization or restoration of the membrane recommences. But, if a second stimulus is applied while a residuum of the original partial depolarization still persists, it may effect a completion of the depolarization. Or perhaps a summation of the effects of three or four stimuli may be required

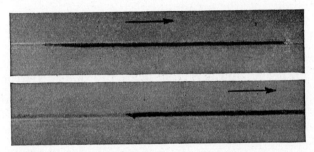

Fig. 142.—A soft iron wire immersed in nitric acid of appropriate concentration conducts visible impulses (*dark segment*) showing many of the characteristics of the nerve impulse. In the lower picture (taken very soon after the upper picture) the impulse has traveled to the right. (From the sound film, *The Nervous System.*)

to complete the depolarization and stimulate the nerve. It is found that, unless three or four subthreshold stimuli summate in this way, any number will fail to do so. The repolarization process apparently occurs at such a rate that, by the time a fifth or sixth stimulus is applied, all effects of the first stimulus have disappeared.

G. The Iron-Wire Model of Nerve

As a final line of evidence supporting the membrane theory of nerve conduction we may cite the experiments of Dr. Ralph Lillie, who has constructed a physical model of nerve with a soft iron wire immersed in nitric acid of the proper concentration. The wire be-

---

building-up of current in the primary and therefore reduces the current induced in the secondary. But, at break of the primary circuit, self-induction in the primary fails to occur because the circuit is not complete, the key having been opened. There is, therefore, an abrupt cessation of current flow in the primary, which induces a current in the secondary greater than at make.

comes coated almost at once with a layer of iron oxide. Such a wire is capable of conducting visible impulses which show many of the characteristics of nerve impulses: the wire can be stimulated electrically, or mechanically; the impulse is accompanied by an action potential; the phenomena of all-or-none action, refractory period, and electrotonus are displayed; the laws of polar stimulation hold. In the iron-wire model it is known that the propagated disturbance is a surface phenomenon and that the impulse rolls along the surface because of local electrical circuits between the oxidized surface and the reduced core of the iron wire. That the wire so closely simulates nerve conduction lends weight to the concept that nerve conduction is similarly a surface or membrane phenomenon, involving local electrical circuits at the membrane of the active fiber.

H. Excitation in Other Tissues

The membrane theory of excitation has been worked out in greatest detail on nerve, and it is from nerve physiology that most of its support is derived. It is probable, nonetheless, that something like this occurs in the excitation of other tissues as well. The finding of currents of injury and action potentials in glands, skeletal muscle, and the heart, exactly as in nerve, suggests that in excitation of all these tissues fundamentally the same changes in polarized surface membranes are taking place.

IX. FUNCTIONS OF THE NEURONE SHEATHS

The significance and functions of the sheaths which invest the nerve fibers proper have excited interest ever since their discovery. There are two of these—the *myelin sheath* and the *neurilemma,* or *sheath of Schwann.*[16] The structure of the sheaths is shown in Figure 143. The myelin sheath is noncellular, composed essentially of fatty materials. Being itself white, it determines the white color of those nerve fibers possessing this sheath. The myelin is not an unbroken tube running the whole length of the axone; its substance is interrupted at intervals called *nodes of Ranvier.*[17]

Surrounding the myelin sheath is the neurilemma, which is cellular and semitransparent. Although no cell boundaries can be distin-

16. After Theodor Schwann, German anatomist. These sheaths are not to be confused with the *membrane* of the nerve fiber. The fiber membrane is the outermost layer of protoplasm in the gelled state, which is common to cells in general.

17. After the French histologist, Louis Ranvier.

guished, there are typical cell nuclei, one being present for each inter-nodal span.

Nerve fibers in the brain and spinal cord possess only the white myelin sheath, lacking the neurilemma. Fibers of the visceral or *autonomic* nerves, though equipped with a neurilemma, have no myelin sheaths and, like naked nerve protoplasm, are therefore gray in color. Ordinary peripheral nerves of the somatic system possess both these sheaths.

What are the functions of these sheaths?

A. The Myelin Sheath

We are at once impressed from the anatomy with the similarity of the myelin sheath to insulation on a wire. Does this sheath in-

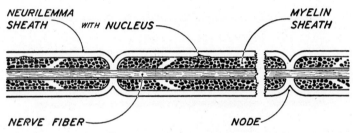

Fig. 143.—The sheaths of nerve fibers. The neurilemma is absent from nerve fibers in the brain and spinal cord. The myelin sheath is absent from peripheral nerves of the *autonomic* system (see p. 406).

sulate the fibers, preventing spread of impulses, largely electrical in nature, from fiber to fiber? In general, the nerves of visceral organs are nonmyelinated. It is true that one's ability to localize accurately the sensations from visceral organs is poor. This would follow if the afferent impulses from these organs should spread from fiber to fiber. It is also true that the motor control of the visceral muscles may not and need not be so precise as the control of skeletal muscle.

But there is no good direct evidence for the insulation hypothesis. The nerve fibers which innervate the wing muscles of insects are non-myelinated, and yet insects are capable of very fine motor control of the wings, which would not be possible if impulses should spread considerably from fiber to fiber. Then also, in mammals, if the fibers of the nonmyelinated cervical sympathetic nerve are gently sepa-rated into several strands, stimulation of any one strand always leads to erection of hairs in exactly the same area, indicating that there is no spread of impulses from one fiber to another.

It has also been suggested that perhaps the myelin sheath is in some way associated with the nourishment of the axone. Possibly the fatlike material of which the sheath is composed is itself a store of energy. However, the nonmyelinated fibers of the autonomic nervous system also carry on metabolism and require fuel and energy, and they certainly get it elsewhere than from myelin.

One undoubted objective finding suggests a possible function. Myelinated fibers, in general, conduct nerve impulses at a more rapid rate than nonmyelinated fibers. An ordinary motor nerve of a mammal transmits at the rate of about 120 meters per second, while a

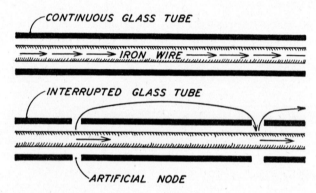

Fig. 144.—If Lillie's iron-wire model of nerve is inclosed in a continuous glass tube, impulses set up in the wire pass slowly along the wire (*short arrows*). If the glass tubing is interrupted with breaks, impulses "jump" from one artificial note to the next (*long curved arrow*), greatly increasing the rate of transmission. It is suggested that a similar phenomenon might be involved in transmission of impulses along myelinated nerves.

nonmyelinated nerve of the autonomic system conducts impulses at about 10–20 meters per second, or even less. Perhaps the presence of the myelin sheath speeds up the conduction of the nerve impulse. But by what mechanism can the sheath effect this?

The only suggestion comes from the behavior of Lillie's iron-wire model of nerve. If the wire is inclosed in several lengths of glass tubing so interrupted at intervals as to reproduce artificial nodes of Ranvier, impulses set up in the wire do not pass straight down the wire longitudinally, as they do if the surrounding glass tube is continuous. Instead, the impulses "jump" from one node to the next, transmitting the impulse at a much more rapid rate (see Fig. 144). Whether or not anything like this occurs in nerve, we cannot say at present. However, it is found that, within limits, the rate of conduction in different myelinated nerves is roughly proportional to the internodal distances, which at least would be in accord with such a manner of transmission.

B. The Neurilemma Sheath: Degeneration and Regeneration of Nerve

The neurilemma sheath is believed to be related significantly to degeneration and regeneration of nerve fibers. If a nerve fiber is cut across, the part of the fiber removed from the nucleus-containing segment will always degenerate, eventually failing to conduct impulses. Finally, in the course of some weeks, the fiber disappears entirely. Death and disintegration are the fate of any fragment of any cell which is cut off from the influence of the nucleus.

When a cut nerve fiber degenerates, characteristic changes occur in the sheaths. The myelin sheath disintegrates into fatty droplets, but in the neurilemma sheath the nuclei proliferate and form a multinucleate tube of tissue around the disappearing axone. In the course of weeks or months the living cut stump of axone commences to grow down through this neurilemma tube. Eventually, such a fiber may grow all the way to the structure innervated originally by the fiber, so that normal function returns. This happens by no means universally when nerve fibers are cut, but it occurs frequently enough so that the physician can often promise the patient a fair chance of recovery from paralysis when a peripheral nerve is cut by accident, provided the cut ends of the nerve are properly sewed together.

We cannot be certain of the role of the neurilemma in this process. It may act mechanically to direct the course of the growing nerve fiber; it may in some way aid in the nourishment of the regenerating fiber. At any rate, there is evidence that, if the degenerated fiber is equipped with a neurilemma sheath, regeneration is more likely to occur. When nerve fibers in the brain or spinal cord are cut, at least in the adult mammal, little if any regeneration occurs. Recovery from paralysis or loss of sensation owing to actual destruction of nerve cells or to cutting of nerves in the spinal cord or brain of man probably does not occur. Whether this failure to regenerate is entirely due to the absence of a neurilemma upon these fibers of the central nervous system, or to some other factor, is still a problem.

X. THE LAW OF BELL AND MAGENDIE

The peripheral nerves and their physiology may be considered independently of the brain and spinal cord as a matter of convenience, but one cannot proceed very far along this line of study before it becomes necessary to take into account the anatomical and physiological relationships of the peripheral nerves to the central nervous system. We may select as typical of this relationship the anatomy of one of the spinal nerves and the nerve roots with which

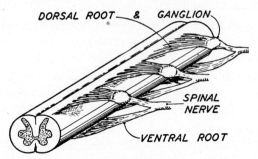

Fig. 145.—Segment of spinal cord showing three spinal nerves and their nerve roots. All spinal nerves are paired. The corresponding three nerves on the left side of the spinal cord are not shown.

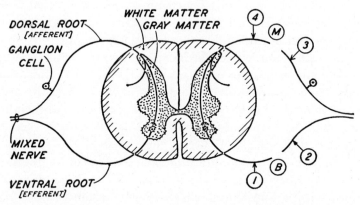

Fig. 146.—Cross-section of the spinal cord showing the relationship of the nerve roots to the cord. The roots on the right are shown cut. Letters and numbers shown here are referred to in the discussion of the law of Bell and Magendie in the text.

it is structurally and functionally connected with the spinal cord. Figure 146 shows the essential anatomical features in a cross-section of the spinal cord taken in such a place as to show a typical spinal nerve.

Each of the thirty-one pairs of spinal nerves[18] has a *dorsal* (in man called *posterior*) and a *ventral* (or *anterior*) root. For convenience,

18. These nerves are grouped as follows: cervical, eight pairs; thoracic, twelve pairs; lumbar, five pairs; sacral, five pairs; coccygeal, one pair.

the diagram shows only one neurone in each root. Actually, there are thousands of such fibers. The cell bodies of which the fibers are outgrowths lie in different positions in the case of the two roots. The cell bodies of the ventral roots lie within the spinal cord in the ventral "horn" of the spinal gray matter, while the cell bodies of the dorsal root lie outside the spinal cord as shown. The cluster of dorsal-root nerve cells produces a visible swelling of the root, referred to as the dorsal-root *ganglion*.[19]

A little over a hundred years ago the Scottish anatomist, Sir Charles Bell, and the French physiologist, François Magendie, demonstrated certain differences in function of the ventral and dorsal nerve roots. Independently of each other, these early investigators opened the bony spinal canal in animals and exposed the spinal cord and its nerve roots.

Bell found that, if he irritated a dorsal root, there was no muscular movement but that, if he irritated the ventral root, "the muscles of the back were immediately convulsed." Magendie observed the effects of cutting these roots. Cutting of the dorsal roots caused a subsequent loss of sensation of the denervated part, without paralysis, while cutting of the ventral roots left sensation unchanged but produced paralysis. From these and other related experiments the conclusion was reached which is now embodied in what is called the *law of Bell and Magendie*, which states that the dorsal roots are sensory (or afferent) in function and that the ventral roots are motor (or efferent).

This law can be demonstrated on an anesthetized animal by exposing the spinal cord and cutting one ventral root (at *B*, Fig. 146) and another dorsal root (at *M*). Stimulation of the central end of the ventral root (at *1*) produces no observable effects because these fibers have no connections with other nerve fibers of the spinal cord to which they can transmit impulses. However, stimulation of the peripheral end (at *2*) always induces muscular movements.

Stimulation of the peripheral end of the dorsal root (at *3*) produces no observable effects because these fibers have no physiological connection with motor structures. Stimulation of the central end (at *4*), which Magendie found to evoke evidences of sensation, will lead only to reflex responses in the anesthetized animal.

The terms "motor" and "sensory" as applied to nerve roots or

19. Any cluster of nerve cells outside the central nervous system is called a *ganglion*, which literally means a "swelling."

fibers, though still used, have largely been replaced by the terms "efferent" and "afferent." These more general terms are better because in the ventral root there are other fibers than those producing motor effects—movements of skeletal muscles. The less conspicuous responses of gland secretion or augmentation of smooth muscle contraction are also elicited from peripheral stimulation of certain ventral roots. In any case, the fibers of this root normally carry impulses away from the central nervous system and are therefore called "efferent," literally meaning "I lead away from."

Many of the impulses in the fibers of the dorsal roots, especially those coming from internal organs, do not reach the sensory areas of the brain. The effects produced by such fibers are reflex in nature. Indeed, stimulation of even those fibers which cause sensations usually produces reflexes also. Normally transmitting impulses toward the central nervous system, the fibers are therefore better called "afferents," meaning "I lead toward."

In these terms we should state the law of Bell and Magendie as follows: The ventral roots of the spinal nerves contain efferent fibers; the dorsal roots, afferent.[20] From the two roots are formed the ordinary spinal nerves, which are referred to as *mixed nerves* because, without exception, they contain both afferent and efferent fibers.

XI. THE AUTONOMIC NERVOUS SYSTEM

A special division of the peripheral nervous system is the autonomic nervous system, which is distinguished by certain definite physiological and anatomical features. The term is employed for those efferent fibers and nerves which supply the innervation of internal or visceral structures, such as the alimentary canal, blood vessels, lungs, heart, uterus, glands, and urinary bladder.

Voluntary or conscious control of internal organs through the autonomic nerves is virtually absent. We cannot voluntarily slow the heart rate by activation of the vagus nerve, or cause the sweat glands to secrete by activation of their efferents, in the same direct way as we can voluntarily contract skeletal muscles.

Although the autonomic nerves are efferents, it is important to recall that there are also afferent nerve fibers included in autonomic nerve trunks. These "visceral afferent" fibers have only very indirect connections with the conscious centers of the brain. Normal stimulation of these afferent fibers, therefore, does not usually cause

20. There is evidence that the dorsal root may also contain vasodilator efferents.

sensations, as contrasted with stimulation of such afferents as those mediating the sense of pain, touch, temperature, or pressure from the skin. We are totally unconscious of the stimulation of the afferent fibers in the lungs which occurs at the end of each normal inspiratory movement in breathing or of the fibers in the arch of the aorta occurring with each beat of the heart.

Yet these and a host of other afferent discharges have very important effects in the many internal reflex adjustments and readjustments that are constantly taking place. In a word, the autonomic nervous system (efferents) and the visceral afferents carry out their functions mainly reflexly and automatically, without its manifold activities reaching consciousness in most cases, or without the possibility of direct voluntary intervention or modification of its activities. For the most part it is this system to which we have devoted our attention in the early chapters of the book in considering the innervation of such organs as the heart, the stomach, and the intestines. Most of the internal organs receive a double efferent innervation from the autonomic system, one set of nerves being excitatory to the organ, the other set inhibitory.

A. The Sympathetic, or Thoracolumbar, System

Anatomically, there are also certain peculiarities of the autonomic system. Its fibers lack a myelin sheath, and, on the efferent side, there is always a two-neurone chain at least[21] between the central nervous system and the structure innervated. In the thoracic region of the spinal cord the autonomic nerves bear the relationships to the cord shown in Figure 147.

The visceral afferent nerve fibers lie along with the other (*somatic*) afferents in the dorsal root, and their cell bodies are located in the dorsal-root ganglion. The ventral-root fibers arise from cell bodies within the spinal cord located laterally in the gray matter. But fairly close to the cord the autonomic efferents separate from the main ventral-root bundle, enter a ganglion, and make functional synaptic connection with the second neurone of the system, whose cell body lies within one of the sympathetic ganglia lying alongside the spinal column. These second neurones may then return to the adjacent peripheral nerve and, along with the afferent visceral

---

21. Because the fibers of the autonomic system may terminate in a peripheral plexus of neurones, as in the heart or the alimentary canal, it is difficult to state whether there are only two neurones in the chain. In some instances there may be more.

fibers, enter into the composition of the mixed spinal nerves, to be distributed, for example, to skin blood vessels, or sweat glands, or hair erector (*pilomotor*) muscles. The fibers coursing from the spinal cord to the ganglia are referred to as *preganglionic;* those whose cell bodies lie in the ganglia, as *postganglionic.*

The ganglia are distributed in pairs alongside the vertebral column, outside the bony canal, and are referred to as the *paravertebral autonomic ganglia.* Some of the postganglionic fibers run upward to join the cervical mixed nerves, and some pass downward to join the sacral nerves. Between the paravertebral ganglia there

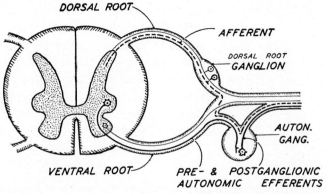

F̲ɪ̲ɢ̲. 147.—Relationships of the autonomic nerves to the ordinary somatic nerves (*latter shown in light lines*) in the thoracic and lumbar regions of the spinal cord. The autonomic efferent shown innervates a blood vessel of the skin. The afferent fiber shown (*dash line*) innervates one of the viscera. (See discussion in text.)

are, therefore, nerve connections, forming with the ganglia the so-called paravertebral *ganglionated chain.*

The anatomical complexities of the autonomic system multiply when we examine more closely into the peripheral distribution of the efferent fibers, especially those of the internal organs (see Fig. 148). Although most of the autonomic ganglia lie alongside the cord as described, several larger ones are located at a little greater distance from the spinal cord. In these ganglia, distributed for the most part in the abdominal cavity, are the synaptic connections between the preganglionic fibers from the spinal cord and the cell bodies of the postganglionic neurones, whose fibers are distributed to the viscera.

Notice, thus, that from the thoracolumbar region of the spinal cord the autonomic efferents fan out to all structures of the body which are innervated by the autonomics, from head to foot. This constitutes but one division of the autonomic system, referred to as the sympathetic, or thoracolumbar, division. Its postganglionic fibers furnish one set of fibers of the double innervation of the autonomically controlled organs and systems.

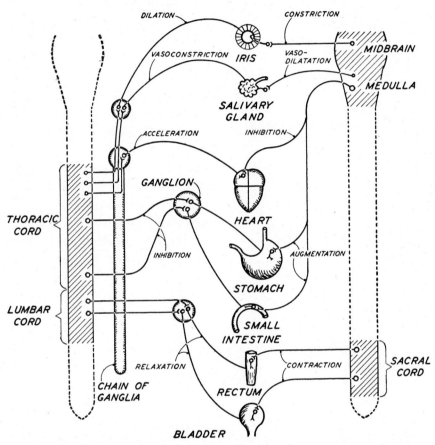

Fig. 148.—The autonomic (efferent) innervation of some of the viscera. The sympathetic autonomic nerves are shown on the left; the parasympathetic, on the right. Note that each organ receives a double innervation and that the action of the two nerves is opposite in each case. The "chain of ganglia" shown is composed of ganglia like that depicted in Fig. 147.

B. The Parasympathetic, or Craniosacral, System

The other half of the autonomic innervation of the viscera, blood vessels, glands, etc., is derived from a second division of the system, originating from the medulla oblongata and midbrain and the sacral portion of the spinal cord. This is known as the parasympathetic, or craniosacral, division of the autonomic nervous system. Its anatomical features are also shown in Figure 148. Note that its efferents are also composed of a chain of two neurones (or more) which synapse in ganglia. The ganglia of this division, in general, lie closer to the organ innervated—sometimes being imbedded in it—than do the ganglia of the thoracolumbar division, which, in general, lie closer to the spinal cord. What may be the significance of this two-neurone efferent system or of the anatomical location of the autonomic ganglia is not understood at present.

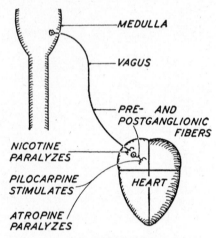

FIG. 149.—Effect of drugs on the vagus nerve endings. After applying *nicotine* to the heart, inhibition can still be effected either by electrical stimulation of the postganglionic fibers of the vagus or by application of *pilocarpine*. At this time stimulation of the vagus trunk (preganglionic fibers) produces no inhibition. After application of *atropine*, the heart cannot be inhibited by electrical stimulation of either the pre- or the postganglionic fibers of the vagus.

Here we have seen, then, the derivation of the double innervation of the various internal organs: the accelerator and inhibitory nerves of the heart, the excitatory and inhibitory nerves of the stomach, the nerves of the dilator and constrictor muscles of the pupil, and the constrictor and dilator nerves of the blood vessels. In each case, one of the nerves is derived from the sympathetic autonomic system, the other from the parasympathetic system.

C. Experiments with Drugs

Many illuminating experiments have been undertaken using various drugs which affect the autonomic nervous system. It is found, for example, that the drug *pilocarpine* stimulates parasympathetic nerve endings generally and that *atropine* paralyzes them.[22] The injection of pilocarpine (0.001 gm.) into an animal will produce effects imitating stimulation of parasympathetic nerves, such as secretion of saliva and slowing of the heart. The injection of atropine paralyzes the nerve endings and renders any subsequent artificial or physiological stimulation of those nerves ineffective. It is therefore possible to stimulate or paralyze the parasympathetic nerves chemi-

22. There are, however, a few exceptions to this generalization.

cally without resort to anesthesia and anatomical exposure of the nerves.

The drug *nicotine* has a highly interesting action. Applied to a ganglion of the autonomic nervous system, where synapses between pre- and postganglionic fibers occur, it prevents impulses from crossing the synapse. It is said to paralyze the preganglionic nerve endings.

Experiments on the vagus nerve illustrate these actions. The application of pilocarpine to the exposed heart stimulates the vagus endings and temporarily slows the heart rate, as if the vagus trunk were electrically stimulated. If nicotine is now applied to the heart, subsequent electrical stimulation of the vagus trunk will be without effect. But, if the postganglionic fibers of the vagus lying in the heart itself are electrically stimulated, the heart is again slowed. Nicotine has paralyzed the preganglionic endings but has not affected the postganglionic fibers. In fact, this effect of nicotine constitutes strong evidence for our concept of preganglionic fibers synapsing with postganglionic fibers in the heart itself.

Finally, if atropine is applied to the heart, subsequent electrical stimulation of either the vagus trunk or the postganglionic fibers will have no effect upon the heart rate. Atropine, we conclude, paralyzes the postganglionic fibers (see Fig. 149).

XII. INHIBITION: NEUROHUMERALISM

This dual innervation of the internal organs poses for us again the general problem of inhibition, which we faced earlier when vagal cardio-inhibition was discussed. How do the vagus nerves, and all the other inhibitory nerves of the autonomic system, exert this action on the structure innervated, despite the fact that the nerve impulses themselves in inhibitory nerves are essentially like those in excitatory nerves?

The experiments of Otto Loewi upon the vagus effect on the heart shed some light on this question. This investigator dissected out the living, beating hearts of two anesthetized frogs and filled each heart with Ringer's solution. He then stimulated the vagus nerve of one preparation, producing the characteristic cardiac inhibition. If he now took some of the fluid from the inhibited heart and placed it into the second heart, it was similarly inhibited, *without stimulation of its own vagus nerve*. Stimulation of the vagus nerve, that is, caused the accumulation in the liquid within the heart of a chemical which had an inhibitory effect upon the second heart.

Loewi called this substance *vagus material*. It was his concept that the material is liberated at the vagus endings and that the chemical and physical organization of the fiber terminations is such that nerve impulses reaching the endings cause the elaboration of the material, in a manner comparable to formation of salivary amylase in the salivary gland when its efferent nerve is stimulated. Crudely stated, the concept is that the vagus endings are really miniature glands— not visible microscopically and not cellular in structure. It is this vagus substance which probably exerts the inhibitory effect, chemically, upon the sinus node pacemaker of the heart.

In Loewi's experiment enough of the material was elaborated so that some of it actually escaped into the chamber of the heart and could be detected there by its effect upon another heart.

It was also soon demonstrated that vagus material obtained from the heart produces excitation of the stomach movements, as does vagus nerve stimulation, and that the vagus substance is also liberated into Ringer's solution circulated through the stomach vessels when the vagus nerve is stimulated.

On the other hand, stimulation of the cardiac accelerator nerves liberates a substance which, upon introduction into a second heart, stimulates it and, when circulated through the stomach vessels, produces the sympathetic nerve effect of inhibition on this organ.

Considerable experimentation has subsequently led to an expansion of Loewi's original notion into the general concept of *neurohumeralism*, which looks upon autonomic nerve action on the peripheral organs, in general, as being mediated by chemicals, which, liberated at the endings of the activated nerves, produce the effects seen. These chemicals are believed to bridge the gap between the nerve ending and the organ innervated. A great deal of experimental evidence supports this view. Stimulation of the nerve causing constriction of the pupil causes material to be liberated into the anterior chamber of the eye, which will duplicate the pupilloconstriction in another eye. The endings of the vasoconstrictor and dilator nerves, and of the secretory nerves to the salivary glands, likewise produce chemicals apparently responsible for the end effects.

Extensive investigation has been directed toward chemical identification of these materials. They are apparently of two general kinds. The vagus material, or, more broadly, the parasympathetic autonomic material, is probably *acetyl choline*. But, if acetyl choline is liberated into the blood of the heart when the vagus is physiologically stimulated, why does it not circulate through the body and produce generalized, chaotic, parasympathetic effects? The answer to this question is apparently quite simple. There is an enzyme in the blood (acetyl choline *esterase*) which very rapidly destroys acetyl choline. This may be the reason why Loewi's original experiments failed to work when blood was used in his experimental frog hearts; the effects described were obtained only with the use of artificial fluids such as Ringer's solution, which contained none of the enzyme which splits acetyl choline. However, it has since proved possible to inactivate this enzyme in blood and to detect an appearance of acetyl choline in the normal circulating stream when parasympathetic nerves are stimulated.

The second of the two kinds of chemicals is liberated by the sympathetic endings and appears to be epinephrine-like in nature. It has been called *sympathin* by its discoverer, Walter Cannon. Unlike acetyl choline, sympathin is not destroyed practically coincidentally with its entrance into the blood stream; only after several seconds is it destroyed by oxidation. In the meantime it may have produced sympathetic effects in regions of the body other than that in which it was formed. Cannon has pointed out that this does not produce the chaotic effects one might expect, because the sympathetic effects in the body are summated quite well into an integrated response complex very similar to that produced by epinephrine (see chap. 12).

CHAPTER TEN

MECHANISMS OF CORRELATION— THE SPINAL CORD AND BRAIN

I. Origin and Significance of the Brain and Spinal Cord
   A. Integrative action
   B. Evolutionary development

II. Reflex Action
   A. Kinds of reflexes
      1. The knee jerk 2. The flexion reflex 3. Crossed extension 4. Complexity of spinal reflexes 5. Classification of reflexes
   B. The stretch reflexes
      1. Muscle tone 2. Maintenance of posture 3. Modulation of movement

III. Conduction in the Reflex Arc
   A. Characteristics of the reflex
      1. Susceptibility to adverse conditions 2. Irreversibility of conduction 3. Long latent period 4. Long afterdischarge 5. Great degree of summation 6. Independent rhythm 7 Great variability 8. Inhibition; reciprocal innervation
   B. Conduction across the synapse
      1. The theory of transmission by chemicals 2. Indirect evidence for the chemical theory 3. Central inhibitory and central excitatory states
   C. Conduction pathways in the spinal cord
      1. Intermediary neurones 2. Spinal shock 3. Gray and white matter 4. The crossing of fibers 5. The specific fiber tracts

IV. The Brain Stem
   A. General features
   B. Reflex centers
   C. Control of muscle tone and posture

V. The Brain Proper
   A. Significance in man and lower animals
   B. Brain size
   C. Significance of the surface convolutions

VI. Structure of the Cerebral Hemispheres
   A. Internal arrangement
   B. The ventricles and the cerebrospinal fluid
   C. Surface configuration

414

VII. LOCALIZATION OF FUNCTION IN THE CORTEX
  A. Motor and sensory areas
    1. Control of the skeletal muscles 2. Muscle sense 3. General body sensitivity 4. Auditory area 5. Visual area 6. Sensation of pain 7. Visceral sensations
  B. The association areas
    1. General functions 2. Interpretation of sensations 3. Interpretation of symbols
  C. The cortex as a complex of reflex centers
  D. Electrical activity—brain waves

VIII. LEARNING
  A. Trial and error
  B. Learning by "getting the idea"
  C. The role of ideation
  D. Conditioned reflexes
  E. The function of the cortex at large

IX. SLEEP

X. FUNCTIONS OF THE CEREBELLUM

XI. ABNORMALITIES OF BRAIN FUNCTION
  A. Destruction of brain tissue
  B. "Functional" nervous disease

I. ORIGIN AND SIGNIFICANCE OF THE BRAIN AND SPINAL CORD

We might speak of the brain and spinal cord as a central clearing-house, or generalissimo's headquarters. "Information" is sent into it via the peripheral afferent nerves from all parts of the body. On the basis of this information, "orders" are sent out along the efferent nerves, causing the various body parts to make the appropriate adjustments. The information does not at all times reach the level of consciousness, nor is there always any conscious direction of the resulting movements. In fact, conscious direction of responses or awareness of sensory stimuli are exceptions rather than the rule. Headquarters is equipped with a staff operating perfectly automatically according to rules and principles which have been inherited from predecessors or acquired by experience.

A. INTEGRATIVE ACTION

This function can perhaps be better expressed in more comprehensive terms without recourse to the foregoing analogy. It is largely through the agency of the central nervous system that all the various tissues and systems of the body are *integrated* into a smoothly operating unit. The billions of cells of the body, which are themselves units in a very real sense, lose their individuality and become the interacting and interdependent parts of a co-operative enterprise. The integrating mechanisms transform what would otherwise be merely a collection of cells and tissues and organs into a single unit—a multicellular individual like man, who can react to his environment as a well-integrated whole.

Even in dealing with the separate organs and systems, we have found it impossible to proceed far without considering these in their relationships to other organs and systems and to the body as a whole. We learn much about the heart isolated from the body, but our information about heart action is woefully incomplete unless the data so collected are regarded as only a beginning. A clear picture emerges only when the heart is considered in relation to the rest of the body—to the blood it pumps and from which it obtains materials necessary for normal function; to the machinery of breathing, which depends upon the pump and which helps make the pumping possible; to the glands of internal secretion, the blood vessels, the skeletal muscles, the digestive apparatus. We have already seen that, to a large extent, these correlations depend upon nerves and reflexes.

Of course, the nervous system is not the only integrating machinery of the body. Many chemical correlations supplement the nervous mechanisms. Hormones, carbon dioxide and other metabolites, and hydrochloric acid and other acids function significantly at some point or other to gear the parts of the machine to one another.

Even such purely physical factors as temperature contribute: the heat produced in muscle contraction helps to accelerate the heart rate and thereby increase the blood flow to the active muscles; it affects the temperature-regulating center of the thalamus, leading to vascular (skin vasodilatation) and secretory (sweating) adjustments which facilitate dissipation of the heat developed.

But the central nervous system seems by far the most important part of the integrative machinery. Despite all the chemical and physical correlating mechanisms, without the nervous system the organism would be little more than a loosely bound together collection of organs, adjusting but poorly to external changes.

A comparison of integration in plants and animals emphasizes this. Lacking a nervous system, plants consist of relatively poorly knit together units. Isolated plant parts—leaves, branches, roots—are usually able to develop new individuals in proper environments, and the loss of parts often affects the rest of the plant organism but little. The death of a branch of a tree may have no effect at all upon the rest of the tree. No single branch is as closely knit into the organization of the whole individual—the tree—as is any single organ in the body of an animal. The high degree of interdependence and unification of the parts of the higher animals contrasts strikingly with the condition in plants.

Even the most complex activities of the brain—conscious processes, emotions, thinking, reasoning—are interwoven with the functions of the lowliest of internal organs. A beautiful symphonic theme may quicken the pulse; anticipation of an examination may flood the blood with sugar released from the liver and accelerate the secretion of urine. On the other hand, the state of the stomach may markedly affect one's "state of mind," depending upon whether the stomach is empty or full, or what it may contain.

B. EVOLUTIONARY DEVELOPMENT

The nervous system had humble evolutionary beginnings. In the two-cell-layered organisms typified by *Hydra*, no *central* clearing agency existed, and the nervous system consisted of a network of nerve-cell fibers, directly connecting the various parts of the organism with one another. In the more advanced type exemplified by the flatworm, with its bulkier body and greater diversification of activity, thanks to the development of mesoderm and muscle, the nerve cells became clustered in centralized masses. These masses of neurones served as intermediaries between stimulus and response. Nerves soon commenced to differentiate into afferent and efferent. In the annelid stage of evolution, resembling the present-day earthworm, anatomical and physiological centralization went still further in the development of chains of ganglia. At this stage there were also the beginnings of concentration of the central nervous elements toward the anterior end of the animal, which comes to exert a greater and greater control over the posterior segments. In the higher animals, including vertebrates and man, further differentiation of this central integrating machinery into the brain and spinal cord took place.

From its earliest beginnings segmentation in the general body plan includes a segmental structure of the central nervous system, with a ganglion or pair of ganglia for each segment. This plan persists in the vertebrates, with their segmentally arranged pairs of spinal nerves. Only in the final stages of brain evolution, with the development of the *cerebral cortex*[1] and the *cerebellum*, is the segmental plan departed from.

II. REFLEX ACTION

A convenient unit for study of the action of the nervous system is the *reflex*. In its simplest form a reflex involves five components: re-

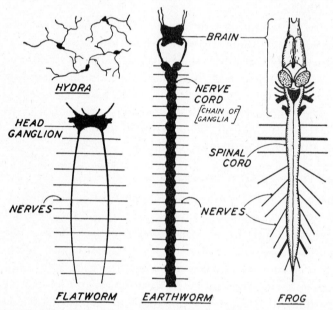

FIG. 150.—Evolution of the central nervous system. In *Hydra* there is a diffuse network of nerves. In the flatworm some of the nerve cells cluster together into a central or head ganglion. The earthworm possesses a definite central nervous system composed of a chain of ganglia, the first of which constitutes a rudimentary brain. The frog has a well-defined brain and spinal cord. In subsequent vertebrate evolution the brain increases tremendously in size and importance.

ceptor, afferent, center, efferent, and effector. *Receptors* are specialized structures which respond to specific kinds of environmental changes. Essentially, a receptor is so organized as to have a low threshold for some specific stimulating agent. The retina of the eye has a low threshold for light waves, the internal ear for sound vibrations, the taste buds for chemical stimuli. Receptors are the sensory end organs of *afferent nerve fibers*, in which nerve impulses are initiated. The afferents transmit the impulses to the central nervous system, where in some reflex *center* or *centers* functional connection

1. The gray covering of the cerebral hemispheres (see Fig. 156, p. 437).

is made between afferent and efferent neurones through synapses. Activation of the *efferent fibers* causes impulses to be transmitted to a structure capable of some such response as movement or cessation of movement, or secretion, known generally as an *effector*.

In the flexion reflex, and in almost all other reflexes, there are one or more intermediary neurones in the central nervous system between afferent and efferent. More or less complicated nerve pathways are usually traversed by impulses entering the central nervous system in afferent fibers, before the efferent fibers are activated.

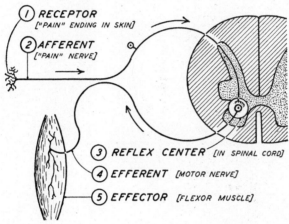

Fig. 151.—The essential components of a reflex, as exemplified by the *flexion reflex*. Arrows indicate direction of transmission of nerve impulses. "Pain" is in quotes because the actual sensation of pain is not essential to the reflex, since the reflex occurs even if the brain is removed. This, and almost all other reflexes, are more complicated than shown. (See discussion in text.)

Although the reflex is a convenient unit of action to study, it should be emphasized that most activity—however simple—of the body or its parts consists of a complex of several or many reflexes occurring simultaneously or in succession. Single reflexes are fragments of functional units isolated artificially to facilitate experimental study.

A. KINDS OF REFLEXES

1. *The knee jerk.*—A typical example of a reflex is the knee jerk. When the muscle tendon at the kneecap is tapped, this tendon is momentarily stretched. Receptors in the muscle, which are sensitive to stretch, are thus stimulated, and there is a reflex activation of the very muscles which were stretched—in this case the extensors

of the lower leg. There results a sudden straightening of the leg. The importance of this kind of reflex we shall consider at a later time.

2. *The flexion reflex.*—Another example, the flexion reflex, is the flexion of a limb produced reflexly by any stimulation of the limb which is of an injurious nature. It is a defensive adaptation of the organism, in which there is an automatic withdrawal of a part from a harmful environmental change. Although this reflex is accompanied

FIG. 152.—The flexion reflex in a frog. The leg is withdrawn (*right*) following stimulation of the leg by application of weak acid (*left*). This automatic response is independent of sensation, since it occurs when the legs are isolated, provided the spinal cord and the nerves of the legs are intact. (From the sound film, *The Nervous System.*)

by pain in the intact animal, the response is independent of the sensation and occurs even if the brain is removed.

3. *Crossed extension.*—Activation of an afferent by such an injurious stimulus produces another reflex, the crossed-extension reflex, simultaneously with the flexion reflex. In this the extensor muscles of the limb on the side of the animal opposite the stimulated side are reflexly activated. Stimulation of the skin of the left leg, that is, causes flexion of the left leg and extension of the right. In conse-

quence, the animal is still adequately supported in the upright position, even though the flexion reflex has withdrawn part of the support. Succinctly put, crossed extension prevents a person who steps on a tack from precipitously sitting on it into the bargain.

4. *Complexity of spinal reflexes.*—Many reflexes are very complex, involving highly integrated activation of many muscles. Movements into which one might easily read planned purposiveness on the part of the animal are often seen to be purely automatic reactions. If we put a drop of mild irritant, such as weak acid, on the skin of a frog's hind leg, a series of complex brushing movements will be performed, as though to rub away the irritant. The movements are so well coordinated and so accurately localized with respect to the exact site of stimulation that feeling and volition—intervention of the brain—seem surely to have played a part. Yet the same reaction takes place in a frog whose brain, or even the entire head, has previously been removed under anesthesia.

If a turtle is suddenly killed, and its entire body cut in two transversely, the posterior segment containing the lower portion of the spinal cord will still display remarkable reflex activity. If the tail is pinched with the fingers, both hind legs will execute a series of movements, pushing firmly against the pinching fingers. There can be no question of the animal "attempting" to get rid of a "pain," because we are dealing with only half an animal, having no brain, with only a part of the spinal reflex machinery intact.

5. *Classification of reflexes.*—Reflexes have been classified in several ways from several points of view. There is the division into the inherited, innate, or unconditioned reflexes, on the one hand, and the acquired, learned, or conditioned group, on the other hand. Our present discussion will deal mostly with the innate reflexes. The conditioned reflexes will be considered in connection with the physiology of learning.

Reflexes have also been classified anatomically, depending upon whether the reflex involves one or more levels of the cord and brain, whether ascending and descending pathways in the central nervous system are involved, and to what extent. This is not fundamentally different from the first scheme of classification, because learned reflexes involve certain anatomical structures, notably the cerebral cortex of the brain. Most reflexes involve some transmission of impulses up or down the pathways of the central nervous system. In perhaps all cases there is at least one intermediary neurone between afferent and efferent fibers.

Physiologically, reflexes have been grouped on the basis of the location and the kind of receptors involved: there are the *entero- ceptive* reflexes, which we have already dealt with, arising in the viscera; the *exteroceptive* reflexes, initiated by stimulation of sense organs on the surface of the body; and the *proprioceptive* reflexes, arising in muscles, tendons, joints, and parts of the inner ear.

Fig. 153.—Complex reflex behavior. A frog decerebrated under anesthesia maintains a normal posture and, when mildly stimulated (*above*) hops away (*below*) exactly as would a normal frog. Here there can be no question of sensation, or voluntary motion. (From the sound film, *The Nervous System*.)

B. THE STRETCH REFLEXES

Muscle tone is intimately related to the proprioceptive reflexes. The latter, of which the knee jerk is an example, depend upon a sud- den stretch of the muscle, which stimulates certain spindle-shaped receptors of the muscle or its tendon. The response is highly local- ized, causing reflex stimulation in the vicinity of just those fibers which are stretched. The knee jerk, as such, and similar tendon jerks are, of course, highly artificial, in the sense that sudden taps on the tendons do not occur physiologically. Muscle stretch of a sustained sort, however, does occur and produces the same kind of reflex. But instead of a sudden, sharp tetanus of short duration, there results the sustained partial contraction we refer to as *tonus*.

1. *Muscle tone.*—At first sight one would be tempted to conclude that tonus is not a reflex because of the sustained character of the weak tetanus. But the fact that cutting either the afferent or the efferent nerve of a muscle abolishes all tone in that muscle clearly demonstrates the reflex nature of the phenomenon.

In the main, the stimulus initiating reflex tone is effected by gravity. For example, gravity tends to cause the lower jaw to drop, the head to fall forward, the legs to flex under the weight of the body, and the back (in quadrupeds) to sag. But note that any tendency for gravity to produce these effects automatically puts the "antigravity" muscles upon a stretch. At once reflex contractions of the muscles offset the effect of gravity. Any tendency for the knees to flex stretches the extensors and reflexly prevents collapse of the limbs. Largely in this way, the tone of the antigravity muscles and the normal posture of the animal are maintained.

2. *Maintenance of posture.*—In posture maintenance the stretch reflexes function steadily and continuously. In contrast to the phasic movements of other reflexes, they are called *continuous reflexes*. Posture represents a sort of base line or foundation upon which actual muscle movements are compounded. Every movement starts from a definite posture and, likewise, ends in a posture.

It is sometimes loosely stated that one of the functions of muscle tone is to "take up the slack," so that activation of the muscle is at once effective in moving skeletal parts. Any such function must be of minor importance. Recall that the latent period in a nerve-muscle preparation devoid of tone is only 0.01 second, which is too brief to be of significance in movements of the skeleton by muscles.

Without muscle tone there could be no inhibition of muscles or *reciprocal innervation*, important phenomena to be considered later.

3. *Modulation of movement.*—Apart from the maintenance of tone, the stretch reflexes are also important in the modulation of muscle movements. Reflex and voluntary movements are normally smoother and less jerky than the same movements induced in muscles whose afferent nerves are cut. In fact, in the ape, cutting the sensory nerves of a limb so seriously reduces the effectiveness of the jerky muscular

movements that the limb is scarcely used voluntarily, almost as if its efferent innervation were destroyed. In man, when proprioceptive reflexes are abolished by syphilitic disease of the spinal cord, which destroys the reflex pathways, there result very sudden, awkward, jerky movements, especially of locomotion.

III. CONDUCTION IN THE REFLEX ARC

We have become familiar with the physiology of certain parts of the reflex arc in our study of conduction of the nerve impulse in peripheral nerve fibers, afferent or efferent. If we examine the conduction of nerve impulses through the reflex arc, certain differences between this and activity in a peripheral nerve fiber alone become at once apparent.

A. CHARACTERISTICS OF THE REFLEX

Conduction through a reflex arc (as compared with a peripheral nerve) is characterized by (1) greater susceptibility to adverse conditions; (2) irreversibility of conduction; (3) a longer latent period; (4) a longer afterdischarge; (5) a greater degree of summation; (6) an independence of the rhythms of electrical stimulus and the response in the efferent fiber; and (7) greater variability.[2] Finally, the puzzling problem of (8) inhibition arises again.

These differences must be due to that added component of the system present in the reflex arc and absent from the peripheral nerve—the reflex center in the central nervous system. Perhaps the conduction of the nerve impulses through nerve cell bodies contributes to these peculiar properties of the reflex arc, but they are probably largely due to the interpolation of synapses in the conduction system—to phenomena associated with transmission of activity from one neurone to the next in the series.

1. *Susceptibility to adverse conditions.*—Reflexes can be elicited only when the blood supply to the spinal cord is adequate, while the nerve-muscle system will respond to stimulation for a relatively long time after cessation of circulation, or even after removal of the preparation from the body. Reflexes are also much more easily abolished by anesthesia or drugs than is activity of nerve fibers or effector organs.

2. We do not refer here to the variability of reflexes induced by training or learning but confine the discussion entirely to the innate reflexes with which mammals and man are equipped by heredity.

This greater susceptibility to adverse conditions of all kinds, to fatigue, lack of blood, or presence of drugs, is dependent upon that portion of the reflex arc lying within the central nervous system—perhaps the synapse, perhaps the cell bodies. We know that the nerve tissue in the central nervous system, composed of fibers plus cell bodies, has a metabolic rate some seventy times that of peripheral nerves, composed only of fibers. We also know that, in general, any part of a biological system having a high metabolic rate, be it part of a cell or a part of an individual, is more quickly affected by adverse conditions than parts with a lower rate of metabolism.

Earlier it was shown (p. 383) that the muscle-nerve junction is more susceptible to fatigue than either the nerve fibers or the muscle. Experiment further reveals that the whole peripheral nerve-muscle complex is far less susceptible to fatigue than are the conduction paths of the central nervous system. If an individual exercises one finger, supporting a weight by it, and lifting the weight repeatedly by flexions of the finger performed in rapid succession, it will be found that the weight will become harder and harder to lift and will be lifted to a gradually diminishing height. Eventually, it can no longer be moved by the utmost voluntary effort. Fatigue has abolished this voluntary action. But the site of *this* fatigue is definitely not in the nerve or muscle or in the junction between the two, because, if now the nerve or muscle is artificially stimulated directly by an electric current, a very powerful contraction of the muscle will lift the weight as high as the very first voluntary contraction. The fatigue is in the central nervous system, somewhere along the pathway from the motor area of the brain to the efferent neurones. Though this activity is not strictly a reflex, the same principle is nevertheless demonstrated—that nerve pathways or connections in the central nervous system are more susceptible to fatigue than are peripheral nerve fibers.

2. *Irreversibility of conduction.*—Nerve fibers can conduct impulses in both directions and do so always when they are stimulated along their course. When the *peripheral end of an afferent* fiber is stimulated, impulses pass peripherally, as evidenced by detectable action potentials, but upon reaching the termination of the fiber—the receptor structure—they are without effect. Or, if the *central end of an efferent* nerve is stimulated, impulses travel "backward" up into the spinal cord. But they do not pass beyond the terminations of the neurone actually stimulated. They cannot cross synaptic connec-

tions in a reverse direction. There is only one-way traffic at each synapse.

3. *Long latent period.*—The time elapsing between the application of a stimulus to a receptor or to its afferent nerve and the reflex response evoked in the effector is much longer than that elapsing between stimulation of an efferent nerve and the effector response. A muscle starts contracting in 0.01 second or less after its nerve is stimulated, but there may be a delay of some seconds between stimulation of an afferent and the reflex muscle response. This difference cannot be accounted for by the longer pathway which must be traversed by the nerve impulse in the reflex. Somewhere in the system, presumably at the synapses involved, there is a considerable retardation of the rate of conduction.

4. *Long afterdischarge.*—In nerve-muscle systems the termination of the stimulation also marks the termination of the effector response. But a reflex muscle response may persist as long as 10 seconds after the afferent stimulation has ceased; that is, the efferent neurone of the arc may continue to be activated for several seconds after transmission of impulses in the afferent has ceased. Presumably, there is a persistence of the effect produced by the afferent impulses in some part of the reflex system—again probably at the synapse.

5. *Great degree of summation.*—In peripheral nerves several weak stimuli in very rapid succession may summate to activate the nerve, though each alone may be quite ineffective in doing so. At most, however, only a few stimuli will summate in this fashion. If four or five subthreshold stimuli fail to excite the nerve, any number of stimuli of the same intensity will also fail.

In reflexes the situation is different. There is a much greater possibility of summation of the afferent impulses somewhere in the arc. If a few impulses in the afferent do not succeed in eliciting a reflex response in the efferent fibers, it is sometimes possible to get a response from the summated effects of many—even hundreds—of afferent impulses.

This seems to indicate that the excitation of a nerve fiber by direct stimulation is in some way different from the excitation of the efferent fiber in a reflex. There appears to be a rapid recovery from those temporary changes produced on a nerve by a subthreshold stimulus—presumably a beginning depolarization. Thus, unless each stimulus of a series is only a little below threshold intensity, the fiber is sufficiently restored toward its resting (repolarized) state even in the short interval elapsing between stimuli, so that the successive stimuli do not progressively depolarize the membrane to completeness.

But, whatever the nature of the change may be which is produced at the efferent nerve by impulses reaching it via the afferent, that change is not so rapidly dissipated. It can be added to by many afferent impulses until, finally, a depolarization and activation of the efferent fiber occur, with elicitation of the reflex response.

6. *Independent rhythm.*—When repetitive stimuli of at least threshold intensity are applied to a peripheral nerve, there will result one impulse per adequate stimulus. The rhythm of the stimuli is reproduced in the rhythm of nerve impulses. This is not always so in the reflex arc. The rhythm of discharge through the efferent system may be independent of the rate set up in the afferent system. For example, when just one or a few impulses are initiated in appropriate afferents in a leg, by application of a single brief stimulus, there results a *volley* of impulses in the efferents to the flexors of that leg, causing not a brief twitch of the muscle but a tetanic contraction. Another volley of impulses of a quite different rhythm is initiated in the efferent fibers innervating the extensors of the opposite leg.

This again means that conduction at some place along the reflex arc is definitely different from conduction along a nerve fiber. The impulses in a given segment of a nerve fiber bear a one-to-one relationship to the impulses at the region of the nerve fiber being stimulated at some distance away. But the impulses in a distant portion of a reflex conduction pathway do not necessarily bear a one-to-one relationship to the impulses at the region of the pathway being stimulated, provided one or more synapses intervene. Impulses are usually not transmitted one by one across synapses.

7. *Great variability.*—In efferent systems, if the condition of the nerve and muscle tissue remains unchanged, the only variability in the response elicited depends upon the number of nerve fibers activated and the number of impulses set up in each fiber—that is, upon the strength of the stimulus to the nerve trunk.

But, in a reflex, even though the state of the nerve and muscle remains constant, other normal conditions besides the strength of the afferent stimulation modify the response obtained. Such modification depends upon the simultaneous occurrence of other reflexes or central nervous system activity normally affecting the same nerve-muscle group.

Consider, for example, the stimulation of certain afferents which play upon the extensor and flexor muscle groups of the *left leg;* tapping the left-knee tendon, producing extension (the knee jerk); pinching the left leg, producing a flexion (the flexion reflex); and pinching the right leg, producing extension (the crossed-extension reflex). If the left-knee tendon is tapped at about the same time that the right leg is pinched, the resultant knee jerk is greatly exaggerated, as though there were an addition of these two effects upon the extensors of the left leg. On the other hand, if the left leg is pinched just as the left-knee tendon is tapped, the knee jerk is diminished or may even be abolished entirely.

This latter effect is not, as one might guess, due to the algebraic summation of contractions of the extensors and flexors of the left leg; it is not a mere balancing of the mechanical effects of muscle groups exerting pulls in opposite directions on the skeletal parts involved. Rather, the extensor muscles effecting the knee jerk are now activated to a lesser degree. This effect occurs in the central nervous system along the reflex pathway. Afferent impulses, which were started by tapping the left knee, fail to activate the extensor efferents if other appropriate afferents are simultaneously activated by pinching the left leg. Transmission of nerve impulses through a reflex arc may thus be modified by the simultaneous occurrence of other afferent stimulation in a manner impossible in transmission of nerve impulses through a peripheral nerve-muscle system.

8. *Inhibition; reciprocal innervation.*—Involved in the foregoing is the phenomenon of inhibition in the central nervous system. Just what is meant by this can best be made clear by analyzing in more detail the events which occur during elicitation of the flexion reflex. Not only are the flexors of the same side activated and, by crossed extension, the extensors of the side opposite, but also there is in each leg an inhibition of the muscle groups whose action is opposite to the action of the activated group. That is, when the pain nerves of the *right leg* are stimulated, the following reflex changes take place:

a) Activation of flexors }
b) Inhibition of extensors } of the right leg

c) Activation of extensors }
d) Inhibition of flexors } of the left leg

How is this inhibition effected? We can dismiss at once, in the light of previous discussions, the possibility of any special kind of nerve impulses in the efferents of the inhibited muscles. There is no such thing as a special kind of inhibitory nerve impulse. If action currents are led off from the efferents involved, it is found that the inhibition is dependent upon a *reduction* in the number of nerve impulses in those efferent nerves whose muscles are being inhibited. Figure 154 shows what happens. At rest (*A*) the "pain" fibers are inactive. The efferents to all four of the muscle groups, however, are normally active, giving rise to the continuous tonic contraction characteristic of all muscle even at rest. At present we need not concern ourselves further with the cause of the tonic activity of the efferents but merely recognize its easily demonstrable presence. Part *B* of the figure shows the changes which occur in the four groups of efferent nerves when an afferent pain fiber of the *right leg* is stimulated:

a) More impulses in the flexor group (causing tetanic contraction)

b) Fewer impulses in the extensor group (causing relaxation)

of the right leg

c) More impulses in the extensor group (causing tetanic contraction)

d) Fewer impulses in the flexor group (causing relaxation)

of the left leg

We encounter here merely a specific example of the general phenomenon of reciprocal inhibition. When any muscle group is activated reflexly or voluntarily, there is normally a simultaneous inhibition of the antagonistic muscle group. The inhibition is accomplished, in the manner described, by means of the nerves of the muscles. Muscle groups, paired into "agonists" and "antagonists" in this way, are said to possess *reciprocal innervation*.[3]

Obviously, the inhibition we here deal with is strikingly different from anything encountered in a peripheral nerve-muscle system. Positive activity in the afferent nerve fibers produces diminished activity in certain of the efferents with which they make physiologi-

3. It should be recalled that the phenomena of activating one group of muscles and inhibiting the antagonistic group by reciprocal innervation apply essentially to overt movements of the body or its parts. In the maintenance of fixed postural positions of body parts, antagonistic groups of muscles may be simultaneously activated through their nerves (see also pp. 441–42).

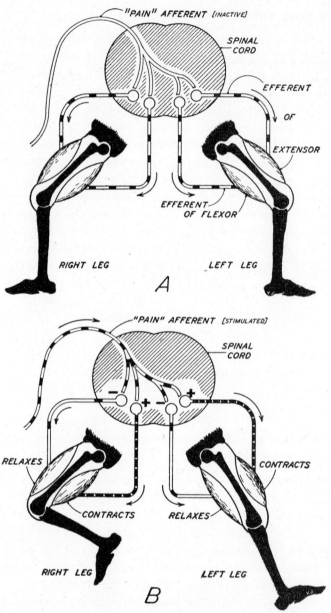

Fig. 154.—Reciprocal innervation of antagonistic muscles. *A:* Flexors and extensors of both legs are shown in a state of tonic contraction produced by a constant stream of efferent nerve impulses (*indicated by shaded segments*). *B:* When a "pain" afferent from the right leg is stimulated, there occur a flexion reflex and a crossed extension reflex, with reciprocal inhibition in both reflexes. Note that nerve impulses (*shaded segments*) in the "pain" nerve produce an *increase* in nerve impulses in the efferents to the right flexor and the left extensor. These same afferent impulses produce a *decrease* in nerve impulses in the efferents to the right extensors and the left flexors. Arrows indicate direction of transmission of nerve impulses. Plus sign indicates a stimulating action; minus sign, an inhibition.

cal connection via synapses. How is this done? Certainly not by any direct transference of activity in one neurone of the reflex pathway to the next. We recognize this as the conclusion which has forced itself upon us repeatedly throughout the discussion of the peculiarities of reflex conduction. How, then, are nerve impulses or effects at neurone endings transmitted across synaptic junctions?

B. Conduction across the Synapse

1. *The theory of transmission by chemicals.*—We can recognize here certain features in common with the problem of inhibition in some of the peripheral systems, such as the vagus-heart system. We saw that this and apparently other inhibitory phenomena in the autonomic nervous system depend at least in part upon the liberation of a material at the nerve endings (see pp. 411–13).

Inhibition or acceleration of the sinus nodal activity depends upon whether inhibitory material (acetyl choline) is liberated at the vagus endings or excitatory materials (sympathin) at the accelerator endings as nerve impulses are set up in one or another of these nerves.

For the moment, let us compare the efferent neurones of the extensor muscles of the right leg with the sinus node, the muscle and tendon afferents from these extensors with the cardioaccelerator endings, and the "pain" afferents from the same leg with the vagus endings (see Fig. 155). Might not the activation of the extensors by the proprioceptive afferents or inhibition by the pain afferents be effected in the same way as the cardiac nerves affect the pacemaker? Perhaps an inhibitory substance is liberated at one nerve ending, or excitatory substance at the other. These substances, in turn, may be responsible for transmission of effects across the synapse to the next neurone in the reflexion chain; they may exert chemical excitation or inhibition upon these neurones.

Looking back to the total effect of an injurious stimulus upon the four sets of efferent neurones involved in the flexion and crossed extension reflexes, including the phenomena of reciprocal innervation and reciprocal inhibition, might we not say that perhaps excitatory substance is liberated by the afferent nerve endings at points plus (+) and inhibitory substance at points minus (−) in Figure 154, *B?*

2. *Indirect evidence for the chemical theory.*—Merely to formulate this hypothesis, as we have done, is no indication of its correspondence with truth, no matter how excellent are our analogies, no matter how attractive the hypothesis may be. Perhaps, with Francis

Bacon, we should view the hypothesis with especial suspicion because it *is* attractive. Is it more than a good guess? What evidence can be unearthed to support or refute it?

First, we should test our hypothesis against known facts. How does it fit in with the peculiarities of conduction through the reflex arc which we have experimentally observed?

Irreversibility of conduction would occur provided only the endings of the *afferent* neurone at the synapse were so organized as to produce excitatory material. When impulses are artificially set up to

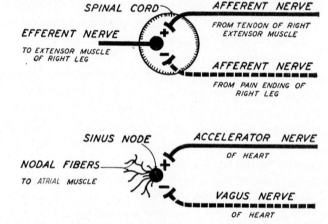

Fig. 155.—Comparison of a reflex center and the sinus node of the heart. The efferent nerve discharges impulses rhythmically, causing muscle tone; the sinus node discharges rhythmically, causing the heart to beat. The rhythmic discharges can be increased or decreased, in either case, by nervous influences: by afferents acting on the reflex center or by the cardiac nerves affecting the sinus. Since the accelerator and vagal influences at the sinus are chemical, perhaps the foregoing afferent nerve endings also produce chemicals at the reflex center. Plus sign indicates a stimulating effect; minus sign, an inhibiting effect.

pass centrally in the efferent fibers, they reach the synapse; but, failing to cause elaboration of the excitatory substance because these endings are not equipped to produce it, they are unable to effect neurones beyond the synapse. There can be only one-way traffic at the synapse.

The long latent period of the reflex arc would be due to the fact that a measurable period of time is required for the elaboration of excitatory substance at the synapse. We know that in peripheral systems like that of the vagus-heart mechanism there is also a much longer latent period than in a skeletal muscle–nerve system. In the former case, time is required for the formation of vagus material; in the latter, the nerve impulses may affect the muscle fibers possibly by some more direct sort of transmission across the nerve-muscle junction. Many peripheral effects involving acetyl choline or sympathin formation take a relatively long time, as also, perhaps for similar reasons, does the reflex.

The long afterdischarge we could ascribe to the slow disappearance of the excitatory state at the synapse between afferent and efferent, just as acetyl choline or sympathin liberated at the sinus node remains for a time to prolong the effect of nerve stimulation beyond the duration of the actual stimulation of the cardiac nerves.

The summation effects described would follow directly from the preceding. Because of the relatively slow disappearance of excitatory material at the synapse, each additional afferent impulse, adding its quota of material, increases the total amount until eventually— perhaps even after scores of impulses—a threshold concentration is reached adequate to effect an activation of the efferent neurone.

The commonly observed independence of the rhythm of impulses in afferent and efferent is also in accord. A one-to-one ratio of impulses in these neurones at all times would be unlikely were the synaptic junctions to be bridged by the intermediary of a chemical substance.

The variability of the reflex response, as analyzed for the knee jerk, we should explain by simultaneous action of excitatory plus inhibitory substance at the synapse, in the case of the inhibiting effect of a flexion reflex upon the knee jerk. Both excitatory substance (from the afferents of the muscle of the same leg) and inhibitory substance (from the pain afferents of the same leg) are produced at the junction with efferents of the extensors of the left leg. If they are produced in amounts of equal effectiveness, no nerve impulses are transmitted to the muscles, and the knee jerk fails to occur. In the same manner properly graded simultaneous stimulation of vagus and accelerator nerves of the heart may liberate acetyl choline and sympathin at the sinus node in such quantities that their effects on the node exactly balance, and the heart rate is unaffected.

No further mention need be made of how this concept helps to account for inhibition, as in reciprocal innervation or elsewhere in the central nervous system. The phenomenon of inhibition was the starting point from which the theory was developed.

3. *Central inhibitory and central excitatory states.*—In the first formulation of this theory by Charles Sherrington, the English

physiologist, he was careful to avoid the use of the terms "material" and "substance." He applied the names "central excitatory" and "central inhibitory" *states* to the highly localized effects produced at the nerve endings of synapses, in general, accounting for trans-synaptic conduction. He postulated that these states might be in the nature of changes of reaction (acid or alkali), physical changes, or, perhaps most likely, the local production of excitatory or inhibitory substances, as we have described.

While this theory is considerably strengthened by the manner in which it explains so many varied peculiarities of transmission across the synapse, we should still be a bit wary of the concept until more direct evidence is available—until the actual presence of these materials is demonstrated at synapses at the time effects are being transmitted across synapses. The difficulties involved in obtaining evidence of this sort are considerable when we remember that, for each efferent pathway, there are several afferents. Synapses are crowded together within distances of microscopic dimensions—some organized to excite, others to inhibit. But, difficult though the problem is, a beginning has been made by investigations of transmission of nervous influences across synapses of the sympathetic ganglia. Although these lie outside the central nervous system, they are probably like the synapses of the brain and spinal cord.

It has been shown that, when nerve fibers on one side of a ganglion are stimulated, causing impulses to be transmitted across the synapses of the ganglia to the fibers beyond, a detectable quantity of a specific material makes its appearance in the ganglion. This presumably corresponds to the central excitatory state, or material. The direct demonstration has not yet been made of the formation of specific chemical substances at synapses within the central nervous system corresponding to the central excitatory state and the central inhibitory state. Until such chemicals have been demonstrated in the synpases of the central nervous system, we shall have to consider this extension of the general concept of neurohumeralism as a theory, for which a great deal of indirect evidence exists, but as a theory nonetheless.[4]

---

4. It must be admitted that mechanisms other than chemical may be involved in transmission of impulses across synapses. For example, the fact that impulses in afferents and efferents sometimes occur in a one-to-one relationship suggests that a more direct mode of transmission may sometimes operate—perhaps physical—in which each afferent impulse stimulates the efferent and initiates a single impulse in the efferent.

C. Conduction Pathways in the Spinal Cord

1. *Intermediary neurones.*—For the purpose of discussion thus far, we have considered the neural elements in a reflex arc to be composed only of afferent and efferent fibers, functionally meeting at a center. In reality, very few reflexes are quite so simple. With the possible exception of such stretch reflexes as the knee jerk, perhaps even the simplest of the reflexes involves at least one or a few intermediary neurones interposed between the afferent and the efferent. All our previous discussion still applies.

In most reflexes there is also a transmission of the activity through nerve pathways in the spinal cord, either upward or downward, or both in succession. Often a stimulus to one region of the body will produce a reflex response in entirely another region, as contrasted with such reflexes as the flexion reflex or the knee jerk. We have encountered the more complex kind of reflexes again and again in our studies upon the internal organs.[5] Obviously, since the centers for many of these reflexes are in the medulla, conduction through the arc must involve pathways in the central nervous system to and from the medulla as well as merely the peripheral efferents and afferents. Many anatomical and physiological studies have been directed toward a determination of just where in the spinal cord all these conducting pathways lie.

2. *Spinal shock.*—The phenomenon of spinal shock aptly illustrates the fact that reflexes which we are likely to think involve only a small part of the spinal cord are considerably modified by, or even dependent upon, more distant regions of the central nervous system. We should certainly expect that destruction of the lower part of the spinal cord would abolish reflexes involving the hind legs by a direct interruption of the reflex pathways. But we are surprised to find that, if the spinal cord is cut across even at its very upper end, there is in most species a temporary abolition of such reflexes—of all reflexes below the cut. This condition we call *spinal shock*. It is as if even the flexion reflex of a hind limb really involves long conduction paths in the spinal cord; the afferent impulses appear not to be relayed at once to the efferent but seem first to pass upward through spinal cord pathways and then down again before completion of the reflex. Perhaps the cutting of the cord interrupts these long reflex

5. In breathing, for example, stimulation of the sensory nerves of the larynx causes reflex effects upon the diaphragm, or, in the circulation, stimulation of the carotid sinus causes reflex changes in the abdominal blood vessels.

pathways. There seem to be potential connections between afferents and efferents of a more direct nature than those used normally. Apparently, these more direct pathways begin to function as the spinal shock gradually disappears.[6]

The duration and the extent of spinal shock—of abolition of reflexes after cutting high in the cord—vary from species to species. In general, spinal shock is more severe and long-lasting in the higher forms. It lasts a few minutes in frogs and hours or days in man.

It is as though in evolutionary development there comes to be an increasing degree of control over lower reflex centers exerted by higher centers in the central nervous system. Therefore, isolation of the lower centers affects spinal reflex function in the higher vertebrates more profoundly than in the lower vertebrates.

3. *Gray and white matter.*—Even grossly the spinal cord is seen to be made of a central gray portion, with surrounding white matter. In a cross-section of the cord the gray matter is roughly of an H shape. We have seen it to be made of nerve cell bodies which display the gray color of their protoplasm. The white matter is made of nerve fibers, so colored because, though the fibers themselves are gray, each is covered with a white sheath.[7] Thus the white matter consists of nerve fibers containing the intermediate nuerones of various reflexes as well as fibers which transmit impulses from all parts of the cord to the conscious centers of the brain and also from the brain to the spinal cord. Extensions into the spinal cord of nerve fibers of the ventral and dorsal nerve roots also contribute to the white matter.

4. *The crossing of fibers.*—In general, it is found that fiber tracts between the brain and spinal cord cross from one side of the central nervous system to the other somewhere between their origin and destination. For example, most of the fibers from a part of the brain to the neurones which innervate the skeletal muscles[8] cross in the region of the medulla oblongata. The remainder of them (see lower right portion of Fig. 156) cross lower in the spinal cord. But all such fibers do cross, so that the left side of the brain controls movement of the right half of the body. The fibers which transmit impulses giving rise to the sense of touch or pressure also cross to the opposite side

6. That there are many more connections between neurones than are normally functional can be shown by injecting *strychnine* into an animal. Even mild stimuli to the body surface may now elicit generalized convulsive responses. Apparently, strychnine "opens up" reflex connections which are always there as potential pathways but are nonfunctional except under abnormal conditions.

7. The myelin sheath.   8. This tract is called the *pyramidal tract.*

of the cord at various levels, so that a stimulus to the right side of the body is experienced in the left half of the brain. No entirely satisfactory explanation of the significance of this crossing has yet been given.

5. *The specific fiber tracts.*—Careful studies have revealed that the fibers are segregated into specific functional groups, some trans-

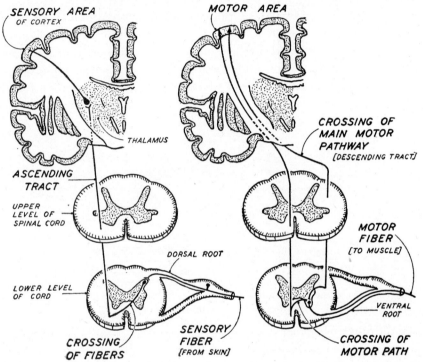

Fig. 156.—Sensory pathway from skin to brain (*left*) and motor pathway from brain to muscle (*right*). The brain is shown in vertical section in the plane of the ears (i.e., in frontal section). Note that both ascending and descending pathways cross the midline somewhere in the central nervous system. The gray matter (*stippled*) lies on the surface of the brain, constituting the cortex. In the spinal cord the gray matter lies inside the white matter (*shown unshaded*). (See also Figs. 157 and 159.)

mitting impulses upward (the *ascending tracts*), others downward (the *descending tracts*).

Many of the anatomical findings have been made because, when nerve fibers are cut, the peripheral end degenerates, and in the central nervous system regeneration rarely if ever occurs. Thus, in a person who has some *lesion*[9] of the central nervous system, produced

9. Anatomical destruction.

either by disease or by accident, the symptoms are carefully noted, and then, after death, the central nervous system is examined by means of dyes which specifically stain degenerated nerve fibers or their modified sheaths to determine what fiber tracts were destroyed. Enough of these observations have been made, supplemented with studies of experimental lesions in animals, to permit us to devise maps showing the location and functions of the various tracts. Figure 157 is a somewhat simplified picture of the main findings.

The functions of some of these pathways will be clear only later, when we have discussed further some of the higher centers and functions of the brain. At present we may point out the function of the fibers in the *posterior columns* of the spinal cord. Impulses which

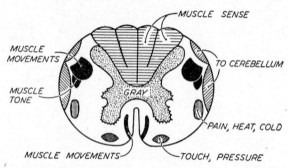

Fig. 157.—The main nerve fiber tracts of the spinal cord. All the tracts shown go to or come from the brain. Much of the spinal cord shown in white contains nerve fiber tracts running up or down the spinal cord for shorter distances. The tracts marked "to cerebellum" are concerned with reflex modulation of muscle movements.

originate in certain sense organs of the muscles, tendons, and joints are transmitted upward in these pathways finally to the brain, producing the so-called "muscle sense." This sense furnishes information of the tension in muscles and positions of the skeletal parts. It is by this mechanism that we are able to tell in what position an arm or a leg or any other part of the body is being held, independently of visual data. When these columns are destroyed, as by syphilitic disease of the spinal cord, muscle sense is lost, and the individual cannot tell in what position his limbs may be unless he can see them.

IV. THE BRAIN STEM

A. General Features

The brain stem may be looked upon as an extension of the spinal cord upward into the skull cavity. The disposition of the white and gray matter here roughly approximates that of the spinal cord, although there is considerably more variation in the internal structure at various levels. On the basis of these structural differences, plus differences in embryonic development and function, we recognize three portions of the brain stem—the hindbrain, the midbrain, and the forebrain (see Fig. 158).

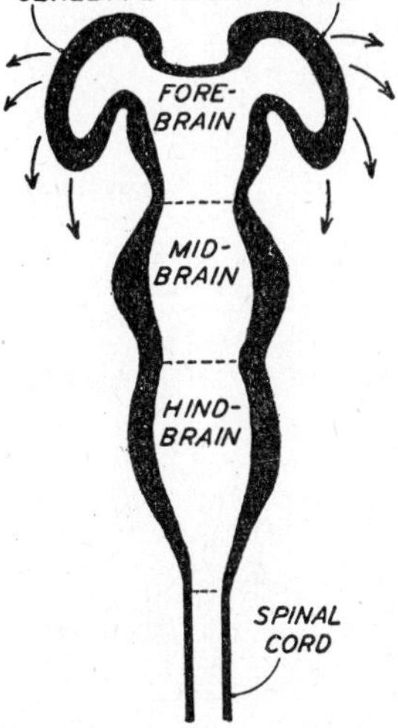

In the white matter of the brain stem are found the extensions of the nerve fiber tracts of the spinal cord running to and from various brain centers, as well as fiber tracts connecting the brain centers with one another.

From the cell bodies of the gray matter spring the efferent components of the twelve cranial nerves.[10] The sensory components of these nerves are composed of fibers whose cell bodies lie just outside the brain stem in ganglia which correspond to (i.e., are serially homologous with) the dorsal-root ganglia of the spinal nerves.

Both the afferent and the efferent components of the cranial nerves participate in a number of reflex actions which have been discussed in preceding chapters. Other reflexes will be described in subsequent sections, as will also certain special sensory features of some of the cranial nerves and their connections. (A brief summary of the main components of the cranial nerves is given in Table 11.)

FIG. 158.—An early stage in the embryonic development of the human brain. The cerebral hemispheres are outgrowths of the forebrain. They grow rapidly to the sides, upward, and backward (*indicated by arrows*) until they almost completely cover the rest of the brain, as seen in Fig. 159. The cerebellum (*not shown*) develops as an outgrowth of the hindbrain. Note that the brain is a tubular structure. The hollow interior persists in the form of liquid-filled ventricles in the adult brain.

B. REFLEX CENTERS

We are already familiar with the fact that the medulla oblongata of the hindbrain contains the respiratory, cardioregulatory, and vasomotor centers. The swallowing center, through which the swallowing reflexes occur, also lies here, for the vagus nerve which constitutes the efferent side of the arc arises in the medulla. It is also known that the reflex center for vomiting as well is located in the

10. As in the spinal cord, these are segmentally arranged, though somewhat less diagrammatically than in the spinal cord. This has led to the term "segmental portion of the brain" to designate the brain stem.

TABLE 11

The Main Structures Innervated by the Twelve Cranial Nerves

Number	Name	Structures Innervated by Efferent Components	Structures Innervated by Afferent Components
I	Olfactory	None	Olfactory mucous membrane of nose (smell)
II	Optic	None	Retina of the eye (vision)
III	Oculomotor	Extrinsic muscles of eye (eye movements, with IV and VI) Muscles of accommodation Iris (constriction of pupil)	Eye muscles (muscle sense)
IV	Trochlear	Extrinsic muscles of eye (eye movements, with III and VI)	Eye muscles (muscle sense)
V	Trigeminal	Some of the muscles used in chewing	Teeth Skin of face Lining of mouth, nose
VI	Abducens	Extrinsic muscles of eye (eye movements, with III and IV)	Eye muscles (muscle sense)
VII	Facial	Muscles of the face Salivary glands (submaxillary and sublingual)	Taste buds of anterior two-thirds of tongue
VIII	Auditory	None	Cochlea (hearing)
	Vestibular	None	Semicircular canals, sacculus, utriculus (senses of movement, balance, rotation)
IX	Glossopharyngeal	Muscles of pharynx (swallowing) Salivary gland (parotid)	Mucous membrane of pharynx (see swallowing reflex, p. 291) Taste buds of posterior one-third of tongue
X	Vagus	Heart (inhibition) Stomach, small intestine (augmentation of peristalsis) Muscles of larynx (speech) Muscles of esophagus (swallowing) Gastric glands (secretory)	Lungs (reflex control of respiratory rhythm; see pp. 237–39) Mucous membrane of larynx (see respiratory reflex, pp. 236–37) Arch of aorta (see pp. 162–63) Stomach (hunger)
XI	Spinal accessory	Muscles of shoulder girdle (shoulder movements)	Muscles of shoulder (muscle sense)
XII	Hypoglossal	Muscles in tongue (tongue movements)	Tongue muscles (muscle sense)

medulla, because cutting the spinal cord across just below the medulla abolishes the vomiting reaction in animals, while a cross-section above the medulla has little or no effect upon it.

The midbrain region mediates certain visual and auditory reflexes, such as the constriction of the pupil when light is thrown on the retina (afferent is optic nerve, efferent is oculomotor nerve), and pricking up the ears in an animal in response to sound (afferent is auditory nerve, efferent is facial nerve).

The thalamus, in the forebrain region of the brain stem, contains the body-temperature-regulating center. Incomplete evidence exists for the presence here of other centers as well which play a part in the regulation of the water balance of the body, the sugar concentration of the blood, fat metabolism, blood pressure, and sleep.

A part of the thalamus is also a great way station for nerve pathways to the sensory area of the cerebral cortex. The sensory pathways of the cord and brain stem synapse in the thalamus with neurones whose fibers make up various tracts in the cerebral hemispheres terminating in the sensory areas (see Fig. 156).

C. Control of Muscle Tone and Posture

Muscle tone, dependent in the last analysis upon the stretch reflexes, is modified greatly by certain of the brain-stem centers, and various related postural reflexes are mediated through them. For example, sense organs in the *labyrinths* (see p. 488) of the internal ear are delicately adjusted to respond to gravitational changes. Shifts in the position of the body, such that the relationship of the body to the earth is disturbed, stimulate these organs and, besides causing sensations of imbalance or falling, initiate reflexes through the medullary centers which automatically adjust to the change. In a large measure these adjustments are tone changes and modifications of posture, such as bracing the body against the tendency to fall. Such reflexes involve afferent fibers from the labyrinths of the internal ear and the medullary centers to which they transmit impulses.

Lying in the midbrain, just above the medulla oblongata, is a cluster of nerve cells—a *nucleus*[11]—which exerts a profound influence, especially in lower mammals, upon the general muscular tone and posture. Strangely enough, this influence seems to be one of inhibition of the spinal centers for tone reflexes. At any rate, when the

---

11. A *nucleus* is any cluster of nerve cells in the central nervous system. It is an anatomical term which is the counterpart of the physiological term "center."

brain stem is sectioned just above the medulla, cutting the descending fibers of this nucleus,[12] a marked exaggeration of muscle tone sets in after some hours or days. All the tendon jerks are exaggerated, and there is an increased tonus of the antigravity muscles, so that what has been called a "caricature of the standing posture" results. The muscles may even become rigid, displaying the phenomenon of *decerebrate rigidity*.

This illustrates further what was pointed out in the discussion of spinal shock—that the higher brain centers have come in evolution to exert a dominating influence over lower centers in the central nervous system, such as those in the spinal cord. This is the physiological counterpart of the anatomical tendency in evolution for the ganglia of the central nervous system to migrate to the head end of the animal—a trend which culminates in the complex brain at the front end of the spinal cord in higher animals.

In monkeys and man this control over the spinal reflex centers for muscle tone has migrated still farther forward. The inhibitory influence of the above-mentioned nucleus in the midbrain is now in large measure taken over by centers in the forebrain—by the cerebral cortex itself. Section of the motor pathways from the cerebral cortex anywhere in the brain stem, therefore, produces definite exaggeration of muscle tone and of the tendon reflexes.

This phenomenon has been of considerable importance to physicians, who can usually tell whether a given paralysis is caused by a lesion in a nerve pathway from the brain or in a peripheral nerve. In the former case, muscle tone and tendon reflexes are increased; in lesions of peripheral nerves, they are diminished or absent.

V.   THE BRAIN PROPER

A. Significance in Man and Lower Animals

There is no semblance of segmental structure in the cerebral hemispheres and cerebellum,[13] which make up the main bulk of the brain. These structures, relatively recently acquired in evolution, develop as outgrowths of the brain stem: the cerebellum from the hindbrain and the cerebral hemispheres from the forebrain. The cerebral hemispheres and their function are of special importance to man because it is the possession and use of these structures which place man

---

12. This nucleus lies near the *red nucleus*, so called because of its reddish color.

13. They are therefore called *suprasegmental* structures.

definitely above the lower animals to a degree exceeding anything we
have encountered so far. In the circulatory, respiratory, digestive,
and excretory systems, man and the lower forms are essentially alike,
part for part, function for function. Even in the nervous system,
until we reach the cerebral hemispheres, there is an essential similarity in the reflex mechanisms, the regulatory centers, and the nerve
pathways. Only in the greater functional plasticity of the cerebral
hemispheres, and especially in the gray covering of the hemispheres
—the cortex—does man seem to be set apart at least in a quantitative sense. Modification of responses to stimuli and of behavior in
general, on the basis of experience and analysis of sensations, are
dependent mainly upon the cerebral cortex.

These modifications can be affected in man to a degree unknown
in lower animals. Yet even here the difference is one of degree. The
capacity to learn—to modify behavior by experience—makes its first
beginnings well down the scale of animal evolution, rudiments being
seen even in forms like the earthworm. It is found throughout the
vertebrate series, and all of us have seen examples of learning in
subhuman mammals.

The varying functional importance of the cerebral hemispheres
from species to species can be most convincingly demonstrated by
studies on animals after surgical removal of the hemispheres or on
man when there is almost complete failure of cortex development
embryologically. The behavior of a frog deprived of its cerebrum is
almost indistinguishable from that of a normal animal. Even in birds
and the lower mammals there is surprisingly little loss of function
from destruction of the cortex. The reflex centers of the brain stem
mediate the vital activities such as breathing, temperature regulation, and circulatory adjustments. A decorticated dog can walk, will
chew and swallow food, and will survive many months if properly
cared for. The care it needs depends upon the fact that decortication
abolishes all responses of the animal acquired by learning, and it will
be able to learn very little, if anything, new. Though it chews and
swallows food, it will not eat from a pan placed before it, and, unless
food is placed in its mouth, it will starve.

Decorticated birds can fly very well and can light and balance upon
a perch. They tend to remain standing quietly for hours, unless disturbed by some potent stimulus from the outside, such as "pain," or
from the inside, such as "hunger" or "thirst." Impelled by such
stimuli, none of which is actually felt, the bird moves about, but in

a perfectly random, unintelligent way. Though hungry, it fails to eat and starves with plenty of food about unless fed artificially.

As we ascend the mammalian scale, we find that the effects of decortication become more striking because a larger and larger proportion of the total normal activity of the higher mammals is based upon cortical control. To a greater and greater extent, this highest center has assumed control over the lower integrating centers.

When the cortex fails to develop in man, the purely vegetative functions are still carried out, but the individual remains an absolute idiot, incapable of all learning and responding to all stimuli essentially as a newborn infant.

B. Brain Size

In a general way there is a correlation between the size of the brain and its functional potentialities—the ability to learn, or "intelligence." A normal man has a brain volume of about 1,500 cc. The highest apes have a brain size of about 600 cc. Skeletal remains indicate that man's immediate ancestors had intermediate brain sizes: the ape man of Java, 940 cc.; Pekin man, 1,000 cc.; Piltdown man, 1,200 cc.; and Neanderthal man, 1,300 cc.

Absolute brain size in itself is not the all-important factor, for, in general, a larger animal has a larger brain, just as it has a larger heart or liver. The rat, having a small brain, seems decidedly more intelligent than the whale, with a brain size of 3,000 cc. But even the ratio of brain size to total body weight is not a good objective index of relative degrees of complexity of function, for in certain vertebrates this ratio is greater than in man.

The most significant size index, so far as correlation with general intelligence is concerned, seems to be the ratio of the brain weight to spinal-cord weight. This would be determined by the degree to which the anterior end of the central nervous system has developed control over the lower levels of the system. This ratio in lower vertebrates is less than 1; in lower mammals it is from 2 to 4; in apes, 15; and in man, 55.

C. Significance of the Surface Convolutions

The convolutions of the surface of the brain are produced by the surface growing more rapidly than the underlying regions and being thrown into folds as a consequence. The complexity of the convolutions is also a general index of complexity of cortical function, from the evolutionary standpoint at least. Lower vertebrates have a smoother brain surface than higher forms. But differences in intelligence of normal individuals of any one species are not sufficiently great to be reflected in significant differences in the degree of complexity of the convolutions. The brain of an absolute idiot might be recognized, but, confronted with the brains of a genius and a man of average, or even low, intelligence, a skilled neurologist cannot tell the difference.

VI. STRUCTURE OF THE CEREBRAL HEMISPHERES

Originating as simple outpouchings of the anterior end of the neural tube, the saclike cerebral hemispheres grow very rapidly—forward, backward, and to the sides. Eventually, they completely cover over the brain stem, hiding it from view except from the underside. Both in evolution and in embryology, development is especially rapid in the gray surface layer—the cortex.

A. INTERNAL ARRANGEMENT

Both gray and white matter are distinguishable in any section made through the cerebral hemispheres. Strangely enough, the disposition of the gray and white elements is the reverse of that in the spinal cord. The gray matter, consisting largely of nerve cell bodies, lies in the convoluted surface, constituting the cortex. Beneath the cortex is the white matter, composed of nerve fibers which connect the cortex with lower centers of the brain and spinal cord, and fibers much more numerous connecting various parts of the cortex with one another. Still deeper in the substance of the hemispheres lies more gray matter, which contains nerve cell bodies—nuclei—and synaptic connections between the neurones of the hemispheres and brain stem. They are relay stations along the course of the nerve pathways to and from the cortex.

B. THE VENTRICLES AND THE CEREBROSPINAL FLUID

Because the central nervous system develops as a hollow tube, there is a cavity, or canal, extending throughout its entire length in early embryology. In the spinal cord this canal becomes almost completely obliterated but remains open in the brain stem. In each of the two cerebral hemispheres is an extension of the cavity referred to as a *ventricle*. The cerebrospinal fluid fills the ventricles and the narrow tubular canal running through the brain stem, besides surrounding the entire brain and spinal cord. This fluid probably furnishes mechanical protection. There is a slow circulation of the fluid from the ventricles where it is formed, down through the brain stem tube, and finally out into the spaces immediately surrounding the central nervous system.

C. Surface Configuration

Among the dips (*sulci*) and bulges (*gyri*) of the cortex there are certain landmarks which anatomically divide each hemisphere into four large areas or lobes—the *frontal, parietal, occipital,* and *temporal* —as indicated in Figure 159. Superficially, the cortex appears to be of the same structure throughout its entire area, but microscopic examination reveals differences in internal structure and in the arrangement of the neurones, which distinguish these major regions from one another. Even within each of the four large areas there are minor local differences; for example, in the motor area of the frontal lobe are found large triangular neurones called *pyramidal cells,* which are to be found nowhere else in the cortex. Are these structural differences correlated with differentiation of activity— with localization of function in the various regions?

The notion that some sort of differentiation of function exists in the brain is very old, finding its first expression in the pseudoscience of phrenology. The phrenologist sought to correlate the size and degree of anatomical development of various parts of the brain with outstanding characteristics of the person examined. From bumps or depressions on the skull he judged the degree of development of the underlying part of the brain. A bump in region *A* came to indicate a development of that part of the brain concerned with business acumen, for example. One endowed with such a bump should go into business. We know now that there is nothing to phrenology except profits for the charlatans who still practice it at the expense of the ignorant. The shape of the skull may vary as much as other parts of the skeleton or the face or hands and is of no more significance, as far as brain function goes. A bump is as likely to be due to an increased thickness of the skull bone at that point as to any peculiarity of the underlying brain tissue.

VII. LOCALIZATION OF FUNCTION IN THE CORTEX

Our modern views of localization of function in the cerebral cortex are based upon several lines of experimental evidence.

First, there is the experimental destruction of various areas. Many experiments have been performed in which one or another cortical area is surgically removed and the effects of the removal upon the behavior of the animal carefully noted. Closely related to this is the second method, the observation of the effects upon man of lesions produced by accident or disease. If the removal or destruction of a specific area regularly causes paralysis of a specific muscle group, we may tentatively ascribe to that area motor control of those muscles.

Third, invaluable data have been obtained by artificial stimulation of various areas in anesthetized animals and by observing the results. The so-called "motor area," which controls the muscle movements of the body, has been mapped out in this way. This method has even been extended to human beings. During the Franco-Prussian War (1870) two physicians directly stimulated the motor area of a soldier whose brain was exposed by shell injury. Since this time, during surgical operations on the brain the motor area has often been stimulated.

Under local anesthesia the various sensory areas have also been stimulated in man, so that, when one region is stimulated, the conscious subject will say, "I see flashes of light"; stimulation of another area will cause sensations of touch or heat or cold in the hand; another area causes sounds to be heard. This method is possible because there is no area of the cortex proper whose stimulation causes pain. It is especially fruitful because it permits co-operation of subject and experimenter; the subject can inform the experimenter of his sensations at each step of the procedure. Studies on the sensory areas of lower animals are often unsatisfactory because the results are difficult to interpret.

A fourth method has been employed extensively—that of observing brain action potentials. By means of electrodes placed upon or into the brain of an anesthetized animal one can determine the course of the nerve impulses from various sense organs to specific brain regions. Light flashed into the eye causes nerve impulses with detectable electrical changes to be set up in the optic pathways of the brain and, finally, in the occipital lobe, their final destination.

Evidence from these different sources indicates a definite degree of localization of function. Certain areas are concerned with specific motor or sensory activities. Figure 159 shows the location of a number of these areas.

A. MOTOR AND SENSORY AREAS

1. *Control of the skeletal muscles.*—Muscular movements are controlled from a region in the frontal lobe just in front of the central fold, or *sulcus*, which separates the frontal and parietal lobes. Detailed studies of localization within the motor area show that, in general, muscles in the upper part of the body are represented in the lower part of the cortex: face muscles are controlled by neurones low in the area, foot muscles by those in higher portions.

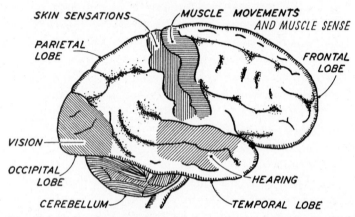

FIG. 159.—The right cerebral hemisphere of man, seen from the side, showing the four lobes and the localized areas concerned with special functions. Association areas are unshaded. "Skin sensations" lie in the parietal lobe; "muscles movements," in the frontal lobe.

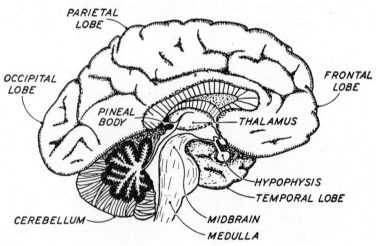

FIG. 160.—The brain as seen in a vertical midline section

The representation is for movements of groups of muscles rather than for single muscles. Stimulation of the motor cortex causes well-integrated body movements, possessing all the characteristics of spinal reflex or voluntary movements (including reciprocal inhibition). The size of the cortical area controlling movements of any muscle group corresponds rather to the complexity of movements than to the bulk of muscle involved. There is a larger area for the hand, with its multiplicity of movements, than for the entire leg group.

The degree of control over muscle groups exerted by this area varies with the evolutionary development of the species. Destruction of part of the area in man causes permanent paralysis of the corresponding muscles, though spinal reflexes involving the muscles still occur. In the dog, and in other lower forms, on the other hand, destruction effects only a transient paralysis, followed by approximately normal function.

2. *Muscle sense.*—Sensory elements are intermingled with the motor neurones of the motor area. They are responsible for *muscle sense* furnishing information of the state of tension in the muscles and the position of the limbs in space. This sense probably also involves the area discussed in the following section.

3. *General body sensitivity.*—The area immediately behind the motor area, in the most anterior folds of the parietal lobe, is responsible for the sensations of heat and cold and touch and pressure, originating from stimulation of receptor structures in the skin. Here again we find lower parts of the body to be represented in regions high in the sensory area and higher parts of the body in lower regions.

4. *Auditory area.*—A large part of each temporal lobe appears to be concerned with hearing. Impulses initiated in stimulation of the *cochlea* of the internal ear by sound are transmitted to this area of the brain via the auditory nerve and nerve pathways.

5. *Visual area.*—The occipital lobes, making up the posterior poles of the hemispheres, are visual in function. The fibers of the optic pathways are easily traceable from the optic nerve to this region. Stimulation of it leads to sensations of light; destruction causes blindness, even though the eyes themselves are entirely sound. It has been said, therefore, that we see not with our eyes but with our brain. This principle applies, of course, to all sensory systems.

6. *Sensation of pain.*—Peculiarly enough, there is apparently no

cortical area concerned specifically with pain perception as such. Any area of the cortex whatsoever can be stimulated, or even cut or removed in a surgical operation, without pain. True, the membranous coverings of the brain are very richly supplied with pain fibers, and surgeons must anesthetize these very carefully when operating on the brain under local anesthesia. But the substance of the cortex is insensitive to pain.

Still other evidence indicates that this primitive sensation is experienced in a less specialized region of the brain—in the thalamus. In that area a primitive sort of unlocalized, undifferentiated pain rises into consciousness. The cortex, via the area of general body sensitivity, apparently functions only to localize the pain—to inform the individual specifically what body area is being painfully stimulated.

7. *Visceral sensations.*—There are no known cortical areas concerned with such sensations from the viscera as hunger and thirst. These primitive, less highly specialized and differentiated sensations may be related to subcortical centers and areas of the cerebral hemispheres. Smell and taste are localized in a fold of the cortex on the undersurface of the brain, near the temporal lobe.

B. The Association Areas

1. *General functions.*—To the rest of the cerebral cortex and, indeed, the greatest bulk of it, no specific, sharply delimited motor or sensory functions can be ascribed. These areas, making up almost all the frontal and parietal lobes and much of the temporal and occipital lobes, are collectively referred to as *association areas.* They are largest in man, relative to the rest of the brain. The retreating forehead and the abrupt downward slope from the top of the skull toward the sides, which we find in prehistoric man, probably indicate less extensive development of the frontal and parietal association areas. These regions are apparently concerned with the "highest" mental faculties: with ideational processes such as memory, imagination, and conception; with learning, intelligence, reasoning, and personality. About some of these little else can be said with certainty. However, certain aspects of the functions of the association areas can be pointed out to give some general notion of the work they do.

2. *Interpretation of sensations.*—The association areas seem to be concerned with what we may call the integration of separate sensations into a meaningful whole. An example will illustrate what is

meant. Let us try to visualize the cortical processes involved in such a seemingly simple experience as identifying a familiar object like an orange. Our inspection of the orange sends information to separated regions of the brain by different pathways. Upon the visual cortex depends our sensing a round, colored image. Sensing of a certain texture, firmness, and a spherical shape depends upon the area of general sensibility. Still other areas of the brain are concerned with consciousness of the smell and taste of the object. But so far we have only a miscellaneous collection of purely sensory data. We do not have the concept *orange* until this information has been compounded into a complex of all the sensory ingredients, each in proper relationship to the others. The bare data furnished by the senses require interpretation, fitting-together. This function is probably carried out largely by the association areas. If we try to be more precise about how this is done, about the specific pathways involved, we come to an abrupt halt. We cannot say explicitly just what takes place. We must agree with Karl Lashley that, in analyzing brain function, we "often have no choice but to be vague or to be wrong."

3. *Interpretation of symbols.*—To a large extent our thought-processes deal with words, speech, language—with symbols which must be manipulated and interpreted, probably in the association areas. Occasionally, disease or destruction of these regions disturbs this function, producing what is known as *aphasia*. In one kind of aphasia the subject may lose the ability to understand written words, though he understands spoken words very well. The inability to interpret visual data which may formerly have been familiar to him is not caused by a defect of vision proper but of interpretation in the association areas of what is seen. The situation is somewhat comparable to the common normal experience of seeing words in a foreign language. A normal individual may be able to see the word *hustru* perfectly well; it "registers" in the occipital cortex but carries no meaning. The association areas cannot interpret it, in this case, not because of any disease of those brain areas but because they lack the proper memory processes necessary to make the interpretation— unless the reader is familiar with a language not unknown to the authors and recognizes this as a foreign word for "wife."

Faulty function of other association areas may prevent interpretation of spoken words, while reading a printed page is done with facility. On the motor side, aphasias may make either writing or speech impossible, although the subject may be fully aware of what

he wants to say or write and suffers from no paralysis of the muscles
of speech or the hands.[14]

C. THE CORTEX AS A COMPLEX OF REFLEX CENTERS

Emphasis upon localized function in the cerebral cortex, upon
motor control and sensations, is likely to engender the notion that
behavior which depends upon the cortex is something unique. There
is a popular tendency to separate responses into the automatic
reflexes, on the one hand, and responses involving sensations, consciousness, and volition, on the other hand. There is perhaps less
justification for such distinction than seems apparent on the surface.
In a large measure, the muscular responses we make voluntarily are
as truly automatic and predictable as is the knee jerk or the flexion
reflex.

The point is that the afferent pathway does not encounter a dead
end at the sensory cortex, or even in the association areas. The process of integration of sensations into a meaningful whole is not the
final stage but leads generally to an intelligent, or at least a fairly
well-integrated, response to the situation. When one sees an orange,
the conception of "orange" does not complete the process; the orange
is eaten or is put aside; if nothing else, the mouth "waters." Probably
in every case the activity associated with sensations is transmitted
eventually, by a complex reflex pathway, through numerous association areas, to motor or secretory or other efferent systems.

Probably, also, motor activity arises *de novo* in the motor cortex, or
even elsewhere in the brain, far less than is popularly believed. To a
large extent it is probably merely the efferent side of a highly complicated reflex, initiated by afferent stimulation somewhere in the body.
True, reflexes through the cortical centers are far more complex,
involving long nerve pathways and highly complicated interconnections between many centers. They are more flexible also, in that they
are modifiable by training. Yet they are largely automatic and predictable. Given a physically, emotionally, and mentally healthy individual with a certain training, we can predict with fair certainty
what many of his "voluntary" responses will be. To the question,

14. There is a localized area of the left frontal lobe—Broca's speech area—whose destruction commonly causes an aphasia in which ideas cannot be expressed in speech, though there
are no actual paralysis of the organs of speech and no interference with written expression.
However, such "pure" aphasias are uncommon. Patients more commonly display a mixture of
more than one type. The lines of demarcation between different varieties of aphasia are not
clearly drawn.

"Which way is Fifty-fifth Street?" almost every person who happens
to know will indicate the direction, just as surely as he would flex his
leg if you deliberately stepped on his toes. In the latter case, one
might safely predict certain specific responses of anger or surprise as
well, via cortical reactions.

Responses of this general sort, which, although acquired, are nevertheless automatic, we designate by the name *habit*. Habits are most
useful to the animal. The dull routine activities of life are carried
out largely automatically. Once walking or writing or talking have
been so well learned as to become habitual, it is no longer necessary
to think about the specific muscular acts involved. They require
scarcely more attention than the automatic beating of the heart or
the movements of breathing. In this way, the brain is largely freed
for the other more intricate tasks of meeting and adjusting to new
environmental situations.

In this discussion it should be clear that there are enormous differences in complexity between a simple reflex such as the knee jerk
and a learned, complex motor response such as driving a car. These
may be alike in that they are equally predictable. But the degrees of
complexity are so great that perhaps they should be considered as
qualitatively different phenomena. Many of the activities involving
the cortex of the brain serve to control or modify the simpler reflexes.
At any rate, all gradations of complexity are to be observed from
simple reflex to complex behavior pattern.

We must also recognize that some activity may arise in parts of
the brain in the absence of afferent nerve activity initiating such
action. We know, for example, that the respiratory center seems to
discharge impulses automatically, apparently from local chemical
stimulation. Perhaps some more complicated activity involving
"thinking" also originates within the brain in the absence of stimulation by afferent nerve pathways.

D. ELECTRICAL ACTIVITY—BRAIN WAVES

We have become familiar with the electrical changes—action
potentials—developed by nerve fibers or nerve trunks in which nerve
impulses are associated with relative electrical negativity. Related
to these comparatively simple electrical changes are those developed
by the entire brain, representing a summation of the action potentials of the many millions of neurones constituting the brain. These
changes may be detected, with a suitably sensitive instrument—the

electroencephalograph—on the surface of the scalp, just as the electro-cardiograph (see p. 148) can detect, on the surface of the body, the electrical changes developed by the muscle fibers of the heart. The electrical changes recorded from the head are commonly called *brain waves*.

These have been intensively studied in man in health and disease. The frequency, shape, and size of the waves are modified characteristically by closing the eyes, by sleep, by fixed mental attention, and by a number of other normal states. Abnormalities of brain function frequently produce modifications in these waves, so that brain-wave tracings may reveal the presence of disease of the brain, such as in convulsive seizures of otherwise unknown origin or certain tumors of the brain or destruction of brain tissue from other causes.

VIII. LEARNING

To a considerable extent the cortical functions of integration and interpretation, discussed above, depend upon learning—upon the modification of behavior by experience. There is little or no learning, or retention of learned material, without the cerebral cortex.

The problems of learning occupy a prominent place in the science of psychology. Here we can deal only with some of the elements of this process. The several factors we shall discuss are not so much distinct components of the learning process as they are lines of approach to the problem. Each represents an emphasis upon one or another aspect of the problem.

Investigations in this field are fraught with difficulties, and the results are often confusing and contradictory. The unknowns at present are many, and the advances which have been made are but feeble beginnings.

A. TRIAL AND ERROR

Certain lines of experiment have emphasized the role of trial and error in learning. Problems are set to animals, often in the form of a box or cage from which they learn to escape by pressing levers or pulling strings. Or the problem may be learning to run a maze or labyrinth. Incentive to learn is afforded by rewarding successes, as with food, and punishing failures, as with a mild electric shock. Important laws of learning in many species of animals have been unearthed by this technique. For example, completely random movements are many, at first, until apparently accidentally the animal performs correctly. Repeated trials on successive days show that the random responses become fewer in number, and the animal comes to perform the required act in less and less time. These findings are embodied in the learning curve (see Fig. 161), which applies fairly generally to learning, especially in animals.

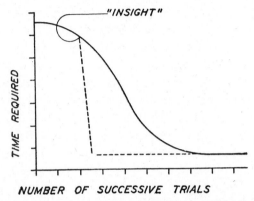

FIG. 161.—The curve of learning. Animals escaping from a problem box take less and less time for a perfect performance in successive trials (*solid curve*). Some investigators think that animals (especially monkeys and man) gain a sudden "insight" into the problem and thenceforth perform with a minimum of waste motion (*dotted portion of curve*).

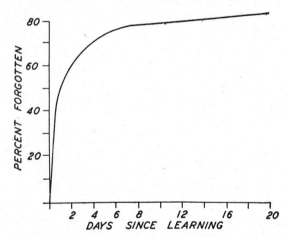

FIG. 162.—The curve of forgetting certain learned material (memorized nonsense syllables) in man. Note that more is forgotten the first day than in the succeeding three weeks.

Forgetting has also been studied in detail. In certain early studies on man it was found that forgetting of memorized material was very rapid on the first day and continued at a gradually diminishing rate, as shown in Figure 162. It has been pointed out frequently that, although forgetting of certain learned material is fairly rapid, and only a small fraction of such learned material is retained, this residue is nevertheless sufficient justification for the pain of learning. Moreover, it has been found that relearning is relatively very rapid as a result of having learned once before.[15]

B. Learning by "Getting the Idea"

As contrasted with the gradual, slow progression to final complete learning, characteristic of lower mammals in the experiments outlined, some of the higher mammals, including apes and man and even dogs, display a type of learning involving what has technically been called *closure*. We may refer to this more simply as "getting the idea." After a few random trials at a problem, an ape or a man will suddenly get the "hang" of the thing, grasp the situation with sudden "insight," and in all trials from then on achieve 100 per cent success. For example, the problem set to a monkey may be to learn to open Door A on one day and Door B on the next. Opening the right door is rewarded with food; opening the wrong door results in mild punishment. After several days, with a number of trials, in which the correct door is at first opened purely by chance, the animal seems suddenly to get the idea of the alternation and from then on opens the right door on the right day without fail. How this modifies the learning curve is depicted in dotted lines in Figure 161.

Such learning is not necessarily of a nature different from that involving trial and error. Some investigators contend that what is termed *insight* is simply a marked acceleration of the learning process —that the higher mammals learn *faster* but in no essentially peculiar manner.

C. The Role of Ideation

In the learning of higher mammals, manipulation of ideas often supplants manipulation of actual objects in the solution of problems. A monkey attempting to reach food which has been suspended overhead beyond reach may aimlessly move an available box about, or unsuccessfully manipulate a short stick which has been placed within reach. After a time, however, he may combine the *ideas* of climbing on the box and reaching with the stick and then successfully perform

15. The rate of forgetting and the shape of the curve vary considerably, depending upon the nature of the learned material and how thoroughly it was originally learned, or "overlearned." Memory—the opposite of forgetting—depends, like learning itself, upon the integrity of the cortex. Memory is considerably impaired by a variety of abnormal cortical states.

the required combination of movements, eliminating further pure trial and error. In a sense, the process becomes one of trial and error of ideas rather than of overt acts. This sort of substitution of ideas for overt activity is characteristic of learning and thinking and reasoning in the higher primates.

D. Conditioned Reflexes

In the foregoing the learning process has been more or less accurately described in terms of the total behavior of the organism. Nothing has been said of the specific neural mechanisms involved. What nerve pathways or brain areas are brought into play, and how? The investigations of the Russian physiologist, Pavlov, have at least made a start toward determining the neural mechanisms involved in the process. Pavlov emphasized the fact that learned behavior is built upon inherited behavior. The inborn reflexes he called responses to *unconditioned stimuli*. The process of learning consists, in large part, of substituting new *conditioned stimuli* for the normal, inborn, unconditioned ones.

The classic example is the conditioning of salivation. Saliva is reflexly secreted normally when food is chewed. Normally, there is no salivary response to a sound like the ringing of a bell. But, if the bell is always rung when food is given, the animal will gradually develop a response to the sound alone and eventually will salivate at the sound even if no food is given. We are prone to say that the animal has learned to associate the bell and the food or that the animal learns that the ringing of the bell *means* that food is forthcoming. Particularly the last statement, however, carries more of interpretation than the experiments warrant. We need not refer to what the animal may be thinking or to what the stimuli may mean to him. This we do not know. We only know that a stimulus which previously was neutral, as far as salivation is concerned, has now become an effective stimulus. We are justified in saying no more than that a new conditioned stimulus has been substituted for an unconditioned stimulus. A *conditioned reflex* has been established.

Pavlov would say of the learning curve that it represents the rate of establishment of the conditioned reflex—that repetition "fixes" the response. Learning by the establishment of conditioned reflexes becomes more complicated as education, in the broad sense, proceeds. One conditioned reflex may serve as the basis for another. The smell of food may become a learned, conditioned stimulus to salivary secretion, by repeated association of smell with taste, which is the unconditioned stimulus. This acquired reflex may then become the basis of a "second-degree" conditioned reflex. Sight of food may be so frequently associated with its odor that sight itself becomes an adequate stimulus. Ultimately, a complicated many-layered series of conditioned reflexes is acquired in this way.

E. The Function of the Cortex at Large

Lashley's work on learning in rats has shown that the cortex as a whole, in a way not hitherto suspected, is involved in learning. The work on localization of function in the cortex tends to foster the idea that learned responses to visual stimuli would involve chiefly the visual cortex; responses to auditory stimuli, the auditory cortex; etc. Lashley confirmed this by finding that removal of the visual cortex abolished the learned responses to certain visual stimuli involving brightness discrimination in rats. However, these rats were now just as capable of learning the very same problem as normal rats. In the absence of the visual cortex, other areas of the brain are apparently fully able to substitute for it in learning.

Any part of the brain, apparently, is capable of this substitution. Far more important than what areas were left behind when the occipital lobes were removed was the total quantity of cortex preserved. The ability to relearn was directly proportional to the total mass of cortex which was left behind in the brain, no matter what areas this included. In the relearning process the entire mass of the cortex seems to be involved.

Though Lashley's experiments are conclusive enough for the rat, it still remains a question as to how extensively these findings apply to human brain function—how much substitution of one part of the brain for another in the performance of cortical activity is possible in man.

IX. SLEEP

We have avoided discussion of the problem of consciousness as beyond the scope of this work. Little more is known about the essential nature of consciousness than that it is dependent upon the cerebral cortex. But we shall deal briefly with a special form of the problem, that of sleep—the normal periodic loss of consciousness of a special nature, characteristic of many animals. Most studies on sleep have dealt with the factors inducing it and the bodily changes resulting from it. Obviously, a complete understanding of the mechanisms involved must await further information about consciousness itself.

The dependence of sleep upon the feeling of fatigue seems at first sight axiomatic, and yet close examination of the relationship reveals certain difficulties. We have already observed the special type of fatigue associated with repeated muscle contractions and the accumulation of waste products (notably lactic acid). We have seen also that the central nervous system is, in general, more susceptible to fatigue than are peripheral systems like the nerve-muscle complex. These findings have led to the erroneous guess that during the waking state, with all its manifold activities, there is a slow accumulation of waste products of a mildly toxic character which especially affect the central nervous system, including the higher centers of the brain. The gradually increasing concentration of these hypothetical products has been thought to induce the state of sleep. During sleep, in turn, the toxins were believed to be disposed of in some way. Except in cases of prolonged sleeplessness, however, direct evidence for the appearance of any such sleep-inducing fatigue products is scanty.

A more important factor, at least in the induction of sleep, seems to be the amount of stimulation to which the sense organs are subjected—the number of impulses in the sensory systems which bombard the sensory centers of the cerebral cortex. Darkness and quiet, which eliminate stimulation of the visual and auditory systems, are conducive to sleep apparently for this reason. Sleep may be interrupted or the onset of sleep delayed or prevented by stimulation of any afferent system which affects the conscious centers: from within by fulness of the bladder or rectum, pain, hunger, or thirst; or from the outside by noise or light. It is possible to fall asleep or to stay asleep quite apart from fatigue, provided such stimulation is absent. On the other hand, if sensory stimulation is sufficient, sleep can be prevented even if fatigue is carried to the point of death.

Yet there is no gainsaying the fact that muscular fatigue in some way favors sleep. What tangible components of fatigue can we fix upon as responsible for this effect?

Contraction of muscle or even the normal state of tone in muscle initiates impulses in the proprioceptive system[16] which "register" in the cerebral cortex as sensations. Relaxation of muscles diminishes this stimulation, and fatigue, it has been suggested, favors relaxation. In fatigue we are less inclined to move our muscles, and even

16. Causing muscle and joint sense (see p. 438).

muscle tone is diminished. Perhaps, then, the role of fatigue is simply the production of a greater muscular relaxation, thereby directly diminishing the stimulation of the afferent proprioceptive system. This would favor sleep in essentially the same way as darkness or quiet.

But, whatever may be the factors which induce sleep, what is sleep itself? A number of the bodily changes that occur in sleep have been elucidated by experiment. There is a general depression of tissue and organ activity, muscle tone is greatly diminished, and the circulation and respiration are slowed. Reduced tissue activity is reflected in a lower metabolic rate, which in sleep is often some 10 per cent below the "basal" level.

The relationship of body-temperature changes to the sleeping-and-waking cycle has long aroused interest. We have seen that the body temperature is lower during sleep than in the waking state, fluctuating through a range of about 2° F. It is lowest in the early morning hours (2:00–5:00 A.M.) and highest in midafternoon. If one regularly sleeps in the daytime and remains awake at night, the temperature cycle may be reversed, so that the high point is now in the early morning hours.

Just what the neural mechanisms of sleep are, and what the fundamental changes in the brain may be, is uncertain. A "sleep center" in the brain stem has been postulated, because an area has been described whose stimulation induces a sleeplike condition. There is more evidence for the existence in the brain stem of a center which continually excites the cortex, keeping it active and the individual awake. Experimentally separating this center from the brain above it causes electrical changes in the cortex characteristic of sleep. Stimulation of the center reverses these changes in animals "asleep" under an anesthetic, so that the electrical activity of the brain resembles that of the waking state. Considerable further study is needed to resolve the problem of the basic nature of sleep.

In a large measure, nocturnal sleep is a habit. The diurnal cycle must be learned by the infant human, which at birth displays a polyphasic cycle with several periods of sleeping and waking throughout the twenty-four hours. Most animals show this same sort of polyphasic cycle, but many of them acquire the diurnal habit.

Much attention has also been paid to the question of the optimum amount of sleep. This is of practical significance in human beings. No very positive statements can be made in this regard. The requirements vary from individual to individual, although about eight hours is a good average figure for most adults. Within limits, habit can again affect the requirements. It is possible to learn to do with somewhat less sleep than eight hours.

We do not know to what the recuperative effects of sleep are due. Contrary to popular opinion, these recuperative effects are not most prominent just upon awakening, or even in the first few hours of the day. The ability to perform acts requiring some skill is generally no better upon awakening than just before retiring. Speed and accuracy in such tests as multiplication, code transcription, and mirror drawing are greatest around noon, or in the early afternoon. Recall that at this time the body temperature is at approximately its highest level for the day.

We are also unable to tell, at present, why sleep is necessary. Experiment demonstrates that it is indispensable for life. An animal kept continuously awake for several days will die.

X. FUNCTIONS OF THE CEREBELLUM

Like the cerebral hemispheres, the cerebellum (see Figs. 159 and 160, p. 448) is not segmented; it is a suprasegmental structure. A relatively late evolutionary acquisition, developing parallel with the evolution of the cerebral cortex, it is roughly proportional in size to the complexity of movement of the skeletal muscles. We shall see that its chief function relates to muscle movements.

The cerebellum is an outgrowth of the hindbrain, remaining connected to the brain stem by large nerve fiber tracts. It is roughly divided into the midportion (the *vermis*) and the two *hemispheres*. The surface is gray, with underlying white fibers. Deep in the substance of the organ are several gray nuclei—synaptic relay stations along the various pathways of the cerebellum.

In internal structure we fail to find the variations in different parts of the cerebellum that characterized the structure of the cerebral cortex. Correspondingly, we discover no high degree of differentiation of function in the different regions.

Consciousness seems not to be experienced in the cerebellum. True, certain afferent pathways converge to the cerebellum, but activity in these tracts never produces actual sensation. Stimulation of the exposed cerebellum of humans causes no sensations, and destruction of it by disease induces no sensory defects.

Electrical stimulation of the surface produces some motor effects, largely in the nature of a modification (augmentation or inhibition) of muscular movements already under way, elicited by other means. There is considerable localization of function: the midportion influences chiefly the muscles of the head, neck, and trunk; each hemisphere affects chiefly the muscles of the limbs on the same side of the animal.

Destruction of the organ experimentally or by disease results in no real paralysis, as does destruction of the cortical motor area; but it is said to cause a generalized diminution in muscle tone and, according to some investigators, in muscle strength as well. The most striking effect, however, is the impaired co-ordination of muscle movements. When the cerebellum is damaged, muscle movements are jerky and decidedly ineffective. A pigeon whose cerebellum is removed is unable to walk or fly, though there is no actual paralysis. When it attempts to fly, the wings simply thrash about in an aimless manner, in marked contrast to the almost normal flying movements seen in a pigeon lacking a cerebral cortex.

Less extensive damage to this part of the brain in man produces less generalized but no less definite defects of muscular movement. The smooth, sweeping, continuous movements of the normal individual are dissociated into their constituent parts. If the hand is raised to the face, it is done jerkily, in stages. The arm may be first flexed at the elbow, then brought forward, and then the entire arm raised. Fine movements, requiring delicate co-ordination, are impossible. If a small object is to be picked up, the hand reaches for it, misses, is pushed past the object, fumbles about, and may ultimately clumsily succeed. Alternating movements, such as closing and opening the fist, can be done only very slowly and with much hesitation. If the muscles of vocalization are affected, speech is thick and slurred. The individual may reel drunkenly when he tries to walk and may fall. He may even be unable to stand quietly without swaying from side to side.

All this signifies that the chief role of the cerebellum is the modulation of muscle motion, lending refinement to the movements and to the tone changes associated with maintenance of posture. Normal voluntary movement, then, requires not only the activity of the motor cortex but modulation by the cerebellum. Between the cerebellum and the cortex, and from the cerebellum to the neurones of the spinal motor nerves, are large nerve fiber tracts involved in carrying out these functions.

XI. ABNORMALITIES OF BRAIN FUNCTION

A. Destruction of Brain Tissue

A variety of factors may operate to destroy or damage brain tissue, producing more or less serious derangement of brain function. There may be faulty embryonic development, such as failure of the cortex to develop, or interference with the drainage of cerebrospinal fluid from the brain ventricles, so that it accumulates under pressure, destroying brain tissue and even causing an enlargement of the skull. A diseased blood vessel of an individual with high blood pressure may suddenly rupture, or a vessel in a normal individual may be ruptured by a blow; the escaping blood may tear and destroy neurones and fiber tracts.

Various infectious agents, notably, syphilis, can produce extensive damage of tissue. Several kinds of tumors, cancerous and non-cancerous, damage the brain substance in their growth or injure it by pressing upon the neurones or their blood supply.

The effects produced depend, of course, upon the extent and the location of the damage. If the cerebral cortex as a whole fails to develop, there is complete idiocy. A tumor of the cerebellum may produce any or all of the serious defects in muscle function described above. An injury to the motor cortex may, by stimulation, cause convulsions and one kind of epilepsy.[17] A ruptured blood vessel may destroy the main motor pathways, producing a paralytic stroke. Syphilis may extensively damage the association areas and effect the marked personality and emotional upsets we refer to as insanity.

B. "Functional" Nervous Disease

The most baffling disorders of the personality are the so-called "functional" disturbances,[18] in which there are serious derangements of function without chemical or structural changes in the brain tissue which have as yet been detected. It is conceded that such changes may exist. However, one working hypothesis is that the parts of the machine are all there but do not operate properly. The manifestations may be identical with those associated with the various kinds of brain damage, but there is no apparent detectable damage.

17. Epileptiform convulsion can be produced experimentally, also, by strong stimulation of the exposed motor cortex.

18. We do not refer here to the malingerer, who believes that he is ill, though in some instances it may be difficult to distinguish true functional abnormalities of the milder types from malingering.

The problem is more than medical; it is of far-reaching social, eugenic, and economic importance as well. Persons with nervous diseases and mental illnesses occupy more hospital beds in this country than do sufferers from all other illnesses combined.

It is granted that the term "mental illness" is an inadequate collective designation of these defects. Varying degrees of disintegration of the personality, of emotional instability, and of inability to make social adjustments are often the most prominent features. An individual who is highly intelligent by all objective standards may be quite unable to integrate his emotional drives in such a way as to enable him to fit himself into the social complex of modern life in a manner satisfactory to himself and useful to others.

The *neuroses* are relatively mild and exceedingly common[19] disorders, with a great variety of manifestations, including anxieties and fears which seem unreasonable and unfounded; excessive shyness or aggressiveness; undue concern about the body and its parts, so that common mild discomforts are sensed as very uncomfortable or even painful. The emotional disturbance may actually affect the body deleteriously, so that there may be digestive disturbances or cardiac irregularities, for example. Characteristically, the patient— and sometimes the doctor—focuses attention upon these outward signs of the disorder, neglecting the true cause. Neurotic manifestations often result from deep-seated inner conflicts, the existence of which may not be realized or even suspected by the disturbed individual. For example, a hostility against someone deserving love (e.g., a parent or sibling or an offspring) may arouse such a sense of guilt that the hostility is deeply repressed. Although repressed and unrecognized, the hostility may be emotionally disturbing and responsible for the neurotic manifestations.

To an extent not popularly appreciated, the more serious disorders of "the mind" and the emotions—the *psychoses*—represent an exaggeration of perfectly normal tendencies. Who among us has not at some time or another experienced the inexplicable swings of mood from depression, when everything seems hopeless, to elation, when all things are possible? In the emotionally unstable these mood swings may be so marked as to produce the picture of the *manic-depressive* psychoses, with months of depression or excitement.

Who has not at times felt himself to be misunderstood, mis-

19. It has been estimated that at least half the people who seek medical advice are in this category.

treated, imposed upon? In some of the mentally diseased this normal tendency may become the dominating influence in behavior—every man's hand is turned against him; friends, relatives, and physicians seek only to injur or kill him; he is ridden with fears and terrors.

The pleasant daydreams we all indulge in may become more real to the sick than the true realities, to which the normal individual can quickly return at will. The dream world, fashioned to the heart's desire, becomes the only reality, and any intrusions into it are resented and rejected. There is a complete escape from the harsh demands of real life into the security of a life which makes no demands, requires no difficult adjustments, expects nothing.

All imaginable degrees of malfunction are encountered, as in any disease: from barely perceptible, mild exaggeration of one or another of these tendencies, offering little or no interference with normal daily activities, to complete disorientation in time and space and in social relationships. There are likewise all degrees of amenability to effective treatment: there may be anything from complete permanent restoration to a normal, useful, happy life, to a progressive decline which cannot be checked in any way.

Numberless difficulties bar the way of the *psychiatrist*, who specializes in mental and emotional disease, in his attempts to understand the underlying mechanisms at fault and to control these conditions. Yet advances, though often slow, are being made on many fronts.

Even the layman is changing his attitude toward people affected by these personality disorders. An intelligent, sympathetic desire to understand the conditions is replacing the sense of shame or derision, not to mention amusement. The words "crazy" and "lunatic" are properly disappearing from the popular vocabulary. More and more, the neuroses and psychoses are considered to be true illnesses, just as are diabetes or heart failure.

CHAPTER ELEVEN

SENSORY MECHANISMS

I. THE SENSES

II. VISION

  A. Physiological anatomy of the eye
    1. Protection of the organ 2. Eye movements 3. Internal structure
  B. The components of vision
  C. Image formation
    1. The lens system 2. Focusing for different distances 3. The mechanism of accommodation 4. Seeing the inverted image upright; projection 5. Errors of refraction; Snellen's type 6. Cataract 7. The iris and its functions 8. Spherical aberration
  D. Perception of distance and depth
    1. The neural basis 2. Judgment of distance 3. Stereoscopic vision
  E. The light receptors
    1. Achromatic vision by the rods 2. Stimulation of the rods and cones by light 3. Color vision by the cones
  F. Injuries and infections of the eyes

III. HEARING

  A. Structure of the ear
    1. The outer and middle ear 2. The cochlea of the inner ear
  B. Stimulation of the organ of Corti by sound
    1. Role of the basilar membrane 2. Discrimination of pitch
  C. Defects of hearing

IV. THE SENSE OF BALANCE AND ROTATION

  A. The sacculus and utriculus
  B. Body-righting reflexes
  C. The semicircular canals

V. RECEPTION OF CHEMICAL STIMULI

  A. Taste
    1. The taste buds 2. Specificity of the taste buds
  B. Smell
    1. The olfactory endings 2. The quality of odors

VI. THE SKIN SENSES

VII. PAIN FROM THE INTERIOR OF THE BODY

VIII. WEBER'S LAW OF SENSATION

I. THE SENSES

Careful consideration reveals the inaccuracy of the popular notion that man is equipped with but five senses—vision, hearing, taste, smell, and touch. Considerably more than touch is experienced from skin stimulation. Sensations of pressure, heat, and cold are also possible from the surface of the body, and pain is likewise a definite sense which may arise here as in many other parts of the body. The "five senses" exclude the important muscle and joint sense—proprioception—which furnishes information about the tension of muscles and the position of the limbs and other body parts. There are also certain senses dealing with the balance of the body and with experiencing rotation. Hunger and thirst are true sensations from the interior of the body, as are also the less readily definable experiences, such as nausea, which may also be initiated from internal organs. The conscious processes induced by tickling have specific characteristics compelling us to speak of a tickle sense, at least provisionally. Similarly, rather distinct and peculiar conscious processes are initiated from some of the organs involved in sexual behavior.

To be stimulated, any *receptor* must be directly acted upon by some energy change. Certain of the more primitive receptors, such as touch, can be stimulated only if the source of the energy is in actual contact with the surface of the body or, at most, only a short distance away, as in sensing an explosion by touch. Other receptors, such as the eye and the ear, are called *distance receptors*, since they can be affected by energy originating at a great distance from the body. Light-emitting objects such as stars can be seen from millions of miles away.

The fundamental difference here is not really in the sense organs themselves. In any case, certain energy changes must be produced at the various receptors. Although they may come from a great distance, light waves must impinge upon the retina to stimulate it. Distance reception depends rather upon the nature of the energy to which the distance receptors are sensitive. Light or sound waves can be transmitted through air for miles, while forms of energy to which certain other receptors are sensitive cannot be transmitted such great distances.

Further, it is only by individual learning that we acquire the ability to judge the source of the stimulus. Presumably, an infant cannot distinguish between the distance from his body of an object he can touch, or one he can smell, or an object—the moon, for example—which he can see. Experience teaches him that the touchable object is close at hand, the moon far away. The aromatic object, he learns, may be a few yards away.

The distance receptors and the ability to interpret sensations received through them endow the higher animals with considerable advantage over more primitive animals. Such receptors enable the animal to obtain far more accurate information about the external world than is otherwise possible.

II. VISION

A. PHYSIOLOGICAL ANATOMY OF THE EYE

1. *Protection of the organ.*—The human eye is well sheltered within its bony socket. Skull bones project forward even beyond the eyeball, above, below, and on the nasal side. Protection is afforded

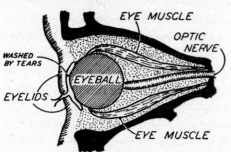

FIG. 163.—The eyeball in its socket, in vertical section. Protection is afforded by bone (*shown black*), the eyelids, tears, and the fat (*shown dotted*) in which the eye is imbedded. The muscles shown are the one which rotates the eye upward and the one which rotates the eye downward. There are four other muscles attached to each eye.

the sensitive anterior surface by the constant flow of tears, secreted by the tear glands lying one above each eye, draining from the eye through a duct into the nasal cavity. The tears tend to wash the eye free from irritating dust particles and to keep its surface moist. The eyelids provide an extra protection. They close reflexly when the eyeball is touched or even when any object is suddenly moved close to the eye.

2. *Eye movements.*—Each eye is equipped with six muscles attached at one end to the bony wall of the socket and at the other to the tissues of the eyeball (see Fig. 163). These muscles can move the anterior pole of the eyeball to the left or the right, up or down, and to a slight degree rotate the eye upon a horizontal, front-to-back axis. The innervation of the muscles is such that the two eyes are normally moved simultaneously and to the same degree, always pointing together in the same direction.

3. *Internal structure.*—The eyeball is approximately spherical in shape (see Fig. 164). Three distinct layers are recognizable in its

wall: the outer layer (the *sclera*), composed of tough fibrous tissue; the intermediate layer (the *choroid coat*), containing pigment and some of the vessels supplying the eye with blood; and the inner coat, the *retina*, containing *rods* and *cones*, which are specialized receptors sensitive to light. The rods and cones connect by intermediary neurones and synapses with the fibers of the optic nerve.

In the anterior portion of the eyeball these layers are modified. The outer coat becomes transparent, constituting the *cornea*. The middle coat forms the pigmented *iris*, with its central orifice, the *pupil*, and the muscle-containing structure (the *ciliary body*), from

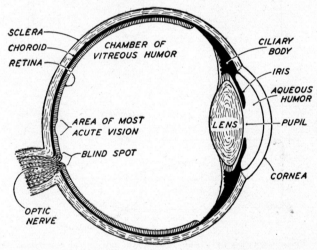

FIG. 164.—Structure of the eyeball, shown in horizontal section. The left eye is shown; the optic nerve lies a little to the nasal side of the midline of the eye (see also Fig. 166).

which is suspended the *crystalline lens*. The retinal layer stops just short of the lens attachment. The large chamber back of the lens is filled with a transparent material of a jelly-like consistency (the *vitreous humor*). The chamber in front of the lens[1] contains a watery fluid (the *aqueous humor*).

B. THE COMPONENTS OF VISION

Vision involves a number of factors—perception of light, form, color, depth, and distance. Some sort of sensitivity to light is a fairly common property of protoplasm. Many of the unicellular forms react positively or negatively to light—moving toward or away from

1. There are really two chambers here—the *anterior chamber* between iris and cornea and the *posterior chamber* between iris and lens attachment. Anterior and posterior chambers communicate through the pupil.

it. Early in evolution, in animals of the flatworm type, special light receptors are encountered. Some perception of form may be accomplished by the compound eye of certain invertebrates, but form perception is probably most accurate in the case of the light-focusing, image-forming eye of vertebrates. Color perception is common to all vertebrates and is found even in some invertebrates, notably, in bees. Some visual perception of depth and distance is

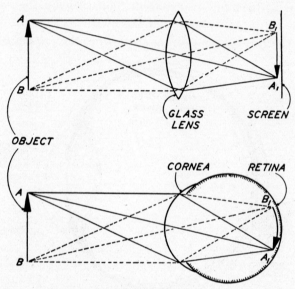

FIG. 165.—Image formation by a glass lens (*above*) and by the eye (*below*). All the light rays from A come together at A_1; all the light rays from B come together at B_1. Hence an inverted image (A_1B_1) of the object (AB) is formed.

apparently possessed by all vertebrates, but this ability reaches its highest development with *stereoscopic binocular vision* in the primates.

C. IMAGE FORMATION

1. *The lens system.*—Form perception is possible because true images of objects looked at are formed upon the retina. This is effected by the lens system of the eye in exactly the same way that any lens will bring rays of light to a focus and form an image of an object emitting light. In Figure 165 the divergent rays of light emitted from point A are bent (*refracted*) by a lens so that all the rays converge to point A_1. Rays from point B are focused at point B_1, and the light from points between A and B converges to points between

A_1 and B_1. As a result, an inverted image A_1B_1 of the object AB is formed upon the screen. In the eye the bending of the rays is effected to a large extent by the surface of the cornea rather than by the lens. The lens plays a minor role in the total refraction. Its importance lies in the fact that it can *change* the focusing power of the eye by changing its curvature.

2. *Focusing for different distances.*—In several respects the eye resembles closely the optical system of a camera. In both there are a lens system whose focus can be changed, a diaphragm which regulates the quantity of light admitted, and a light-sensitive plate upon which the image is formed.

In a camera changes in focus are effected by moving lens and plate closer together when distant objects are photographed and farther apart for near objects. Conceivably, the eyeball might have been similarly constructed—to lengthen or shorten, depending upon the distance of the object viewed. In fact, in some birds there is apparently just such a mechanism. But, for the most part, and entirely so in man, adjustments are made by changing the focusing power of the lens—by changes in its thickness and curvature, making it a "stronger" or "weaker" lens, as the case may be. The closer the object, the thicker the lens becomes. Changes in the lens thickness, called *accommodation*, are effected by the action of tiny muscles (the *ciliary* muscles) lying in the middle coat of the eye at the place where the lens is attached.

3. *The mechanism of accommodation.*—The ciliary muscles are so arranged (see Fig. 166) that their contraction loosens the fibers supporting the lens, permitting that structure to bulge of its own elasticity. Relaxation of the muscles causes tension to be placed upon the fibers, and the lens flattens, when distant objects are viewed.

The degree of accommodation which is possible depends upon the contractile power of the ciliary muscles and the elasticity of the lens. In the young it is possible to cause such a bulging of the lens that sharp images can be formed of objects at a distance of only a few inches from the eye. This distance we term the *near point* of distinct vision. The eye is unable to accommodate for objects closer to the eye than this, and only blurred images are formed. With increasing age, the elasticity of the lens decreases. Contraction of the ciliary muscles, loosening the tension on the lens, now fails to result in as great a bulging of the lens. Stated otherwise, the near point recedes with age. For this reason, an older individual may require convex glasses (i.e., thicker at the center than at the edge) for near vision—in reading—artificially lending aid to the relatively inelastic lens system.

4. *Seeing the inverted image upright; projection.*—Invariably the question arises: How is it that, though the retinal image is inverted, we nevertheless "see" objects looked at as upright? The problem is really one of cerebral interpretation of sensory data. In general terms, we interpret stimulation of the upper part of the retina as

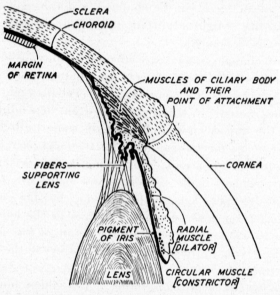

Fig. 166.—Section of the eye showing the relationships of the ciliary body to lens and iris. The muscles of the ciliary body contract in accommodation. This pulls the ciliary body toward the fixed point of attachment of the muscles, which, in turn, releases the tension upon the fibers supporting the lens. The lens then bulges of its own elasticity. The arrangement of the muscles of the iris which dilate and constrict the pupil is also shown (see also Fig. 164).

being produced by an object or a point in the lower field of vision. Light falling upon the left side of the retina is interpreted as coming from the right field of vision. This can be demonstrated strikingly by a simple procedure known as *Scheiner's experiment.*[2]

The point of a pin is looked at through *two* pinholes placed close together on an opaque card held just in front of one eye, the other eye being closed. When the point of the pin is accommodated for through the two pinholes, a single point is visualized. But when one focuses for an object at some distance *beyond* the point of the pin, and

2. After Christopher Scheiner, early-seventeenth-century German physicist.

on a line with it, the rays of light from the pin which get through the two holes will now tend to come to a focus *behind* the retina (see Fig. 167). Two *blurred* images of the pin will be seen because light from the pin is made to form two images upon the retina, at P_1 and P_2. If now the left-hand hole in the opaque card is covered, image P_2 will, of course, no longer stimulate the retina. But the cutting-off of this image on the *left-hand* part of the retina will cause the subjective disappearance of the *right-hand* pin, rather than the left, of the two previously seen out in space. This means that light falling upon the retina at P_2 is interpreted as coming from the *right-hand*

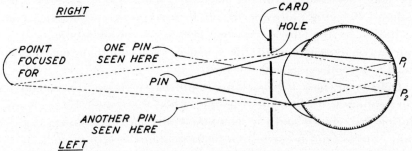

FIG. 167.—Projection, as demonstrated by Scheiner's experiment. If one focuses for a distant point and observes a pin (placed closer to the eye) through two holes in an opaque card, two pins will be seen, since blurred images of the pin are formed at two points (P_1 and P_2 on the retina. When the *left* hole is covered, image P_2 is no longer formed on the retina, and the *right* pin disappears. This indicates that images formed on one side of the retina are interpreted as being caused by objects in the opposite field of vision. Dotted lines show path of light from distant point to retina. Solid lines show path of light from pin to retina. Dot-and-dash lines show the direction of "projection."

field of space, actually, from along the extension of a line connecting the point of the retina stimulated with the very central point (the *nodal point*) of the refracting system of the eye.

This interpretative phenomenon is referred to as *projection;* we "project" images formed upon the retina out into the field of vision in this manner. The capacity to project correctly retinal images or points to the actual source of stimulation out in space is acquired by training and practice. We learn by experience that, when an image is formed on the left side of either retina, we must reach to the right to grasp that object or walk to the right to approach it. We learn to project visual stimuli properly just as we learn to project other subjective sensory data; for example, we must learn that stimulation of a specific sensory area in the cerebral cortex means that the index finger of the left hand is being touched.

If the light rays are experimentally tricked by the use of special glasses to focus in such a way that upright retinal images are formed, the whole world seems upside down at first. But, in time, the experimenter learns how to project correctly the images formed under the new conditions. Visual data are again interpreted properly. If now the special inverting glasses are removed, the process of relearning, this time to the normal manner of projection, must be repeated.

5. *Errors of refraction; Snellen's type.*—In the normal eye the length of the eyeball and the refractive power of the cornea and lens are such that when the muscles of accommodation are relaxed, and the lens therefore maximally flattened, images of objects at a distance of 20 feet or more are sharply focused on the retina. No accommodation by the lens is required, that is, to see such objects. They can be seen clearly with the muscles of accommodation relaxed. Such a normal eye is called *emmetropic.*

Defects in these relationships requiring correction by glasses are very common. In farsighted people, when the muscles of accommodation are relaxed and the lens maximally flattened, images of distant objects tend to come to a focus behind the retina,[3] making the retinal image blurred. The farsighted individual can correct this condition by contracting his muscles of accommodation, making the lens more convex, or "stronger." This means that the person will have to accommodate continuously. The mechanism of accommodation can never rest except when the eyes are closed, and "eyestrain" is a frequent consequence. Glasses made of convex lenses correct the situation, bringing light from distant objects to a focus without accommodation. The muscles of accommodation will now have to be used only when objects closer than 20 feet are being observed, just as in the normal eye (see Fig. 168).

In the nearsighted subject light rays from distant objects come to a focus in front of the retina when accommodation is relaxed.[4] There will be a blurring of the images of all objects located beyond a critical distance from the eye—the *far point* of distinct vision, which is infinity in normal and farsighted eyes. In this case, accommodation, which increases the lens thickness and therefore its refracting power, cannot correct the situation. Only by the use of concave lenses (i.e., thicker at the edge than in the center) can distant objects or objects beyond the far point be seen clearly. Such lenses bend the light rays

3. Called *hypermetropia*, commonly resulting from the eyeball being too short for the refractive system.

4. *Myopia*, commonly due to the eyeball being too long for the refractive system.

so that, when they strike the cornea, they are divergent or more so than before and are now brought to a sharp focus on the retina by the refracting system of the eye.[5]

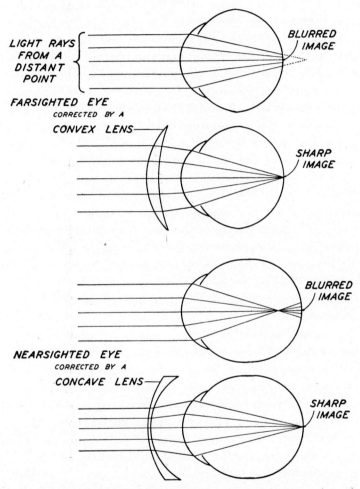

FIG. 168.—Errors of refraction and their correction. In the farsighted eye (*top figure*) with relaxed accommodation light rays from points 20 feet or more away (i.e., parallel rays) tend to focus *behind* the retina. In the nearsighted eye (*third figure*) such rays come to a focus *in front* of the retina. Either condition can be corrected (*second and fourth figures*) with suitable glasses.

These abnormalities of refraction can be detected and measured by the use of Snellen's test type.[6] The letters on a Snellen chart are of

5. Variants of farsightedness or nearsightedness, such as *astigmatism*, are even commoner than the simpler defects described. In astigmatism the refractive power of the cornea or lens is different in different meridians.

6. After the Dutch ophthalmologist, Hermann Snellen.

such sizes that they can easily be read by normal individuals at the distance specified on the chart for each line of type. In practice the eyes to be tested are rendered unable to accommodate,[7] and the reading is done at a distance of 20 feet. The normal eye should be able to see and read the appropriate line of type at this distance without accommodating. If the eye is myopic, the light from the chart will focus in front of the retina, and the blurred retinal image will render the letters illegible. If the eye is hypermetropic, the images tend to form behind the retina. In this latter case, unless accommodation is paralyzed, the subject can bring the letters to a focus by unconsciously accommodating, and the existence of the hypermetropia would not be detected.

In either case, the placing of the proper "trial lenses" before the eyes—convex in hypermetropia, concave in myopia—enable the subject to see clearly. The kind as well as the degree of abnormality are thus determined, and the proper glasses prescribed.

6. *Cataract.*—Any opacity of the lens interrupting the pathway of light to the retina is called a cataract. Cataract is fairly common in old age. The only measure which will restore the failing eyesight is to remove the opacity or the entire lens surgically. This enables the subject to see again, but, of course, renders him totally unable to accommodate.

Treatment of cataract has been a source of tremendous profit to the unscrupulous. There are "eye specialists" who capitalize on the fact that, though cataract becomes progressively worse, it may be characterized by brief remissions during which eyesight is temporarily improved. Such improvement is credited to magic drops[8] instilled by the charlatan. The only magic in the transaction is the rapid transference of currency from patient to quack. There are no "drops" which are known to dissolve cataracts—or gallstones or kidney stones, for that matter.

7. *The iris and its functions.*—The iris is a thin sheet of pigmented tissue suspended in front of the lens, having a hole—the pupil—in its center through which light is admitted to the eye. By means of muscles arranged circularly about the pupillary margin of the iris, the pupil can be made smaller by a sphincter-like action. Muscles dis-

---

7. Instillation of *atropine sulfate* temporarily prevents the use of the ciliary muscles.

8. Sometimes atropine is instilled. This dilates the pupil widely and may improve vision by admitting more light to the eye—until the effect of the atropine is gone. This drug does nothing to the cataract itself.

posed radially, like the spokes of a wheel, dilate the pupil when they contract (see Fig. 166). The quantity of light admitted to the eye is regulated by this mechanism, after the fashion of the diaphragm of a camera. In the eye the diaphragm is adjusted automatically by the pupillary reflex, the optic nerves being the afferents in this adjustment. The greater the intensity of light striking the retina, the greater is the reflex contraction of the muscles constricting the pupil.

Pupillary changes are also geared to the accommodation machinery, so that in accommodation for near objects the pupil becomes small; when accommodation relaxes, the pupil enlarges. This serves the purpose of admitting a maximum of light to the eye from a distant object, facilitating clearer vision; it admits less light when less is required in near vision.

8. *Spherical aberration.*—Constriction of the pupil in near vision serves the additional function of causing the retinal image to be more sharply focused. This depends upon the phenomenon of spherical aberration: all light rays from any point are brought to a focus at not exactly the same point, either by the eye or by other ordinary lenses. Rays passing through the margin of the lens come to a focus sooner than those passing through the central areas of the lens because of certain purely physical properties of the lens. These rays at the margin are said to be aberrant. Pupilloconstriction tends to cut them out, and only the rays passing through the center of the lens, all of which come to a sharp focus, are admitted to the eye.

D. Perception of Distance and Depth

Binocular vision in man is more than simply the possession and simultaneous use of the two eyes. In addition, images of a given object are formed upon the two retinas at the same time, and these two images are *perceived in consciousness as one*. This is impossible in the lower vertebrates such as the fish, whose eyes cannot be simultaneously directed to the same object. Such binocular vision is first encountered in the higher vertebrates and reaches its highest development in man and other primates.

1. *The neural basis.*—Neurologically, the basis for man's binocular vision is found in the arrangement of the optic nerve fibers and the optic tracts, connecting retinal units with specific regions of the visual cortex. Activation of any point on the right-hand side of either retina is transmitted to the right occipital cortex (see Fig. 169). Furthermore, for every point on one retina (P_1), there is a so-called

corresponding point upon the same side of the other retina (P_2). Simultaneous stimulation of the two points gives rise to a single sensation. Presumably, this is because the optic nerve and tract fibers from these two points terminate in the same point in the occipital cortex, where only the single sensation rises into consciousness. A flat object, then, made of innumerable points emitting light, will stimulate a pattern of corresponding points on the two retinas, and the whole object will be perceived as single.

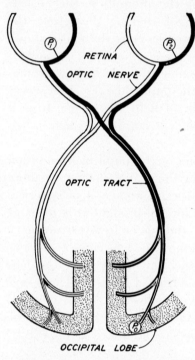

It is possible to cause images of a single object to form upon noncorresponding points of the two retinas, by "looking cross-eyed," for example. Double vision is the consequence because now, presumably, two separate—though perhaps overlapping—areas of the occipital cortex are activated by the images in the two eyes.

2. Judgment of distance.—Binocular vision in man is an important aid in accurate judgment of distance. Its efficacy in this regard is made apparent by a simple experiment. Hold a pencil, point up, at arm's length. With one eye closed (monocular vision), try to bring the tip of the index finger of the free arm from a position at the side, directly to the point of the pencil, in a rapid, accurate movement. Alternate a few times with the left and right hands. Then repeat the experiment with both eyes open and note with how much greater precision and rapidity the movement can be performed.

Fig. 169.—The optic pathways as seen in a horizontal section through the head. The right half of each retina sends fibers (*shown in black*) to the right occipital cortex. P_1 and P_2 are shown as a pair of *corresponding points*. Fibers from these two points presumably go to the same point in the visual cortex (P_0), since simultaneous stimulation of these two points causes a single sensation. If images of an object are formed on noncorresponding points, the object is seen double.

This improvement in distance judgment in binocular vision depends upon *parallax*—upon the fact that the two eyes lie several inches apart. Therefore, any object is seen from two angles—from two vantage points separated by the distance between the two eyes. It is a utilization, on a smaller scale, of the principle of triangulation in surveying. We do get fairly accurate clues of distance in monocular vision, but our chief basis for judgment is the degree of convergence of the eyes—the extent to which the eyes must be turned inward for both to be pointed directly toward the object looked at.

3. *Stereoscopic vision.*—Distance judgment in binocular vision is also the basis for depth perception, or stereoscopic vision. We judge

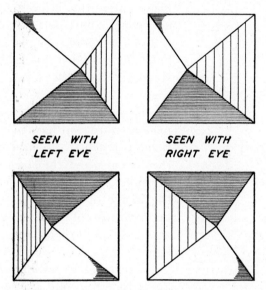

FIG. 170.—Stereoscopic vision. *Above:* Images of a pyramid seen with the two eyes, when the apex of the pyramid points toward the eyes. *Below:* Images which would be seen if the apex of a hollow pyramid were pointing away from the eyes. Such differences in the two retinal images enable us to perceive depths of objects.

a solid object to have depth even though the images formed on the retinas are flat, partly because binocular vision enables us to determine that certain parts of the object are farther from the eye than other parts. In addition, stereoscopic vision depends upon the fact that each of the two eyes gets a slightly different view of a close object. The right eye sees a little more of the right-hand surface of the object. The left eye sees a little less of this surface but perhaps more of the left surface. Experience has taught us that, when the images on the two retinas differ in this regular fashion, the object seen is three dimensional, possessing depth (see Fig. 170).

E. The Light Receptors

The retina is made up of receptors specialized for light reception. There are two kinds—the rods and the cones—differing in shape (see Fig. 171) and, to a certain extent, in function. They are more densely packed together in the center of the retina, where vision is most acute. They are less concentrated at the periphery of the retina, where vision is poorest. At the point where the optic nerve enters the retina there are no rods and cones. This portion of the retina, called the *blind spot*, is insensitive to light. Its presence can easily be demonstrated: Close the left eye and hold Figure 172 before the open right eye. *Fixing the gaze on the cross*, move the page forward and backward until a distance is reached (about 10 inches from the eye) at which the black spot becomes invisible. With the figure in this position, the image of the black spot is formed upon the area of the retina devoid of rods or cones and, therefore, insensitive to light.

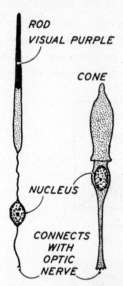

ROD
VISUAL PURPLE
CONE
NUCLEUS
CONNECTS WITH OPTIC NERVE

Fig. 171.—The light receptors of the retina. Rods function in colorless (achromatic) vision, and vision in dim light. Cones function both in color vision and in achromatic vision. The periphery of the retina contains only rods. The area of most acute vision (*fovea centralis*) contains only cones. Elsewhere on the retina rods and cones are intermingled, except in the blind spot (see Fig. 164, p. 469), where neither is found. Connections of the rods and cones with the optic nerves are indirect, with intervening neurones between these receptors and the optic nerve proper.

1. *Achromatic vision by the rods.*—We know that the rods are concerned with achromatic or colorless vision because, in areas of the retina where only rods are present (at the periphery), there is no color perception. If one directs the gaze of one eye straight ahead, and gradually moves two colored objects into the peripheral field of vision from the side, it will be noted that the objects can be seen as objects, though indistinctly, before their colors can be distinguished or differentiated. Not until the objects are brought closer to the direct line of vision, so that their images fall upon cone-bearing portions of the retina, closer to its center, do their colors become apparent. Also, in dim light, when cone vision is known to be almost absent, colors cannot be distin-

guished, and all objects, seen by means of the rods, appear in shades of gray.

2. *Stimulation of the rods and cones by light.*—How does light effect its stimulation of the retinal receptors? What changes occur in them, initiating the nervous activity which is transmitted to the brain via the optic nerves and tracts? The changes taking place in the rods are much better understood than those in the cones. The rods contain a material called *visual purple*, which is bleached to a yellowish color under the action of light. Changes from purple to yellow and back again can be induced in this substance even outside the body by alternate exposure to light and dark. When light strikes the retina, then, it produces a chemical change in the visual purple which is thought, in turn, to set up nerve impulses in the optic system.

The common experience of gradual improvement of vision for a time after one enters a dark room (*dark adaptation*) seems to depend

Fig. 172.—Demonstration of the blind spot on the retina. (Use of the figure is described in the text.)

upon this behavior of visual purple. In the dark, visual purple is gradually restored from its bleached condition. After one remains in the dark for several minutes, more of the pigment is now present in the purple form, rendering the rods more sensitive to what little light is available.

This capacity of the rods for dark adaptation renders them more sensitive than the cones in dim light. And, since in man there are no rods at the very center[9] of the retina, it follows that the keenest vision in dim light is just off the center of the retina, in the region where the rods—dark adapted—are most abundant. It is for this reason that one is often unable to see a dimly lighted object like a distant star by looking straight at it; but, if the gaze is directed slightly to one side of the star, it becomes visible. Perhaps this same phenomenon is sometimes involved in seeing "ghosts" at night. The ghost—any dimly lit object—is visible off to the side but mysteriously disappears when the eyes are shifted in an attempt to see it better.

9. The *fovea centralis.*

In the reformation of visual purple, vitamin A is required and apparently enters directly into the chemical structure of visual purple. One of the manifestations of inadequate vitamin A intake is "night blindness," or impaired vision in dim light. It is caused by an inadequate resynthesis of the visual purple so necessary for vision in the near-dark.

3. *Color vision by the cones.*—Perception of color seems to depend upon cone function, since there is no color discrimination on the periphery of the retina where there are no cones. Nor is there color discrimination in dim light when the cones play but a small role in vision. Of course, the cones are also responsive to white light of sufficient intensity. In bright light the keenest vision, both achromatic and chromatic, is in the center of the retina, which is rod-free.

The mechanism of color vision has puzzled physiologists and psychologists for many years. Very fundamental neurological problems are involved. In general, a more intense stimulus of any kind produces a stronger sensation because more nerve fibers are involved or because more nerve impulses are initiated in each fiber. Perception of different *qualities* of sensation, however, depends upon separate sensory systems: separate receptors, afferent nerves, fiber tracts in the central nervous system, and (it is also believed) discrete though tiny brain areas. This is plainly the case where widely different senses are concerned, such as vision, hearing, or taste, each of which has its specific peripheral sense organs, nerves, conduction pathways, and cerebral areas.

As far as all our information goes, this principle probably applies also in the case of qualitative differences in sensation within a given sense. Nerve fibers are apparently not equipped to conduct different kinds of impulses. In vision it is highly improbable that a given receptor or neurone transmits one kind of impulse for red and another for green or blue. Distinct neural mechanisms are apparently necessary for each qualitative difference not only in vision but also in pitch discrimination in hearing and in differentiation of various qualities of taste in the gustatory sense.

Elaborate theories of color vision have been developed which seek to account for the objective facts. All the theories start from the knowledge that all colors, including white, are combinations of red, green, and blue. Any color can be duplicated by an appropriate mixture of the proper shades of these three *fundamental colors*. Common to all theories of color vision is the notion that there are at

least three different kinds of cones, each of which is in some way
activated by one of the three fundamental colors. Sensations of the
intermediate colors would result from simultaneous but usually un-
equal stimulation of two types of cones. White would be caused by
equal stimulation of all three kinds. Since the central area of the
retina is responsive to all colors, it would be necessary that all three
types of receptor structures be represented in each tiny fraction of
this area.

Evidence has been found for the existence of several kinds of
cones, each kind responding to light in a narrow range of wave
length. This has led to the theory that, instead of three kinds of
cones, there are several, corresponding to the several main colors
of the spectrum.

Most of the evidence favoring the various theories of color vision
is indirect; each accounts for some but not all of the data experimentally verifiable.

F. Injuries and Infections of the Eyes

Well protected though the eyes are by the bones of the cheek, nose
and forehead and by the lids, which close reflexly when the cornea is
merely touched, they are, nevertheless, subject to mechanical injury.
Usually, the most serious consequences of abrasions or cuts is the
development of an opacity of the cornea. This produces blindness in
the same way as does cataract—by interrupting the pathway of light
to the retina. But, unlike cataract, corneal opacities are difficult or
impossible to deal with effectively. Attempts to remove the opacity
simply cause a further extension of it. And, if an injured cornea heals
without an opacity developing, irregularities in the healed corneal
surface may so interfere with sharp-image formation as to render the
eye useless for clear vision.

Of the infections to which the cornea is subject, one of the most
important is *gonorrhea*, which is responsible for a high percentage of
blindness. The organism of gonorrhea has a special predilection for
the genital passages and the cornea, in which regions it seems to
thrive best. Consequently, many infants have been doomed to a life
of blindness by infection acquired at birth. The incidence of blindness from this cause has been greatly reduced since the enactment of
laws in many states requiring instillation of silver nitrate into the
eyes of all newborn infants. In proper concentrations, this compound kills the infectious organisms without injuring the eyes.

III. HEARING

A. STRUCTURE OF THE EAR

The ear is divided into three portions—the outer, middle, and inner ear. It is the *cochlea* of the inner ear which contains the nervous elements—the sound receptors. The outer and middle ear are conducting mechanisms which convey the sound waves to the cochlea.

1. *The outer and middle ear.*—The outer ear consists of the cartilaginous skin-covered organ (known as the *pinna*) and the external auditory canal, leading from the outside to the eardrum, which separates the outer from the middle ear.

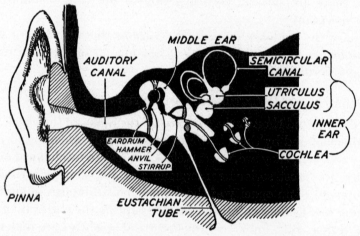

FIG. 173.—Structure of the ear, shown schematically in section. Bone is shown in black. Figures 173–78 should be studied as a group. Appreciation of any one of these calls for reference to some of the others.

The "ears" (the pinnae) of man are of some slight aid in hearing, by directing the sound waves into the external canal. But man's ears cannot be turned to the sources of the sound, as the ears can in the horse, for example. The muscles for accomplishing this are all there, but they are rudimentary and of significance only as evolutionary vestiges, except in the few accomplished individuals who are gifted with the ability to wiggle the ears.

The air-filled middle-ear cavity is a modified gill pouch of the embryo—a heritage from a stage of evolution when these outpouchings of the pharynx extended entirely through the pharyngeal wall to the exterior as gill slits, whose respiratory function in fish we have already mentioned. As such, the middle ear retains its connection with the pharynx through the *Eustachian tube*. The frequency of middle-ear infections depends upon this open channel through which organisms living in the mouth or pharynx can rather easily reach the middle ear, especially in childhood, when the tube is short and rather widely open.

The discomfort or even pain in the ears associated with rapid ascents or descents in the air depends upon undue stretching of the eardrum. In ascending in an airplane the pressure in the auditory canal diminishes, and the eardrum is stretched outward by the higher pressure of the air trapped in the middle ear. Relief is experienced as air bubbles pass from the middle-ear cavity through the Eustachian tube into the pharynx. As a plane descends, reverse pressure changes occur, with the eardrum at first pressed inward by the rising pressure in the auditory canal. For the normal balance to be established, air must pass from the pharynx into the middle ear. This is usually more difficult than is passage of air in the reverse direction, so that people may experience no appreciable ear discomfort in ascending but considerable pain in descending.

These distortions of the eardrum may also temporarily impair hearing, because the abnormal stretching and bulging of the eardrum may interfere with its vibrations in response to sound waves.

Sound waves which enter the external auditory canal, setting the eardrum into vibration, are transmitted across the cavity of the middle ear by the three little ear bones.[10] They bridge across the middle-ear cavity from the eardrum, or *tympanic membrane*, to the membrane (of the oval window) which separates the middle from the internal ear.

2. *The cochlea of the inner ear.*—The cochlea is a membranous structure lying imbedded in bone. It is spiral shaped (see Fig. 174), broader at the base than at the apex. Its internal structure can best be understood by uncoiling it—stretching it out straight. We see it to be made of three tubes or canals in the shape of gradually tapering narrow cones, coming to a point at the apex of the cochlea. At the base of the *vestibular canal* is the membranous *oval window*, to which the third bone of the middle ear, the stirrup, is attached. At

---

10. These bones are called the *auditory ossicles;* from their shape they derive the names of hammer (*malleus*), anvil (*incus*), and stirrup (*stapes*). These bones constitute a lever system whose long arm is attached to the eardrum, the short arm to the oval window. It is thought that this lever system diminishes the amplitude of the vibrations transmitted to the cochlea but increases their force. Such a system serves admirably to transform the vibrations occurring in the air of the outer ear into vibrations of liquid in the inner ear.

the base of the *tympanic canal* is the *round window*, closed off from the middle ear by a membrane. The vestibular and tympanic canals communicate at the apex of the cochlea. Vestibular and tympanic canals are filled with fluid (the *perilymph*). When the vibrations of the stirrup are communicated to the oval window, the whole fluid system vibrates.

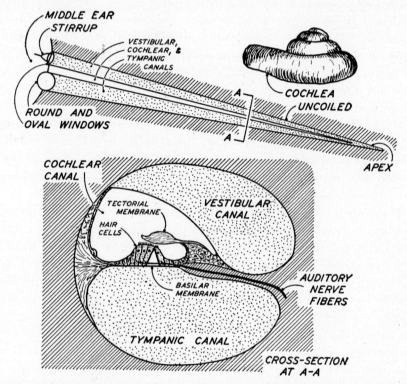

FIG. 174.—The organ of hearing. "Uncoiling" the spiral-shaped cochlea and splitting it open reveals the relationships of its three canals. Vibrations of the stirrup set the perilymph (*shown stippled*) into vibration. These vibrations are communicated to the basilar membrane and the organ of Corti, which rests upon it. The organ of Corti is shown much simplified, structures other than the hair cells being omitted. The hair cells, innervated by fibers of the auditory nerve, are the receptor cells for hearing. The entire cochlea is imbedded in bone (*shown crosshatched*).

Between these canals, separated from them by membranous partitions, is the liquid-filled (*endolymph*) *cochlear canal*, which contains the organ of hearing—the *organ of Corti*.[11] The details of its structure are shown in Figure 174 in the cross-section taken through the entire

11. After the Italian anatomist, Alfonso Corti, its discoverer.

cochlea. In it are found the "hair" cells,[12] which are really ciliated cells. They are arranged in rows which extend up and down the cochlea throughout its entire length. The cells rest upon the basilar membrane, and their cilia are in contact with an overhanging structure (the *tectorial membrane* or "roof" membrane).

B. STIMULATION OF THE ORGAN OF CORTI BY SOUND

The problem facing the physiologist is the correlation of this complex anatomical structure with function. How are the sound waves converted into nerve impulses in the auditory nerve fibers after being transmitted across the middle ear by the ear bones and after setting up waves in the fluid of the canals? It seems plain that the hair cells are directly involved, for they are anatomically in connection with the actual nerve fibers. These cells are apparently the sensory end organs of hearing. Just how are they stimulated?

1. *Role of the basilar membrane.*—Much attention has been paid to the basilar membrane upon which the hair cells rest. This is composed essentially of transverse connective-tissue fibers attached firmly at either end and stretched taut, bearing some resemblance to the strings of a harp. If these fibers should be set into vibration by the pulsations of the fluid in the cochlear canal, the hair cells resting on the fibers would also move. The cilia, in contact with the overhanging tectorial membrane, might be moved in a manner perhaps adequate to stimulate the cells and thus initiate impulses in the auditory nerve fibers. This concept is incorporated into the Helmholtz theory of hearing.

2. *Discrimination of pitch.*—Any theory of hearing which we may tentatively accept must account for pitch discrimination—for our ability to distinguish tones of different vibration frequencies. The Helmholtz basilar membrane theory does this very nicely. The fibers of the membrane are of different lengths at different levels of the cochlea. They gradually become longer toward the top of the spiral cochlea.[13] The whole resembles a musical instrument equipped with strings of graded length and, therefore, of graded intrinsic frequencies. If a note is sounded by a tuning fork, the string of such an instrument possessing the same inherent frequency as the tuning fork will

12. In some insects "hearing" also depends upon "hair" cells on the surface of the body, which are apparently directly stimulated by sound vibrations in the air.

13. This seems paradoxical, for the cochlea as a whole is widest at the base and narrowest at the top. The fibers are shortest at the base, however; here the points of attachment of the ends of the fibers project farther into the cavity than at the apex.

be set into sympathetic vibration. Similarly, it is thought that specific regions in the basilar membrane are set into vibration by sounds of specific pitches.

The concept is, then, that sounds of different vibration frequencies stimulate different hair cells in the cochlea: low notes stimulate those near the apex, where the fibers of the basilar membrane are longest; high notes stimulate hair cells near the base, where the fibers of the membrane are shortest. Experiment bears out this view to the extent that, if the hair cells of a given restricted region of the cochlea are destroyed, there results a deafness for a given restricted range of pitches. It has further been shown that damage to the cochlea near its base, by sounding intensely loud high-pitched sounds into the ear of an experimental animal, causes deafness for tones of high frequency. Injury to the apex, where the fibers are longest, causes deafness for low tones.

Varying volumes of sound would, of course, stimulate the specific hair cells for that pitch more or less intensely. A loud note would initiate more impulses in a given set of nerve fibers than a soft note of the same pitch.

C. Defects of Hearing

Deafness of any degree can be caused by damage either to the *sound-perceiving* apparatus in the inner ear or to the *sound-conducting* mechanisms. The latter is more common. Infections of the middle ear may damage the ear bones, or interfere with their motion, or fill the middle-ear cavity with pus. In other instances the infection may injure or perforate or entirely destroy the eardrum.

A fairly common cause of slowly progressing deafness in young adults apparently depends upon a gradual "fixing" of the ear bones. New bone formation interferes with their free vibration.

In the above hearing difficulties there is no known defect of the inner ear—of the organ of Corti. The sensory endings are present and intact but do not function because sound waves fail to reach them.

IV. THE SENSE OF BALANCE AND ROTATION

Besides the cochlea, the inner ear contains other structures—the tiny chambers called the *sacculus* and *utriculus* and the *semicircular canals* (see Fig. 175). These, together with the cochlea, constitute the labyrinths of the inner ear, so called because of the labyrinthine intricacies of their structure. The cochlea, or acoustic labyrinth, is a rather recent evolutionary acquisition. It is entirely absent from fishes, whose "hearing" mechanisms are probably quite different from those of the higher vertebrates. The remainder of the labyrinths, the nonacoustic portions, have nothing to do with hearing but contain the sensory end organs concerned with equilibrium. They are innervated by the nonacoustic (or *vestibular*) fibers of the auditory or eighth cranial nerve.

Destruction of these organs produces profound disturbances of equilibrium. A pigeon so operated on can neither stand nor fly, al-

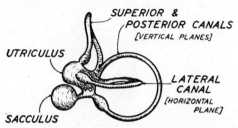

Fig. 175.—The nonacoustic labyrinths of the internal ear (*schematic*). Each of the semicircular canals lies in a plane perpendicular to the planes of each of the other two canals.

though there is no actual paralysis. In time the pigeon relearns how to maintain its equilibrium, mainly by the use of vision. Blindfolding such an animal brings on the equilibrium disturbances anew.

A. The Sacculus and Utriculus

The sacculus and utriculus are apparently concerned with static equilibrium, with the relationship of the body—really the head—to the pull of gravity. They inform us more or less accurately which way is up or down when other clues are lacking, as in the case of swimming under water, or flying in an airplane in complete darkness. The end organs, consisting of hair cells in contact with fibers of sensory nerve, are diagrammatically represented in Figure 176. Resting upon the cilia of each of the four groups of hair cells are concretions of calcium carbonate ($CaCO_3$) called ear stones, or *otoliths*. The manner in which these receptors seem to be stimulated by changes in the gravitational pull is well illustrated by the functions of a similar but less complex organ found in certain crayfish (see Fig. 177).

This crayfish organ is saclike, made of cells whose cilia project into the hollow interior. In the interior, resting against the cilia, is a calcareous particle. An ingenious experiment has been reported in which

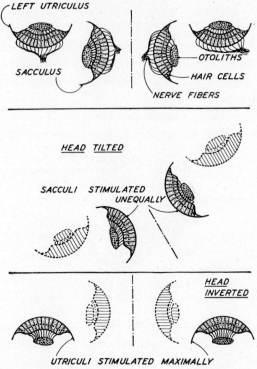

FIG. 176.—The sense organs of static equilibrium. The adequate stimulus to these sense organs (the *maculi* of the utriculus and sacculus) seems to be a *pull* upon the "hairs" of the hair cells, exerted by the action of gravity on the clusters of otoliths to which the hairs are attached. Note that, when the head is tilted to an asymmetric position to the left, the hair cells of the left sacculus are pulled, and those of the right sacculus tend to be compressed. When the head is tilted forward or backward, there is no change in the stimulation of the saccular maculi, but there is a pull upon the hair cells of the utriculi.

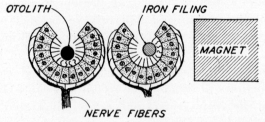

FIG. 177.—Equilibrium organ of the crayfish. *Left:* The otolith is pulled downward by gravity, pressing upon the lower hair cells of the organ, when the animal is upright. *Right:* Replacement of the otolith with an iron filing. Upon bringing a magnet near, the filing is pulled against the hair cells of the side of the organ, causing the animal to act as if its equilibrium were disturbed.

these ear stones of crayfish were replaced by particles of iron. Such an iron particle could be pressed against the hair cells of any part of the sphere by placing a magnet near the animal. If the magnet was held below the animal, the filing was pulled against the lower hair cells, exactly as the normal otolith would have been by gravity. The animal maintained its upright position. If the magnet were held to the right of the upright animal, simulating the gravitational pull if the animal were lying on its right side, it made "righting" movements, which, of course, only upset the postural equilibrium of the already upright animal. Only when the lower hair cells of the organ were stimulated by the pressure of the filing did the animal behave as though it were in the normal upright position.

In mammals, including man, the otolith hair-cell complex seems to function in similar fashion. Any disturbance, by an altered position of the head, of the normal gravitational pressure directly down upon the utricular hair cells, or of the shearing pull of the saccular otoliths, is sensed in consciousness as an upset of the normal, symmetrical, upright posture of the head. Of course, this sense is normally supplemented by visual stimuli. But, if an animal whose labyrinths are removed is also blindfolded, it fails to "right" its head. Its head tends to hang limply downward no matter into what position it may be rotated. The animal seems not to "know" that its head is in an abnormal position and does nothing about it.

It must not be supposed that the readjustments in head position of the normal animal, which occur in situations of disturbed equilibrium, depend upon actual consciousness of the disturbed state any more than the flexion reflex depends upon the attendant pain. The responses are mainly reflex in nature and will occur in animals in which all sensory pathways from the internal ear to the conscious centers of the brain are destroyed by section of the brain stem in the region of the thalamus.

B. Body-righting Reflexes

We see, then, how visual and labyrinthine reflexes induce a righting of the head itself when the relationship of the head to gravitational pull is disturbed. But what of the righting of the rest of the body? Instructive in this respect is the well-known experiment of holding a cat dorsal surface down and then letting it drop a distance of about 3 feet. Here not only the head but the whole body rights itself in the air, and the cat usually lands upon its feet. Motion pictures have demonstrated that the head rights itself first, the body following. The visual and labyrinthine reflexes described right the head. Experiment has shown that the subsequent righting of the trunk depends upon the twisted position of the neck produced by the initial righting of the head. Sensory endings in the neck muscles are stimulated by the twisted position, initiating a reflex rotation of the trunk. The upright position is assumed, then, by a sort of spiral rotation, commencing with righting of the head, followed by righting of the body progressively from front to back, the whole reaction depending upon a chain of reflexes.[14]

C. The Semicircular Canals

There are three semicircular canals in each labyrinth, looping through a short part-circle from a starting point in the utricle back again to the utricle. They are so arranged that one pair (one canal from each labyrinth constituting a pair) lies in each of the three planes of space, each plane lying at right angles to the plane of each other pair (see Fig. 175). Each liquid-filled canal is equipped with a receptor structure innervated by fibers of the eighth cranial nerve. The evidence indicates that the nerve endings are stimulated by movements of liquid inside the canal. These movements are due to inertia, or a lag in the movement of the liquid when the body or head is moved either in a straight line or in rotation in any plane.

Rotation of an animal to the left, for example, will cause an initial lag in the fluid of the left horizontal semicircular canal, producing a relative flow of fluid to the right into the swelling of the canal containing the sensory receptor. Such motion stimulates the end organ (see Fig. 178) and produces not only consciousness of the rotation but reflex adjustments to it, such as eye and head movements in a direction opposite to the rotation.[15]

Evidence for this concept comes from experiments in which the fluid of the semicircular canals is artificially caused to move without actual rotation. This can be effected in man by irrigating the external auditory canal with warm or cold water. Convection cur-

14. Invertebrates, lacking labyrinths, are nevertheless equipped with equally effective mechanisms of other kinds for righting the body and maintaining it in normal relationship to gravity. What these mechanisms are, especially in the flying insects, is not clear at present.

15. This is called *nystagmus*, of which the "slow component" is described.

rents are set up in the fluid of the canals. Sensations of rotation—dizziness—are experienced, and all the reflex adjustments attendant upon rotation itself occur under these artificial conditions.

Straight-line movements cause inertial lags (fluid movement) in the semicircular canals of the two sides symmetrically and equally. Most translational movement in land animals is in the horizontal plane. Man is unused to the kind of stimulation of the canals which

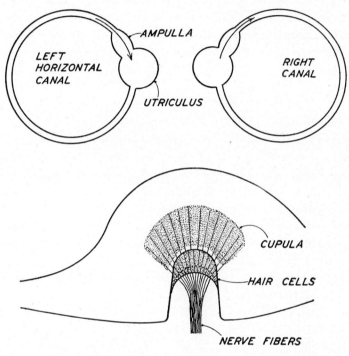

Fig. 178.—The sense organs of dynamic equilibrium. *Above:* Horizontal semicircular canals. Rotation of an animal to the left causes movement of fluid (*endolymph*) in these canals to the right (*indicated by arrows*), because of inertia of the fluid. Movement of fluid in this direction stimulates the sensory end organ of the *left* canal. *Below:* End organ (the *crista*) of the ampulla of a semicircular canal.

occurs when the body is moved upward or downward in the direction of the long axis of the body. For some poorly understood reason, this relatively unusual kind of movement produces sensations of nausea and even vomiting. This is a large factor in seasickness, where the motion of the boat caused by the waves is in a vertical plane. If one lies down, these vertical movements now stimulate the semicircular canals in a fashion resembling the more normal stimulation occurring when one moves on a horizontal plane, as in walking. Thus, if one remains recumbent, the movements of the boat seem markedly reduced, and the gastric disturbance is less likely to occur.

On the same principle, the queer sensation vaguely localized in the pit of the stomach, experienced when an elevator suddenly starts or stops, can be greatly lessened by flexing the head forward to a position at right angles with the axis of the trunk. The line of movement of the elevator car now bears the same relation to the semicircular canals as does the line of movement of an automobile on a level road. In consequence, the sensations in the elevator will be no more disturbing than those experienced in a rapidly accelerating or decelerating automobile.

V. RECEPTION OF CHEMICAL STIMULI

The sense organs of taste and smell are *chemoreceptors*, so called because they are normally stimulated directly by chemicals in solution. We are reminded here of the direct effect of certain chemicals upon cell organisms such as the ameba, which responds in characteristic ways to chemical stimulation by foods or irritants. In the cells of the taste and smell receptors the effect of the chemical stimulation is one merely of initiation of nerve impulses in the gustatory and olfactory nerve fibers. To be effective, the chemicals must be in solution in the saliva or the secretions of the nasal cavities. A perfectly dry tongue is incapable of tasting; a dry nasal cavity cannot be stimulated by aromatic materials.

These senses have extremely low thresholds. They are stimulated by chemicals in great dilution. Quinine in water can be tasted in concentrations as low as one part in two million. Olfactory sensations arise from some materials even more dilute. We can smell musk in a dilution of one part in eight million.

Taste and smell are often confused subjectively. The oral and nasal cavities, freely open to each other at the back of the mouth, are often subjected to stimulation by the same chemicals. The extent to which smell contributes to what we loosely call "taste" is evident from the "tastelessness" of foods eaten when one has a nose cold, when the olfactory endings are rendered relatively nonfunctional by the accumulation of mucus in the "stopped-up" nasal cavity. On the other hand, there are certain smells, such as the "sweetish" smell of chloroform, which owe the sensation they produce to stimulation of the taste rather than the olfactory endings.

A. TASTE

Taste per se is also confused with other properties of substances in the mouth. Hot and cold coffee are chemically alike and therefore stimulate the taste mechanism itself in the same way. The difference in what we call the taste is really a difference in stimulation of temperature receptors in the mouth. Texture and consistency of foods also modify their "taste." A crisp cracker does not really taste different from a soggy one; the difference is in the consistency and the feel of the cracker against tongue and teeth and cheeks.

In fact, the qualitative differences in foods or chemicals which are detectable by taste alone are few. Observers are fairly well agreed that there are only four fundamental tastes—salt, sweet, bitter, and

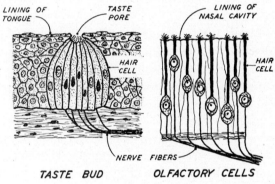

FIG. 179.—The chemical receptors: sense organs of taste and smell

sour. These may occur in combinations, of course. Some people apparently lack the ability to taste bitter substances.

1. *The taste buds.*—The receptor structures for taste, known as taste buds, are distributed chiefly to the tongue but also to the pharynx, the roof of the mouth, and the laryngeal region (on the epiglottis). Each bud consists of a cluster of cells imbedded in the mucous membrane (see Fig. 179). Peculiarly enough, these receptor structures are ciliated, like the receptors of the labyrinths, although the significance of this structural feature is unknown.

2. *Specificity of the taste buds.*—Again, because differences in kinds of nerve impulses do not occur, we should expect that different taste buds are concerned with qualitatively different taste sensations, just as different cones of the eye respond to different fundamental colors and different hair cells in the cochlea are stimulated by sounds of different pitches. Some evidence for this is found in the topograph-

ical distribution of the four fundamental taste sensations over the tongue. The tastes of salt and of sweet predominate at the tip of the tongue, bitter is localized at the base of the tongue posteriorly, and acid (or sour) along the borders. Presumably, this corresponds with the distribution of at least four specific varieties of buds, each able to give rise to only one kind of sensation.

More direct experiments support this conception. Localized stimulation of circumscribed clusters of taste buds indicates that different buds respond to different kinds of chemical stimuli. Highly interesting is the report that certain chemicals (e.g., Na_2SO_4) will produce a sweet taste when applied to one region and a bitter taste in another. That is, the "sweet endings" can give rise only to sweet sensation, and the bitter endings, to bitter sensation, no matter what the nature of the stimulating agent.[16]

In further support of this concept is the finding that the effect of the local anesthetic, cocaine, is not the same for all taste sensations. It affects bitter before sweet, sweet more than sour, and salt scarcely at all, indicating that these four sensations depend upon different taste buds having different resistances to cocaine.

To a very limited degree it is sometimes possible to anticipate, from the physical and chemical properties of a substance, what its taste properties will be. We know, for example, that acidity produces a sour taste, and alkalinity, bitter. But, in general, there are no such clear-cut objective criteria for classification of qualitatively different tastes as there are in vision or hearing, where colors or pitches can be expressed in terms of wave length or frequency.

B. SMELL

Olfaction is thought to play a much more important part in the life of the lower mammals than of man. In many species it forms perhaps the chief source of information about the external world, especially as regards food, danger, and mating. A large portion of the head of an animal like the dog is given over to the nasal cavity in the snout, which contains the olfactory receptors, and a fairly large proportion of the brain of lower vertebrates such as the shark appears to be devoted to responses based upon olfaction.

In the primate line of evolution, tree-dwelling is thought to have placed a premium on good eyesight and to have selected evolutionary

16. Needless to state, chemicals of this category, able to stimulate more than one kind of taste bud, are exceptional.

improvements in this sense at the expense of olfaction, which has steadily declined in importance in this evolutionary series. In man olfaction plays a very minor role in furnishing information about the external world, and relatively few reflexes are built around it.

1. *The olfactory endings.*—In each nasal cavity in man there are about 5 sq. cm. of olfactory epithelium, consisting of ciliated cells. The receptor structures are even less highly differentiated than the taste buds, which are, in turn, considerably less complex than the organs of vision or hearing. The ciliated cells are the end organs of nerve fibers which pierce the skull and pass back to the brain as the olfactory nerve or tract (see Fig. 179). Like the peripheral neurone system of the visual apparatus, the olfactory "nerve" is also composed of a series of neurones and is not entirely a true nerve but partly a fiber tract of the brain.

The olfactory epithelium is bathed in liquid, and stimulation depends upon solution of the chemicals in this fluid. The moist receptors lie in the upper portion of the cavity, off the direct pathway of the stream of respired air. This tends to prevent drying of the surface and is responsible for the necessity of "sniffing" to bring a maximal concentration of chemicals in the air into contact with the sensitive cells.

2. *The quality of odors.*—There has been no satisfactory classification of qualitatively different odors comparable to colors in vision, pitch in hearing, or the four primary components of taste. Odors are characteristically described with reference to the name of some specific aromatic substance. We say, "This smells like camphor, or like balsam, or like roses." We are apparently confronted with the existence of almost as many kinds of odors as there are kinds of aromatic substances. At once the question arises as to the degree of differentiation of the receptor structures. Are there as many kinds of receptors as there are kinds of odors? This seems improbable, although no satisfactory evidence on this point is available at present.

The easy fatigability of the olfactory receptors is of some interest and seems to shed some light on this problem. The intensity of any odor, as experienced subjectively, falls off very rapidly in just a few minutes of olfactory stimulation. Everyone has repeatedly observed that odors in a room go quite unnoticed to one remaining in the room but are often evident to a marked degree to one just entering the room, whose olfactory apparatus is not fatigued. Bearing upon the question of specificity of individual smell receptors is the fact that this fatigue of the olfactory mechanism for one odor does not greatly affect the sensitivity for other odors. This suggests a differentiation of receptors into very many specific types.

A degree of similarity of olfaction to certain aspects of visual perception has been described. There are said to be "complementary" odors, which neutralize one another just as complementary colors (e.g., yellow and blue) give a visual sensation of white or gray. Proper concentrations of iodoform and Peru balsam, or of carbolic acid and the odor of putrefaction, are examples. Whether such effects depend upon the character of the receptor units and the manner of their stimulation, as is the case in vision, or upon actual chemical combination of the aromatic molecules before they affect the receptors, is quite unknown. In fact, our knowledge of the peripheral mechanisms in this relatively primitive sense of smell is far less complete and satisfactory at present than is our understanding of the visual and auditory senses, with their much more complex receptor organs.

VI. THE SKIN SENSES

Upon superficial consideration, the sensitivity of the skin to stimulation by touch, pressure, pain, heat, or cold seems to be uniformly distributed, but experimentation reveals the fact that this is not so. Certain areas are far more sensitive to a stimulation than other areas. The fingertips are more sensitive to a touch than the back of the hand; the lips are more sensitive to heat than the hands. Nor are all areas of the skin sensitive to all these kinds of stimuli. It can also be shown that in any given skin region there are discrete areas, often of the dimensions of a point, which are capable of mediating only one kind of sensation. Touch spots, or heat or cold spots, can be fairly clearly mapped out in any given region (see Fig. 180). A moderately warm object will stimulate only the heat spots; a moderately cold object will activate only the cold spots.

This depends upon the presence in the skin of special receptors for each kind of stimulation. Each of the senses, apparently, has its own peculiar receptor structure especially adapted to respond to the peculiar kind of environmental change with which it is associated. Some of these receptors are schematically represented in Figure 181. These structures possess the same degree of specificity in their relationship to the effective stimulating agent as do the rods or cones for light and the cochlear hair cells for sound.

Again we encounter the principle that a given peripheral system can give rise to only one type of sensation, no matter how it may be stimulated. A pin point applied lightly to a touch spot (i.e., stimulating a touch receptor) gives rise to a touch sensation; applied to a pain spot, it causes pain—just as any kind of stimulation of the optic nerve system causes a sensation of light, and of the auditory system, sound.

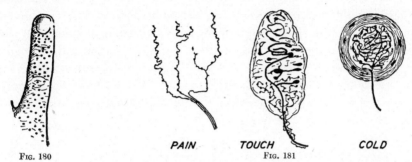

FIG. 180   PAIN   TOUCH   COLD
                                            FIG. 181

FIG. 180.—Cold spots on the back of the little finger. Heat spots or touch or pain spots could be mapped out similarly. The locations of these spots vary considerably in different individuals. Their spacing varies widely in different parts of the body.

FIG. 181.—Some sensory nerve endings in the skin.

Combinations are possible here also, of course, as in taste or any other sense. Very hot objects stimulate not only the heat receptors but the pain elements as well, causing a combination of the two sensations. No special receptor seems to be involved in pain sensibility. In the pain spots we find only free nerve endings.

VII. PAIN FROM THE INTERIOR OF THE BODY

Though afferent nerves of the internal organs are very active in the many reflex adjustments of the body, there is but little sensation from these regions. Some of these sensations we have dealt with—muscle and joint sense (see Fig. 182), hunger and thirst, and fulness of the bladder and rectum. Besides these, there is normally little else.

Abnormally, the sense of pain may be experienced from the interior of the body. There are certain differences between this pain sense and that of the skin, however. Internal pain is usually less definitely localized. Furthermore, the way in which internal pain nerves are stimulated is not clear. It is possible, for example, to cut, tear, pinch, crush, or burn almost any of the internal organs without producing pain. This has frequently been done in the course of surgical operations on patients whose abdominal wall only is locally anesthetized.

Yet internal pain does occur from many regions, as all have experienced. It has been suggested that perhaps the pain fibers are fewer in number in the internal organs and that such stimuli as those mentioned are not distributed over a sufficiently large area to stimulate many of the nerves.

Observers are generally agreed that distention of hollow internal organs produces pain, especially if this is associated with spasms of the muscular wall. Distention of the colon with gas, or spasms of the bile duct or ureter containing a stone, may be acutely painful. Any mother will bear witness to the pains associated with the powerful contractions of the uterus at childbirth. Spasms of blood vessels may be involved in the excruciating pain associated with disturbances of circulation through the heart muscle.

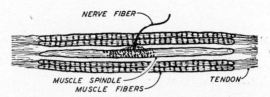

Fig. 182.—The end organ of "muscle sense." These sensory structures, together with similar receptors in tendons, furnish information as to the tension of skeletal muscles and the position of the limbs in space. The spindles are stimulated by muscle tension.

Chemical irritation is believed to be a large factor in the production of the pain of stomach ulcers, which are bathed by hydrochloric acid. The inner surface of the abdominal wall is far more sensitive to chemical irritants than are the viscera themselves. In fact, this surface is sensitive to almost all the types of stimuli which act upon the pain endings of the skin. Much abdominal pain, such as that associated with infections of the appendix, is thought to be due to irritation of this sensitive surface.

VIII. WEBER'S LAW OF SENSATION

Everyone has experienced certain relationships between changes in stimulus intensity and changes in the resulting sensations: The light from a candle is far more noticeable when added to a faint source of light in a dimly lighted room than when the candlelight is added to that of a strong electric bulb. Growth of a few inches is easily detected in an infant, though the same quantity of growth in a boy of sixteen may pass unnoticed. One pound added to a five-pound weight makes a greater difference than one pound added to a fifty-pound weight.

Struck by these commonplace facts, the German physiologist, Ernst Weber, designed experiments to determine how much the intensity of a stimulus need be increased to be just detectable. He found that the required increase was not always the same but depended upon the intensity to which the increase was added. If an individual was able to judge a weight of 31 gm. to be greater than 30 gm., he was not able to detect the difference between weights of 300 and 301 gm. but would judge 310 gm. to be just heavier than 300. The same relationships apply in the judgment of lengths and distances and in detecting differences in intensity of many kinds of stimuli involving a number of sensations.

Weber's law states that a barely perceptible difference in intensity of stimulus bears a constant relationship to the intensity of the original stimulus. In the case of judgment of weight differences he found that the barely detectable added weight had to be about one-thirtieth of the original weight, as in the example cited above. The law is somewhat limited in its application, holding only over a fairly small range of stimulus intensities.

CHAPTER TWELVE

CHEMICAL CORRELATION: THE GLANDS OF INTERNAL SECRETION

I. THE STUDY OF THE ENDOCRINE GLANDS
  A. General functions of the endocrine glands and the hormones
  B. Principles of experimentation
  C. Data from human abnormalities
  D. Early experiments

II. THE PANCREAS
  A. The pancreatic islets and diabetes mellitus
    1. Experimental diabetes   2. Function of the islets   3. Discovery of insulin
  B. Functions of insulin
    1. Combustion and storage of glucose 2. Relationship to fat metabolism
  C. Insulin excess
  D. The fundamental cause of diabetes

III. THE THYROID GLANDS
  A. Deficient function in adults
    1. Myxedema   2. Effect on the basal metabolic rate
  B. Cretinism and its control
    1. The condition in man   2. Experimental cretinism
  C. Goiter
    1. Compensatory thyroid enlargement   2. Simple goiter   3. Relation of the thyroid to iodine
  D. The thyroid hormone
    1. Chemical composition   2. Control of hormone secretion
  E. Hyperthyroidism
    1. The occurrence of goiter   2. Effects of overactivity   3. Evidence for excessive hormone production
  F. Relationship of the thyroid to body heat
  G. Unsolved problems

IV. THE PARATHYROID GLANDS
  A. Indispensability of parathyroids for life
    1. Fatal result of extirpation   2. Parathyroid tetany
  B. Regulation of the blood-calcium level
    1. Effect of calcium on muscle irritability   2. Control of tetany by calcium administration
  C. The parathyroid hormone

D. The parathyroid glands in man
  1. Hypofunction   2. Hyperfunction   3. Control of calcium balance

V. The Medulla of the Adrenal Glands
  A. Epinephrine
    1. Discovery and synthesis   2. Action of epinephrine
  B. Function of the gland
    1. Epinephrine and the sympathetic nervous system   2. The secretion of epinephrine   3. Significance to the organism   4. Emergency function
  C. Summary

VI. The Cortex of the Adrenal Glands
  A. Pioneer observations
    1. Addison's disease   2. The experiments of Brown-Séquard   3. Independence of cortex and medulla
  B. Insufficiency of the adrenal cortex
    1. Effects upon experimental animals   2. The search for the hormone   3. Chemical studies   4. Usefulness of cortical extracts   5. Control of hormone secretion   6. Salt balance of blood

VII. The Gonads
  A. Reproduction
    1. The primary sex organs   2. Production of sperm   3. The reproductive cycle in the female
  B. Endocrine control of the female reproductive cycle
    1. Chemical control of uterine changes   2. Control by the ovaries
  C. The follicular hormone
    1. Effect on the uterus   2. Cyclic changes in hormone concentration of the blood
  D. The corpus luteum hormone
    1. Function in the sexual cycle   2. Function in pregnancy
  E. Additional functions of the ovarian hormones
  F. Disorders of the menstrual cycle
  G. The testicular hormone
    1. Effects of castration   2. Endocrine portion of the testes   3. Functions of the hormone

VIII. The Pituitary Gland or Hypophysis
  A. Structure and development
  B. The posterior lobe
  C. Control of growth
    1. Growth abnormalities in man   2. Experimental dwarfism and gigantism   3. The growth-promoting hormone
  D. Control of gonad function
    1. Effects of removal of the hypophysis   2. Artificial-substitution experiments   3. The gonad-stimulating hormones
  E. Interactions of the hypophyseal and ovarian hormones
    1. Stimulation of the follicles and the corpus luteum   2. Pregnancy tests   3. Reciprocal action

F. The control of lactation
G. Hypophyseal control of other endocrine glands
IX. OTHER ENDOCRINE STRUCTURES
    A. The stomach and duodenum
    B. The liver
    C. The thymus (?)
    D. The pineal body (?)
X. CONTROL OF THE ENDOCRINE GLANDS
    A. Nervous control
       1. The adrenal medulla   2. The pancreas   3. The thyroids
    B. Chemical factors
       1. Secretin   2. Rate of insulin production
    C. Hormonal interrelations
       1. The hypophysis: the "master-gland"   2. Endocrine interactions in
       diabetes

I. THE STUDY OF THE ENDOCRINE GLANDS

All glands are organized groups of cells, frequently columnar in shape, which possess the property of extracting certain materials from the blood and lymph and either concentrating those materials or manufacturing from them some new specific chemical compound or compounds. We have seen how the sweat glands, the salivary glands, the pancreas, and the liver do these things. In these instances the glandular products are passed by the gland into ducts which lead to the surface of the body or to the alimentary canal.

A. GENERAL FUNCTIONS OF THE ENDOCRINE
GLANDS AND THE HORMONES

There is another kind of gland in the body—the *endocrine* or *ductless glands*. They are true glands in that they also manufacture complex compounds from ingredients which they get from the blood or lymph. These compounds, again, are usually highly specific. Each endocrine gland manufactures its own characteristic material or materials with specific chemical, physiological, or pharmacological properties. However, instead of emptying its product into a duct, to be transmitted to a region where it may function, an endocrine gland, having no duct, liberates its secretion directly into the blood stream, or into the lymph, whence it soon enters the blood stream. In general, these glands are richly supplied with blood vessels which carry away their products. The adrenal glands, for instance, have a richer blood supply and receive more blood per gram of tissue per second than most other tissues in the body, including even the very vascular tissue of the central nervous system.

Once in the blood, these endocrine products are carried along in the general circulation and are distributed to all parts of the body. They may exert a specific influence, therefore, upon cells, tissues,

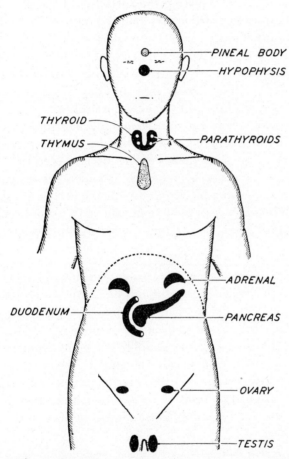

FIG. 183.—Locations of the glands of internal secretion. It is uncertain whether the *pineal body* or *thymus* are endocrine glands. The liver (perhaps a gland of internal secretion) and the stomach (which secretes *gastrin* into the blood) are not shown.

organs, or systems distantly removed from their site of origin. For example, some of the products of the *pituitary* gland, lying within the skull cavity, exert a specific influence upon the *gonads*, located in the lower part of the abdominal cavity. On the other hand, these endocrine products may exert an effect upon several different organs or even upon all the cells of the body. The secretion of the *thyroid* glands apparently modifies oxidative activities in perhaps every cell in the body.

We recognize a certain basic similarity in the general role of the endocrine glands and the nervous system. Both systems function as integrating machinery, knitting the activities of the various parts of the body into a unified whole. We pointed out earlier that the nervous system is aided in this function by certain chemicals, such as the carbon dioxide produced in muscular activity. This product helps to integrate the breathing movements, the beating of the heart, and the caliber of the blood vessels into an adaptation best adapted to increase the oxygen supply of the active muscle and to remove waste from it. To such chemical-correlating mechanisms we must now add the endocrine glands and the hormones produced by them.

The specific active ingredients of the internal secretions are called *hormones*. They have also been called "chemical messengers," which name implies something in common with the nervous system, which also transmits "messages" influencing the activity of organs located at some distance from the original source of the message.

B. Principles of Experimentation

Simple and straightforward as this description of endocrine glands may seem, it has been very difficult in many cases to determine whether or not an organ is really an endocrine gland. It is still uncertain whether the *pineal body* in the skull cavity or the *thymus* in the chest really are endocrine glands. The establishment of the internal secretory function of all the glands whose true endocrine nature we are now certain about involved extensive experimentation. Specific criteria must be fulfilled by any body structure suspected of having an endocrine function before we can be sure of its true endocrine character:

1. Surgical removal or destruction of the gland should lead to certain measurable effects, since the body would then be deprived of its normal source of the hormone
2. It should be possible to counteract the consequences of removal by administering the hormone of the gland
3. The hormone should be present in the blood and obtainable by extraction of the blood
4. The hormone should be present in the gland itself and obtainable from it by extraction

The foregoing really constitutes an outline of the methods used in studying the glands of internal secretion experimentally. There are many ramifications to these methods of experimentation. For example, when a glandular product affects many organs in a number of ways, the problem of determining the effects of gland extirpation may be exceedingly difficult and may involve a variety of special techniques.

C. Data from Human Abnormalities

Parallel with studies of this kind go the important observations of physicians upon human beings. Unfortunately, nature sometimes destroys or injures endocrine glands in man. Through disease, accident, or defective heredity, a gland may function subnormally and produce an insufficient quantity of its secretion or no secretion at all. The physician studies such abnormal people. If and when the patient dies, and if the relatives are intelligent and possess in some degree a sense of responsibility to humanity, the physician may be permitted to examine the body to ascertain just what structures have been destroyed or modified by the disease process.

Once the relationship between a series of symptoms and the destruction or injury of a specific gland has been established, it may be possible to resort to the same sort of substitution for glandular deficiency described in the case of experimental animals. Artificially prepared gland extracts, obtained from animals at slaughter-houses for instance, may replace the lacking secretion.

This deficiency type of endocrine disorder is called *hypofunction*, or *hypoactivity*. In addition, there may be a derangement called *hyperfunction*, or *hyperactivity*, in which the gland secretes an excess of its normal product. Here nature again performs experiments from which valuable physiological data may be obtained, for, in this case, it is possible to observe directly the physiological effects of endocrine secretions. Hypersecretion in man corresponds to the injection into a normal animal of an excess of the hormone from the gland in question.

D. Early Experiments

The term "internal secretion" was coined by Claude Bernard about a century ago. He observed that, in fasting, the blood leaving the liver is richer in sugar than the blood entering the liver. He concluded from this and from other studies that, between meals, the stored carbohydrate of the liver—namely, glycogen—is converted into glucose and is passed into the blood. Though Bernard's term

still persists, this secretion of glucose by the liver is no longer considered a true endocrine activity, because glucose is not a peculiar
product of the liver. Glucose, that is, is not a true hormone.

Some of the earliest experimental studies upon the endocrine
glands proper were done sometime later by the French physiologist
and neurologist, Charles Brown-Séquard, who extirpated in dogs
the paired glands lying near the kidneys in the abdomen called the
adrenal glands. His operative methods were so crude, however, that
he erroneously concluded that the destruction of only one adrenal
gland was necessarily fatal. We know now that removal or destruction of only one of the glands produces but little effect, while absence
of both glands is invariably fatal in untreated animals. Brown-
Séquard, likewise, was one of the first to study the effects of administration of glandular products upon man. He injected preparations made from the male gonads into himself and described vague
beneficial results. He "felt stronger and more vigorous." Here again,
though Brown-Séquard pointed the way to an important line of
investigation, his specific conclusions were incorrect, for subsequent
work has demonstrated that the injection of testicular products is
without such definite effects upon bodily well-being.

The secretory product of an endocrine gland was given the name
"hormone"[1] by the British physiologists, Sir William Bayliss and
Ernest Starling in 1902. They applied the term to *secretin*, which is
an endocrine product secreted by the cells and tissues of the intestinal wall (see p. 280).

II. THE PANCREAS

The pancreas lies in the upper abdominal cavity, stretching along
the lower border of the stomach. At one end it is fairly firmly attached to the tissues of the duodenum. The general appearance of
this gland is familiar to all who have seen fresh sweetbreads, which
may contain pancreas.[2] We have already noticed that the pancreas
is a gland of external secretion manufacturing an important digestive
juice which it pours into the upper intestine through two ducts leading from the gland into the duodenum (see pp. 275–76).

If the gland is examined microscopically, one observes that it resembles the salivary glands. Millions of typical secretory cells are
arranged along and about the terminal ramifications of the ducts,

1. From the Greek *hormōn,* "to excite."
2. Sweetbreads consist mostly of thymus tissue.

into which the digestive juice is secreted. However, one also sees another kind of cell unlike those of the main mass of cells. Scattered throughout the entire organ are isolated clusters of cells which stain differently from the digestive-juice-manufacturing cells. The gland is dotted with these clusters of special tissue called *islet tissue*.[3] The role of the islet tissue remained unknown for a long time after its anatomical existence had been demonstrated.

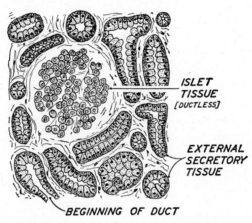

ISLET
TISSUE
[DUCTLESS]

EXTERNAL
SECRETORY
TISSUE

BEGINNING OF DUCT

FIG. 184.—Section of the pancreas showing the two kinds of secretory tissue: the external secretory portion, which secretes digestive enzymes into ducts, and the islet tissue, which secretes insulin into the blood stream. Figure 92 (p. 267) shows the location of the pancreas near the stomach.

A. THE PANCREATIC ISLETS AND DIABETES MELLITUS

1. *Experimental diabetes.*—The chief interest of physiologists in the pancreas seems to have been in its digestive functions. In 1889 two German physicians, Johann von Mering and Oscar Minkowski, decided to observe the digestive upsets which might result from extirpation of the pancreas.[4] One would expect a decided interference especially with fat digestion in such animals, since the only important fat-splitting enzyme of the body is manufactured by the pancreas. So depancreatized dogs were prepared, and the digestive disturbances observed.

But another thing happened to these dogs. It is said that the

3. These clusters of cells are also referred to as *islets of Langerhans*, after their discoverer, Ernst Robert Langerhans, the German histologist.

4. This operation is called a *pancreatectomy*. The same terminology is used also for excision of other glands or organs. Hence we may speak of *thyroidectomy*, *adrenalectomy*, etc. An animal whose pancreas has been removed is said to be *pancreatectomized*, or *depancreatized*.

animal caretakers observed that ants collected about the cages of those dogs from whom the pancreas had been removed. The ants were attracted to the urine of the operated animals. The discovery soon followed that the urine contained great quantities of sugar. Von Mering and Minkowski extended their experiments and demonstrated that, invariably, when the pancreas was removed, urine was secreted in great volumes and was always "loaded" with sugar. The condition was fatal in ten to thirty days.

These dogs thus resembled persons afflicted with severe diabetes mellitus. This disease had been known in man for centuries, although its cause was unknown and its control quite inadequate. A rare opportunity now presented itself. If the true nature of this diabetes in dogs could be ascertained, perhaps an understanding of human diabetes might also follow. If the experimentally induced disease could be controlled, perhaps the human disease could likewise be controlled. And so many investigators set to work.

2. *Function of the islets.*—It was first necessary to establish the fact that the effects noted in the dogs were entirely due to the absence of the pancreas, for, after all, when the pancreas is removed surgically, many things are necessarily done besides depriving an animal of the organ. The animal is subjected to more or less prolonged ether anesthesia; there is an incision in the abdomen, and nerves and blood vessels are cut; there may be considerable hemorrhage. The organs in the upper abdomen are necessarily injured somewhat by manipulation, and the structures immediately adjacent to the pancreas are considerably damaged. In addition to all this, the dog is deprived of its most important digestive juice and of the pancreas. What justification is there for thinking that only one of these—the loss of the pancreas—causes diabetes? Only further experimentation could provide the answer.

Control experiments were done—"dummy operations" performed —in which exactly the same procedure was followed, involving the same degree of anesthesia, trauma, and hemorrhage; but the pancreas, though traumatized, was not removed. The abdominal incision was closed, and the dogs were treated exactly like the others postoperatively. These dogs failed to develop diabetes. Unquestionably, the lack of the pancreas was the specific cause.

In mammals the pancreas is really two organs, although in its gross structure it appears to be a single gland. There is this division, microscopically, into tissue which secretes digestive juice and islet tissue. Which of these is necessary to the animal to prevent the onset of diabetes? It would be impossible to dissect out one kind of tissue alone and remove it. However, if the ducts of the pancreas are tied off, it is found that the digestive portion of the gland degenerates in time; only the islet tissue remains. And, again, in such dogs there is no diabetes. Thus, absence of the whole gland causes diabetes, but absence of the digestive portion does not. Therefore, absence of the islet tissue alone causes diabetes. In support of this is the frequent finding of degenerative changes in the pancreatic islets in human diabetes. The whole gland may also be diminished in size. There is on record the case of a woman whose pancreas at autopsy was found to weigh only 15 gm. as compared to the normal of about 90 gm.[5]

3. *Discovery of insulin.*—How does absence or deficiency of the islet tissue cause diabetes? One hypothesis—the correct one—was that the islet tissue normally produces a material which is secreted into the blood or lymph and that this material is responsible for the prevention of diabetes. Investigators reasoned that, if pancreas tissue were fed by mouth, perhaps the material manufactured by the pancreas—the hormone—might be absorbed from the alimentary canal in quantities sufficient to replace the lacking hormone of the depancreatized dogs. But this was not so. If any of the hormone had been stored in the glands, it was apparently destroyed by the digestive juices before absorption could take place. The diabetes of the operated dogs fed pancreas was just as severe, and they died just as soon as operated dogs fed an ordinary diet. The hormone—if there was a hormone—would have to be injected under the skin or into the blood.

Numerous investigators tried to extract the hormone from great quantities of pancreas tissue, empirically using many kinds of solvents, both acid and alkaline in reaction. But none of the extracts, upon hypodermic (i.e., under the skin) or intravenous injection, had any appreciable effect upon experimentally diabetic dogs. Many of these extracts were actually injurious. Twenty years of repeated failure passed.

Then Dr. Frederick Banting, a young Canadian physician, finding this problem more interesting than the practice of medicine, got an idea which he presented to Dr. J. J. McLeod, at the University of Toronto. Possibly the clue to the whole difficulty lay in the fact

5. Such findings are not always observed. In some instances little or no visible change is detectable in the pancreas of diabetics.

that the pancreas is two glands. Perhaps the digestive enzymes of the pancreas itself actually destroy such hormone as may be present in the tissue, so that, when a mash of the whole pancreas is made, the hormone is destroyed before the chemical extractions can be made. Banting wanted to extract islet tissue alone, freed from the destructive enzyme-producing cells. Furthermore, he knew where he could get such islet tissue.

In embryonic development the islet tissue of the pancreas develops before the rest of the gland produces digestive enzymes. So, with the collaboration of McLeod, Charles Best, and J. B. Collip, Banting extracted enzyme-free embryonic pancreas and injected his extracts into diabetic dogs. The diabetes became less severe.

Since this work in 1922 the problem has been one of obtaining more and more highly purified extracts—preparations containing the highest concentration of the hormone. Today a highly purified preparation is available. It is called *insulin*, from the fact that it is the hormone of the islet tissue. Insulin will keep depancreatized dogs alive and in relatively good health for months. Obviously, the hormone must be injected at rather frequent intervals—sometimes, two or three injections a day—and even then there is but a roughly adequate substitution for the more or less continuous liberation of insulin occurring in the normal animal. A few hours after insulin injection is stopped, of course, the depancreatized dog becomes severely diabetic and dies in about two weeks.

Ultimately, the hormone preparations were sufficiently free from injurious impurities to warrant attempting their use in human diabetics. The effect was exactly the same as on diabetic dogs; the diabetes was practically completely controlled. Today there are hundreds of thousands of diabetics who owe their comparative good health, and even life itself, to insulin. Unfortunately, no insulin preparation has been developed which is active when taken by mouth, since it is apparently destroyed by the digestive enzymes. Hypodermic injection must always be used, but patients soon learn how to calculate the required dosage of insulin and how to inject it. As long as proper self-administrations of insulin are given, and a certain dietary regime adhered to, diabetics lead practically normal lives.

B. Functions of Insulin

Why is insulin necessary for life? What specifically is the function of the hormone? An intimate relation to sugar metabolism is suggested by the invariable presence of sugar in the urine of diabetics. The concentration of sugar in the blood is also elevated. Instead of the normal concentration of about 0.1 per cent by weight, there may be ten or twenty times this amount.[6] It is the elevation of the blood sugar which is the direct cause of the appearance of glucose in the urine.[7] Now, the presence of sugar in the urine is not in itself harmful. A normal individual may excrete some sugar, without even being aware of it, after a meal rich in carbohydrate, or in a period of emotional stress. The sugar in the urine and the high blood-sugar

Fig. 185.—Injection of insulin. Diabetics, even children, can be taught to give themselves subcutaneous injections of lifesaving insulin, at intervals and in amounts which they themselves learn to compute. Repeated checks by the physician are also required. (From the sound film, *The Endocrine Glands.*)

concentration in the diabetic are symptoms of the disease and not the disease itself.

1. *Combustion and storage of glucose.*—Many studies suggest that the completely diabetic animal or the severely diabetic human being[8] is unable to store and to burn sugars in the normal manner. There is apparently considerable interference with the deposition of glucose as glycogen in the liver. The abnormal glucose combustion is more

6. Called *hyperglycemia.*

7. Called *glycosuria.*

8. It is improbable that complete diabetes in man ever exists. The pancreas of diabetics probably produces some insulin, the quantity varying with the severity of the diabetes.

serious than is apparent at first because, of the amino acids absorbed from protein digestion or liberated from protein breakdown in the organism, over one-half are normally converted into glucose. In the diabetic this glucose also is improperly metabolized and becomes added waste.

2. *Relationship to fat metabolism.*—A striking consequence of insulin deficiency is the rapid withdrawal of fat from the storage depots. The blood may become so rich in fat that the plasma appears milky. It is thought that some of this fat is converted into glucose and that this, coupled with an increased glucose formation from proteins, results in an "overproduction" of glucose in the body. The hyperglycemia, then, is ascribed not only to a defective utilization and storage of sugar but also to an abnormally great formation of sugar in the body from proteins and fats.

The disturbance of fat metabolism is one of the most serious defects in diabetes. In the normal combustion of fats the molecule is broken down ultimately to carbon dioxide and water. Intermediate products in this oxidation are certain acid bodies, related to acetone,[9] which is also produced. In normal metabolism, when much of the required energy of the organism is derived from glucose combustion, the oxidation of these acid intermediate products of fat metabolism is complete, and there is no accumulation of the acids. But, in diabetes, when carbohydrate metabolism is impaired, and when the body derives an unusual proportion of its energy from fat oxidation, the intermediate products of fat breakdown cannot be entirely oxidized to carbon dioxide and water. These products accumulate in the body and account for much of the damage occurring in diabetes.

Uncontrolled diabetes is fatal, then, partly because of interference with energy liberation in the body and depletion of the carbohydrate and fat reserves. The diabetic becomes extremely emaciated and is especially subject to infections. Equally important in determining a fatal outcome is the accumulation of acid products.

Insulin injections tend to rectify all this. Sugars are again normally metabolized and can be stored as glycogen. Excessive glucose production from proteins and fats ceases. The fats are completely oxidized to carbon dioxide and water, and the acid intoxication disappears.

9. Because acetone is volatile, its odor can be detected on the breath. The related acid bodies are *acetoacetic acid* and *beta-hydroxybutyric acid*.

C. Insulin Excess

Insulin administration is not without its dangers. If one injects more than is needed to effect the oxidation or storage of just the quantity of glucose eaten, or if excess insulin is injected into a normal individual, serious consequences follow. The concentration of sugar in the blood and body tissues falls. Instead of the normal level, there may be only 0.06 or 0.05 per cent. For some unknown reason, certain nerve cells in the brain are unable to function normally when the concentration of sugar in their environment is lowered to this point. They become highly irritable, and when the level of sugar in the blood reaches about 0.04 per cent, a little less than half the normal concentration, convulsions may set in, soon followed by unconsciousness and death.

A condition simulating this occurs "spontaneously" in those rare cases where the islet tissue becomes hyperactive. A tumor of the pancreas may so increase the quantity of islet tissue that abnormally large quantities of insulin are produced. This condition, the opposite of the hypoinsulinism of diabetes, is known as *hyperinsulinism*. The acute symptoms can be relieved by ingesting or injecting glucose, which promptly but temporarily elevates the blood-sugar concentration. The more permanent control of spontaneous hyperinsulinism, as in the case of hyperthyroidism, is effected by the surgical removal of the excess glandular tissue.

D. The Fundamental Cause of Diabetes

Although diabetes can be fairly well controlled with insulin, the fundamental problem is by no means solved. Entirely unanswered as yet are the questions: What starts the diabetes, and what causes the pancreas to become defective and secrete inadequate quantities of insulin? Diet is apparently not an important factor, and there is but little experimental justification for the popular belief that a high carbohydrate diet induces diabetes. There is only the fact that, once the disease has established itself, it is sometimes desirable to reduce the carbohydrate intake. There seems to be some hereditary basis for the condition, for, while there is no evidence for an infectious origin, it may make its appearance in several members of a family and tends to "run in families" in some instances.

Perhaps the initial defect is related to the hypophysis rather than the pancreas (see pp. 574–75), since injections of extracts of the hypophysis not only produce some of the features of diabetes but also cause degeneration of the pancreatic islet cells, at least in some experimental animals. But, if the hypophysis causes the pancreas defects, we are still left with the problem of what causes the abnormal activity of the hypophysis.

Obesity seems to bear some relationship to diabetes, since the disease occurs much more frequently in the obese than in those of normal weight. What the causal relationships might be is not clear.

Far from answered also is the problem of just how insulin functions to make possible the burning and storage of sugar. And to what degree and why is the burning of fat dependent upon carbohydrate combustion? Just what are the chemical, enzyme, and hormone relationships here? These and many others are questions which only future experimentation can answer.

III. THE THYROID GLANDS

The thyroids are paired glands lying alongside the trachea in the neck. In man the two structures are attached by a narrow band, or *isthmus*, of thyroid tissue which passes in front of the trachea. Embryologically, the gland[10] develops as a single midline outpouching of the primitive alimentary canal near the base of the tongue in the back of the mouth. In normal development the duct connecting the gland with the primitive digestive tract disappears.

Microscopically, the gland is seen to contain groups of cuboidal cells arranged in many one-cell-thick irregular spheres, the interior cavities of which normally contain a colloidal material which we now know contains some of the stored hormone which the thyroid cells have manufactured. Between these adjacent spheres (or *follicles*) are found connective tissue and a rich network of blood vessels into which the hormone is secreted.

A. DEFICIENT FUNCTION IN ADULTS

Our first knowledge of the function of the thyroids came before the endocrine nature of the glands was known. The beginnings were made, as in the case of the pancreas, through observations upon underactivity, or hypofunction. In 1874 Sir William Gull associated specific physiological defects in certain of his patients with degeneration of the thyroids of the afflicted individuals. Today hypothyroidism is still often designated as *Gull's disease*. Within a decade surgeons had unwittingly produced Gull's disease in man by surgical

10. The singular term *thyroid* is used frequently to designate all the thyroid tissue collectively—the paired lobes plus the isthmus.

removal of *goiters*—enlarged thyroids. Hypothyroidism has since been effected in a variety of animals. There is a close parallel between the picture of a human being with spontaneous hypofunction of the thyroids and animals with extirpated thyroids.[11]

1. *Myxedema.*—Some of the effects of hypothyroidism are very apparent. Gull described a great reduction in mental and physical vigor, a loss of hair, and peculiar thickening of the skin, which resembled the appearance of skin containing an abnormal quantity of water, that is, edema. This skin picture gave the name *myxedema* to the condition of hypothyroidism in adults.

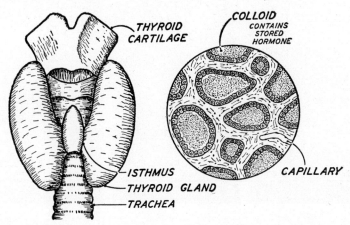

Fig. 186.—The thyroid gland. *Left:* Location and gross appearance seen from the front. *Right:* Microscopic structure. The gland cells are arranged in spheres, whose interiors contain colloid. The interior of the sphere is not to be confused with a duct, which is lacking in the thyroid gland of adults.

2. *Effect on the basal metabolic rate.*—A still more important observation was made some years later. The heat production of people with Gull's disease was found to be considerably less than normal, sometimes by as much as 40 per cent. A similar lowering was soon described in animals deprived of the thyroids. This led to the concept, amply confirmed today, that the thyroid in some way regulates the oxidative rate and, therefore, the heat production and energy liberation of probably all the cells of the body.

Many of the effects of hypothyroidism probably bear a significant causal relationship to the generally lowered oxidative rate. The de-

11. Confusion resulted at first because, in some experimental animals (e.g., dogs and cats), the *parathyroid glands* lie imbedded in the thyroids. Consequently, the results of thyroid removal were complicated by the simultaneous occurrence of experimental hypoparathyroidism.

creased energy liberation in specific organs and systems probably is important in producing some of the specific defects. The mental sluggishness, lowered heart rate, and decreased sex drive may represent decreased metabolic activity of the nervous system, the heart, and the gonad system. There may also be a gain in weight. The food eaten, being burned relatively slowly, is stored in greater quantities than normal.

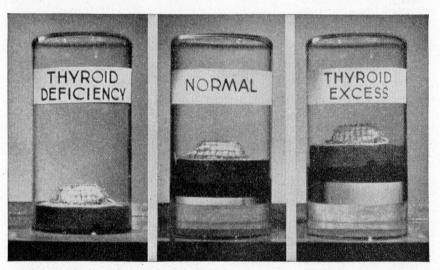

Fig. 187.—Effect of the thyroid hormone on oxygen consumption. Each of the three chambers is so arranged that, as oxygen is used up by the rat within, the platform upon which the rat rests rises in the oxygen-filled chamber. Each rat is restrained in a small wire cage, to reduce muscular movements and therefore approximate "basal" metabolic conditions. Note that the rat with thyroid deficiency (thyroid glands removed) uses less oxygen than the normal rat and that the rat with thyroid excess (from feeding thyroid hormone) uses more oxygen than the control animal. At the start of the experiment (not shown) all three platforms were at the same height (i.e., each chamber contained an equal volume of oxygen), and all rats weighed the same. (From the sound film, *The Endocrine Glands.*)

It was soon demonstrated that the glandular defect in myxedema or experimental hypothyroidism could be substituted for, and normality more or less adequately restored, by the administration of thyroid substance. The difficulties met with in research on diabetes were not encountered here, for oral administration of thyroid-gland products was found to be quite effective. The thyroid hormone, unlike insulin, is highly resistant to the action of the digestive juices. The whole gland may be eaten, or, as is the practice today, a dried, powdered preparation of whole gland may be taken in capsules.

It seemed unquestionable, therefore, that the thyroids were glands

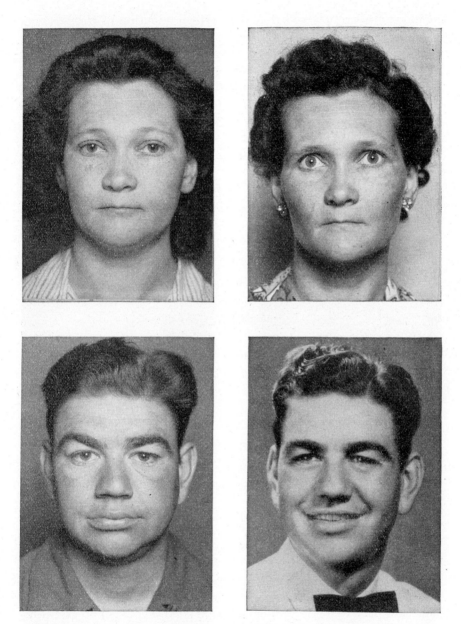

Fig. 188.—Hypothyroidism in the adult, before (*left*) and a few weeks after (*right*) daily administration of thyroid-gland hormone. Note especially the before-and-after differences in alertness, skin puffiness, and hair texture.

of internal secretion and that defective production of the hormone could readily be balanced by thyroid substance taken by mouth.

B. Cretinism and Its Control

1. *The condition in man.*—Persisting hypothyroidism beginning in infancy presents a more striking picture and adds considerably to our knowledge of the function of the thyroid hormone. There is a marked

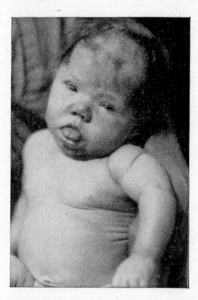 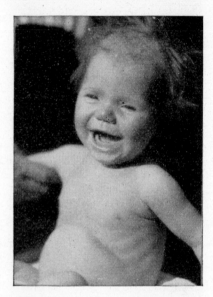

Fig. 189.—Hypothyroidism in the young. A cretin infant, before and after administration of thyroid hormone. The sluggish, lethargic child (*left*) would have become a mentally retarded dwarf. Administration of thyroid hormone for six weeks transformed the child (*right*) toward responsive normality. Continuous administration of the hormone has resulted in a subsequent relatively normal life.

retardation in mental, physical, and sexual development. This condition is called *cretinism*, and the individuals *cretins*. Cretins are dull mentally and, if untreated, may never exceed the level of intelligence of a child of five years. Bodily growth is retarded and a permanent dwarfism is established. Not all cretins are sterile, but they display a delay in sexual maturity and a lack of normal sexual vigor. At autopsy the thyroid glands are seen to be quite abnormal or very small.

2. *Experimental cretinism.*—All these results can be paralleled by removal of the thyroids in young animals, so that there can be no question of the causal relationship of cretinism to the thyroid glands.

As might be expected from what has been said about myxedema, this condition also can be fairly successfully alleviated by eating dried thyroid preparations. This applies equally well to "spontaneous" and experimental cretinism. Obviously, such control of the condition must be instituted at an early age. A human cretin of twenty years of age—a dwarf, and perhaps an imbecile—cannot be restored to normal. The developmental period is nearly over and cannot be reinstituted to any great extent.

Cretins display the same low heat production encountered in myxedema. Probably the defects in development are in large part secondary to this lowered rate of metabolic activity.

Fig. 190.—Normal and cretin rabbits. The two small rabbits, weighing about 800 gm. each at twelve weeks of age, had their thyroid glands removed soon after birth. The large rabbit, weighing 1,600 gm., is a litter mate of the others. Its thyroid glands are intact. All three rabbits were fed the same diet. (From Basinger, courtesy of *Archives of Internal Medicine.*)

From these data upon the effects of hypofunction,[12] we may conclude that in the normal individual the thyroid hormone helps to regulate (*a*) rate of cellular oxidations and heat production—basal metabolic rate; (*b*) physical growth; (*c*) mental development; and (*d*) attainment of sexual maturity.

C. Goiter

Goiter is an enlargement of the thyroid gland. One might guess that an enlarged thyroid would necessarily indicate a hyperactive thyroid, but this is by no means always the case. There is a relatively

12. Anemia also occurs in hypothyroidism. This defect of bone marrow activity is probably related to the general decrease in tissue activity.

common form of goiter associated with hypothyroidism. Even cretins may have goiter.

1. *Compensatory thyroid enlargement.*—The relationships here are believed to be about as follows: a defect of thyroid physiology occurs in which there is an inadequate output of the thyroid hormone; secondary to this the thyroid enlarges. More thyroid cells are produced, which makes up in part for the deficient activity of each cell. This is a compensatory reaction whose machinery is poorly understood; it is a specific example of the general tendency for the organism to compensate for defects. If part of an organ is destroyed experimentally or by disease, the remaining organ or tissue may enlarge or increase its activity in such a way as to tend to offset the loss.[13]

2. *Simple goiter.*—Goiter formation may or may not compensate for the original defect. Each thyroid cell may be so hypoactive that, even with a greatly increased number and size of cells, the total hormone output of the enlarged gland may still be insufficient. On the other hand, presumably if the defect of each cell is relatively slight, the increased number of cells may bring the total hormone production up to normal. We should then have essentially a normal individual, with a normal thyroid hormone output, a normal metabolic rate, and normal growth and mental and sexual development, plus an unsightly swelling of the neck. Such a goiter is called a *simple* goiter.

3. *Relation of the thyroid to iodine.*—The search for the cause of the original defect of the hypoactive thyroid cells has led finally to the isolation, purification, and even the artificial synthesis of the thyroid hormone, besides yielding a comparatively effective control of goiter, whether or not it is associated with hypofunction.

Investigators for a century have been struck by the fact that goiter, myxedema, and cretinism have been especially common in certain geographical regions. These regions are the Himalayan Plateau in Asia, the Alps, Pyrenees, and Carpathian mountain regions in Europe, the Andean Plateau in South America, and the St. Lawrence and Great Lakes regions in North America. For the most

---

13. It should be realized that to call a reaction such as goiter formation in hypothyroidism a "compensatory" phenomenon is purely descriptive and tells nothing about the physiological machinery involved. It still leaves the question: Specifically, what machinery is set into operation, and how, which results in the compensatory adjustment?

part, the goitrous areas either are mountainous or are valleys and plains whose soils were leached by the last glaciation. But what have mountains or leached soils to do with goiter?

About a century ago the observation was made that the soil, water, and vegetation of these regions were relatively poor in iodine

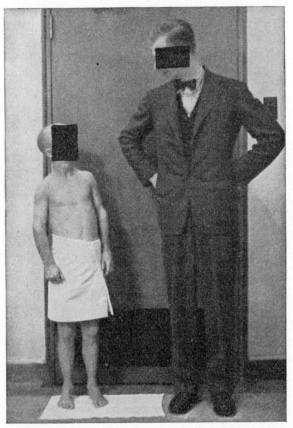

Fig. 191.—An adult cretin compared with a normal man. (From the sound film, *The Endocrine Glands*.)

and a possible causal relationship was suggested. It now seems strange that so little attention was paid this observation, especially since it had also been demonstrated that iodine-feeding had some value in the treatment and control of goiter. These observations began to take on meaning at about the turn of the century, when it was conclusively shown that iodine is to be found in the thyroid gland, in relatively large quantities, in a firm organic combination.

Further work amply confirmed the tentative hypothesis which could now be formulated: The thyroid hormone contains iodine as an important atom of the molecule. This iodine must eventually come from foods and drinking water. When these are deficient in iodine (e.g., because of soil leaching), the thyroid cells are unable to manufacture the hormone, and hypothyroidism results, sometimes partly compensated for by goiter formation. It became clear, therefore, that the rational control of simple goiter was to furnish iodine to the organism artificially. Today this is done on an extensive scale, in goiter districts, commonly by the addition of iodine in the form of sodium or potassium iodide to table salt. Simple goiter, cretinism, and myxedema have been greatly diminished as a result. The incidence of goiter in one canton in Switzerland declined from 88 per cent of the population to 13 per cent in a period of three years.

Nevertheless, the problem of hypothyroidism is by no means wholly solved. Defective hormone output may still occur even though adequate iodine is supplied to the gland cells. Although this condition may be controlled by feeding thyroid-hormone preparations, the question still remains: What causes the failure of the thyroids to manufacture and to secrete a sufficient quantity of the hormone?

D. The Thyroid Hormone

1. *Chemical composition.*—There have been great forward strides in our knowledge of the chemistry of the thyroid hormone. It is found in the cells and in the stored colloid of the gland and is called *thyroglobulin.*[14] Investigators have succeeded in chemically splitting off the protein part of the thyroglobulin molecule, leaving a crystalline product which has been given the name *thyroxin.* This product, when fed or injected, produces essentially all the physiological effects that have been described for whole dried thyroid. Upon analysis, thryoxin was found to have the formula $C_{15}H_{11}O_4NI_4$.

A further important chapter in thyroid-hormone chemistry was written when thyroxin was synthesized in the test tube. Artificially prepared thyroxin apparently possesses the chemical and physiological properties of natural thyroxin. Since the structural formula of the synthetic substance is known, it seems likely that this is also the

14. The globulins comprise one group of proteins.

structural formula of the thyroid hormone. Thyroxin is really an amino acid. The structural formula is as follows:

```
      I  H              I  H            H   H
      |  |              |  |             \ /
      C—C               C—C          H   N      O
     ╱    ╲            ╱    ╲         |   |      ‖
HO—C        C—O—C          C—C—C—C—OH
     ╲    ╱            ╲    ╱         |   |
      C=C               C=C          H   H
      |  |              |  |
      I  H              I  H
```

The importance of iodine in thyroxin is indicated by the fact that, if two of the iodine atoms are replaced by hydrogen, the substance loses most of its stimulating action on cell metabolism; if all four iodine atoms are replaced by hydrogen, all its activity is lost.

2. *Control of hormone secretion.*—Apparently the rate of secretion of the thyroid hormone is under the direct control of the hypophysis (see also p. 566). Hypothyroidism essentially identical with that induced by thyroid-gland destruction can be produced in an animal with an intact thyroid gland by removing the hypophysis. In such animals normal thyroid function can be restored by administering a chemical derived from the hypophysis, called *thyroid-stimulating hormone.*

It appears likely that hypothyroidism may be due mainly to deficient hypophyseal secretion of this hormone rather than primarily to a defect of the thyroid gland.

E. HYPERTHYROIDISM

In man spontaneous hyperactivity is far more common with the thyroid gland than with the pancreas. This may be a matter of each cell producing a great excess of the hormone or of an actual enlargement of the gland. At any rate, there is an abnormally large quantity of thyroxin produced and secreted into the blood stream.

1. *The occurrence of goiter.*—The occurrence of goiter (Fig. 192) in both hypothyroidism and hyperthyroidism is at first sight very confusing. The enlargement in the two cases is seemingly quite different and is induced in different ways. That in hyperthyroidism is apparently the original defect, resulting from some abnormal stimulation of unknown nature and origin. This directly causes the excessive

hormone production. On the other hand, the goiter of hypothyroidism is probably secondary to defective hormone production. It is a reaction of compensatory adjustment by the animal to the abnormally small secretion. It may partly correct the defect. But the primary difficulty is the hypoactivity of the thyroid cells. These differences may be represented schematically, as in Figure 193.

2. *Effects of overactivity.*—Some of the effects of hyperthyroidism could be predicted from the knowledge we have of the action of the thyroid hormone gained from studies upon spontaneous or experimental hypothyroidism.

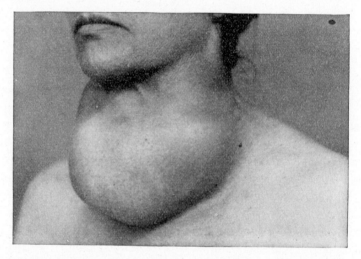

Fig. 192.—An unusually large goiter. Many goiters associated with serious abnormal thyroid gland function are so small as to be unnoticeable upon casual inspection of the neck.

In hyperfunction there is an elevation of the rate of heat production—an increase in the metabolic rate of all the cells of the body to 30 per cent or more above normal. The cells are overstimulated by the excessive quantity of thyroid hormone in the blood and body fluids. Certain consequences of the increased fuel combustion naturally follow. The hyperthyroid individual keeps warm easily, characteristically uses less clothing than a normal person, and is too warm in a room which is comfortable for others. The increased heat production stimulates the physiological machinery which dissipates heat; consequently, profuse sweating is common. There may also be an increase in appetite, and the individual may eat ravenously. Yet he not only fails to gain weight but usually actually loses. The added food ingested is not stored but serves as fuel for the metabolic flame fanned by the excess of thyroid hormone. Lacking reserve fuel, wastefully squandering energy, the hyperthyroid lacks vigor and finds it more difficult to perform any sustained physical work than does the normal individual.

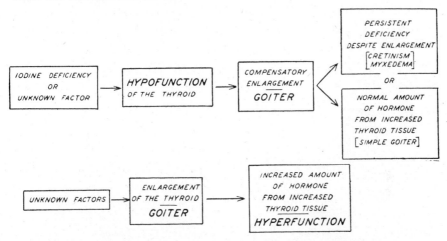

FIG. 193.—Causal relationships of goiter to hypo- and hyperthyroid function. Arrows indicate "followed by" or "causes." Hypo- or hyperthyroidism may also occur without goiter.

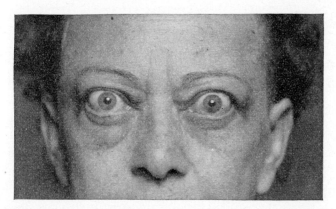

FIG. 194.—Involuntary bulging of the eyes—exophthalmos—frequently seen in hyperthyroidism.

In contrast with the sluggishness and apathy of myxedema, there is in hyperthyroidism a marked, subjective nervous tension as well as overt nervousness and heightened excitability. The metabolism of the central nervous system is whipped up, along with other cells of the body, and often this nervousness is the most outstanding complaint. The heart also is stimulated, so that characteristically there is an accelerated heart rate.

Because growth is impaired in hypothyroidism, one might expect that gigantism would result from an excess of thyroid hormone. But hyperthyroidism usually manifests itself in the young adult, so that growth effects are not seen. Nor can giants be produced experimentally with an excess of the hormone. However, it is possible to

TABLE 12

EFFECTS OF UNDERACTIVITY AND OVERACTIVITY
OF THE THYROID GLANDS

| EFFECTS ON | HYPOTHYROIDISM | | HYPERTHYROIDISM |
	Cretinism	Myxedema	
Metabolic rate (cell oxidations)	Decreased	Decreased	Increased
Body weight	Poor growth Dwarfism	Increased, sometimes	Decreased, usually
Mentality	Sluggish Feeble-minded, sometimes	Sluggish	Normal Usually slight to extreme nervousness and excitability
Sex	Poor sexual development	Decrease in sexual vigor	Normal
Heart rate	May be normal	Slow	Fast
Skin	Dry	Puffy, thick Sweating decreased	Moist from excessive sweating
Sensitivity to cold	Increased	Increased	Decreased
Other effects*		Anemia Hair thin, coarse	Damage to heart Gastrointestinal upsets Bulging of eyes

* See pp. 521 and 525 for discussion of goiter.

accelerate growth in certain animals. Tadpoles living in water in which the hormone is added become mature frogs much more quickly than untreated tadpoles.

3. *Evidence for excessive hormone production.*—That all these effects, and others,[15] are dependent upon an excessive hormone production is proved by the following observations.

15. Hyperthyroidism often results in a peculiar bulging of the eyes, called *exophthalmos* (see Fig. 194). There sometimes are gastrointestinal upsets, such as diarrhea. There may be injury to the heart muscle which may dangerously weaken that organ.

First, experimental hyperthyroidism can be produced easily in man or animal by feeding the thyroid hormone to normal individuals. An elevated metabolic rate, increased appetite, loss of weight, decreased sensitivity to external cold, and nervousness—all these can be artificially induced by administration of the thyroid hormone.

Second, the hyperfunction as it occurs in man can be controlled by reducing the hormone output of the gland. No practical way is known of permanently decreasing the hormone secretion of the gland cells themselves, but it is possible to reduce the number of cells and, therefore, the hormone secretion by the entire gland. The most effective manner of accomplishing this is the surgical removal of part, or sometimes nearly all, of the overactive gland. Various nonsurgical methods of destroying gland cells have also been attempted; for example, by radiation with X-rays, which is usually not effective.

Since the development of the atomic bomb and our vastly increased knowledge of the structure of the atom, a most ingenious technique for destroying thyroid-gland tissue has been evolved. It is now possible to convert ordinary atoms into "radioactive" atoms which emit rays somewhat like those produced by naturally occurring radium. While acquiring radiation properties, such atoms, in addition, retain most if not all of their former chemical properties. This is true, for example, of iodine. Converted into radioactive iodine, this substance, when fed, is deposited primarily in the thyroid gland, just as is ordinary iodine. Normally, this atom is employed in the gland for the manufacture of the thyroid hormone. However, the radioactive iodine, in addition, destroys thyroid-gland tissue, much as if the gland were exposed to X-rays or radium, except that the destructive action is apparently localized to the gland, in which about 80 per cent of iodine eaten localizes.

A final evaluation of this procedure remains to be made, since the degree of gland destruction achieved is difficult to control, and occurrence of possible delayed injurious effects elsewhere in the body (exposed to the 20 per cent of absorbed iodine not deposited in the gland) has not been determined. The method has been especially useful in instances of hyperthyroidism occurring in individuals whose general physical condition is very poor, thus making an operation especially dangerous. Such therapy has also proved promising in efforts to destroy the tissues in cancer of the thyroid gland.

F. Relationship of the Thyroid to Body Heat

While it is quite clear that the thyroid controls the rate of cell oxidations and, therefore, of heat production, there is no convincing evidence that the body heat production can be changed *rapidly* by variations in hormone production when changes in the environmental temperature occur. The automatic increase in body heat production which occurs in a cold environment, and helps maintain a constant body temperature, is probably not induced by any temporary physiological increase in thyroid-hormone output. The metabolism-stimulating action of the thyroid apparently occurs too slowly to be of significance in this situation, for, if thyroid hormone is artificially administered, the increased heat production fails to manifest itself until some twenty-four or thirty-six hours have elapsed. Possible adjustments to external cold by an increase in hormone secretion and body heat production over a longer period of time have been discussed earlier (see pp. 332–33).

G. Unsolved Problems

Many questions of normal and abnormal thyroid physiology remain to be answered by future investigators. What other factors, controllable or otherwise, besides iodine intake, regulate the function of the thyroid? What is the underlying cause of hyperthyroidism? How does hypothyroidism induce the compensatory goiter formation? What initiates the goiter responsible for hyperthyroidism? Is thyroid malfunction essentially a defect of the hypophysis?

IV. THE PARATHYROID GLANDS

A. Indispensability of Parathyroids for Life

1. *Fatal result of extirpation.*—Our knowledge of parathyroid physiology commenced with an accidental discovery and an erroneous conclusion. Investigators who were interested in the relationship of the thyroids to myxedema and cretinism removed the thyroid glands of dogs. There resulted a striking series of events, culminating almost invariably in death within a few days. The observers concluded that the thyroid extirpation was responsible and that, at least in dogs, the thyroid glands are necessary for life. However, no such fatal effect follows thyroid extirpation in rabbits. The significant difference was soon found to be due to no difference in thyroid function in these two species of animals but to the fact that in dogs the small structures now called parathyroids lie imbedded within the thyroid glands, whereas in rabbits they do not. Consequently, when a thyroidectomy is done on a dog, the parathyroids are also removed. In rabbits at least some of the parathyroids are undisturbed by thyroidectomy.

2. *Parathyroid tetany.*—The effects of parathyroid extirpation have been studied in detail. Within a day or so after the operation there are involuntary muscular twitchings, which become more and more severe, amounting finally to convulsions. These attacks are called *tetany*, or, more specifically, *parathyroid tetany.*[16] The animal

may die during the first convulsive seizure. If not, there is a spontaneous remission, followed sooner or later by another attack. Few dogs survive three attacks, and all are sure to die in a few days if the entire parathyroid tissue has been removed.

What causes the tetany? There is a demonstrable increased irritability of muscles and nerves. If one applies an electrical stimulus over the muscles[17] of a dog lacking his parathyroids, it is found that the muscles will twitch in response to stimuli much weaker than those necessary to stimulate the muscles of a normal

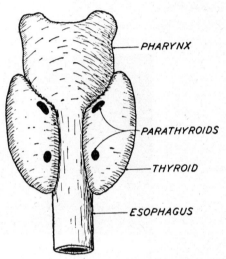

FIG. 195.—The parathyroid glands, located on the posterior (dorsal) surface of the thyroid glands. Sometimes the parathyroids are imbedded in the thyroid substance. Parathyroid tissue is composed of a dense mass of cells, quite unlike the arrangement of thyroid cells.

dog. In the parathyroidectomized animal, then, many of the internal and external stimuli which are subthreshold for the normal tissues evoke the muscular twitchings and convulsions of tetany (Fig. 196).

B. REGULATION OF THE BLOOD-CALCIUM LEVEL

But why does parathyroid removal increase muscle irritability? The answer to this important question was undoubtedly arrived at

16. This is not to be confused with *tetanus* of skeletal muscle or with the condition produced by the *tetanus bacilli*.

17. Skeletal muscles are readily stimulated in the intact animal without discomfort. Testing the electrical irritability of the intact muscles of man is often done in certain nervous diseases or in mild human parathyroid deficiency.

more quickly because of a piece of research of the type which many are inclined to regard as utterly useless, since its immediate application to practical problems is not at once apparent.

1. *Effect of calcium on muscle irritability.*—Jacques Loeb was far more interested in the physiology of cells than in endocrinology. He experimented upon cells of all kinds, including nerve and muscle cells. Empirically, and without the slightest thought of the parathyroid glands, Loeb observed the effect of various ions upon the irritability of excised muscles and nerves. Among other things he learned that, when muscles are placed in a fluid lacking or deficient in calcium ions, there is a greatly increased irritability of the muscles,

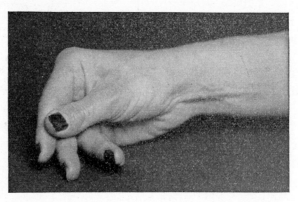

Fig. 196.—Parathyroid tetany. Involuntary contractions of hand and arm muscles resulting from insufficient secretion of parathyroid hormone and a consequent lowering of the blood calcium.

which may manifest itself in "spontaneous" twitchings. If a little calcium is added to the fluid, the twitchings cease, and normal irritability is restored.

2. *Control of tetany by calcium administration.*—This work served to focus attention upon the calcium concentration of the blood and tissue fluids in parathyroidectomized animals. The blood-calcium concentration was found to be uniformly lower than normal.[18] If the tetany were significantly related to this fact, which Loeb's work might suggest, then it should be possible to relieve or prevent the tetany by artificially restoring the blood-calcium concentration to its normal level.

The experiment was tried with dramatic results. If calcium salts

18. Other changes in the blood chemistry also occur; for example, there is always an elevation of the blood phosphorus.

are injected into the vein of an animal in tetany, the convulsions cease within a minute, and the animal recovers from the attack. Further attacks can be prevented by calcium injection. It is also possible to stop or prevent the attacks, and keep the animal alive, by administering adequate quantities of calcium by mouth. In any case, the calcium administration must be repeated frequently because, some hours after injection or ingestion, the blood calcium returns to its former dangerously low level.

Here we encounter another specific example of the fact that the cells of the body are very nicely adjusted to just the proper kind and concentration of materials in their environment. Among other things, just the proper concentration of calcium ions in the fluid environment is indispensable for the normal functioning of the cells, particularly of muscle and nerve cells. This concentration is small, for we find only one part of calcium in ten thousand parts of blood. If this quantity is reduced to one-half, increased irritability, tetany, and death result.

We conclude thus that the outstanding function of the parathyroids seems to be the control and maintenance, directly or indirectly, of the proper calcium-ion concentration in the blood. Death is the consequence of a breakdown of this control.

C. The Parathyroid Hormone

The parathyroids exert this control by the secretion of a hormone. It has been possible to extract from the gland a material called *parathormone*, which will prevent tetany and maintain life when injected into parathyroidectomized dogs. The hormone preparation must be administered by injection, for it is ineffective when fed, presumably because it is destroyed by the digestive juices. Its tetany-preventing action is apparently due to its effect upon the blood calcium. Within a few hours after injection into either a normal animal or a parathyroidectomized animal, the blood calcium rises to a higher level and remains so for some hours.

D. The Parathyroid Glands in Man

1. *Hypofunction.*—Thus far nothing has been said of spontaneous malfunction of the parathyroids in man. Most of our knowledge of the physiology of these glands preceded an awareness of even the existence of malfunction. Even today we know but little about hypoparathyroidism in man. Of course, hypofunction follows the occa-

sional accidental removal of some of the parathyroids during surgical excision of a hyperactive thyroid. Usually the parathyroids lie in the dorsal portion of the thyroid glands of man; consequently, they are ordinarily undisturbed by a thyroidectomy, in which only the ventral portions of the gland are removed. But sometimes the parathyroids chance to lie somewhat more ventrally than normal. This condition is impossible to detect by gross examination of the thyroid, and a mild parathyroid deficiency may follow thyroidectomy. This can be controlled by giving parathormone, or calcium, or more commonly both, just as experimental hypofunction is controlled.

No other instances of undisputed hypofunction of the parathyroids in man are known. It has been suggested that convulsions of various kinds may be manifestations of parathyroid deficiency. Parathyroid hormone or calcium administration often improves such conditions, but the relationships have not been clearly established.

2. *Hyperfunction.*—However, hyperfunction of the parathyroids is known to occur in human beings in association with enlargement or tumors of the parathyroids, and the condition can be duplicated in dogs by injecting large amounts of parathormone into normal animals. The blood calcium is abnormally high, and, quite significant for our consideration of the function of the glands, there is a withdrawal of calcium from bones, causing cyst formation and weakening of the bones.

3. *Control of calcium balance.*—We arrive at the view, then, that the primary function of the parathyroids is the regulation of the balance between the calcium in the blood, calcium excretion, and calcium in the bones.[19] Too little parathormone causes a decrease of calcium in the blood, and too much results in a withdrawal of calcium from the deposits in bone and elevates the concentration in the blood. We note here a direct relationship of the hormone to bone and tooth growth and maintenance. Indirectly, through its effect on the blood-calcium concentration, the hormone also regulates nerve and muscle irritability.

We have indicated that the parathyroid glands are necessary for life. This statement needs some qualification. If a dog, deprived of its parathyroids, is maintained in comparative good health by cal-

---

19. This regulation may be rather indirect, via a primary control of the blood level of phosphates, whose blood concentration usually varies inversely with the blood concentration of calcium. Blood phosphates are elevated in hypoparathyroidism, and they decrease when the blood calcium is elevated.

cium or parathormone administration for several months, it will be found that the amounts necessary to prevent tetany gradually diminish until, finally, the animal remains tetany-free under ordinary conditions without special treatment. The reason for this is not entirely clear. It is reasonable to suppose that accessory parathyroids, known to be frequent, would grow larger and gradually be able to replace in function the parathyroids which have been removed. In many instances it seems that a new readjustment occurs between cells and the environment, with its low calcium content.

V. THE MEDULLA OF THE ADRENAL GLANDS

The paired adrenal[20] glands derive their name from the fact that they lie near the kidneys. Few organs, if any, have a richer blood supply than this endocrine gland. The mammalian adrenal is really a combination of two glands, which are quite different embryologically, anatomically, and physiologically. The inner core of the gland is called the *medulla* (or "marrow"). Completely surrounding this and comprising the external surface of the gland is the *cortex* (or "bark"). The only known relationship between the two is this intimate anatomical apposition in most adult vertebrates. They develop from entirely different embryological precursors.

If the gland is cut in half, the two portions can be readily distinguished with the naked eye, and, upon microscopic examination, a decided difference in the staining qualities and the shape and arrangement of the cells is to be seen. In a few of the lower vertebrates (e.g., the shark) the two structures remain anatomically separate even in the adult. Studies on these forms have amply confirmed the conclusions arrived at in studies on the higher vertebrates as to the physiological independence of the two parts. Our account of adrenal physiology is therefore treated under two separate headings—the medulla and the cortex.

A. Epinephrine

1. *Discovery and synthesis.*—The place of the adrenal medulla in the animal economy is anomalous. Our information of the chemistry and physiological action of its hormone is more extensive than that available for any other hormone with the possible exception of thyroxin. Furthermore, the hormone is used extensively and effec-

20. From the Latin *ad*, "near," and *ren*, "kidney." The term *epinephrine* is derived from the Greek *epi*, "upon," and *nephros*, "kidney." The term *adrenalin* has also been used for the product of the adrenal medulla.

tively as a drug in modern medical practice. And yet the true significance of the gland and its hormone and the exact role it plays in the organism are still the subject of some dispute.

About a hundred years ago it was shown that the blood passing from the adrenal gland gave a chemical reaction which indicated the presence in it of a peculiar product, now known to be the hormone from the medulla. This material was not found in the blood of the artery entering the gland. Years later (1895) the nature of the material was more fully investigated. Two Englishmen, George Oliver and Edward Schafer, made an extract of the adrenal medulla and ob-

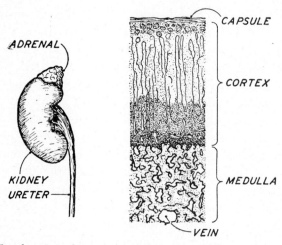

FIG. 197.—The adrenal gland, showing its location (in man), capping the upper pole of the kidney. The microscopic section (*right*) shows the division into cortex and medulla. The section is not magnified enough to show individual cells.

tained a product which, upon injection into an animal, produced rather remarkable effects, the most striking of which was a considerable elevation of the arterial blood pressure. No such product was obtainable from other tissues, even from the adrenal cortex. The material was given the name epinephrine. It was soon prepared in a form pure enough to enable chemists to determine its specific structural formula.

No longer is it necessary to extract ox or sheep glands to obtain epinephrine; accurate knowledge of its chemical structure has enabled chemists to synthesize the product in the test tube. The synthetic product has all the properties of the naturally occurring hormone, and, since its formula is definitely known, we have excellent confirmation of the formula arrived at by chemical analysis of the hormone itself.

2. *Action of epinephrine.*—The physiological actions of epinephrine which result from its intravenous injection[21] are readily demonstrable. It causes a rise in arterial blood pressure, an acceleration of the heart rate,[22] an elevation of the glucose content of the blood, and, according to Cannon, an increase of muscular power and resistance to fatigue as well as an increased rate of blood coagulation.[23]

B. Function of the Gland

1. *Epinephrine and the sympathetic nervous system.*—At first sight, the effects of epinephrine appear to be unrelated—a heterogeneous list of physiological responses. However, it was at once noted that nearly all these effects are duplications of the responses which may be induced by stimulation of the sympathetic (thoracolumbar) division of the autonomic nervous system. Stimulation of these nerves causes a vasoconstriction and an accelerated heart rate, which together cause an elevation of the arterial blood pressure, a liberation of glucose from the liver, a delayed fatigue of the skeletal muscles, an erection of the hairs, a dilatation of the pupil, and an inhibition of intestinal peristalsis. In general, it may be said that the injection of epinephrine duplicates, or mimics, the effects of a rather generalized stimulation of the sympathetic nervous system. Because of this relationship, epinephrine has been called a *sympathico-mimetic* substance.

Because of these and other [24] effects, epinephrine has come to be a useful drug in the hands of physicians. It can be used, for example,

21. Like many hormones, epinephrine is inactive when taken by mouth.

22. Epinephrine itself exerts a chemical stimulation upon the pacemaker of the heart. However, when *large* quantities are injected, there is usually a slowing of the heart. This is a secondary effect entirely dependent upon a reflex initiated by the blood-pressure elevation (see p. 201).

23. Various effects are seen upon nonstriated muscle, such as contraction of the muscles which dilate the pupil and the muscles which erect the hairs, or produce "goose flesh" in man, where the body hair is sparse. There is also an inhibition of the gastrointestinal movements.

24. Epinephrine is sometimes an effective agent in the control of certain asthmatic attacks, partly by causing vasoconstriction and a reduction in the engorgement of pulmonary blood vessels, but also by reducing a spasmodic contraction of the smooth muscles encircling the finer air passages of the lungs. The narrowed caliber renders breathing difficult. Stimulation of the sympathetic nerves of these muscles causes them to relax. Such stimulation is difficult to effect in the intact animal, but the action is readily "mimicked" by epinephrine injection. The resulting sudden alleviation of the respiratory embarrassment in an acute asthmatic attack is most startling.

to elevate the blood pressure—though temporarily—in collapse of the circulatory system during surgical procedures.

2. *The secretion of epinephrine.*—Without question we may say that epinephrine is secreted physiologically. It has been identified in the adrenal-vein blood, using both chemical and physiological tests. The rate of secretion has been determined to be about 0.00000025 gm. per minute per kilogram body weight. Experimentally, this quantity may be increased by stimulation of the efferent nerves of the adrenal glands, or the secretion of epinephrine can be stopped by cutting the nerves of the glands. The direct control of an endocrine gland by nerves is most apparent and most easily demonstrated for the adrenal medulla. Such control is either lacking or difficult to demonstrate in the case of other glands of internal secretion.

3. *Significance to the organism.*—From all this evidence one might conclude that epinephrine is a very important hormone. And yet, when an animal is deprived of all adrenal-medullary function, nothing startling, or even clear cut, results. It is possible to remove the medulla of each gland experimentally by exposing it, slitting it open, and spooning out the medullary core, or practically all epinephrine secretion can be stopped by denervation of the gland. In either case, there is no significant disturbance of function of the cortex, removal of which is invariably fatal. In such a demedullated animal there is no particular disturbance of blood pressure, heart rate, blood-sugar concentration, coagulation time, resistance of skeletal muscle to fatigue, or any other effect which might be expected from our knowledge of the continuous secretion of epinephrine normally and the physiological effects of injection of the hormone.

If epinephrine does anything important in normal life, its role is apparently easily assumed by other mechanisms. In fact, we have already indicated the sort of substitution which may operate. Nervous stimulation, by way of the autonomics, can produce effects upon internal organs and systems identical with those of epinephrine.[25]

4. *Emergency function.*—Some investigators have pointed out that most of the effects of epinephrine would be of utility to an organism in times of stress—during "flight or fight" or fear. A blood-pressure

25. Recall that sympathin is liberated by autonomic nerves (see p. 413) and that sympathin closely resembles epinephrine in its actions.

elevation by vasoconstriction in the abdominal blood vessels would accelerate the blood flow through the heart, the brain, and the skeletal muscles. An increased heart rate would accentuate this flow; an elevated blood sugar would make more fuel immediately available; a greater resistance to fatigue would allow for more intensive muscular effort; and the increased blood coagulability would serve to produce a quicker staunching of blood flow from a wound. Now, although all these effects, except perhaps the last, can be induced in "emergency" situations by mechanisms other than epinephrine liberation, the theory postulates that epinephrine aids the nervous mechanisms in producing these effects.

This hypothesis is very attractive, but attractiveness and apparent conformity with logic are not proof of conformity with nature. Granted that epinephrine secretion would be useful to the hardpressed animal—that it would render him capable of more effective flight or combat—it had to be demonstrated that epinephrine is actually liberated in emergency situations in quantities sufficient to produce these effects.

C. Summary

In summary, we may say this of the adrenal medulla: It secretes epinephrine continuously in quantities too small to be of *observable* physiological significance. Under emergency conditions, in times of stress or danger, the secretion may be raised to such a level as to cause increased circulation, some delay of fatigue, an elevation in the blood sugar, and a heightened coagulability of the blood. We still lack conclusive evidence that epinephrine is important for the existence of man or any other animal or for adaptation to emergency situations. Apparently most, if not all, of the actions attributable to epinephrine secretion by the adrenal glands can equally well be carried out by a generalized stimulation of the sympathetic division of the autonomic nervous system. Epinephrine secretion may merely augment the functions of sympathetic nerves.

VI. THE CORTEX OF THE ADRENAL GLANDS

A. Pioneer Observations

1. *Addison's disease.*—About the middle of the last century (1855) a British physician named Thomas Addison described a disease of human beings which is still known as *Addison's disease.*

He observed a chronic low blood pressure, a marked muscular weakness, digestive upsets, general apathy, and a peculiar gradual bronzing of the skin as the outstanding characteristics. He found the disease to be invariably fatal in a few months or years. Groping empirically for causal relationships, he performed as many autopsies as he could on patients who had succumbed to the malady. In nearly every case he found a degeneration of the adrenal cortex, usually of tuberculous origin.

Thus, before any clear concept of endocrine glands had been formulated, Addison discovered a causal relationship between destruction of the adrenal cortex and a fatal disease of man. The concept of the adrenal cortex preventing the disease through liberation of a hormone came many years later. Addison's original observations and conclusions have withstood the test of time and have repeatedly been confirmed by physicians and investigators.

2. *The experiments of Brown-Séquard.*—The first experimental work on the adrenal cortex was done by that cosmopolitan pioneer in physiology, Brown-Séquard, who surgically removed one of the paired glands from dogs. His work was of value in the path of the investigation he opened up rather than in the specific conclusions his results yielded, for he concluded that removal of one adrenal gland was invariably fatal. Today we know that, though removal of both adrenal glands is always fatal in a few days, an animal can live out his usual life-span, and remain apparently perfectly normal, with either of the two glands intact. This is partly due to the fact that, in accord with the general principal of "margins of safety" in the animal organism, the normal animal is provided with more adrenal tissue than is needed at any one time. In an emergency the animal organism can spare some of almost any tissue it possesses. Especially is this true in the case of the paired organs, such as the kidneys and lungs, where one of the pair can often do the work of both if necessary.

In addition, when one adrenal gland is removed from a normal animal, the other usually increases in size and presumably in function. This has been amply confirmed in experimental animals and furnishes another example of the sort of compensatory response of an animal to a deficiency, described earlier in the chapter in the discussion of goiter.

3. *Independence of cortex and medulla.*—Early workers failed to distinguish clearly between the cortex and the medulla, so that, when

an extract of the adrenals—epinephrine—was found to have a marked blood-pressure-elevating effect, it was thought that perhaps Addison's disease, with its characteristic low blood pressure, was due to a deficient production of epinephrine. But this hypothesis was soon proved to be untenable for the following reasons.

Epinephrine was found to be a product of the adrenal medulla, and extirpation of the medulla of both adrenals in animals never produced Addison's disease or any other demonstrable abnormality.

Epinephrine injection in persons with Addison's disease, or in experimental animals with both adrenals removed, though it did result in a transient elevation of the blood pressure, had no appreciable effect upon the course or the fatal outcome of the condition.

Thus it became clear that the adrenals are two separate organs, that Addison's disease is related to a defect of the adrenal cortex, and that the adrenal medulla has little if anything to do with the effects of bilateral extirpation of the adrenals. However, this is far from saying that the adrenal cortex is an endocrine gland and that Addison's disease is really caused by hyposecretion of a hormone.

B. INSUFFICIENCY OF THE ADRENAL CORTEX

1. *Effects upon experimental animals.*—Only after some time was it possible to conclude that the cortex secretes a hormone. For many years all attempts to employ the substitution method, so effective in diabetes or hypothyroidism, were unsuccessful in controlling experimental or spontaneous Addison's disease.

In the meantime, adrenal research yielded a large amount of detailed information about the exact nature of the defects in adrenalectomized animals. Back of all these studies was the desire not only to control Addison's disease but also to learn more about the physiology of the adrenal cortex.

Besides the blood-pressure lowering, the muscular weakness, and (in man) the skin pigmentation, several other systems were found to be involved. First, there are profound changes in the blood. Sodium and chlorides diminish in quantity because of an increased elimination of these elements in the urine, and the potassium is greatly increased. There is also a loss of water from the blood and tissues, so that the formed elements and most of the dissolved substances in the blood become more concentrated, with a reduction in blood volume. The tissues become dehydrated.

Second, renal function is impaired, so that certain metabolic wastes accumulate in the blood. There is a marked increase in the total quantity of urea, creatinine, phosphate, and sulfate of the blood.

Third, carbohydrate metabolism is interfered with, and there is a decrease in the concentration of the blood sugar, as well as a diminution of the stored carbohydrate, especially in the liver. Even though large quantities of glucose are injected into an adrenalectomized animal, the stores of glycogen of the liver cannot be appreciably increased.

2. *The search for the hormone.*—The first successes in the long search for the hormone came when it was reported by several investigators in the late 1920's that extracts of the cortex could be made which, upon injection, would prolong the life of adrenalectomized animals from the usual average of about seven days to about twenty days. Later it was possible to prolong life to a month, and now extracts are available which will keep experimental animals alive and apparently well for years.

These results indicate conclusively that the adrenal cortex is an endocrine gland; the defects which appear when an animal is deprived of the normal source of the hormone can be controlled by substitution—by injection of the hormone extracted from the adrenal glands of other animals.

Unfortunately, the control of adrenal insufficiency in man has not met with the complete success that might be expected. Some patients with Addison's disease fail to respond to cortical-hormone administration. Is this because Addison's disease is more than an adrenal deficiency? We do know that frequently there is also a tuberculous infection present. Or is our failure due to an inability to furnish the patient with adequate quantities of the hormone? Does the hormone as available at present lose its activity by the time it is used on patients? There is some evidence that this occurs.

3. *Chemical studies.*—Intensive studies have revealed that several distinct substances, presumably hormones, can be extracted from the adrenal cortex. Their chemical formulas have been determined, and their actions upon the animal organism tested experimentally. One group of these substances influences primarily the metabolism of carbohydrates, and another, the salt balance of the blood.

4. *Usefulness of cortical extracts.*—The major importance of corti-

cal extracts is their employment in the control of such seriously deficient hormone secretion as occurs in Addison's disease. One of the extracts—*cortisone*—has proved to be useful in the control of one kind of *arthritis*, a crippling disease of the joints. Whether such treatment represents a true replacement therapy, substituting for defective adrenal cortical function in people with arthritis, is not clear.

5. *Control of hormone secretion.*—Adrenal cortical function provides another example of the control of one endocrine gland by another. Several effects of adrenal gland hypofunction can be produced by destruction of the hypophysis (as was also seen in the case of hypothyroidism [p. 525]). Some of these defects—for example, that of carbohydrate metabolism—can be corrected by administering extracts of the hypophysis.

It has also been shown that extirpation of the hypophysis causes degeneration of the adrenal cortex, although not sufficient to cause death. Such degeneration can be prevented by suitable injections of certain extracts of the hypophysis.

This has led to the concept that the hypophysis (anterior lobe) secretes an *adrenocorticotrophic* (i.e., nourishing or stimulating) *hormone*, called *ACTH*, reminiscent of the thyroid-stimulating hormone.

6. *Salt balance of blood.*—Considering the various effects of adrenalectomy upon lower animals or destruction of the adrenal glands in man, it seems likely that the fatal outcome may be closely related to the diminution of blood sodium and chloride and the elevation of blood potassium. This concept is strongly supported by attempts to control adrenal insufficiency by experimentally administering diets rich in sodium chloride and low in potassium. Animals deprived of their adrenals have been kept alive and in fairly good condition for many weeks by such diets, with no administration of hormone except in the acute stage of deficiency immediately following the operation.

Perhaps the adrenal cortex hormone is concerned essentially with the control of the sodium, potassium, and chlorine content of the blood, just as the parathyroid hormone controls the calcium level. In these two glands, then, we should have the machinery controlling the blood concentration of sodium, potassium, and calcium, which exert such a profound influence upon cellular activity in general.

VII. THE GONADS

A. REPRODUCTION

1. *The primary sex organs.*—The primary reproductive structures in animals are the *gonads*—the male *testes* and the female *ovaries*. These produce the reproductive units, the single-celled sperms and eggs. In many animals there is but little else to the reproductive system; very slight specialization of accessory or secondary sexual structures has occurred. The eggs of the female fish or amphibian are passed into the water, and the sperm are ejected upon the eggs. Fertilization of the eggs and development of the new individual occur more or less independently of the parental bodies.

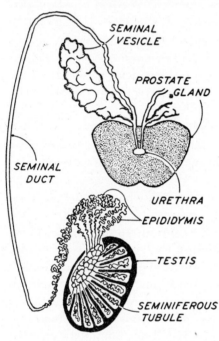

FIG. 198.—The genital system of the human male.

In the higher vertebrates a number of accessory structures have been evolved, particularly in the female. All these we may look upon as facilitating fertilization and providing more adequately the requirements of the developing embryo, nutritional and otherwise. Such elaborate provisions for the well-being of the embryo are characteristic of both the higher plants and the higher animals. So highly specialized have these mechanisms become that reproduction is impossible in the mammals, for example, without the accessory sex structures. Even though the ovary of the human being may continue to produce perfectly normal eggs, reproduction cannot occur unless the uterus and other accessory structures are also present and functioning normally.

The production of sperm and eggs is itself a nonendocrine function. It is an example of specialization of tissues for the production of a peculiar kind of cell. The bone marrow produces red blood cells, the skin produces special cells at the roots of the hairs, and the gonads produce reproductive cells. None of these is an endocrine function.

However, the specialized cell activities of egg and sperm production, especially in higher vertebrates, are under the control of the endocrine glands and their hormones. The activities of the accessory structures as well, upon which depends the successful development of the embryo, are also dependent upon hormones. First among the endocrine glands concerned are the gonads themselves.[26] Hormones as well as reproductive cells are produced by the gonads. An understanding of the functions of these hormones can only follow an appreciation of the physiology of reproduction itself, since the major functions of the hormones are related to the activities of reproduction. We commence our discussion of the sex hormones, therefore, with a consideration of the main outlines of the reproductive processes in mammals.

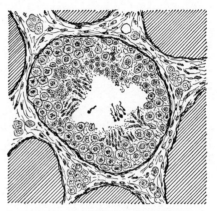

FIG. 199.—Genesis of sperm in the human testis as seen in a section through a seminiferous tubule. Several developmental stages are shown. A few mature sperm (see Fig. 208, p. 584) are present in the lumen of the tubule. Crosshatched areas are adjacent tubules. Between the tubules is tissue which probably elaborates the male sex hormone or hormones.

2. *Production of sperm.*—Beginning with puberty, sperm in all stages of development, from a somewhat primitive type of cell to the mature form, can be seen making up the walls of the many tiny ducts or *seminiferous tubules* of the testes. The more mature cells lie closest to the lumen of the tubules, and the adult forms lie free within the ducts. This, then, is the primary male reproductive activity—production of sperm. In man and a few other mammals it is a continuous process, though ejection of sperm to the exterior is not. In most mammals sperm production is cyclic, varying with the seasons of the year.

Besides these sperm-producing units there are testicular cells quite different in appearance and structure. They communicate with no duct, although each individual cell resembles in appearance the cells of glands in general. It is these cells, we shall see, which produce the male sex hormone.

The other structures of the male sexual system serve important accessory functions in reproduction. (*a*) Storage of sperm occurs in

26. See also the discussion on the hypophysis, p. 563.

the coiled tube which makes up the *epididymis*, into which the seminiferous tubes empty. (*b*) The *seminal duct* serves to transport the
sperm from the testis to the urethra, through which the sperm pass
out through the *penis*. (*c*) The *seminal vesicles* and the *prostate
gland* pour into the seminal duct or into the urethra secretions
which serve mainly as a vehicle of transportation for sperm. The
sperm plus these secretions constitute the *semen*. (*d*) The penis serves
as the organ of intromission for depositing the sperm-containing
semen into the female reproductive passages.

   3. *The reproductive cycle in the female.*—Starting at puberty, the
ova develop in the interior of the ovaries, which lie in the pelvic
portion of the abdominal cavity. Microscopic examination reveals
ova in various stages of development. The mature eggs lie at the
surface of the ovary. At this stage the ovum is contained within a
fluid-filled cavity called a *follicle* and is attached to the wall of the
cavity in the manner shown in Figure 201. At fairly regular intervals
one or more of the ova become mature, the wall of the follicle ruptures, and the egg escapes along with the follicular fluid. This
liberation of the mature egg is called *ovulation*.[27] The egg passes into
the funnel-shaped termination of the *oviduct*, or *Fallopian tube*.
Through this it reaches the uterus, and, if it is not fertilized, it
disintegrates or is passed to the exterior.

   Sperm which have been introduced into the vagina at copulation
migrate upward through the female genital passages partly by virtue
of their own motility. They "swim" by means of a lashing, whiplike
action of their "tails." The sperm traverse the uterine cavity and
move up into the Fallopian tubes. Should a sperm encounter an egg
here, fertilization might occur. If it does, the fertilized egg, or *zygote*,
passes down the Fallopian tube and becomes imbedded in the uterine
wall, where further development takes place.[28]

   Occasionally, the fertilized egg may imbed itself in the wall of the
Fallopian tube. Or it may even escape from the funnel-shaped opening of the oviduct into the general abdominal cavity, to become
lodged in the lining of the abdominal wall. Again, a mature egg
may fail to leave the ruptured follicle; and, if it is fertilized here,

   27. The interval between ovulation periods varies greatly from species to species and, to a
lesser extent, from individual to individual within a species. In the human female the period is
approximately twenty-eight days; in the dog it occurs twice annually; and in the rat, every
five days. In addition, in some species ovulation seems to be induced by copulation. This
is not true of man.

   28. Subsequent details of embryonic development may be found in chap. 14.

development of the embryo may commence within the substance of the ovary. Fertilization of ova in any of these abnormal sites usually leads to disorders which call for surgical intervention.

After ovulation has occurred, structural changes are undergone by the cells comprising the collapsed wall of the ruptured follicle. The

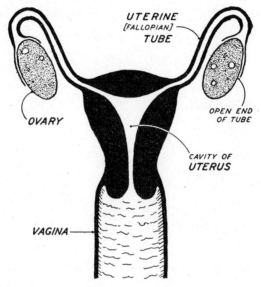

Fig. 200.—The genital system of the human female. Several follicles (see Fig. 201) are shown in each ovary.

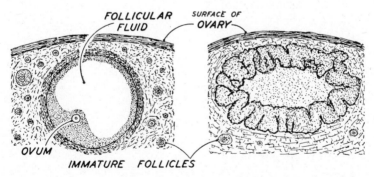

Fig. 201.—A mature follicle (*left*) and a corpus luteum (*right*) at the surface of the ovary

entire structure becomes a compact mass of cells, lacking the fluid-filled cavity of the follicle. It is yellowish in appearance and therefore has been called the yellow body, or *corpus luteum*. The corpus luteum remains as a definite ovarian structure for a time after each ovulation. It may suffer either of two fates, depending upon whether or not fertilization has occurred. If there has been no fertilization, the corpus luteum gradually decreases in size and disappears in about two weeks. If fertilization takes place, the corpus luteum remains as a definite structure during the entire period of pregnancy. We know

TABLE 13

THE REPRODUCTIVE CYCLE IN THE HUMAN FEMALE

DAYS AFTER MENSTRUATION CEASES*	OVARIES		LINING OF UTERUS
	Follicle and Ovum	Corpus Luteum	
1st to 7th........	Gradual ripening of ovum and increase in quantity of follicular fluid	Absent	Resting condition, then increasing thickness of uterine lining with increased vascularity and gland formation
About 8th........	OVULATION and passage of ovum into the Fallopian tube (where fertilization may occur)†	Forms from the cells of the ruptured follicle	Uterine glands begin to secrete a viscid fluid
9th to 14th.......	Arrival of ovum in uterus	Grows	Ready for reception of fertilized ovum
ALTERNATIVES: (a) If Fertilization Has Not Occurred			
14th to 24th......	Ovum can probably no longer be fertilized and disintegrates	Gradually disappears	Secretion subsides
24th to 28th......	New follicle begins to develop	Absent	MENSTRUATION: uterine lining sloughs off, with moderate bleeding
(b) If Fertilization Has Occurred			
14th to 280th.....	No follicle formation or ovulation during pregnancy	Remains during period of pregnancy	Fertilized ovum imbeds itself in uterine wall and grows

* These figures are approximate; they vary from individual to individual.

† Most investigators think that sperm introduced into the female passages two or three days before ovulation may remain and fertilize the ovum after ovulation has occurred.

now that the follicle and the yellow body are important endocrine structures having related though different functions and playing distinct parts in the regulation of the reproductive cycle and in the development of the embryo.

The ovarian rhythm culminating periodically in ovulation is accompanied by rhythmic changes in the uterus and vagina and, in lower animls, in the intensity of the sex urge. The period of most intense sex drive in animals is called the *estrous period*, and the whole cycle of changes the *estrous cycle*. In this cycle, periodically the uterine lining proliferates, becomes thicker and softer, more vascular and glandular.[29] The height of the proliferation, vascularity, and secretion is reached a few days after ovulation and represents a periodic adaptation of the uterus for the reception of a fertilized egg.

If fertilization and implantation of the zygote into the uterine wall fail to occur, the uterine lining, in lower mammals, gradually subsides to its previous less vascular condition, with some sloughing-off of the uterine lining. In some primates, including man, the sloughing is accompanied by moderate hemorrhage. Fragments of the uterine mucosa, plus blood, continue to leave the uterus by way of the vagina for a period of about four days. This process is called *menstruation*.

The estrous cycle repeats itself, running again and again through the elaborate process of preparation for pregnancy and demolishing the preparations if no pregnancy occurs. But it is found that menstruation and ovulation do not occur simultaneously. Elaborate studies have shown that ovulation occurs about midway between the *onset* of two successive menstrual periods (see Table 13).

B. Endocrine Control of the Female Reproductive Cycle

Our account of the ovaries has so far been purely descriptive. The bare sequence of cyclic changes has been outlined. These in themselves are complex enough, but an understanding of the underlying controlling mechanisms constitutes a real challenge to the physiologist seeking to untangle the interacting physiological forces involved. To find an explanation of the machinery regulating the cycles, of the physiological devices which correlate so perfectly the rhythmic activities of the ovaries and uterus, has been no easy task. It is still not completed, and many of the conclusions arrived at must be considered tentative, pending the accumulation of further experimental data.

1. *Chemical control of uterine changes.*—From the start we can be sure that the factors which effect the estrous cycle are chemical in

29. The deep glands finally secrete a viscid, translucent material, most of which does not escape into the uterine cavity but is retained for a time in the wall of the uterus.

nature. If a bit of the uterine lining of an animal is removed and placed into the anterior chamber of the eye of the same animal, the transplanted tissue will grow in a fair proportion of the cases. In this new location the tissue can be observed directly. Since such tissue is obviously devoid of innervation, any changes observed in it must be ascribed to chemical influences reaching it by way of the blood stream and body fluids. Uterine transplants in the eyes of monkeys undergo the same changes known to take place in uterine tissue in its normal location, including even the bleeding, which corresponds to menstruation and which occurs at the same time the uterus of the animal is menstruating.

2. *Control by the ovaries.*—It is also clear that the source of this chemical influence is the ovaries. If the ovaries of an animal are removed, obviously there is no further egg production, and the animal is henceforth sterile. But other effects are also produced. Even though the ovariectomy has left all the accessory sex organs intact, their behavior is markedly altered. All the cyclic changes cease. There are no estrous periods, no periodic changes in the uterus, no menstruation. This effect ensues both in experimental animals and in the human female when disease of the ovaries necessitates their removal. Some effect from the ovaries obviously controls the estrous cycle.

If the ovaries of the human or subhuman mammal are removed during the early stages of gestation, moreover, the pregnancy is soon terminated. The fetus dies, the embryonic uterine contents are reabsorbed, or an abortion occurs. The ovaries, then, are also necessary for the maintenance and development of the embryo, particularly in the early period of pregnancy.

We have seen not only that the ovaries are complex structures, histologically, but that the internal structure varies with the estrous cycle: the follicle is transformed into the corpus luteum, and then a new follicle develops, in a regularly repeated alternation. The cycles of the uterus are apparently related in some way to the follicle and corpus luteum cycle of the ovaries. We may formulate the tentative hypothesis that the follicle and the yellow body are endocrine glands, each secreting a specific hormone or hormones which are absorbed into the blood vessels of those structures and are transported in the circulating blood to the uterus, where they exert their effects.

The logical experiment to test this hypothesis would be to extract from the follicle or the follicular fluid, and from the yellow body, specific materials having precise actions upon the uterus. This has now been done successfully. Although the cells of the yellow body are the very same cells which formerly made up the follicle, the products obtained from the extraction of the two are definitely different in action though not unrelated chemically.

What actions do these products—presumably hormones—have? This question can be answered by the expedient of substitution experiments. We can remove the ovaries of an animal and determine which of the effects produced will be corrected by injection of extracts, first of the follicle, then of the yellow body.

C. THE FOLLICULAR HORMONE

1. *Effect on the uterus.*—An animal deprived of its ovaries has no estrous cycles. The uterus remains in a dormant, "resting" condition. If follicular extract is injected, the changes characteristic of estrus are restored. The animal displays sexual behavior and will accept the male in copulation. The uterine lining becomes more vascular and glandular as injections are continued day after day. Then, upon cessation of the injections, the uterus subsides to the resting stage. Or, if the experiment is done upon monkeys, or upon human females without ovaries, a more or less typical sloughing of the uterine lining, with bleeding and menstruation, will occur when the injections are stopped.

Therefore, it appears that the ovaries manufacture a hormone— *estrogen*—which (supplemented with the corpus luteum hormone) induces changes in the uterus and prepares that organ for reception of the fertilized egg. Menstruation occurs when these hormone secretions subside.

2. *Cyclic changes in hormone concentration of the blood.*—We now have data which indicate that the follicular hormone is produced in quantities which vary cyclically. It is possible to detect the quantity of follicular hormone in either blood or urine not by chemical tests but by observing the effects produced by injection of extracts of measured quantities of blood or urine into animals deprived of ovaries. Such experiments reveal the fact that after a menstrual period there is a gradual increase in the hormone content of the blood and urine, which reaches its maximum just before menstruation, when the proliferation of the uterine lining is maximal. Then just before menstruation there is a drop in the hormone concentration of the blood.

The source of all the follicular hormone of the blood is not entirely clear. Of course, much of it comes from the growing ovarian follicle, and, with the increase in growth and size of a follicle following menstruation, we may conclude that the hormone production likewise increases. But recall that ovulation occurs about midway between menstrual periods. Yet the production of "follicular" hormone continues to increase until the following menstrual period. Where does this hormone originate? Does it come from other immature follicles of the ovary, or does it come from some other structure entirely? We do know that a product having all the properties and actions of the follicular hormone can be obtained from other sources. The yellow body itself yields "follicular" hormone, or estrogen, upon extraction and may possibly be the chief source of that hormone immediately after ovulation. In pregnancy, also, the concentration of the estrogen in the blood rises far above that of the usual menstrual cycles, and the placenta, amniotic fluid, umbilical cord, and fetal blood contain large amounts of it, although follicle formation is in complete abeyance during pregnancy.

D. THE CORPUS LUTEUM HORMONE

1. *Function in the sexual cycle.*—We have seen that uterine growth changes are induced by follicular hormone, whether it is secreted by the follicle or by some other structure or is injected into a female possessing no ovaries. This growth change is in the direction of a preparation of the uterus for the reception of the fertilized egg. But in the ovariectomized animal the preparation is not complete if only the follicular hormone—estrogen—is injected. The mucosa becomes thick, vascular, and glandular. But the glands do not secrete their viscid product, which secretion normally completes the preparation of the uterus. This product presumably furnishes nutritive substances for the embryo before blood-vessel connections are established through the placenta.

Experimentally, extracts of the corpus luteum can bring about this final premenstrual change even in human females deprived of their ovaries. This last uterine change of the cycle is called the *progestational* change (or "change before pregnancy"), and the luteal extract has, therefore, been given the name *progesterone*. In the normal female, as yellow-body formation follows follicle eruption, the luteal hormone progesterone is presumabely secreted and effects the final uterine changes. Examination of the urine for progesterone verifies this conclusion. It begins to make its appearance here shortly after the time of ovulation and reaches a maximum shortly before menstruation. Progesterone has not yet been demonstrated in the blood, however.

2. *Function in pregnancy.*—We must look for some additional function of the corpus luteum and its hormone—progesterone. What is the significance of the persistence of the corpus luteum during pregnancy? We have indicated that the continuation of pregnancy is dependent upon the ovaries being intact. If they are removed, especially in early pregnancy, without directly interfering with the embryo and uterus, death of the embryo follows. Exactly the same effect is produced if, instead of removing the entire ovary, only the corpus luteum (or corpora lutea if there is a multiple pregnancy) is removed. Implantation of the embryo in the uterus and its early development depend upon the presence of the yellow body; they are contingent upon continued progesterone production by that structure, for, if the ovaries are removed from an animal after the onset of pregnancy, injections of progesterone continued for the duration of the pregnancy will prevent a premature termination.

Hence, it seems clear the progesterone functions in two ways: it induces the final premenstrual changes in the uterus preparatory to pregnancy and it makes possible the nourishment and development of the embryo in the early stages of its growth.

This hormone inhibits uterine contractions, probably important in preserving pregnancy, and also seems to be responsible for the enlargement of the mammary glands late in pregnancy.

We shall see that ovarian endocrine function, that is, the secretion of estrogen and progesterone, are directly under the control of the hypophysis (see pp. 563–64). Removal of the hypophysis causes cessation of ovarian hormone secretion and of the phenomena normally induced by secretion of estrogen and progesterone. Subsequent injection of appropriate hypophyseal hormones restores ovarian function.

E. Additional Functions of the Ovarian Hormones

Less complex but also important are a number of other actions of the female sex hormones. During the period of body growth there is a direct effect of the ovarian hormones upon the growth and development of the uterus, oviducts, and vagina. If the young animal is ovariectomized, these organs will remain in an infantile condition throughout life. Estrogen injections can substitute for the experimental defect and will support normal development of the accessory sex organs, though obviously an ovariectomized animal will remain sterile. The induction of precocious development of the sex organs has also been achieved in experimental animals by injections of sex hormone preparations.

The development of the mammary glands and their maintenance in the adult also depend upon the sex hormones, for removal of the ovaries leads to their atrophy. This consequence of ovariectomy can also be prevented by suitable injections of estrogen.

The relationship of the sex drive to the ovarian hormones is still largely an open question. Psychological elements play a large role, and in human beings the problem is difficult to deal with objectively. In experimental animals absence of the ovaries leads to a cessation of the characteristic sex behavior, and the male is not accepted in copulation. In some animals—rats and mice—it has been possible to restore mating activity in ovariectomized females by estrogen injections.

Certain secondary sex characters, not directly related to the mating act, such as the plumage of fowls, are also dependent upon ovarian hormone production.

We can summarize the functions of the ovary as follows:[30]

a) Production of ova (nonendocrine)
b) Production in the follicles of the hormone estrogen, which
  1. Induces the growth, vascular, and glandular changes in the uterus during each menstrual or estrous cycle
  2. Supports the development of the uterus and other accessory reproductive structures during the growing period and maintains them in the adult condition
  3. Supports the growth of the mammary glands and maintains them in the normal adult condition
  4. Contributes to sexual behavior and the sex urge
c) Production by the corpus luteum of progesterone, which
  1. Completes the uterine changes started by estrogen, preparatory to pregnancy, during each estrous period
  2. Makes possible the imbedding of the fertilized egg in the uterus and the maintenance and continuation of pregnancy in the early months
  3. Stimulates mammary-gland enlargement late in pregnancy
d) Production of other hormones by the remainder of the ovary (?)

---

30. The reader should at this point refer again to Table 13 and reconsider the uterine and ovarian changes of the menstrual cycle in the light of this later material on the actions of estrogen and progesterone.

Many problems remain unsolved in this field. We need to know more about the rate of secretion of the hormones, and the change in rate of secretion during the cycle, and more about their intricate interaction. What other sex hormones, besides those mentioned, may there be? What is the significance of the high concentration of "follicular" hormone in the blood in pregnancy, and where does it come from? What endocrine role, if any, is played by the temporary structures of pregnancy, such as the placenta, or the embryo itself? What is the relation of the sex hormones to sterility? How are they related to the sex urge? What controls the cyclic changes in rate of secretion of estrogen and progesterone? We shall partially answer this last question in the section on the physiology of the hypophysis (p. 564). The very rapid strides being made in our understanding of this complex field of physiology warrant the hope that solutions to these problems may soon be found.

F. Disorders of the Menstrual Cycle

Disorders of the cyclic uterine changes are frequently encountered. Painful or difficult menstruation, scanty flow or absence of flow, as well as excessive bleeding, are sometimes caused by hormone imbalances, though some of these defects are often of emotional origin. The cessation of menstrual cycles—the *menopause*—occurring in women at about forty-five to fifty years of age normally marks the termination of the reproductive period. At this time ovulation ceases, and the uterus and mammary glands undergo atrophy. Frequently, women experience untoward vasomotor, neuromuscular, and psychic symptoms at this time. Sterility in women may also conceivably occur, even though ova are produced, through defects in the normal cyclic functioning of the uterus.

In what ways can our knowledge of the sex hormones and their functions be of service here? The administration of estrogens has been beneficial in alleviating some of the symptoms which may occur at the end of reproductive life. For other derangements of the reproductive cycle, attempts at alleviation by substitution therapy have not been very successful. One of the most serious difficulties lies in the fact that the production of hormones by the ovaries is cyclic; the concentration of the follicular or luteal hormones in the blood varies in regular fashion but in a rather complicated interrelationship. By injections of the hormones it is still difficult or impossible to reproduce this normal hormone balance.

Psychological elements probably play an important role. There is a common notion among the uninformed that the cessation of menstruation, even that which occurs normally in middle age, is harmful because of retention of poisons in the body. This belief is without foundation.

It is possible to reproduce cyclic uterine changes, culminating in a more or less normal menstrual flow, by proper variations in quantity and alternation of follicular and luteal hormone injections. But this is expensive and difficult and not always successful. Of just what value it is, besides the psychological effect it produces, is problematical. In middle-aged or older women, perfect physical well-being is quite compatible with absence of cyclic uterine changes. In the young female, absence or cessation of menstruation may be of minor significance in itself, though in other cases it may be an indication of some serious general systematic disease, such as tuberculosis.

G. The Testicular Hormone

The principles underlying the proof that the testes are endocrine glands are the same as those applied to the study of other endocrine structures. What happens when the testes are removed? Can these changes be prevented or reversed by the administration of some product or products known to be elaborated specifically by the testes?

1. *Effects of castration.*—Of course, the removal of the testes—*castration*—causes a permanent sterility. No more sperm can be produced, and nothing can be done to the animal to restore this non-endocrine function.

Some of the secondary effects, now known to be related to the male sex hormone, have been familiar for centuries. Man has castrated his domestic animals for almost as long a time, probably, as domestication has been practiced. A noteworthy effect of this has been to produce somewhat larger and fatter animals. Just how much of this effect is due to the decrease in general bodily activity, such as the pursuit of the female, is not clear. Some of the increased growth is perhaps more closely related to the absence of the hormone itself, for in the bone growth of castrates the union of growing portions of bone to one another is delayed, which results in larger and longer bones.

Certain secondary sex characters associated with the attainment of sexual maturity are also dependent upon the intactness of the testes. The growth and distribution of body hair and the voice and stature changes characteristic of puberty in man fail to develop in the absence of the testes. Once these have made their appearance in the normal individual, however, they are unaffected by subsequent castration. In fowl such secondary sex characters as plumage and comb growth are also dependent upon intact, functioning testes.

Many experiments have demonstrated that perhaps the most significant effects of castration are upon the accessory reproductive structures which have to do with the storage of sperm, their transportation to the exterior, and the insemination of the female. The seminal vesicles, prostate gland, and seminal ducts either fail to reach the mature state, if castration is done on immature animals, or undergo atrophy and fail to form their characteristic products in castrated adults.

2. *Endocrine portion of the testes.*—These effects can be ascribed to absence or hypofunction specifically of those cells of the testes which are not concerned with sperm formation. Nature performed the experiment, proving this before man thought of it. If, in the course of the development of the individual, the testes fail to descend from the site of their original differentiation in the abdominal cavity into the *scrotal sac*, there is an almost complete degeneration of the sperm-producing cells, leaving the rest of the testis intact. Such men or other animals fail to display the effects of castration described, but these soon make their appearance if the testes, now composed mainly of the endocrine cells, are subsequently removed.

That we are dealing here with an endocrine activity of these testicular cells has been demonstrated by observing the effects upon castrated animals of injecting testicular extracts. All the effects described (except sperm genesis) can be prevented by injections of such extracts containing the testicular hormone. Secondary sex characters and accessory reproductive structures can be made to attain their full maturity in the young castrate, and regressive changes are prevented in the adult castrate.

Moreover, the hormone—*testosterone*—has been detected in the blood of certain animals and in the urine of others, including man, so that there can be no doubt of the existence of a hormone mechanism. The chemical nature of the hormone is fairly well known. Chemically, it is closely related to one of the female sex hormones.

Both the endocrine and the sperm-forming activities of the testis are under control of hypophyseal hormones, as is indicated by a cessation of these activities after hypophysectomy and by their resumption upon subsequent administration of appropriate hypophyseal extracts.

3. *Functions of the hormone.*—Thus far it has not been possible to put this knowledge we have of the testicular hormone to much practical use. The hormone, of course, has no effect upon sterility caused by failure to form sperm. As a matter of fact, the hormone has if anything an inhibitory effect (indirectly) upon sperm production.

There exist many erroneous notions about the role of the male sex hormone and the gonads in general bodily well-being. Brown-Séquard gave impetus to such ideas many years ago by administering testicular extracts to himself and by concluding that his vigor and strength were increased. In the opinion of some persons, grafting of testes has led to a regenerating, invigorating effect. But, in the first place, grafted testes or any other tissue can survive only a very short time, especially if they are interspecies grafts. In the second place, even if the grafts should "take" and produce the sex hormone for an indefinite period, there is no evidence that the effects desired could ever be accomplished. A rejuvenated, invigorated feeling is difficult to measure objectively, and castration in animals has never been shown to decrease the life-span. At least from the point of view of the animal breeder, the effects of castration upon general bodily well-being are rather on the credit side of the ledger.

Just what is the role of the hormone in characteristic sex behavior? This also is not entirely clear. Male guinea pigs castrated even before reaching sexual maturity will exhibit pursuing tendencies toward females for months after becoming mature, and certain domestic animals display some mating behavior after castration. But, ultimately, sexual behavior diminishes and, therefore, may be said to depend largely upon the male sex hormone.

In the human species the problem is made more complex by the psychological elements present. Much of the sexual behavior of the human male is undoubtedly independent of the sex hormone, although fertility is dependent upon normal testicular function.

We may summarize the functions of the testes in the following outline:

a) Production of sperm (nonendocrine)
b) Production of the male sex hormone, which
    1. Causes (but is not always necessary for the maintenance of) certain anatomical changes in nonsexual structures, characteristic of sexual maturity
    2. Stimulates the development and functioning of the accessory sex organs and maintains them in the adult stage
    3. Contributes to sexual behavior and to the sex urge

VIII. THE PITUITARY GLAND OR HYPOPHYSIS

A. STRUCTURE AND DEVELOPMENT

The pituitary gland or hypophysis lies in a little depression in the floor of the skull. It is of dual origin embryologically, part of it arising as an upward outpouching of the roof of the primitive mouth, and part being a downward growth from the brain itself. Throughout life the two portions are distinguishable microscopically and are called the *anterior lobe*—from the roof of the mouth—and the *posterior lobe*—chiefly from the brain.

The old embryonic connection of the anterior lobe with the rear of the mouth cavity disappears early in development, but the attachment of the posterior lobe with the base of the brain persists throughout life. The hypophysis of the adult human being, then, is a small structure, weighing about a half-gram, attached to the base of the brain by a narrow stalk.

B. THE POSTERIOR LOBE

The physiological role of the posterior lobe is rather obscure. Studies upon this structure remind one of the kind of difficulty encountered in the earlier studies upon the adrenal medulla. From the posterior hypophysis it is possible to extract materials which have remarkable and undisputed physiological actions, which were well known even before the actions of the secretions of the more important anterior lobe were understood. The general name for such extracts is *pituitrin*.

The outstanding effects of pituitrin are upon the smooth muscle of the arterioles and of the uterus of late pregnancy and upon the volume of urine secreted. The action of pituitrin upon uterine muscle has made it a useful drug in the hands of physicians. When properly used, it facilitates the uterine contractions which cause expulsion of the products of conception from the uterus in labor.[31]

One might entertain with logical satisfaction the view that the normal uterine contractions of labor are due to increased secretion of pituitrin at the time of childbirth. This would very nicely explain much of the mechanism underlying the induction of labor. However, there is no evidence that pituitrin plays such a physiological role, for all the phenomena of labor occur quite normally even if the hypophy-

31. Pituitrin has been separated into two components. One fraction, called *oxytocin*, or *pitocin*, acts upon the uterine muscle. The other, called *pitressin*, affects mainly the arteriolar muscle and urine secretion.

sis has been removed shortly before labor sets in. Nor do any changes in blood pressure follow such an operation.

One fraction of pituitrin has been of value in the control of a human disorder known as *diabetes insipidus*,[32] which is characterized by a profuse secretion of dilute urine (not containing sugar) and an excessive thirst. Injections of a pituitary extract reduce the urinary flow to normal. It was once thought that this was simply a drug effect, in which certain chemicals extractable from the posterior lobe happen to exert a beneficial effect upon a disorder not related to any pituitary dysfunction, but now we have evidence that diabetes insipidus may be caused by a deficient secretion of a substance which normally prevents excessive urine secretion.

This we learn from animal experimentation. Investigations show that destruction or removal of the main parts of the posterior lobe of the hypophysis in dogs or rats produces a state closely resembling human diabetes insipidus. The profuse secretion of urine in the experimental animals is effectively controlled by injections of pituitary extracts.

Thus, although it is as yet uncertain whether there are posterior lobe hormones which control blood pressure or labor, it does seem that there is a hormone which controls the volume and concentration of the urine secreted.

C. Control of Growth

1. *Growth abnormalities in man.*—Our first knowledge of the function of the hypophysis came about when peculiar disorders of human growth or body size were correlated with hypophyseal malfunction. Excessive growth was associated with enlargement of the hypophysis. The stimulating effect of enlargement of the structure was recognized as being of two kinds. When the enlargement occurs during growth, there is a general growth acceleration, producing a condition of *gigantism*. Most of the circus giants are of this type (see Fig. 202). If the enlargement of the hypophysis occurs after the period when the long bones have stopped their growth in length—that is, in an adult —the condition called *acromegaly* results. Here there is no gigantism, but the growth phenomena are restricted to overgrowths and enlargements of parts of the skeleton. The bony ridges over the eyes, the cheek bones, and the jawbone enlarge, producing a physiognomy

32. Causally unrelated to this is the more common form of diabetes controlled by insulin, which we have already discussed. Pancreatic diabetes is called *diabetes mellitus* (meaning "honey" or "sweet" diabetes) because of the sugar in the urine.

quite characteristic of the derangement. The bones and soft parts of the hands and feet also enlarge greatly.

But the demonstration of the relationship of these conditions to enlargement of the hypophysis did not prove that the structure was an endocrine gland or that these defects were really expressions of hypersecretion of a growth hormone.

Fig. 202.—Pituitary gigantism in man compared with normal stature. Excessive secretion of the growth-promoting hormone of the anterior lobe of the hypophysis caused the gigantism.

2. *Experimental dwarfism and gigantism.*—Such proof came years later from several sources. When a German worker (Aschner, 1909) operated on puppies, removing the hypophysis, there was marked retardation of growth and development (see Fig. 203). The same phenomenon has been produced repeatedly in rats. Afterward, investigators were successful in applying the substitution principle. They

prepared hypophysectomized rats and then each day implanted under the skin of the rat pieces of hypophysis freshly removed from recently killed animals. The dwarfism was prevented. The rats without hypophyses grew as well as normal control litter-mates.

Gigantism was also produced by the simple expedient of implanting a larger amount of fresh hypophyseal tissue daily. This could be done equally well on normal or on hypophysectomized animals. The treated animals attained a much greater size than the untreated controls. The gigantism was real, involving the skeleton as well as the soft tissues, and was not simply a laying-on of fat. It has also been

Fig. 203.—Experimental dwarfism. The dogs are litter-mates and were fed the same diet. The small dog had its hypophysis removed soon after birth. Such dwarfism can be prevented by administering the growth-promoting hormone of the hypophysis. (From Aschner.)

possible to reproduce experimentally some of the characteristics of the hyperpituitarism which develops in the adult human being. Acromegaly in dogs can be achieved by injection of hypophyseal extracts to adult dogs.[33]

3. *The growth-promoting hormone.*—Attempts at isolation of the growth-promoting principle from the anterior lobe of the hypophysis have not met with complete success. Extracts of beef anterior lobe are available, which duplicate in animals the growth-stimulating action of the fresh gland implants used by the earlier workers, but highly purified preparations capable of complete replacement in the hypophysectomized animal have not yet been made. However, some of these extracts are of sufficient purity for injection in cases of

33. At present, the only method by which hyperpituitarism can be successfully controlled is by destruction of part of the hyperactive gland surgically.

human dwarfism caused by hypophyseal hypofunction.[34] Some improvement of growth is effected.

The growth-promoting action of this hypophyseal hormone recalls the role played by thyroxin in growth. Does the hypophysis control growth in any way similarly to thyroxin? Apparently not, for, although the hypophysis seems to have some effect upon tissue oxidations, thyroxin cannot replace the hypophyseal hormone in hypophysectomized animals.[35]

D. Control of Gonad Function

1. *Effects of removal of the hypophysis.*—The absolute dependence of gonad function upon the hypophysis is striking. If the hypophysis is removed in the male or female before puberty, the gonads remain infantile and will never produce gametes. The endocrine activities of the gonads do not develop, and, consequently, all the accessory reproductive structures and the secondary sex characters fail to reach the adult state.

If hypophysectomy is done on the adult male, there is a striking atrophy of the testes, which may shrink to one-tenth their former size. Genesis of sperm ceases, and there is no further hormone production. In the adult female, ova fail to mature, no follicles form, and no hormone is produced, and the phenomena of estrus and the menstrual cycle disappear. Similar conditions obtain in man with defective hypophyseal function. Hypopituitarism is usually associated with sexual infantilism.

If the hypophysis is removed early in pregnancy, abortion follows, and it can be shown that the corpus luteum of the ovary, characteristic of pregnancy, has disappeared. Consequently, we must look upon all the phenomena which are under control of the gonad hormones to be dependent indirectly upon normal hypophyseal function.

2. *Artificial-substitution experiments.*—Substitution for the removed hypophysis by the injection of proper hypophyseal extracts has succeeded in preventing all these defects. Injections of extracts into hypophysectomized females have restored the sexual cycles and supported normal pregnancy. There can be no question but that this action is by way of the gonads, for, if the gonads are removed, no

34. Such dwarfs are rare.

35. In this connection see also the discussion of the thyroid-stimulating hormone of the hypophysis (p. 566).

amount of injected hypophyseal extract will induce an estrous cycle or support pregnancy.

3. *The gonad-stimulating hormones.*—These gonad-stimulating hormones have been prepared in fairly pure form and have actually been detected in the blood and urine of normal animals and man, so that there seems to be no question of their true hormone nature. Hence, we have gonad-stimulating hormones as well as a growth-promoting hormone of the anterior lobe of the hypophysis.

E. INTERACTIONS OF THE HYPOPHYSEAL AND
   OVARIAN HORMONES

1. *Stimulation of the follicles and the corpus luteum.*—A more detailed knowledge of the intricate interrelations of the ovarian and hypophyseal hormones has resulted from extensive work of the last decades. At least two gonad-stimulating hormones have been differentiated in the female. It will be recalled that there are at least two ovarian hormones secreted, respectively, by the follicle (estrogen) and by the yellow body (progesterone). Correspondingly, investigators have isolated a follicle-stimulating hormone, which is responsible for maturation of the ova and development of the follicle with hormone production by that follicle. This hormone has no effect upon the corpus luteum of pregnancy.

A second hypophyseal hormone is without effect upon the follicle, but it will stimulate growth of the corpus luteum and maintain it intact. This is the luteinizing hormone.

2. *Pregnancy tests.*—During pregnancy, particularly in the early weeks, the ovary-stimulating hormones of the hypophysis are present in relatively large concentrations in the blood and can be readily detected in the urine. Injections of such urine, or simple urine extracts containing the hormones, into immature female mice or segregated adult female rabbits cause characteristic changes in the follicles of the ovaries which can be detected by the naked eye. This reaction constitutes a reliable test for pregnancy in the human female before other indications of pregnancy are apparent.

3. *Reciprocal action.*—There is considerable evidence for a reciprocal action of gonadal and hypophyseal hormones. While the hypophyseal hormones stimulate production of the ovarian hormones, these latter, in turn, seem to inhibit production of the gonad-stimulating hormones by the hypophysis. These facts have led to the formulation of a hypothesis which attempts to account for the

baffling cyclic nature of the changes in the female genital system, including, of course, the cyclic changes in hormone production.

In the first place, cyclic changes in the production of gonad-stimulating hormones by the hypophysis have been demonstrated indirectly. This would account, then, for the fact that the ovarian hormone secretion, in turn, is cyclic. But why is the hypophyseal secretion cyclic? The theory is this: After a menstrual period the hypophysis commences its secretion of gonad-stimulating hormones. This effects follicle growth and the production of the follicle hormone. However, when the follicular hormone has reached a certain concentration in the blood, it exerts its inhibitory action upon the hypophysis. Hence, when the follicular hormone is most concentrated in the blood, the hypophysis has been almost entirely inhibited. Then with menstruation, and the drop in the follicular hormone content of the blood, the hypophysis is released from inhibition and increases its hormone output and thereby stimulates the progress of the follicle development of the next menstrual cycle.

F. THE CONTROL OF LACTATION

Even after the period of pregnancy is normally terminated by the birth of the young, the hypophysis still plays a part in the reproductive process. A substance independent of the other hormones of the hypophysis has been isolated from the gland which, upon injection, stimulates the milk-secreting mammary glands. This effect is particularly striking upon the mammary glands at the time of labor and immediately after, though it can be induced to a lesser extent independently of pregnancy.

In animals hypophysectomized just before delivery of the young, milk secretion starts anyway but ceases in a few days. The continuation of the secretion seems dependent upon the intact hypophysis and, presumably, upon a specific hormone.

Nevertheless, other factors operate here also, for it is well known that continued removal of the formed milk from the mammary glands by the suckling young is also necessary for further secretion. Failure to remove the milk as it is formed leads to premature cessation of lactation.

G. HYPOPHYSEAL CONTROL OF OTHER ENDOCRINE GLANDS

The effects of the hypophysis upon other glands of internal secretion do not end with the ovaries and testes. In fact, the control of the hypophysis over other endocrines appears to be so prevalent that it has been called the "general headquarters" of the endocrine system (see also pp. 574–75).

In the hypophysectomized animal degenerative changes occur in the thyroids and in the cortex of the adrenal glands. Injections of appropriate hypophyseal extracts prevent such degeneration and, in the normal animal, may cause an increase in size and secretory activity of these endocrines.

The adrenocorticotrophic hormone (ACTH), already mentioned, exerts considerable control over hormone secretion by the adrenal cortex, so that certain defects characteristic of hypofunction of the adrenal can also be produced by hypophysectomy and can be corrected by administration of ACTH.

There is also a thyroid-stimulating hormone distinct from the other hormones which is secreted by the hypophysis. It is therefore possible to reproduce the picture of hypothyroidism by removal of the hypophysis, or of hyperthyroidism by injection of the thyroid-stimulating hormone of the hypophysis, at least temporarily.

One might be led to believe that the growth-promoting action of the hypophysis operates entirely through stimulation of the thyroids, since the thyroids are also known to affect growth. The relationships here are not fully worked out, but apparently the growth-promoting action of the hypophysis is more or less independent of the thyroid, as evidenced by the fact that the dwarfism of experimental hypopituitarism cannot be corrected by injection of the thyroid-stimulating substance but only by the injection of the specific growth-promoting hormone.

It seems fairly conclusively demonstrated that the anterior lobe of the hypophysis secretes at least six specific hormones, which have these actions:

a) Promotion of general bodily growth, especially of the skeleton
b) Stimulation of growth of the gonads to the mature condition and maintenance of them; regulation of the female sexual cycle, involving estrus and ovulation; maintenance of the corpus luteum, which is necessary for the existence of the embryo (at least two hormones are believed to be involved in these gonad-stimulating actions)
c) Stimulation of milk secretion by the mammary glands
d) Stimulation of the thyroid gland
e) Stimulation of the adrenal cortex

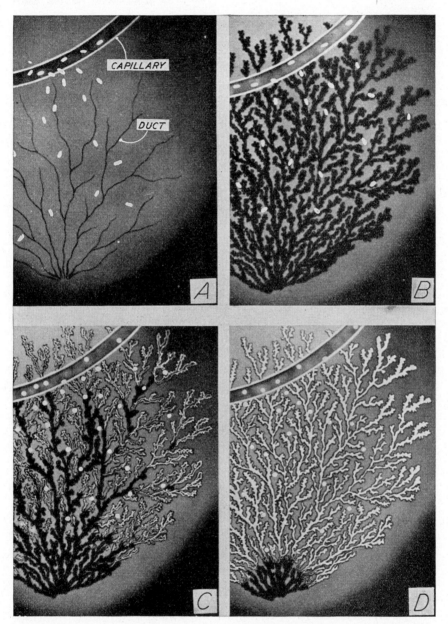

Fig. 204.—Hormone control of milk secretion. In *A* the ducts of the mammary gland are thin, and no milk is produced. Ovarian hormones are carried to the gland in the blood. Molecules of the hormone (*white ovals*) are seen diffusing from a blood capillary into the mammary-gland substance, which is stimulated to grow (*B*) and develop milk-producing cells. In *C* another hormone (*white circles*), produced by the hypophysis, stimulates the newly developed mammary-gland cells to produce milk (*shown in white*). In *D* milk has nearly filled the ducts of the gland. (From the sound film, *The Endocrine Glands.*)

IX. OTHER ENDOCRINE STRUCTURES

A. The Stomach and Duodenum

The discovery by Bayliss and Starling that the application of acid to the duodenal mucosa causes, physiologically, the liberation of a specific material which stimulates secretion of pancreatic juice has already been discussed (see p. 280). It has been convenient to

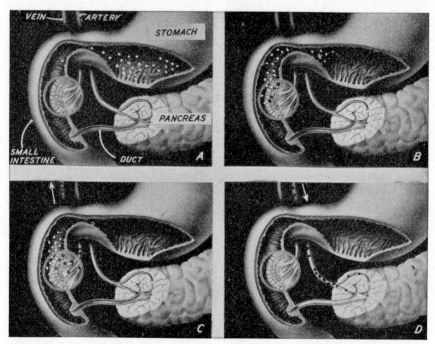

Fig. 205.—Control of the pancreas by secretin. In *A* the stomach is shown secreting hydrochloric acid (*white dots*) which is moved on into the small intestine. In *B* some of the acid has penetrated into the cells lining the small intestine (*circular insert*), where the acid stimulates the cells to form secretin (*black dots*). In *C* secretin formation continues and some secretin has entered the (duodenal) vein and is carried back to the heart. In *D* some of the secretin is carried in the (pancreatic) artery to the pancreas, which is stimulated to produce pancreatic juice. The juice (*shown in white*) fills the pancreatic duct and flows into the small intestine. Arrows indicate the direction of blood flow. (From the sound film, *Digestion of Foods.*)

discuss the physiology of this hormone, called *secretin*, in the chapter on digestion. Secretin is merely mentioned here because it is apparently a true hormone, and the cells of the duodenal mucosa, which produce it, are therefore true internal secretory structures.

The evidence that a true hormone is liberated by the gastric mucosa, having a stimulating action upon the gastric secretory cells, is less convincing. A more detailed discussion of gastrin and its actions was also given in the chapter on digestion (see p. 284).

B. THE LIVER

Whether or not the liver secretes a hormone or hormones is not yet definitely known. The commonly employed method of extirpation of the organ is not available in investigations on the liver. The liver is a large organ—the largest gland of the body—and it has at least a score of known important functions nonendocrine in nature. So diversified and important are these functions that death ensues shortly after liver extirpation. Withdrawal from the body of these nonendocrine functions is quite sufficient to cause death, regardless of whether or not the gland may in addition secrete a hormone. We might compare liver extirpation with removal of the medulla of the brain. The latter is fatal within a few minutes, from respiratory failure, and obviously no extract of the medulla would avail to prevent the death.

However, one can obtain from the liver an extract which has important properties and has served a very useful purpose in the hands of physicians. We refer to the liver extract which controls the condition of pernicious anemia and prevents death from that disease. Unquestionably, an antianemic principle is to be found in the normal liver. Apparently, this substance is not a product of the liver itself, not an internal secretion, but is simply stored in the liver. The antipernicious anemia factor in liver is probably mostly if not entirely vitamin B_{12}.[36]

C. THE THYMUS (?)

The thymus and the pineal body are two structures which have also been classed as endocrine glands. In the adult the thymus lies within the chest cavity in the midline, just above the heart. Embryologically, it develops from the same pharyngeal clefts which give rise to the parathyroids, and reference has already been made to the fact that accessory parathyroid tissue frequently lies imbedded within the thymus. Its internal cellular structure is compatible with an endocrine function, although much of the thymus tissue is lymphoid in nature, resembling the internal structure of the lymph nodes.

The thymus is mentioned most frequently in association with growth, and it has been thought to elaborate a growth-stimulating

36. See discussion on pp. 102–4.

hormone. This idea was first suggested because the thymus is largest during the growing period and regresses after puberty. But this is a far cry from demonstrating even a significant relationship to growth, let alone possible elaboration of a hormone.

Evidence for the existence of such a growth hormone has been repeatedly sought. If it does exist, it is apparently not essential to growth, for removal of the thymus in young or old causes no clear-cut disturbance of any phase of growth or development or anything else. It has been reported that repeated injections of extracts of the thymus accelerate growth and the attainment of sexual maturity. The material appears to have a cumulative action, for, if the injections are continued from generation to generation in rats, it is reported that full growth and sexual maturity are attained, to a certain limit, earlier and earlier in successive generations.

Is this extract an endocrine product, or is it a growth-stimulating drug which is formed as an artifact during the process of extraction? At present the weight of the evidence seems to be against the hormone interpretation, although the final answer must await further experimentation.

D. The Pineal Body (?)

The pineal body is a small structure projecting from the brain substance upward between the two cerebral hemispheres. Lying as it does in almost the geometric center of the brain cavity, it has been thought also to be figuratively at the center: Descartes believed it to be the seat of the "soul," or the "self," but there exists no reliable evidence for this hypothesis. Evidence for any endocrine function of the structure is just about as scant. Extirpation leads to no constant specific effects which can be ascribed to the absence of the pineal body per se. In some animals the location of the structure makes surgical approach to it without injury to the adjacent brain tissue almost impossible, and undoubtedly many of the effects ascribed to removal are the result of brain injury. No extracts having any kind of consistent physiological action have ever been isolated from the body. Negative findings warrant the tentative hypothesis that the pineal body is not an endocrine gland, nor is any nonendocrine function ascribable to it. It is homologous with the third or pineal eye of primitive extinct vertebrates and in the present-day mammal is significant probably only as a museum piece, a relic of bygone evolution—and of the soul-seeking yearnings of man.

X. CONTROL OF THE ENDOCRINE GLANDS

One important aspect of endocrine physiology has been referred to repeatedly in the preceding discussions. What controls the rate of hormone production by the endocrine glands in the normal individual? By what machinery is rate of activity correlated with the other bodily processes? This question the student and investigator must ask repeatedly no matter what field of physiology is being considered. Our knowledge of the salivary glands, for example, is incomplete even if we should know fully about the formation, chemistry, and actions of saliva. In addition, we must understand at least the mechanisms which effect salivary secretion when food is ingested —that a reflex is initiated by the presence of food in the mouth.

Again, to know in great detail all the complicated chemical and physical phenomena of the heartbeat, and the intricate details of its pumping action, is fragmentary knowledge which must be supplemented by a knowledge of the chemical, nervous, and physical influences which play upon the heart and modify its pumping action, correlating the rate of its activity with the varying rates of activity of the body and its parts. Likewise we must ask about the hormone whose chemistry and effects we may know in some detail—what controls its secretion rate? In many instances our answers to this question will have to be more tentative than many of the conclusions we reached in the preceding sections on endocrinology.

A. NERVOUS CONTROL

1. *The adrenal medulla.*—Are the endocrines under the control of nerves? Do we have a situation analogous to salivary secretion, where the gland is paralyzed if denervated, and normal secretion occurs only by way of stimulation of the nerve of the gland? Apparently, this is not generally true of the endocrines. However, the adrenal medulla has been shown to be under nervous control. Epinephrine secretion practically ceases when the nerves of the gland are cut, and variations in the rate of secretion are readily induced by stimulation of the efferent nerves. But under what physiological conditions the adrenal nerves may be stimulated is debated. There is some evidence that a lowered blood-sugar concentration stimulates them. If this mechanism really operates physiologically, its utility in maintaining a constant blood-sugar concentration is plain, for epinephrine undoubtedly causes the liver to liberate glucose into the blood stream. The adrenal nerves are stimulated in emotional

states and also reflexly when painful injury is inflicted upon the animal. Epinephrine secretion reproduces and reinforces the various actions[37] of the sympathetic nervous system which are aroused in such stress situations as "fight and flight."

2. *The pancreas.*—There is some evidence, too, that insulin production is under nervous control, for the secretion of insulin has been reported to increase when the vagus is artificially stimulated. It has also been claimed that insulin secretion, occurring when the sugar concentration of the blood is elevated, depends on the vagus nerves being intact. The facts are not conclusive, however. If such nervous control exists, it apparently is not of prime importance, for the denervated pancreas, or even a transplanted pancreas, apparently produces insulin at normal rates.

3. *The thyroids.*—Evidence has been repeatedly sought for a nervous control of the thyroid glands. There is no question but that the nerves of the gland are capable of modifying the blood supply of the gland by constriction or dilatation of the vessels and that this might conceivably modify the hormone output indirectly. But the presence of specific secretory nerves has never been conclusively demonstrated. If such nerves do exist, they are probably not of great importance; for, if a fragment of the thyroid is transplanted to some new region of the body, it functions entirely normally despite the absence of all nerves. It also enlarges or atrophies parallel with the thyroid tissue remaining in its normal location with intact innervation. Furthermore, destroying all the nerves of the thyroid gland in animals does not seem to alter the rate of the thyroid hormone production.

B. Chemical Factors

Chemical changes in the composition of body fluids exerts a degree of controlling influence upon some of the endocrines. We have mentioned the theory of elevated blood sugar stimulating the nerves of the pancreatic islets and lowered blood sugar stimulating the nerves of the adrenal medulla. In still other instances the cells of the endocrine glands are directly activated or inhibited by changes in the chemical composition of the body fluids in the immediate environment of the hormone-producing cells themselves.

1. *Secretin.*—An excellent example of this is seen in the production of secretin by the cells of the duodenal mucosa. The most potent

37. See discussion of the physiological effects of epinephrine, p. 537.

stimulus to secretion of this hormone is the presence of hydrochloric acid in the intestine and, by diffusion, in the walls of the intestine. Secretin is produced, circulates in the blood to the pancreas, and there causes a stimulation of the enzyme-producing cells of that gland. As a consequence, there is a copious secretion of pancreatic juice. Here is a beautiful example of the operation of physiological machinery which correlates rate of hormone secretion with bodily requirements. Pancreatic juice is especially useful just when partially digested food passes from the stomach into the intestine; that is, secretion of the hormone is desirable just when food enters the intestine. The machinery is such that secretin formation is brought about at precisely this time, for, normally, the stomach contents are passed on into the intestine when they are definitely acid in reaction. And acid is the stimulus to secretin production. Hence we see that the very agent which sets the stage for an adaptive response— namely, entrance of gastric contents into the intestine—initiates a physiological reaction the end result of which is an adaptation useful in such a physiological setting.

2. *Rate of insulin production.*—Insulin secretion seems to be controlled essentially by the glucose concentration of the tissue fluids in the pancreas, which is, of course, determined by the glucose concentration of the blood. The insulin output of the pancreas must be measured indirectly. Ideally, we should like to be able to analyze the blood of the pancreatic vein for insulin by some quantitative chemical test. This cannot be done at present, and indirect measurements must be made. These tests depend upon some physiological reaction of insulin, such as the effect it has of lowering the sugar concentration of the blood. Insulin concentration must be measured crudely by injecting insulin-containing blood into an animal and noting the extent of diminution of blood-sugar concentration produced. Blood from a fasting animal has but little of this action. Blood from an animal recently having eaten carbohydrate has a distinct blood-sugar-lowering effect, presumably because of an increase in insulin content.

Although some question still remains as to the correct interpretation of these results, undoubtedly there would be considerable utility in such a mechanism. Insulin would be especially useful when the blood-sugar concentration is elevated, as after a meal, in facilitating combustion and storage of the sugar. Apparently, the blood-sugar elevation itself is the adequate stimulus to increased insulin output, which, in turn, decreases the blood-sugar content. Assuming these conclusions to be correct, we understand something of the machinery for keeping the blood-sugar concentration at a fairly constant level through the effects of changes in that concentration indirectly upon epinephrine production and directly upon insulin production.

C. HORMONAL INTERRELATIONS

Our discussions of the individual endocrine glands suggests that one of the most important factors in the control of the rate of secretion of one endocrine is stimulation or inhibition of that gland by the hormones of other glands. Very complex hormonal interrelationships occur, whose intricacies are far from being fully understood. Great strides are constantly being made in this field, however, and certain problems of endocrine interrelations are well on their way toward a satisfactory solution.

1. *The hypophysis: the "master-gland."*—The hypophysis comes to mind at once, and we have seen how extensive is the control exerted by the several hypophyseal hormones upon other glands of internal secretion. We saw that secretion of the sex hormones is dependent upon stimulation of the gonads by hypophyseal hormones and that even the phasic nature of the female sex hormone production might be accounted for by a mutual interaction of hypophysis and gonads upon each other via their respective hormones.

We noted evidence also for thyroid secretion being under control of the hypophysis. The thyroid tends to atrophy after hypophysectomy and increases its secretion of thyroxin when the proper hypophyseal hormone is injected. Hypophyseal enlargement has long been known to be common in hyperthyroidism, and the question arises: Are not hypothyroidism and hyperthyroidism essentially dependent upon hyposecretion or hypersecretion of thyroid-stimulating hormone by the hypophysis?

Similarly, the ACTH of the hypophysis largely controls the secretion of the adrenal cortical hormones. This control is apparently not complete. If the hypophysis is removed, the adrenal cortex continues to function sufficiently to preserve life. Removal of the cortex of both adrenal glands is soon fatal.

2. *Endocrine interactions in diabetes.*—Some rather startling results have been obtained in experimental studies upon the interrelations of the hypophysis and the pancreas. The hypophysis apparently exerts some antagonistic action upon insulin, possibly by

means of a specific hormone. The blood-sugar-lowering effect of insulin is decreased when hypophyseal extracts are administered, and, if the hypophysis is removed, the animal becomes very sensitive to insulin. Small quantities of insulin now lower the blood sugar to a dangerous or even fatal level.

The administration of suitable hypophyseal extracts to normal animals leads to some of the signs of diabetes, including a high blood-sugar concentration and the production of acid end products in the metabolism of fats, even though the pancreas is intact; also, diabetes is fairly common in marked hyperactivity of the hypophysis in man.

Most striking of all is the finding that the removal of the hypophysis markedly alleviates the diabetes produced by extirpation of the pancreas in animals. A hypophysectomized animal may survive a subsequent removal of the pancreas for six months to a year without insulin treatment; a pancreatectomized animal with an intact hypophysis will expire in about three weeks if untreated. At any rate, hypophysectomy greatly diminishes the amount of insulin required to keep the experimental diabetic animal in fairly normal health. Injections of hypophyseal extracts into animals deprived of both hypophysis and pancreas promptly increases the severity of the diabetes and of the insulin requirements.

Whether this anti-insulin action of the hypophysis is exerted by a specific hormone is not known. Purified extracts of the hypophysis free from other recognized hypophyseal hormones, yet possessing this anti-insulin action, have never been prepared. If such a hormone does exist, the question arises whether diabetes in man might not be primarily a hyperactivity of the hypophysis. But there is at present no clear-cut evidence of this in most patients afflicted with diabetes.

CHAPTER THIRTEEN

BODY DEFENSES
AGAINST DISEASE

I. HEALTH AND DISEASE
   A. Emergency functions of the body machinery
   B. Margins of safety
   C. Replacement of damaged tissue
   D. The function of pain

II. PARASITIC INFECTIONS
   A. Bacterial disease
   B. Species and tissue immunity

III. BARRIERS AGAINST ENTRANCE OF BACTERIA INTO THE BODY
   A. Mechanical factors
      1. The skin   2. Secretions
   B. Ciliary motion
      1. Ciliated cells in nature   2. Cilia in mammals
   C. Reflexes
   D. Chemical factors

IV. LOCALIZATION OF INFECTIONS
   A. Inflammation
   B. Phagocytosis
   C. Lymph vessels and nodes
   D. Destruction of bacteria after spread

V. ACQUIRED RESISTANCE TO INFECTION
   A. Pasteur's experiments with anthrax
   B. Antigen-antibody relationships
   C. Active and passive immunity

VI. TREATMENT OF INFECTIONS WITH CHEMICALS
   A. The sulfa drugs
   B. Penicillin
   C. Limitations of treatment with chemicals
      1. Specificity of chemical agents   2. Injurious effects upon man's body
      3. Development of resistance by bacteria

VII. BIOLOGICAL WARFARE

I.   HEALTH AND DISEASE

The normal span of life is determined partly by heredity both in the individual and in the species. The animal machine operates for a limited time. Its parts are susceptible not only to constant wear and tear but to damage which may more or less seriously impair the function of the whole organism. No tissue or organ or system is exempt from this. It is true of all plants and animals—of all living things. Success in the struggle for life depends in a large measure, therefore, upon the evolution of mechanisms which weight the balance between health and disease in favor of health, at least until the individual has contributed to the perpetuation of the species by reproduction.

A. Emergency Functions of the Body Machinery

The study of normal and abnormal physiology, of the conditions of health and disease, necessarily go hand in hand. Often the very same mechanisms which are always functioning in the normal animal also operate as defensive mechanisms in the emergencies of disease. Identical mechanisms are involved in the constant replacement of red blood cells normally destroyed every day and in the replacement of red cells destroyed in excessive numbers in disease. The preservation of the normal water balance between blood and tissues and the speedy restoration of fluids to the blood when blood is lost from a vessel both depend upon the same mechanisms. The stimulating effect of diminished oxygen tension in the blood upon the bone marrow and the interacting forces of filtration and osmosis in the capillaries are not merely cogs in the physiological machine; they are mechanisms of defense against injury to the machine.

In a sense, all physiological mechanisms constitute defenses against disease, for they function in the preservation of a constancy in the internal environment, a balance between factors which, if unopposed, spell disease and death. Claude Bernard said, "All the vital mechanisms, however varied they may be, have only one object, that of preserving constant the conditions of life"—and, he might have added—of health. In the words of Frederique, "Each disturbing influence induces by itself the calling forth of compensatory activities to neutralize or repair the disturbance . . . [tending] to free the organism . . . from unfavorable influences and changes in the environment."

B. Margins of Safety

An important bulwark against serious or fatal consequences of tissue or organ damage lies in the fact that there is usually a wide margin of safety between the normal quantity of tissue in the body and the minimum quantity necessary for health. In general, we are equipped with more of every specific kind of tissue than we actually need. An animal can get along very well with considerably less than half his liver or pancreas, provided the remaining tissue is normal. In the case of a paired organ, one can usually do the work of both if necessary. An animal can exist fairly well with one lung or one kidney, so that considerable damage may often be sustained without impairment of health. The extra supply of liver and pancreas tissue and the extra lung and kidney represent a reserve supply of tissues; they furnish a rather wide margin of safety.

We encounter the same principle when we examine the blood supply of most organs. In general, they are equipped with more than one artery and more than one vein, so that, if one artery or vein is damaged, the blood supply and drainage of an organ or part of an organ are not seriously impaired. There are important exceptions to this rule. A given area of the heart muscle, or of the brain, has but one nutrient artery. If this is damaged, the tissue supplied by it dies. If the medulla, with its vital centers, is so affected, death is certain and sudden.

C. Replacement of Damaged Tissue

We have seen how tissues which are moderately overactive tend to increase in size and functional capacity. Just what the mechanism of this effect is, we do not know. But we do know that physical exercise in some way stimulates increased muscle growth and, hence, muscle power. Similar effects on the heart constitute another defensive mechanism. When, for example, abnormalities of the blood vessels cause a gradually increasing blood pressure, a strain is placed upon the heart; blood is now pumped against the high pressure with increased difficulty. The heart is enabled to meet the extra load partly because in some way the added work it does stimulates an increase in size and strength of the organ.

This phenomenon plays a role also when the margins of safety are encroached upon by disease. The tissue which survives becomes overactive and, like muscle, grows. When one adrenal gland is removed or destroyed by disease, the other gland tends to enlarge, presumably also increasing its hormone-producing capacity.

Such growth may even replace the lost tissue entirely. Lost blood is replaced by an equal quantity of normal blood, damaged skin by new skin. New nerve fibers can regenerate when peripheral nerves are destroyed. There are, of course, definite limitations here. Some tissues can reconstitute themselves only to a limited degree or not at all. Little if any replacement of damaged kidney tissue occurs, except by a fibrous connective-tissue scar, which leaves the kidneys minus some secreting elements. Little if any replacement of neurones occurs in the central nervous system, or of skeletal muscle fibers, at least in adult mammals.

D. The Function of Pain

Pain appears to be primarily an adaptation which protects against injury. This is obvious in the case of those reflexes which are initiated by injurious stimuli, such as flexion of the limb when pain fibers are stimulated and blinking of the eye when the cornea is touched. But, though normally associated with pain, these reflexes are not the result of it. They occur even if no pain is experienced, in an anesthetized man or animal. However, pain does afford an added protection. Because of it, we may decide to move away from the harmful agent, or we learn to prevent a recurrence of the injury.

Of equal importance is the function of pain in informing us that something is amiss. Slight pain warns of slight damage, which may then be attended to before more serious harm results. Pain calls attention to a sliver in the finger; the sliver may be removed before serious infection sets in. Pain tells us that a tooth is decaying, and we attend to it before an abscess forms. Pain not only warns of trouble but it often tells just where the trouble is. It not only says, "You have cut your finger," but it tells on which finger, and just where, the cut is. It is one of the most important symptoms of disease, and the physician often gets much information from knowing exactly what kind of pain his patient suffers and just where the pain is localized.

Another important function of pain is its indirect aid in the repair and replacement of damaged tissue. Rest is probably the best *single* aid to healing. Whether or not the physician uses drugs or serums or diet or other special treatment, he always prescribes rest as well. Pain enforces rest. If it is severe enough, it may impose absolute rest in bed. Or the pain may merely enforce rest of the injured part. Pain immobilizes a broken arm, facilitating the growth of new bone which heals the fracture. The surgeon merely aids in the enforcement of rest of the arm by splinting it.

In a world in which danger of injury by accident and infection threaten on all sides, pain is therefore of considerable adaptive value. The task of alleviating pain is one rather of eliminating the causes of pain than simply treating the pain itself. Often, of course, the pain itself must be treated in the sick, but only incidentally and after it has served the function of telling where the trouble lies.

II. PARASITIC INFECTIONS

Our main concern in this chapter will be with the mechanisms of defense against infectious agents. Man is host to infestations by animals,[1] by plants,[2] and, of most concern to us here, by bacteria of all kinds.[3]

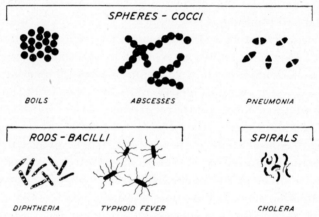

FIG. 206.—Some bacteria which produce disease in man, illustrating the three main forms of bacteria.

A. BACTERIAL DISEASE

Our accurate information about the role of bacteria in disease dates from the researches of such men as Louis Pasteur in the latter half of the last century. In the face of intense opposition on the part of a conservative medical world, these men demonstrated that many

1. Examples are intestinal worms, such as the tapeworm and the hookworm, and protozoa, such as the malarial parasite and the ameba of amebic dysentery.

2. An example is the fungus which causes "athlete's foot."

3. The morphology and biology of the bacteria cannot be delt with here in detail. We may state that, on the basis of structure, these organisms are classified into the *cocci*, which are of spherical shape; the *bacilli*, which are rod-shaped; and the *spirella*, which are tiny spirals. A number of infectious diseases, such as smallpox, are produced by agents which in many respects are like bacteria but are invisible even under the microscope and can pass through filters of the finest mesh. These agents are called *filterable viruses*.

diseases are caused by bacterial infection. They proved that a specific bacterium causes tuberculosis and that another produces anthrax. The German bacteriologist, Robert Koch, formulated several principles which must be adhered to in proving the causal relationship between a specific form of bacterium and a specific disease.[4] The discoveries of these men and of their contemporaries and students constitute a thrilling chapter in the history of man's conquest of disease, but it is beyond the scope of this book.

Of course, not all bacteria are disease-producing. Many are harmless. Some—those which figure in the cycle of matter and energy, causing such phenomena as decay—are even indispensable for animal and plant life on this planet.

B. SPECIES AND TISSUE IMMUNITY

In their environmental relationships all forms of life display a high degree of specificity. This applies fully to the bacteria, including the disease-producing varieties. The ability of bacterial parasites to grow and produce damage varies considerably from species to species. The tubercle bacillus grows poorly in dogs, very well in the tissues of man, and better still in guinea pigs. There is apprently something about the composition of the dog's tissues which renders them unsuitable for the best growth of the tubercle bacilli. We commonly express such relationships by saying that the dog has an inherent resistance to this disease, or that it is relatively *immune* to tuberculosis, and that the guinea pig has a low degree of natural immunity to tuberculosis.

A great many specific immunities of this nature exist. Each species is subject to infections to which it is especially susceptible. Each species has an inborn relative immunity to other infections. In reality, of course, this immunity is also a defensive mechanism of a sort. Whatever may be its nature, there are factors present which protect the animal more or less successfully against infection by certain particular parasites.

A further refinement of host-parasite specificity is found in the differences in susceptibility of the various tissues of the body of a

---

4. Koch's postulates state that, to prove a specific bacterium causes a given disease, one must: (*a*) always find the organism in the blood or tissues of individuals having the disease; (*b*) cultivate the organism in an artificial medium, excluding all other forms of life; (*c*) produce the disease in typical form by inoculation of organisms from the culture into an experimental animal—man or beast; and (*d*) find the same organism in the blood or tissues of the experimental animal.

given animal to an infection. The organisms of infantile paralysis (a virus) and tetanus (or its toxins) have a special predilection for nerve tissue, for example. Other tissues of the body are less susceptible, for some reason, to injury by these agents. They possess a degree of natural tissue immunity.

III. BARRIERS AGAINST ENTRANCE OF BACTERIA INTO THE BODY

The mechanisms of defense against bacterial disease are organized along several lines. In the first place, there are devices which guard against the entrance of bacteria into the body tissues. Of course, ordinary common sense figures here. We avoid contact with people known to have contagious infections as well as with objects which may harbor harmful bacteria. Sanitary measures of all kinds, which have been so effective in reducing the death rate from infectious diseases and in preventing the spread of infections,[5] are common-sense applications of our knowledge of infectious diseases. But the body defenses do not rely too much upon common sense. They are inborn and function purely automatically.

The barriers against entrance of bacteria are especially massed about the external openings into the interior of the body—the mouth and pharynx and respiratory passages. These natural openings are always potential *portals of entry* for bacteria.

A. Mechanical Factors

1. *The skin.*—Some of the primary defenses are of a purely mechanical nature. The skin plays a major role here. Ordinarily, bacteria cannot penetrate this layer, whose outer surface is composed of dead cells. They do sometimes gain entrance into the sweat glands and produce local infections, but they usually cannot reach the deeper tissues, except when a break or tear occurs in the skin, affording a portal of entry. Oil secreted by tiny glands at the hair roots serves to keep the outer layers of dead skin cells freely flexible, so that cracks affording openings for bacterial entry do not result from drying of the skin.

2. *Secretions.*—Secretions upon other surfaces help prevent bacterial penetration in another way. The linings of the alimentary

---

5. Sanitary measures such as the proper disposal of sewage and protection and disinfection of the water supply (along with vaccination) have been largely responsible for great reductions in the annual typhoid-fever death rate. In 1893 the death rate in Chicago was 170 per 100,000 population. It is now less than 1 per 100,000.

canal, the nasal cavity, and the upper lung passages are coated with a thick slimy secretion called *mucus*, which entraps and arrests the migration of bacteria on the flypaper principle. The lining cells themselves also constitute a mechanical barrier much as the skin does. Swarms of bacteria are always present in the lower small intestine and colon, yet normally they do not get into the tissues of the intestines, or into the blood, in significant numbers.

The flow of secretions is often an effective device. It mechanically retards the passage of bacteria against the moving stream. In

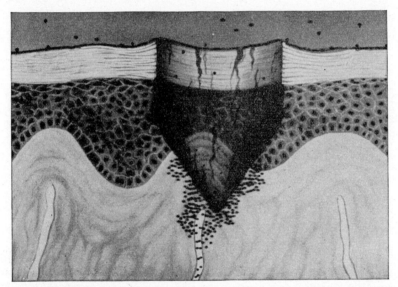

Fig. 207.—Injury to the skin providing a portal of entry of bacteria. The skin constitutes a mechanical barrier to entrance of bacteria into the body. When the skin is broken (here, by a nail), bacteria (*black dots*) enter the tissues and may multiply rapidly (see Fig. 211). (From the sound film, *Body Defenses against Disease.*)

this way bacteria are fairly effectively prevented from migrating upward through gland ducts to infect the gland itself, whether this be a sweat gland, a salivary gland, the pancreas, the liver, the kidney, or any other gland of external secretion. Tears continually wash the cornea free from bacteria-laden dust particles.

B. Ciliary Motion

Still another mechanical factor is ciliary motion, particularly in the upper-lung passages of mammals. Ciliated cells are common in nature.

1. *Ciliated cells in nature.*—Cilia are extremely fine protoplasmic

extensions of cells. Though sometimes described as hairlike, they
are not to be confused with true hair. Ciliated cells are widespread
in nature. A whole class of Protozoa[6] is known as the Ciliata. *Paramecium*, a member of this group, is ciliated over its entire surface.
Other Protozoa possess but a single or few cilia which, though essentially like cilia, are called *flagellae*. In all these forms the cilia (or
flagellae) are used in locomotion. Each cilium is flipped rapidly in
one direction and then slowly returned to its original position. The
beating movement serves very effectively to propel the animal
forward.

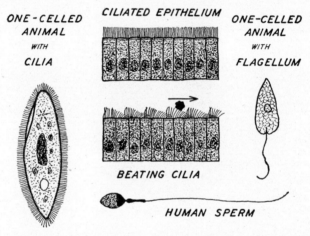

FIG. 208.—Ciliated cells. Cilia are found on some one-celled animals, where they function in
locomotion. In higher animals the cilia of stationary cells function to move particles (indicated
by arrow). Human sperm resemble certain one-celled animals in their manner of locomotion,
employing a single large cilium or flagellum.

In the higher animals ciliated cells are also common. Here, however, the cells are not able to move about freely, and the ciliary
movement is not used for locomotion.[7] The ciliated cells in vertebrates are parts of a tissue, many cells being attached together, lining a duct or cavity, with the cilia projecting into the cavity. Here
the concerted action of the cilia produces movement of fluids or particles which may come into contact with the ciliated surface. Thousands of cilia beating in unison in such tissue, when viewed under
the microscope, present an appearance resembling a wheat field
"waving" in the wind.

6. Unicellular animals.

7. Except for the flagellum constituting the "tail" of the male reproductive cell, which
propels it along through the female genital tract.

What controls ciliary action? What mechanical changes occur in the cells to effect the lashing movements? The answer is by no means clear. It is known that at the termination of each cilium, imbedded or rooted in the cell proper, there is a granular swelling which in some manner controls the movements. Granules have been dissected from the roots of cilia under the microscope with the result that the cilia cease beating. Also, the cilia of fragments of ciliated cells will continue to beat for a time, provided these granules remain attached. Apparently, some sort of activity of the protoplasm at the root of each cilium is involved, a rapid contraction on one side causing the quick flip of the cilium, and a relaxation producing the slower return movement, both being under control of the granule.

The mechanisms by which the many cilia are correlated with one another in their synchronized activity are also puzzling. In forms like *Paramecium* there is apparently a sort of primitive nervous system which extends as a protoplasmic network throughout the cell to each cilium. Exactly how this effects correlation is not known. In the case of the frog's gullet or man's breathing passages the problem is even more perplexing, for here there are many cells involved, each possessing a great many cilia. Whatever may be the synchronizing influence, it extends across cell boundaries, affecting several adjacent cells in a well co-ordinated manner.

2. Cilia in mammals.—The gullet of the frog is lined with cilia which beat in such a manner that food particles are swept downward from the mouth to the stomach. This function is reminiscent of the action of the cilia in the gullet of *Paramecium*,[8] which sweep microscopic food particles to the base of the tube into the interior of the cells.

In man there are no cilia in the gullet. Other mechanisms operate

8. Though the term "gullet" is used for both animals, it is to be remembered that, while these structures are of similar function, they are not *homologous*. The gullet of a *Paramecium* is part of a cell. The gullet of a mammal is a large multicellular structure, composed of several kinds of tissues.

Given structures occurring in more than one species of animals are said to be *homologous* when they have a common embryonic and evolutionary history. The arm of a man, the flipper of a whale, and the wing of a bird are homologous because they are constructed on a similar plan; they develop in the same manner embryologically; and they represent further specialization of an appendage evolved by common ancestors—the primitive fish. Homologous structures need not have identical functions.

Structures not related *phylogenetically* (i.e., in evolution) but having a similar function are said to be *analogous*. Examples are the lungs of man and the gills of a fish; the wings of birds and the wings of insects.

to move food from mouth to stomach. But the main breathing passages of mammals contain this type of epithelium. Here, however, the cilia move so as to sweep particles upward, from the finer branches lower in the lungs, into the back of the mouth or pharynx. As dust particles are breathed in with the air, they come into contact with the ciliated epithelium, stick to its moist surface, and are swept slowly upward. This constitutes a protective mechanism which not only prevents much irritating dust from reaching the delicate tissues of the lung but also guards against infection by bacteria which may be lodged upon the dust particles.

The hairs of the nasal cavity, not to be confused with cilia, serve to strain out bacteria-laden dust particles which happen to be present in the inhaled air.

C. Reflexes

The cough reflex functions in expelling infected particles, or masses of mucus which may contain bacteria, from the pharynx, larynx, trachea, or even the smaller respiratory passages. It serves not merely in defense against bacterial infection but likewise in the prevention of mechanical obstruction of the breathing passages when solid particles accidentally gain entrance. Should an infection of the deeper lung passages occur, the increased quantities of mucus secreted by the epithelium of the respiratory tubes will also initiate the cough reflex by mechanical stimulation, and bacteria-laden quantities of mucus are blown from the lungs.

The corneal reflex, in which there is an involuntary blinking of the eye when the cornea is touched or irritated, helps to prevent bacteria from growing locally. The reflex inhibition of respiration during swallowing usually prevents food, which is nearly always contaminated with bacteria, from being drawn into the lungs.

In a sense, even the flexion reflexes constitute a defense against infection, for without them injuries to the skin would be more common and more severe, providing portals of entry for the ever present bacteria.

D. Chemical Factors

In general, the body secretions themselves are only mildly if at all injurious to bacteria. A notable exception is the gastric juice, which by virtue of its high acid content destroys bacteria chemically or at least retards bacterial growth. As a consequence, even though bac-

teria are always present in the mouth and saliva and in the foods
swallowed, very few bacteria are found in the materials emptied from
the stomach into the small intestine.

IV. LOCALIZATION OF INFECTIONS

A. INFLAMMATION

Should bacteria penetrate these barriers against entrance into the
body, a number of devices operate in localizing the invading infection. Among these is the phenomenon of inflammation, which is
exemplified by a boil. This involves an augmented blood supply to
the infected area effected by a dilatation of the arterioles and capillaries. Fluids escape from the blood into the region, diluting the
bacteria and their products. This plasma fluid contains the various

FIG. 209.—Behavior of white blood cells (*leucocytes*) in defense against certain infections.
Left: Migration through a capillary wall. *Center:* Movement toward the invading organisms
by ameboid locomotion. *Right:* A cell containing several bacteria which have been ingested
by phagocytosis.

blood-clotting elements, and coagulation takes place, perhaps initiated largely by the thrombokinase liberated from cells destroyed by
the bacteria. The whole locally infected area is in this way converted
into a jelly-like mass, enmeshed with strands of fibrin. Fibrous connective tissue forms around the area, helping to wall off the infection.
Sometimes, when this occurs in the skin, man interferes with nature
at this point by squeezing the inflamed area to "let the pus out." He
incidentally also may break down the limiting barriers of the coagu-
lated mass, inviting spread of the infection. But, ordinarily, inflammation effectively limits the spread of the infection.

B. PHAGOCYTOSIS

Meanwhile white cells of the blood have been collecting in the
area, and many of the bacteria are destroyed by phagocytosis. Leucocytes are called *microphages*, or small phagocytes. Large phagocytes, or *macrophages*, come to their assistance. These cells are found
ubiquitously distributed throughout the tissues, through which they

slowly migrate. Like the white blood cells, they normally collect in great numbers in infected areas.[9]

C. Lymph Vessels and Nodes

Still further barriers to generalized spread are encountered. If the bacteria gain entrance into the lymphatic vessels, profusely distributed to all parts of the body, they are filtered out in the lymph nodes in great numbers and ingested by the large phagocytic cells found in the meshes of the nodes. Lymph nodes and vessels are especially abundant in regions in which bacteria are most likely to

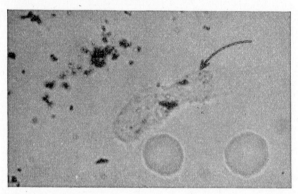

Fig. 210.—Photograph of a large phagocyte (*indicated by arrow*) which has engulfed a number of bacteria (*shown in black*). Phagocytes engulf particles by extending pseudopods ("false feet") which flow around the particles (see also Fig. 29, p. 119). (From the sound film, *Body Defenses against Disease.*)

gain entrance—about the pharynx and respiratory and gastrointestinal passages.

The large phagocytic cells found in the linings of the blood channels of the liver and spleen function in similar fashion, except that it is from the blood that they remove bacteria. They also engulf and destroy bacteria by digesting them.

Often so many bacteria are filtered out in the lymph nodes along a vessel draining an infected area that the node itself becomes swollen and tender. Many persons have experienced the painful swellings in the neck during a siege of "sore throat" or in the armpit after an infection of the hand or arm. These swellings are lymph nodes which have filtered out and are destroying the infectious agents.

9. Belonging also to the macrophage group are those cells in the spleen, liver, and lymph nodes which possess the capacity for ingesting and destroying bacteria.

D. Destruction of Bacteria after Spread

When the bacteria are numerous and highly virulent, and the agencies for the limitation of spread fail to localize the infection, the combat between invader and defenses continues in all parts of the body to which the organisms happen to spread. We cannot sharply delimit all these factors from the preceding group. Bacteria which have been generally disseminated are engulfed by the white blood

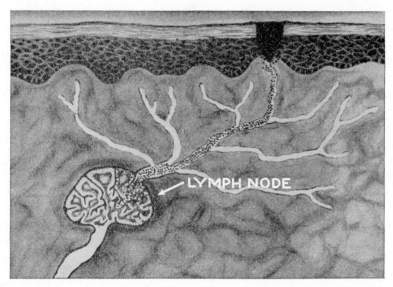

Fig. 211.—Prevention of spread of bacteria. In the tissues, bacteria (*black dots*) enter the lymph vessels (*white channels*), and many are filtered out and destroyed in the lymph nodes. (From the sound film, *Body Defenses against Disease.*)

cells. The large phagocytes found throughout the body tissues, but especially numerous near the blood vessels, carry on their activities everywhere. The phenomenon of inflammation, which originally failed to prevent the spread, is repeated in the many new regions where bacteria have now lodged. Multiple areas of inflammation and abcess formation may occur throughout the tissues of the body.

Finally, the processes dependent upon acquired immunity serve to combat the onslaughts of the disseminated bacteria.

V. ACQUIRED RESISTANCE TO INFECTION

The presence of an infection within the healthy body tends to increase the effectiveness of those defensive mechanisms which destroy bacteria and neutralize their injurious effects. This phenom-

enon was recognized many years before the nature of the defense was
known. Persons who recovered from smallpox seldom contracted the
disease a second time. This indicated that the original infection
changed the body in some way so that it could more effectively combat or entirely prevent a second infection. The individual had acquired an enhanced immunity to the infection.

Of especial interest here is the fact that in many cases we have
learned how to effect this enhanced immunity without subjecting an
individual to the dangers of an actual infection. In 1796 Edward
Jenner, in England, discovered that the same high degree of immunity to smallpox could be produced by artificially infecting an individual with an extremely mild form of the disease—so mild as to be
practically symptomless and without danger. He instituted the
practice of *vaccination* based upon this principle.

A. Pasteur's Experiments with Anthrax

Perhaps the most dramatic demonstration of acquired immunity
was made by Louis Pasteur in his investigations of the disease
anthrax, which was decimating the herds of cattle not only in France
but in all Europe and even fatally attacking man as well. Pasteur first
demonstrated, at least to his own satisfaction, that the disease was
caused by rod-shaped bacteria—the bacilli of anthrax. He then produced weakened strains of the organism by cultivating them for a
time at fairly high temperatures. After a few days, a drop of culture
injected into ten sheep killed only four or five instead of all ten. After
a few more days, an injection killed none at all; it merely made the
sheep slightly ill. But the important finding was that these weakened
microbes did produce an enhanced immunity, which caused the sheep
to resist successfully a subsequent injection of enough of the anthrax
rods to have killed several ordinary sheep. His conclusions were received with skepticism and even hostility by the conservative physicians of the day. But, sure of his repeatedly verified results, Pasteur
arranged a public demonstration to convince his skeptical contemporaries. The great of French medicine and science attended.

Fifty sheep were gathered in a single inclosure. Half of them, selected at random, were injected with broth cultures teeming with
rods which had been greatly weakened by heat. The twenty-five injected sheep were marked to distinguish them from the others—the
controls—which received no injection.

Twelve days later, before the reassembled audience, the marked
sheep were reinjected, this time with rods which were also weakened
by heat but not so much as those of the first injections.

Still later the whole lot of fifty marked and unmarked sheep were injected with a deadly dose of rods. The marked sheep even received triple doses. On the following day eighteen of the unmarked sheep were dead, while some of the vaccinated sheep developed only a mild fever.

On the day of the final phase of the demonstration, twenty-two of the untreated sheep lay lifeless; the remaining three were breathing their last. Every one of the twenty-five vaccinated sheep was in excellent health—a perfect experiment by "the most perfect man who has ever entered the kingdom of science."

B. ANTIGEN-ANTIBODY RELATIONSHIPS

Subsequent investigations have elucidated many features of this antigen-antibody phenomenon. It is a very generalized reaction, restricted in no way to bacteria alone, their proteins, or their products. The introduction into an animal of almost any protein which is "foreign" to the animal induces in that animal the development of agents tending to destroy that protein.

We have seen that the body fluids of any dog will hemolyze the red blood cells of any rabbit. But, if a dog is *injected* with a rabbit's corpuscles, we find after a period of time that the dog's body fluids are rendered many times more hemolytic to the rabbit's corpuscles. The presence of the rabbit's corpuscles within the tissues stimulates the production in the dog of large amounts of substances which destroy the rabbit's corpuscles. Any such foreign proteins as those of red blood cells which are introduced into an animal's body are called *antigens;* the destructive agents formed are *antibodies.* Antigen-antibody relationships are highly specific. Any antigen, whether it be bacteria, bacterial products, or proteins, induces the production of antibodies which affect only that specific antigen. Most, if not all, antibodies are proteins.

The exact mechanism of antibody production is obscure, but we do know what constitutes the stimulus for their production and something also about what the antibodies do. Some of them, the *opsonins*, enhance the phagocytic activity of the phagocytes; others, the *bacteriolysins*, destroy bacteria by dissolving them; the *agglutinins* act by causing bacteria to clump together, which prevents spread and makes more certain that they will be caught in the filtering meshes of the lymph nodes.

Other antibodies, the *antitoxins*, neutralize the poisons produced by bacteria, which are often more injurious than any direct tissue damage produced locally by the growing bacteria themselves. Some of these antitoxins, the *precipitins*, act by precipitating the toxins.

Still other forms of increased immunity, of a less well understood nature, may be produced. In tuberculosis the relative immunity takes the form of an enhancement of the inflammatory reaction that occurs when the organisms invade the immunized tissues.

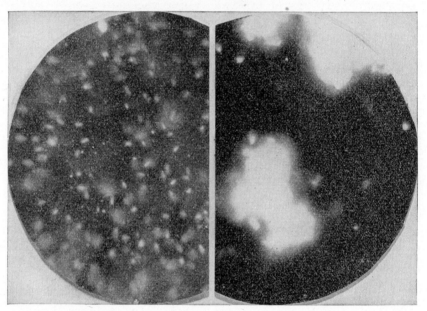

Fig. 212.—Clumping of bacteria by antibodies. Antibodies called *agglutinins* facilitate destruction of bacteria (*white spots*) by causing them to clump together (*right*). Ordinarily, the bacteria remain dispersed (*left*) (see also Fig. 19, p. 89). (From the sound film, *Body Defenses against Disease.*)

C. Active and Passive Immunity

Animals whose own tissues have produced the antibodies are said to possess *active immunity*. This may be acquired by the individual actually having had the disease and having recovered from it, or it can be induced by the artificial inoculation of weakened or dead bacteria or bacterial products. This latter is called *vaccination*. Artificial immunization against smallpox or typhoid fever are notable examples.

In many cases it has been possible to stimulate antibody production in an animal and collect the blood serum which contains the antibodies and transfer them into another animal or, in actual practice, into man. Their effectiveness persists for a time in the new animal. A man so inoculated is said to possess a *passive immunity*, because the antibodies of his blood and tissues have been produced not by himself but by another animal. In general, passive immunity is not quite so effective or so long-lasting as active immunity, although the protection it affords commences at once following injection.

An interesting and important discovery involving active and passive immunity in poliomyelitis was made in 1952. There is evidence that most adults have had poliomyelitis, usually in such mild form as to have been indistinguishable from a cold but sufficient to stimulate the formation of some degree of active immunity.

A fraction of the protein—*gamma globulin*—from the pooled blood of many "normal" adults seems to be or to contain the antibodies. In the 1952 experiment thousands of children were injected with gamma globulin, while other thousands in the same area—the controls—received injections of a neutral inert substance. It was found that there was significantly less poliomyelitis among the children receiving the globulin as compared with the controls. Apparently, the injection of such gamma globulins into children provided significant although temporary protection against poliomyelitis, presumably by conferring a passive immunity.

It is sometimes found effective to combine artificial passive and active immunity. In diphtheria epidemics children are injected with both a modified diphtheria toxin and its antibody in a toxin-antitoxin combination. The resultant passive immunization confers a resistance to the infection which is immediate. It tides the child over the period during which its own tissues are producing antitoxin, after which there is an active immunity which is of higher degree and longer duration than the passive.

Unfortunately, immunity of a high degree and long duration is not characteristic of all infections or of artificial immunization against infections. There are wide variations of effectiveness. Active immunity against smallpox is nearly 100 per cent effective and, once established, is likely to persist throughout life. The common cold produces only a slight degree of immunity of brief duration. The effectiveness of typhoid vaccination is intermediate in duration, persisting for from five to eight years.

A number of infections cannot yet be controlled at all by artificial immunization. Nevertheless, the high degree of success in combating many of the infectious diseases with these weapons constitutes one of the most gratifying and encouraging successes in man's fight against disease.

VI.   TREATMENT OF INFECTIONS WITH CHEMICALS

An important function of the physician and of the science of medicine is to assist, strengthen, and supplement the natural reactions of the body to deleterious influences. In the repair of a broken bone by the body, it is necessary that there be virtually no movement at the break. Pain reduces such movement, but the cast applied by the physician practically eliminates it. The transfusion of blood by the physician aids the red bone marrow to provide an adequate number of red blood cells in an anemic individual. In a similar manner efforts have been numerous in the history of medicine to discover and employ chemicals harmful to bacteria but harmless to man in the control of bacterial infections in attempts to reinforce the natural defenses of the body which have been described.

Many chemicals are known which destroy bacteria, but unfortunately most of these damage the cells and tissues of the body also, so that they cannot be employed in combating infections. The search has been for chemicals which destroy or harm bacteria but do not injure man. Such substances, furthermore, must not be rendered inactive by chemical reactions in the body and should be of such a nature that they reach the bacteria of the infected organ or part via the blood stream, after administration by mouth, by vein, or under the skin.

Perhaps the best-known such chemical, until recent years, was the famous "606" (*Salvarsan*, or *arsphenamine*), said to be the six hundred and sixth chemical compound in a series synthesized and tested by Paul Ehrlich in attempts to control infections with the organism producing syphilis, against which the body defenses are relatively poor.

A. THE SULFA DRUGS

Tremendous impetus was given to research upon and testing of antibacterial drugs by the necessities and opportunities of World War II. The need for such chemicals sprang from the desire to reduce or eliminate infections of wounds, and the opportunity to test and show the effectiveness of promising substances (already proved to be at least harmless) was provided by the large number of men under arms. This was the case with the sulfa drugs, which came into prominence during the war.

The sulfa compounds had been known a long time before their medical usefulness was demonstrated. *Sulfanilamide*, as an effective agent in the treatment of certain infections, was the product of thirty years' investigations by many scientists. First manufactured in 1908, sulfanilamide was originally important only as a dye component. Its antibacterial action was noted upon several occasions but was not convincingly demonstrated until a German scientist reported (in 1935) that mice given a sulfanilamide compound failed to die when infected with otherwise fatal doses of certain bacteria.[10] This epoch-making discovery stimulated numerous experiments in Europe and America which demonstrated the effectiveness of this substance and its chemical relatives, called collectively the *sulfa drugs*, in controlling certain infections in man as well as in the experimental animal. Employment of this new weapon against disease and death in the wounded of World War II was an important factor in winning the war; although the chemical originated in Germany, the development and extension of its effective use took place primarily in the laboratories and hospitals of the United States and England.

B. Penicillin

Penicillin, the product of a lowly mold, has become an even more important bacteria-combating agent than the sulfa drugs. Alexander Fleming, its English discoverer, was conducting experiments on the growth of various bacteria on artificial culture plates in his laboratory. The magnitude of his discovery was by no means apparent from his simple description of his original observations, published some twenty-five years ago:

> While working with staphylococcus variants a number of culture-plates were set aside on the laboratory bench and examined from time to time. In the examinations these plates were necessarily exposed to the air and they became contaminated with various micro-organisms. It was noticed that around a large colony of the contaminating mould the staphylococcus colonies became transparent and were obviously undergoing . . . [destruction].

The destructive action was shown to be due to a substance produced by the mold, a purely accidental contaminating agent on the culture plates. Although Fleming suggested that penicillin "may be an efficient antiseptic for application to, or injection into, areas in-

10. The streptococci, which are cocci that grow in chains, as depicted in Fig. 206 (p. 580).

fected with penicillin-sensitive microbes," it remained for other investigators to nurture Fleming's discoveries to full fruition. Extensive experiments were conducted to test possible injurious action of
the substance on animal tissues. The efficacy of penicillin in combating certain infections in man, especially with pus-producing
organisms, was not convincingly demonstrated until eleven years
after Fleming's work, by another Englishman, Howard Florey, and
his co-workers.

To those uninitiated into the difficulties besetting the scientific investigator, this seems an unwarrantedly long time. One major
obstacle to rapid progress in this instance was the very small amount
of bacteria-destroying penicillin produced by even large cultures of
the mold. It was difficult to secure enough of the substance to test
it properly. Indeed, even after penicillin was known to possess antibacterial action in some human infections, quantities sufficient to
meet the demands of the wounded in World War II were not available in the early years. A little-known chapter in the history of the
war was the nation-wide collaboration of the government, scientists,
and drug manufacturers in a remarkably successful campaign to push
penicillin production to the required levels. Initially, the drug was
rationed for civilian use, although production increased so astonish-
ingly fast that there soon was enough for use by both military and
civilian physicians.

Penicillin belongs to the general category of *antibiotics*, of which
many have been developed. They are formed in or by living organisms and kill or inhibit the growth and reproduction of many kinds
of infectious agents.

C. Limitations of Treatment with Chemicals

Since the advent of the first sulpha drugs and penicillin, literally
dozens of new chemical compounds and various types of molds and
their products have been developed, tested for antibacterial action in
the test tube, screened for possible injury to the tissues and organs
of the body, and finally employed in man in attempts to combat
human infections.

The fact that such search and research continues arouses suspi-
cions of the efficacy of chemical control of harmful infections by
"wonder drugs." Effective though such control has been in many
kinds of infections, serious obstacles remain.

1. *Specificity of chemical agents.*—It has been found that an anti-bacterial agent effective against one organism may be quite harm-less against other kinds of bacteria. No one or even a few chemicals is injurious to more than one or a few kinds of bacteria. Since large numbers of organisms infect and injure and even kill man, it appears that perhaps almost equally large numbers of bacteria-injuring agents must be discovered by painstaking research if the chemical control of the infections of man is to be broadly effective. For many infections—the virus producing poliomyelitis is an example—there has been limited promise of success by chemical attack upon the organism.

2. *Injurious effects upon man's body.*—There are many chemicals which kill bacteria quickly. However, most of these kill the tissues and organs of man as well. To discover chemicals which kill or injure bacteria without harming the human being harboring the organisms is the problem. As might be expected, absolutes are rare here as elsewhere. Usually, the best that can be achieved is to find an agent which is relatively very harmful to infectious organisms and rela-tively harmless to man himself. An infection may be controlled by a chemical agent, but severe nausea and vomiting may also occur, or the skin and other tissues may become waterlogged and swollen.

3. *Development of resistance by bacteria.*—Perhaps the most frus-trating findings in this area of research are the repeated observations that a chemical originally very effective in combating an infection gradually loses its effectiveness. Actually, it is the infecting organ-isms which change.

Streptomycin—a mold product—is extremely injurious to tubercle bacilli and relatively harmless to man so that its administration may initially retard tuberculosis. However, continued administration be-comes decreasingly effective or even useless. Eventually it is found that tuberculosis organisms in the treated patient are extremely resistant to the chemical which was once deadly to the organisms. It is probable that, initially, only a relatively few tuberculosis germ⁵ are so constructed as to resist the harmful agent. But these few multiply, producing myriads of resistant offspring, replacing the many drug-susceptible organisms originally involved. The disease is still tuberculosis. At first, perhaps 95 per cent of the organisms were susceptible to destruction by streptomycin, perhaps 5 per cent were resistant, so that administration of the substance produces spec-

tacular benefits. Later, the 5 per cent resistant organisms multiply
rapidly so that the ratio becomes 95 per cent resistant and 5 per cent
susceptible. Streptomycin is now virtually ineffective.

Such development of resistance is the rule rather than the exception and has intensified the search for still other chemical agents
whose administration will destroy infecting bacteria without foster-
ing the development of bacteria resistant to the injurious agent.

One conclusion emerges clearly. Self-administration of such agents
or employment of them for minor infections is risky. The immediate
difficulty may be apparently resolved, but the development of resistant strains of bacteria may be fostered, so that subsequent treatment of serious infections may be dangerously impaired.

VII. BIOLOGICAL WARFARE

Biological warfare is a term applied to the employment of bacteria
or other infectious agents or their poisonous products to injure or
kill military or civilian personnel of the enemy or damage crops and
domestic animals. There have been gross exaggerations of the potentialities of such attacks. There are as yet no supergerms or ultra-
poisons which can be effectively disseminated.

Major and unsolved problems for the offenders in such warfare are
to effect a wide distribution of the injurious agents and to introduce
them into the bodies of victims. The natural spread of artificially
introduced infections is slow and uncertain. The natural body defenses limit the disease and its spread. Epidemics can frequently be
effectively controlled by well-known methods. As an example, when
one man with smallpox was discovered in New York City in 1947
(not as a result of biological warfare), over six million people in the
area were immediately vaccinated. A total of only twelve cases
eventuated.

A major problem in employing biological warfare is that, if such
warfare is successful, and if the conquered population becomes diseased and its domestic animals and crops destroyed, the victors
acquire a dependent, starving people, whose disease may be equally
hazardous to occupation forces and the vanquished.

Although the dangers of biological warfare are not so tremendous
as the imagination might picture, organized measures for dealing
with the problem have been developed under the auspices of the
federal Civil Defense Administration.

CHAPTER FOURTEEN

REPRODUCTION AND EARLY DEVELOPMENT

I. Growth and Cell Division
   A. In ameba and man
   B. Adaptive value of cell division
   C. Mechanisms
      1. Lower plants   2. Mitosis   3. Unknowns
II. Forms of Reproduction in Multicellular Organisms
   A. Vegetative reproduction
   B. Reproduction by spores
   C. Sexual reproduction
      1. Male and female gametes; fertilization   2. The biological value of sexual reproduction
III. Evolutionary Progress in Sexual Reproduction
   A. In lower plants and animals
   B. Invasion of the land
   C. Mammalian advances
      1. Fertilization and implantation   2. Embryonic membranes   3. The placenta
IV. Parental Contributions to the Young
V. Germ Layers and Organ Differentiation
   A. The three primary germ layers
   B. The nervous system
   C. Control of embryonic differentiation
VI. The Recapitulation Principle
VII. Birth of the Young
   A. Uterine contractions
   B. Straining
   C. Changes at birth
   D. Controlling mechanisms
VIII. Nutrition of the Infant

I.   GROWTH AND CELL DIVISION

A. In Ameba and Man

Specific synthesis is a basic life-process in which each kind of protoplasm in each living cell increases in quantity, using the food materials at hand. This process involves both growth and reproduction in an intimate interrelationship. Synthesis of ameba protoplasm by the one-celled ameba first causes cell growth; the animal becomes larger. Still further synthesis is accompanied by reproduction—an orderly division of the cell into two daughter-cells, each of which continues and repeats the processes.

During the growing period single cells in man's complex body undergo similar changes. A cell increases in size, then divides. But here though *cells* have reproduced, the *individual* has not. Cell reproduction, following upon cell growth, has caused growth of another kind; the multicellular individual has become larger by virtue of an increase in the number of cells.

B. ADAPTIVE VALUE OF CELL DIVISION

The adaptive value of the phenomenon of cell division in either unicellular or multicellular animals depends in part upon the fact that exchanges of materials (foods, oxygen, wastes) between interior and exterior of cells occur largely by diffusion. In most instances diffusion can occur rapidly enough to support metabolism over only a limited distance. Now, as cells enlarge, the volume of the protoplasm increases faster than the area of the surface through which the materials exchanged must diffuse. The critical distance from the source of food and temporary depository of wastes just outside the cell to any bit of protoplasm within the cell is kept small by the division of cells which have grown to a certain size.

The usefulness of the process is therefore clear. No plant or animal group has achieved conspicuous success without cell division. Indeed, without the evolution of cell division, life as we know it would probably not have developed.

C. MECHANISMS

1. *Lower plants.*—In a few simple one-celled plants which possess no organized nucleus, cell division seems to be relatively simple. The dividing cell usually elongates, a constriction appears and deepens, and the cell is "pinched" in two (Fig. 213). Observing this process, one sees nothing to indicate that there is an orderly halving of the cell contents. Yet it is probable that there is an accurate division of at least certain cell constituents (i.e., the genes [see next section]) and an equal distribution of these elements to the two new daughter-cells.

2. *Mitosis.*—Most cells, from an ameba to the units constituting man's body, possess nuclei. The nucleus seems to exert the main control over the process of cell division and displays a characteristic and complicated behavior (see Fig. 214). At first the dark-staining portions of the nucleus become visible threads, called *chromosomes*, which soon line up at the equator of the cell. Each thread duplicates itself, lengthwise, and the daughter-chromosomes separate. Actually, under the microscope, it appears that they are *pulled* apart. The daughter-chromosomes move in opposite directions, and each group reassembles to form a new nucleus. Lastly, the rest of the cell also divides into two usually equal parts.

This process, called *mitosis*, seems to be a mechanism which insures an exact and equal division of the chromosomes between the two daughter-cells. There is irrefutable evidence that the chromosomes contain the hereditary units, called *genes*, which largely determine the characters of the daughter-cells. Genes are reponsible for the new cells resembling one another and the parent-cell, anatomically and physiologically, when the cell environment remains constant.

3. *Unknowns.*—Great as is our knowledge concerning mitosis and cell division, we are largely ignorant of the stimulus for, and many of the factors controlling, this process. Does the size of the growing cell in some way determine when mitosis shall occur? How could cell size effect such control? Are there chemical stimuli to mitosis, produced by changed metabolic reactions incident to the decrease in the proportion of cell surface to cell volume, which occurs in growth? We should like to know the answers to these questions not only to understand normal growth and reproduction but to understand, and perhaps control more effectively, the abnormally rapid and frequent mitoses that occur in the cells constituting the cancers.

II. FORMS OF REPRODUCTION IN MULTI- CELLULAR ORGANISMS

In the multicellular plants and animals, where mitosis alone cannot produce new individuals, reproduction of the organism takes various other forms.

A. Vegetative Reproduction

Some multicellular organisms reproduce *vegetatively*. A part of the original plant or animal, separated or removed from the main body, can sometimes in a suitable environment produce a new individual.

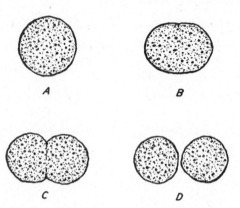

FIG. 213.—Cell division in nonnucleated cells. The parent-cell elongates and divides into two daughter-cells (see also Fig. 214).

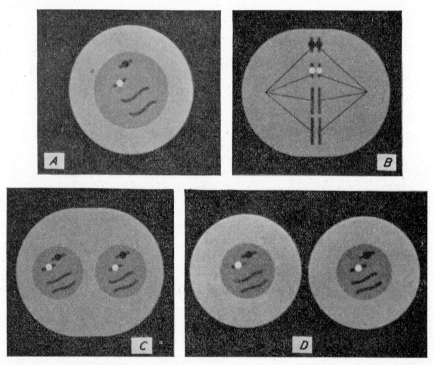

FIG. 214.—Cell division in nucleated cells—*mitosis*. In *A* is shown a cell whose nucleus possesses four chromosomes. Each chromosome contains many genes, which are the units of heredity. In *B* the chromosomes have lined up at the equator of the cell, and each chromosome has split in two lengthwise. In *C* the chromosomes have pulled apart, and two new nuclei are formed. In *D* the cytoplasm has also divided, and two daughter-cells have been formed, each having a hereditary composition exactly like the parent-cell in *A* (see also Fig. 213). (From the sound film, *Heredity*.)

Thus the twig of a tree may form an entire new, independent being—a new tree. Or each of the two halves of a bisected flatworm may produce an entire new animal. In the main, however, vegetative reproduction is limited to plants and the lower multicellular animals. Except in the very early stages of embryonic growth, in higher animals portions of the body isolated from the whole lack the capacity to produce other parts or a new individual. In fact, such isolated parts usually die. We relate this to the high degree of interdependence of organs and systems upon one another in the closely integrated bodies of the higher animals.

B. REPRODUCTION BY SPORES

A second form of reproduction, by *spores*, occurs in many multicellular plants. Spores are single cells produced in large numbers, each possessing the capacity to produce a new individual in the appropriate environment. Spores are disseminated through the water or air, reproducing the plants and widely distributing the species over areas providing suitable environments. Nature is prodigal in the production of spores; the vast majority die, for they do not encounter favorable growing conditions. For example, the air commonly contains thousands of spores of bread mold, but only an extremely small proportion of these happen to light upon a piece of bread or similar substrate to produce new mold plants.

C. SEXUAL REPRODUCTION

1. *Male and female gametes; fertilization.*—Sexual reproduction occurs in practically all multicellular plants and animals, including those plants which also reproduce by spores. The higher animals can reproduce only sexually. Unlike a spore, the cell unit in sexual reproduction, called a *gamete*, cannot alone develop into a new individual. Gametes are characteristically of two kinds: the male gamete, or sperm, and the female gamete, or egg. Only after the union of a sperm with an egg—*fertilization*—can further development occur.[1]

2. *The biological value of sexual reproduction.*—The almost universal occurrence of sexual reproduction in the higher plants and animals raises the question of the relative effectiveness of this manner of re-

---

1. Normally, only one sperm fertilizes an egg. Usually, after one sperm has fused with the egg, a protective membrane is formed around the zygote, preventing the entrance of further sperm. A few animals display a modified form of sexual reproduction, in which sperms do not fertilize eggs. The eggs develop into new individuals independently without fertilization. This process is called *parthenogenesis*.

production as compared with the widely occurring spore process. Certainly as a method of reproduction, sex has definite biological disadvantages. Zygotes and spores share the hazard of encountering unsuitable growing conditions, but in sexual reproduction there is an additional hazard not encountered by spores: ordinarily, a sperm must come into contact with and fertilize an egg before a new individual can develop. Although spores can effectively reproduce the individual, disseminate the species, and preserve the race, yet sex has evolved in both animals and plants. Does sex possess some less obvious advantage over asexual reproduction? Does it have some definite survival value?

Apparently there is but one clearly established advantage: *sex tends to multiply variations.* Thus, for example, an individual plant possessing the characters R (e.g., redness) and T (e.g., tallness) could asexually produce only RT offspring.[2] Another plant with the characters r (e.g., whiteness) and t (e.g., shortness) could produce only rt offspring. But if these two individuals are male and female, mating could yield not only the RT and rt kinds of offspring but also (in the second generation) Rt (e.g., red, short) or rT (e.g., white, tall) as well—offspring different from either parent, possessing some characters of each.

In evolution there is a constant selection of those forms best adapted to the environment. Less well-adapted forms are likely to perish and leave no descendants. Perhaps for a long time the environment is such that the original RT and rt plants do equally well. Then a change occurs in the environment, such that the new Rt or perhaps rT plants have the edge on their competitors for a livelihood. Barring the rare phenomenon of mutation, spores can never produce such superior plants, but sexual reproduction can. It seems that sex and sexual reproduction have meaning and value only in a world in which new combinations of old characters[3] are of survival value— that is, a world in which changing environment, a constant struggle

2. Unless the comparatively rare phenomenon of *mutation* occurs. Mutations are sudden changes in genes. Since genes largely determine the characters of the cells possessing them, a gene change can produce a new character in a cell, plus those daughter-cells, and granddaughter-cells, which are produced in mitosis. Such new characters may be either beneficial or harmful to the organism.

3. Note that sex itself produces no *new* hereditary unit. Almost all *new* hereditary units occur as sudden changes—gene mutations. By means of sex, nature tries out new *combinations* of genes already in existence, which combinations may endow the animal or plant with new qualities or capacities.

for existence, and the survival of those forms best adapted to the environment are the order of life.

III. EVOLUTIONARY PROGRESS IN SEXUAL REPRODUCTION

Advances in sexual reproduction in the higher plants and animals can be summed up as follows: as evolution continues, better and better care and nourishment of the young are provided, both before and after birth.

A. IN LOWER PLANTS AND ANIMALS

At the outset of sexual reproduction (as in the alga *Ulothrix*) eggs and sperms are apparently alike. In fact, one cannot properly designate these seemingly identical cells as eggs and sperms; but in higher species there is a differentiation. Sperms remain small and actively motile, helping to insure contact with eggs, which become quiescent and much larger, containing food stored against the needs of the zygote or embryo.

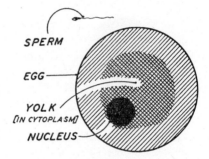

Fig. 215.—Human sperm and egg, showing the relative sizes of these cells. Most of the difference in size is due to the large amount of yolk stored in the cytoplasm of the egg.

Even fairly complex mating acts occur in some of the fishes and amphibians, but in most of the lower vertebrates[4] the female simply lays her eggs in the water, and sperms are poured over the eggs by the male. In most instances after fertilization has occurred, the new individual, nourished at first by food stored in the egg, must shift for itself, with relatively little further attention or care from the parents.

B. INVASION OF THE LAND

An event of the most far-reaching consequence in the evolution of the vertebrates was their successful invasion of the land. The amphibians made a start in this direction, for many of them are primari-

4. The vertebrates (subphylum Vertebrata of the phylum Chordata) comprise the following groups (or classes): (a) jawless and limbless vertebrates (fishlike forms); (b) primitive fishes (possess skeletons of cartilage; include sharks, skates, rays); (c) bony fish (possess skeletons of bone; include most modern fish); (d) amphibians (live on land except in early developmental stages; include frogs, toads, salamanders); (e) reptiles (first to lay eggs on land; include snakes, crocodiles, turtles); (f) birds (closely related to reptiles); and (g) mammals (possess mammary glands, suckle the young).

ly land dwellers, equipped for land locomotion and for obtaining oxygen from air instead of from water. Yet amphibians were never completely emancipated from the water, for most of them go back to the water to reproduce, and the young are water dwellers, more like fish than land forms.

Freedom from a watery external environment came only with the evolution of the land egg[5] in reptiles. Fertilized inside the maternal body by the introduction of sperm in the act of copulation, the reptilian egg contains a great deal of stored nutriment—the *yolk*—which provides food for the young in the early critical stages of growth.

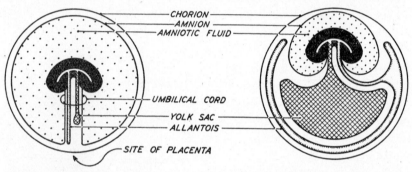

CHORION
AMNION
AMNIOTIC FLUID

UMBILICAL CORD
YOLK SAC
ALLANTOIS

SITE OF PLACENTA

FIG. 216.—Embryonic membranes in man (*left*) and the land egg of reptiles and birds (*right*). The embryo proper is shown in black. Note that both embryos are surrounded by amniotic fluid and that corresponding membranes are present in each egg. In the land egg the yolk sac is large, containing much stored food, and the allantois extends under much of the egg surface. The allantois, containing embryonic blood vessels, effects exchanges of oxygen and carbon dioxide between embryonic blood and the outside air. The shell of the land egg is not shown.

Equally important, the developing egg is protected by membranes (see Fig. 216) which are thin enough to permit exchanges of oxygen and carbon dioxide with the outside air through the outermost shell, yet sturdy enough to prevent an injurious amount of evaporation of water. Within the membranes is a fluid bathing the developing embryo.

It may be said that in the land egg of reptiles and birds and, indeed, even in the embryonic mammals, including man, the growing embryo is provided with a private pond whose consistency somewhat resembles blood plasma. All these forms, then, start life in the

5. The term "egg" may mean simply the female gamete—a cell. Or, as in this instance, may mean this cell plus the stored food, the membranes, and the shell. This latter is, of course, the popular conception of an egg.

water, like fish and amphibians, except that now the water is inclosed within a membrane of the egg.

So successful was the land egg that it enabled numerous reptiles—such as the dinosaurs—to spread over the surface of the earth in the fabulous Age of Reptiles millions of years ago. And the reptilian land egg is maintained practically unchanged in the likewise very successful present-day birds. Yet the land egg is not perfect. It is thought that the dinosaurs disappeared from the earth largely because of a change in climate. The earth became colder, vegetation diminished, and these huge animals starved to death. One may well imagine another contributing factor: that, at this time, stealing eggs for food in lieu of vegetation was not merely a popular pastime of animals but actually a matter of life or death.

C. MAMMALIAN ADVANCES

In reptiles the care of the young by the parents lasts but a short time. At a relatively early age, with critical stages of development still ahead, the young are deserted by the parents. This is no longer the case in birds and mammals, both of which groups evolved from reptiles. In birds postnatal care by the parents often continues a long time. The evolution of mammary glands and the suckling of the young in mammals established a still greater intimacy in the parent-offspring relationship and helped to prolong the period of parental protection and care. Mammary glands first occurred in egg-laying forms (represented today by the duckbill of Australia). But early in the evolution of mammals, organs and mechanisms developed which provided for a rather extended growth of the young inside the maternal body.[6]

1. *Fertilization and implantation.*—Fertilized in the uterine tube, the fertilized egg—now a zygote—commences cell division at once, and by the time the young embryo[7] reaches the uterus proper it is several cells in size. It becomes imbedded or implanted in the richly vascular uterine wall, where it derives nourishment from the capillaries of the maternal uterus in a manner not unlike the nourishment of the uterine muscle cells.

6. At this time it will be well to review the anatomy and physiology of the reproductive organs of mammals (see pp. 544–58 and 563–65).

7. In mammals embryonic life lasts from the time of fertilization of the egg until birth. When the embryo begins definitely to resemble the parental form (in man, at about the third month of pregnancy), it is usually called a *fetus*.

2. *Embryonic membranes.*—The growing embryo develops membranes which are essentially like the membranes of the reptilian and bird egg. There are the *amnion*, which incloses the embryo within the pond of amniotic fluid, and the outermost *chorion.* As in the land egg, there are two additional membranes—the *yolk sac* and the *allantois*—outgrowths of the primitive intestine of the embryo and lying between the amnion and chorion (see Fig. 216).

As we might guess, the yolk sac is very small compared to that of the land egg, since the quantity of yolk in the mammalian egg is not great. Little or no yolk is required here, for food is now derived from the mother. The persistence of the vestigial and functionless yolk sac speaks strongly for the evolution of the mammalian membranes from the reptilian egg and, indeed, for the evolutionary descent of mammals from reptile-like forms. The allantois carries blood vessels from the embryo to the uterine tissue, where, together with the chorion and the lining of the uterus, it forms a structure found in all but the most primitive mammals—the *placenta.*

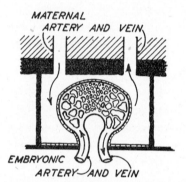

MATERNAL ARTERY AND VEIN

EMBRYONIC ARTERY AND VEIN

FIG. 217.—A small portion of the placenta (*diagrammatic*). Note how the maternal blood, containing oxygen and food, bathes the embryonic capillaries. The placenta is the source of food and oxygen and the depository of wastes for the embryo. Arrows indicate directions of blood flow.

3. *The placenta.*—The placenta is the site of exchange of all materials between maternal and embryonic blood. It is here that oxygen and food taken in by the mother pass into the embryonic circulation; here embryonic wastes, including carbon dioxide, enter the blood stream of the mother, to be excreted by her.

There is no actual mixing of the blood of the mother and the blood of the embryo, for the embryo makes its own blood and pumps it with its own heart through its own closed vascular system. The exchange of materials between the two blood streams is effected mainly by diffusion.

The two blood systems, however, come very close together in the placenta. The embryonic capillaries dip into tiny pools (sinuses) of maternal blood. Hence, to move from one blood system to the other, molecules of food or wastes need only pass through a thin layer of cells consisting of little more than the embryonic capillary walls (see Fig. 217).

IV. PARENTAL CONTRIBUTIONS TO THE YOUNG

A knowledge of the foregoing facts and principles of early development is a relatively recent acquisition of man. Once it was vaguely stated that in some mysterious manner the uterus conceived an embryo much as the brain conceived a thought—however that might be. Even after they knew about cells, including sperm, eggs, and zygotes, scientists thought also that a tiny embryo was simply a preformed miniature of the animal-to-be, with all its organs and parts, and that in the uterus these parts simply enlarged, or unfolded, much as a rosebud becomes a rose. We now know this to be erroneous.

Some misconceptions still persist. Many persons are not aware that the father and mother have made their total hereditary contribution to the child when a sperm fertilizes an egg. At that time the hereditary units—the genes of the parental gametes—have already determined the sex of the child, approximately its future stature, its eye and hair color, as well as many other characters. Even the future intelligence of the new individual (barring disease or accident) is largely delimited at the time of fertilization.

After fertilization the mother contributes nothing but a suitable environment for the embryo, supplying food and oxygen and removing wastes. It follows, therefore, that if the mother eats food poor in iron, the embryo or child may be anemic. If the mother has defective kidneys, waste elimination is impaired, and the embryo may suffer. Or, if the mother is syphilitic, the infection may be transmitted to the offspring through the placenta. But such influences are limited essentially to the physical factors involved in growth and nutrition. We know that an intense preoccupation with beautiful thoughts by the mother will in no wise affect the child's disposition, nor will diligent pursuit of music—concert-going or piano-playing—influence the child's musical ability. A sudden, intense fright will not produce a birthmark (though, coupled with physical injury, it may predispose to premature birth of the young).

V. GERM LAYERS AND ORGAN DIFFERENTIATION

A. The Three Primary Germ Layers

Soon after the fertilized egg has begun to divide, when the embryo consists of relatively few cells (in man, perhaps a few dozen), the cells become arranged into two distinct, so-called "germ layers": an outer *ectoderm* and an inner *endoderm*, roughly corresponding to the two layers of cells in the body wall of adult *Hydra* (see Fig. 12, p. 64). At this stage the future of these layers has already been determined. For example, the ectodermal layer will give rise to the skin and the endodermal layer will produce the inner lining of most of the alimentary canal.

All animals above *Hydra* in the evolutionary scale possess, in addition, a third embryonic layer, the *mesoderm* (see flatworm, Fig. 12, p. 64), which develops, as the name suggests,[8] between ectoderm and endoderm. All the organs of the body develop or "germinate" in an orderly fashion from these three primary germ layers. The great

TABLE 14

MAIN DERIVATIVES OF THE THREE EMBRYONIC GERM LAYERS

Ectoderm	Endoderm	Mesoderm
Skin (including hair, nails, sweat glands)	Inner lining of alimentary tract	Skeleton
Nervous system (including brain, spinal cord, nerves, ganglia)	Digestive glands and inner lining of ducts	Heart, blood vessels, blood
Lining of part of mouth, nose, and anus	Respiratory passages and lungs	Muscle, smooth and striated
	Inner lining of most of urinary bladder	Lymph nodes and vessels
		Alimentary tract (except for inner lining)
Eye		Kidneys
Enamel of teeth		Reproductive organs
		Teeth (except for enamel)

bulk of the bodies of higher animals is derived from the mesoderm (see Table 14).

B. The Nervous System

One might have guessed that the lining of the alimentary canal is derived from endoderm, skeletal muscle from mesoderm, and skin from ectoderm. Not quite so obvious is the origin of the nervous system from ectoderm. Fairly early in development the ectoderm along the back (dorsal) surface of the embryo begins to thicken, forming what is called the *nerve* (or neural) *plate*. This plate then folds inward (see Figs. 218 and 219) until a tube is formed, which later is pinched off from the overlying ectoderm. Outgrowths from the tube form the nerves, and the tube itself becomes the brain and spinal cord. Recall that the brain retains a tubular structure even

8. Ectoderm literally means "outer skin"; endoderm, "inner skin"; and mesoderm, the "between layer."

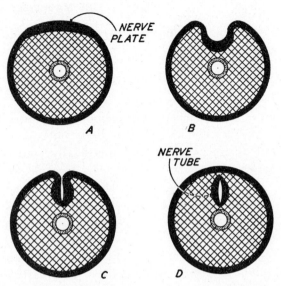

Fig. 218.—Cross-sections of the embryo, showing formation of the nervous system from ectoderm. Ectoderm is shown in black, mesoderm crosshatched, endoderm dotted. The ectoderm on the back (dorsal) surface of the embryo thickens into the nerve (or neural) plate (*A*) and begins to dip inward (*B*). The infolding increases (*C*) and pinches off (*D*), so that the ectodermal body surface is separate from the nerve (or neural) tube. Brain, spinal cord, and nerves develop from this tube (see also Fig. 219).

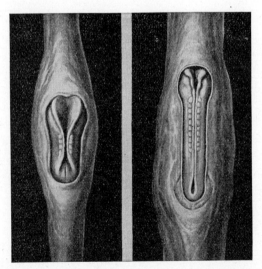

Fig. 219.—Formation of the neural tube in the pig. The uterus has been opened to show the back (dorsal) surface of the embryo. On the left is a fourteen-day-old embryo, showing the beginning of infolding of the nerve (or neural) plate. On the right is the same embryo twenty-four hours later. Here the infolding and nerve tube formation are nearly complete (see also Fig. 218). (From the sound film, *Reproduction among Mammals*.)

in adult life, containing the fluid-filled ventricles in its interior (see p. 445 and Fig. 158, p. 439).

C. Control of Embryonic Differentiation

One of the most fascinating problems—and difficult of solution— in embryology is the nature of the mechanisms controlling the development of specific cells into specific body parts or organs. Early in development all cells look alike, superficially, and seem to have equal developmental potentialities, at least in some species. In certain starfish, at an early stage when the embryo consists of about eight cells, one can carefully separate the cells and place them in an appropriate nutrient medium. Almost every cell will grow and will produce a complete individual. But, as growth proceeds, the individual cells quickly lose their generalized potentialities, and the subsequent developmental pattern becomes fixed. Very early in the human embryo a given cell group can produce only kidneys, for example. Other cells soon lose their kidney-forming potentialities and are destined to produce other organs. What starts some cells irrevocably along the "to-kidney" pathway and blocks this road to all other cells, which in their turn become committed to the development of heart or spleen or muscle?

A few modest beginnings have been made in our attempts to solve these problems. We know that the skin in the eye region—indistinguishable from skin elsewhere—develops into the lens of the eye. Why? Apparently, at least in one species of frog, the adequate stimulus comes from the underlying retina, because if the growing retina is removed before the lens has started to form, and is successfully transplanted beneath the skin in certain other body regions, a lens will fail to form at the normal site but will form instead over the transplanted retina.

There are numerous examples of this sort, where one organ determines the developmental course of adjacent tissue, but still our knowledge remains fragmentary. What makes the *retina* develop where it does in our frog? How does the retina effect its control upon the overlying skin? And why are the controls apparently so different even in closely related species? We know that in another species of frog normal lens will develop from skin in the normal site despite previous removal of the retina. Intensive work in this field is going on, and we may hope that many of these questions will be answered eventually.

VI. THE RECAPITULATION PRINCIPLE

Animals commence life as a single cell. In the course of their life-time some animals never go beyond this one-cell stage. In other animals there are varying degrees of multicellular complexity. Some forms are but a hollow ball of cells; others, like *Hydra*, consist of two layers of cells—the ectoderm and endoderm. In flatworms a meso-derm appears between these layers. Now, it is an interesting observation that in development a hollow-ball form starts its existence as a single cell; the more complex *Hydra*, also a single cell at first, passes through a hollow-ball stage before it becomes a two-layered animal; a flatworm, still higher in the scale of evolution, repeats these stages in order, besides lastly adding the mesoderm. Similarly, all forms above the flatworms in the evolutionary scale pass through the same embryonic stages, plus also a flatworm-like stage. Man also begins as one cell, becomes a modified hollow ball,[9] and then develops endo-derm, ectoderm, and, finally, mesoderm. In embryonic development and growth each individual tends to pass through, or recapitulate, the successive stages of the evolutionary past of that individual (see Table 15).

Often these embryonic reminiscences of evolutionary past seem to be without function in the embryo. Some of them may even persist and remain functionless, or nearly so, in the adult. A few examples of this in man are the old tail muscles, the vermiform appendix, some of the skin muscles, the muscles of the external ear, and the body hair.

Still other embryonic structures linking man with his far-distant past may be observed to change in embryonic growth and to be made over gradually, until they ultimately come to serve some function quite different from that of the corresponding (i.e., "homologous") structure from which they were derived in evolution. In fish, for example, there are several pairs of openings—the pharyngeal gill slits—connecting the pharynx with the outside. These openings are useful to the adult fish, for oxygen-laden water is forced through them, releasing oxygen to the blood in the capillaries of the gills. Embryonic man possesses somewhat similar outpouchings of the pharynx, which are utterly useless in aerating blood. But the first (most anterior) pair of these outpouchings is made over into the middle-ear cavity and the Eustachian tube, structures of importance in hearing (see Fig. 173, p. 484).

9. The early human embryo bears only a rough resemblance to the hollow-ball stage in the development of lower animals.

Between the pharyngeal slits of adult fish are pairs of pharyngeal arches consisting of flesh-covered cartilage or bone (see Fig. 220). They contain arched blood vessels, supplying blood to the gills, which are aerated by water moving through the pharyngeal slits. These arches are also present in the human embryo, but most of them disappear—the fate of a number of these embryonic vestiges.[10]

TABLE 15

THE RECAPITULATION PRINCIPLE

In its own embryonic development each animal (*left column*) tends to repeat or recapitulate the stages of evolution (*listed across top*) through which the species has passed.

Animal	One-Cell Stage Occurs In:	Hollow-Ball Stage Occurs In:	Two-layered Body (Ectoderm and Endoderm) Occurs In:	Three-layered Body (Mesoderm Added) Occurs In:	Could be
Ameba	Entire life-time				continued further for
Volvox	Beginning of each individual	Adult form			segmentation of body, de-
Hydra	"	Early stage of embryonic development	Adult form		velopment of one-way
Flatworm	"	"	Later stage of embryonic development	Adult form	alimentary canal, etc.
Man (and many forms intermediate between flatworm and man)	"	"	"	Still later stage in embryonic development	

The skeletal parts of the first pair of pharyngeal arches, however, are converted into the jaws.[11]

This repetition or recapitulation in embryonic development is not nearly so faithful as our account might imply. Some evolutionary changes are passed over very quickly by the embryo, and stages

10. Witness the pronounced tail of the human embryo (see Fig. 221, p. 616), which later almost entirely disappears.

11. In fact, the process of reworking the pharyngeal arches begins in the cartilaginous fishes. These animals—the first with jaws—derived their jaws from the first pair of pharyngeal arches of their evolutionary predecessors, the jawless and limbless vertebrates.

which definitely followed one another in evolution often overlap in embryonic development.

It follows from the law of recapitulation[12] that, the more closely animals are related in evolution, the longer will the embryonic development of those animals remain similar. The embryonic development of man is like that of the flatworm for only a brief period. But the human embryo resembles the embryo of another placental mammal for a much longer time. Figure 221 shows how similar are human and pig embryos for a relatively long duration. Only in the second picture from the bottom (C) is it clear which embryo is human.

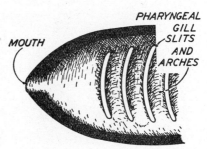

FIG. 220.—Anterior end of a vertebrate embryo, split open lengthwise, and viewed from within the pharynx, showing the pharyngeal gill slits and arches.

VII. BIRTH OF THE YOUNG

All mammals bear living young. The embryo, well nourished via the placenta, passes through the critical early stages of growth in the well-protected environment of the maternal uterus. Then occurs one of the most remarkable events in biology—the phenomenon of birth.[13]

A. Uterine Contractions

At the end of the gestation period (i.e., "at term") the tissues of the mouth of the uterus have become soft and relaxed, and the uterus commences to contract rhythmically. The contractions become gradually more powerful and more painful. Pressing with the fingertips upon the abdominal wall of a pregnant woman at term, the physician can feel the organ become intermittently very hard, at which time the mother experiences pain. Uterine contractions are "labor pains" to the pregnant woman.[14] The contractions become increasingly frequent and powerful as labor proceeds. So strong are

12. Formally stated, this law says that *ontogeny* (the embryonic development of the individual) tends to recapitulate *phylogeny* (the evolutionary development of the species).

13. Technically called *parturition*. The interval between fertilization and birth of the young —called the period of *gestation*—varies with different species. In the rat it is 3 weeks; the dog, 9 weeks; the pig, 17 weeks; man, 40 weeks; the horse, 48 weeks; the elephant, 86 weeks.

14. The stretching of the *birth canal*, as the fetal head passes through, is also painful. The birth canal is the passage through which the fetus passes; it includes the neck of the uterus and the vagina.

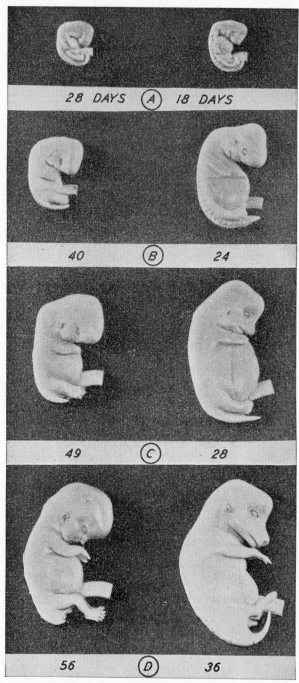

FIG. 221.—Successive stages (A to D) in the embryonic development of the human (*left*) and pig (*right*) embryos. In each instance the figure given refers to the age of the embryo in days. Note the similarity of the two embryos, which are practically identical in appearance in A and B; note also that the human embryo possesses a tail in all the stages shown. (From the sound film, *Reproduction among Mammals*.)

they that, in the rare instances when the birth canal is obstructed by some abnormality, the uterine wall may actually rupture. Normally, these contractions gradually expel the fetus (see Fig. 222), usually head first.

B. STRAINING

During the final stages of labor, when the mouth of the uterus has been fully dilated and the child begins to be forced further along the

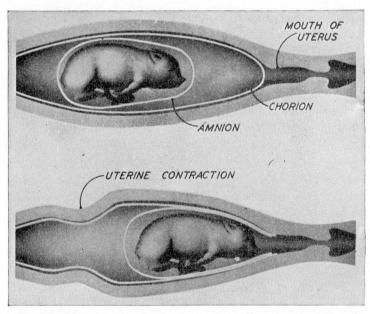

FIG. 222.—Birth of the young in the pig. *Above:* The uterus, slit open lengthwise, contains the young pig, inclosed by the embryonic membranes. *Below:* Uterine contractions have started, forcing the young animal toward the mouth of the uterus (and later, through the dilated birth canal). Note (*lower picture*) that the membranes burst before the animal is born, liberating the amniotic fluid (see Fig. 216, p. 606). (From the sound film, *Reproduction among Mammals.*)

birth canal, the mother usually assists the uterus by "straining"; that is, she compresses the abdominal contents by taking a deep breath (diaphragm descends), closing the glottis (so that air cannot escape from the lungs, and the diaphragm remains "fixed"), and contracting the abdominal muscles (further compressing the abdominal contents, including the uterus). Recall that straining also aids in the evacuation of the bladder in urination and of the rectum in defecation (see also pp. 247–48 and 298).

C. Changes at Birth

After an infant is born, the umbilical cord still connects the child with the placenta, which remains attached to the uterine wall. The umbilical cord should be tied and cut at once. The infant is now "on its own," which event it usually punctuates with a wail of protest with the first breath of air. A few minutes to a half-hour later the placenta separates from the uterine wall and is expelled as the after-birth through the birth canal.

At birth the normal infant quickly makes several remarkable adjustments to its new environment. Hitherto, oxygen was supplied, and carbon dioxide excreted, through the placenta. Now these exchanges are at once effected through the lungs.

Associated with this drastic change are circulatory readjustments.[15] During intrauterine life very little of the carbon dioxide–laden, oxygen-poor blood returning to the right heart is pumped to the nonfunctional lungs. Such blood as may be pumped into the pulmonary artery is shunted almost at once into the aorta, through the *ductus arteriosus*, an embryonic vessel connecting pulmonary artery and aorta. Much of the blood entering the right atrium passes immediately into the left atrium, through a hole in the wall between the two atria, called the *foramen ovale*, or oval opening. Thus the partially "impure" blood returned to the right atrium is pumped on unchanged into the systemic circulation, only later to find its way into the placental vessels, where aeration occurs.

All this is changed at birth. Instead of passing through the two shunts past the lungs (i.e., the ductus arteriosus and foramen ovale), practically all the blood entering the right atrium now flows on through the lungs, where it is aerated.

What effects this drastic and necessary change at just the right time? Unquestionably the new environment of the infant provides the stimulus, for the same changes occur at birth in infants born any time from thirty weeks (or even somewhat less) to the normal forty weeks after fertilization.[16] The sequence of events is thought to be as follows: the first deep breath of the infant,[17] stimulated perhaps by

15. The circulatory changes can best be visualized by studying Fig. 223 with the text.

16. Infants born earlier than forty weeks after fertilization are called *premature* babies.

17. It has been shown that shallow breathing movements commence even before birth, producing a tidal flow of amniotic fluid into and out of the lungs. Of course, this effects no aeration of the fetal blood.

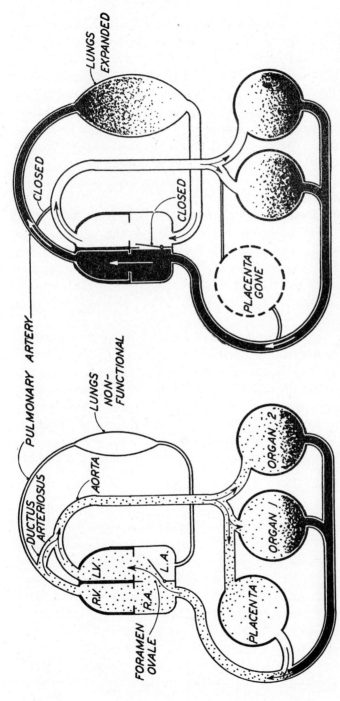

FIG. 223.—Circulatory changes at birth. Aerated blood is shown in white; unaerated blood, black; a mixture of aerated and unaerated blood, dotted. Arrows indicate the direction of blood flow. The embryonic circulation is shown at the left. Note that the aerated blood from the placenta mixes with the blood from various organs and flows to the right atrium (*R.A.*). This blood either passes through the foramen ovale into the left heart (*L.A.* and *L.V.*) or, entering the right ventricle (*R.V.*), soon passes through the ductus arteriosus. Thus practically all the blood entering the right atrium soon finds itself, unchanged, in the aorta, little or none of it flowing through the nonfunctional lungs.

The changes at birth are shown at the right, where absence of the placenta, expansion of the lungs, closure of the foramen ovale, and degeneration of the ductus arteriosus establish the kind of circulation which persists throughout life. (See discussion in text.)

an accumulation of carbon dioxide in the blood after the umbilical cord is tied and cut, expands the lungs widely, including not only the air passages and air sacs but also the lung blood vessels, which were previously partially collapsed. The venous blood now pumped into the pulmonary artery by the right ventricle flows on into the expanded lung vessels instead of short-circuiting through the shunt (i.e., the ductus arteriosus). Consequently, a good deal of blood now enters the left atrium *from the lungs*. This tends to close the valvelike opening between left and right atria (i.e., the foramen ovale), because the increased pressure in the left atrium forces shut, at least partially, the flaps of this opening.

Moreover, that blood which had formerly passed from right to left heart through the foramen ovale is now moved on through the lungs, which still further increases the flow from lungs to left atrium, closing the oval orifice even more tightly. Normally, all this is effected in a few minutes or less. Eventually the flaps of the opening actually grow together, and, in addition, the shunt from pulmonary artery to aorta degenerates.

If the foramen ovale or the ductus arteriosus fails to close properly, there may be an inadequate aeration of the blood: a "blue baby" results, unless death is immediate. One of the triumphs of modern surgery has been the operative closure of the ductus arteriosus in such cases. In fact, it is now possible to operate with considerable success upon the heart and large blood vessels near the heart to correct a number of developmental defects. Such operations were made possible by practice upon experimental animals, especially dogs.

D. Controlling Mechanisms

The preceding account of childbirth is mainly descriptive: a statement of empirically observed events in proper sequence. At many points little or nothing has been said of the *mechanisms* involved. This is because we are largely still ignorant of the driving forces, and many problems remain to be solved by further investigation. What adjusts the onset of uterine contractions to the state of development of the fetus? What stimulates the uterus to undergo these rhythmic contractions? What factors other than the mechanical ones mentioned earlier are responsible for the circulatory changes at birth? What causes the placenta to separate from the uterine wall? What prevents excessive uterine bleeding when this occurs? These last two phenomena may be related causally. It is thought that perhaps the placenta is torn loose from the uterine wall by strong contractions of the nearly empty organ and that the contraction of the uterus to a small size probably mechanically occludes the uterine vessels which supplied the placenta with blood. But what causes the uterus to contract down to this small size?

And why do the rhythmic uterine contractions cease soon after the fetus and placenta are expelled? Why are the uterine contractions painful? Probably this last is related to an interference with the blood supply of the uterus by compression of the capillaries. Marked impairment of blood flow in skeletal and cardiac muscle may also be painful. But why this is so we do not yet understand.

VIII. NUTRITION OF THE INFANT

After birth food and water are no longer available from the placenta and must be taken by mouth. The normal food of infants is mother's milk. We have seen that growth and enlargement of the mammary glands and at least the beginning of secretion of milk are probably stimulated in the mother by hormones of the ovary and anterior lobe of the hypophysis (see pp. 565 and 567), so that normally food is immediately available to the newborn infant.

Milk is an excellent food, containing proteins, carbohydrates, and fats of the proper kinds and in the right proportions. It contains several vitamins and nearly all the required salts. The infant needs no other food during the first month of life. But even normal milk is deficient in some respects, and, unless the milk diet is supplemented rather early, dietary deficiencies will develop before the time of weaning.

The most notable defect in milk is its low iron content (see also p. 101). Since red blood cells are continuously destroyed in the infant as well as in the adult, and since iron is indispensable for the manufacture of the hemoglobin in the red cells replacing those normally destroyed, an adequate source of iron must be available to the busy red bone marrow. During intrauterine life the embryo stores iron in the liver, provided the iron intake of the mother is adequate. These reserves supply the needs of the bone marrow in early infancy, but, unless the iron-poor milk diet is soon supplemented with some rich source of iron, a severe anemia will develop. Perhaps the best source of iron for infants is egg yolk. Even later in life this is a superior source of iron; we may be fairly sure that an egg a day will keep iron-deficiency anemia away.

The vitamin C and vitamin D content of milk also is usually insufficient for the needs of the growing infant, and the milk diet should be supplemented with orange or tomato juice, for vitamin C, and cod-liver oil, or some other source of vitamin D, when the infant has reached an age of only a few weeks.[18]

18. The *pediatrician*, who is a specialist in the nutrition and abnormalities of children, thinks it best to begin adding these vitamins when the child is one month old. Usually iron, in the form of egg yolk, is added at about four months of age.

CHAPTER FIFTEEN

CONCLUSION

In this presentation of the machinery of the body our efforts will have been well worth while if the reader has gained some insight into the principles and problems of physiology in health and in disease. The need for further experimentation should stand out clearly from these pages. The past achievements of research in the science of life are ample justification for further investigation. They are so numerous as to remove all doubts of the worth-whileness of investing all the brains, money, and time possible in the quest.

Painstaking research has been exceedingly fruitful, both in the laboratories of peacetime and on the proving ground of wartime emergency, not only in man's efforts to conquer disease but also in his eternal quest better to understand the nature of life and his position in the physical world. Enough has been learned to indicate that man and other animals must be studied by the same scientific methods and with the same objectivity as are employed in unearthing the secrets of the atom or the universe.

We have attempted to demonstrate how such methods have led us to a better understanding of the manner in which cells and tissues, organs and systems, and the body as a whole carry out their exceedingly complex and beautifully integrated functions.

In our presentations of abnormal physiology at no time has it been our aim to equip the reader with more than an understanding of some of the principles involved. The diagnosis and treatment of disease are jobs for the well-trained physician. But understanding based on reliable information is an insurance against one's own folly as well as against fraud on the part of one's fellow-men. When disease threatens man's happiness or life, an understanding of the principles of health and disease is the best guide to sources of scientific counsel. The informed layman will not expect miracles from such counsel in the face of nature—"red in tooth and claw"—and of the many unknowns in biology still obscuring the path. But, from past progress in the control of disease, man may take courage in the probability of still better days to come for the human species.

SELECTED REFERENCES

BIOLOGY

MARSLAND, DOUGLAS. *Principles of Modern Biology*. Rev. ed. New York: Henry Holt & Co., 1951.

WELLS, H. G.; HUXLEY, JULIAN; and WELLS, G. P. *The Science of Life*. New York: Garden City Publishing Co., Inc., 1939.

PRINCIPLES OF EVOLUTION

MOODY, P. A. *Introduction to Evolution*. New York: Harper & Bros., 1953.

EVOLUTION AND EMBRYOLOGICAL DEVELOPMENT OF MAN

ROMER, A. S. *Man and the Vertebrates*. 3d ed. Chicago: University of Chicago Press, 1941.

PROTOPLASM AND THE CELL

MITCHELL, P. H. *Textbook of General Physiology*. 4th ed. New York: McGraw-Hill Book Co., 1948.

SCHEER, B. T. *General Physiology*. New York: John Wiley & Sons, 1953.

HEART AND CIRCULATION

HARVEY, WILLIAM. *Motion of the Heart and Blood*. Reprint with annotation by CHAUNCEY D. LEAKE. 3d ed. Springfield, Ill.: Charles C Thomas, 1949.

NERVOUS SYSTEM

BEERS, C. W. *A Mind That Found Itself*. New York: Doubleday, Doran & Co., 1935. (Reissue 1948.)

GARRETT, H. E. *Great Experiments in Psychology*. 3d ed. New York: Appleton-Century Co., 1951.

HERRICK, C. J. *Introduction to Neurology*. 5th ed. Philadelphia: W. B. Saunders Co., 1931.

PHYSIOLOGICAL CONSTANTS

CANNON, W. B. *The Wisdom of the Body*. New York: W. W. Norton & Co., 1939.

HISTORY OF PHYSIOLOGY AND MEDICINE

FULTON, J. F. *Selected Readings in the History of Physiology*. Springfield, Ill.: Charles C Thomas, 1930.

HAGGARD, H. W. *Devils, Drugs and Doctors*. New York: Harper & Bros., 1945.

BACTERIA AND INFECTION

BAYNE-JONES, STANHOPE. *Man and Microbes*. Baltimore: William & Wilkins Co., 1932.

VALLERY-RADOT, R. *Life of Pasteur*. New York: Garden City Publishing Co., Inc., 1937.

TEXTBOOKS OF PHYSIOLOGY

BEST, C. H., and TAYLOR, N. B. *The Physiological Basis of Medical Practice*. 5th ed. Baltimore: Williams & Wilkins Co., 1950.

HOWELL, W. H. *Textbook of Physiology*, ed. JOHN F. FULTON. 16th ed. Philadelphia: W. B. Saunders Co., 1949.

MACLEOD, J. J. R. *Physiology in Modern Medicine*, ed. PHILLIP BARD. 9th ed. St. Louis: C. V. Mosby Co., 1941.

STARLING, E. H. *Human Physiology*, ed. C. LOVATT EVANS. 11th ed. Philadelphia: Lea & Febiger, 1952.

PLANTS AND LOWER ANIMALS

BUCHSBAUM, RALPH. *Animals without Backbones*. 2d ed. Chicago: University of Chicago Press, 1948.

COULTER, MERLE C. *The Story of the Plant Kingdom*. Chicago: University of Chicago Press, 1935.

WARTIME ADVANCES IN MEDICINE

TALIAFERRO, WILLIAM H. (ed.). *Medicine and the War*. Chicago: University of Chicago Press, 1944.

The authors of this book have welcomed their self-imposed task of explaining to human beings how their bodies function. The book itself grew out of their experience as teachers to undergraduates of the fundamentals of physiology. They both have insisted that knowledge of the workings of one's own body is at least as important to an individual as any other branch of knowledge. As well as teaching, the authors pioneered in the production and use of sound films in the classroom. In addition, both authors have studied physiology through controlled experiment; neither has hesitated to take his place "on the table" when volunteers were needed for crucial experiments. Carlson starved himself to study hunger and was the third human being to inhale the anesthetic ethylene. Johnson voluntarily subjected himself to insulin overdosage when the effect of the drug which has prolonged the lives of hundreds of thousands of diabetics was still imperfectly known.

Until he was almost thirty, the life of Anton J. Carlson, the senior author, seemed unconcerned with any science. Born in Svarteborg, in southern Sweden, on January 29, 1875, he herded goats as a farm boy and followed an elder brother to the United States at the age of sixteen. Working as a $1.25-a-day carpenter on Chicago's South Side, he accepted some help from his fellow-Swedes and went to Augustana College at Rock Island, Illinois, to study for the Lutheran ministry. His employer assured him that he was "making the greatest mistake of his life."

At Augustana, Carlson studied everything, including philosophy and psychology, winning a B.A. degree in 1898 and an M.A. in 1899, and casually substituted for a Latin teacher when the latter was ill. Then, black-haired, ruddy-cheeked, intense, he was sent to Anaconda, Montana, to take charge of a Lutheran church. Already dedicated to his philosophy of work, he lectured to young people's groups on philosophic subjects, read and climbed mountains in his spare time.

That year marked an intellectual turning point. At its end he headed west and entered the still new Leland Stanford University to study biology. In California he met his second real scientist, Jacques Loeb, the great zoölogist. (The first had been an Augustana geologist.) He began his study of the mechanism of nerve impulses there

and won his Ph.D. degree in' 1902. His first important discovery
showed that the speed of a muscle contraction activated by a motor
nerve depended on the speed of the impulse through the nerve. He
settled a three-centuries-old argument about the nerve-versus-muscle
origin of the heartbeat in three days at the marine laboratories at
Woods Hole, Massachusetts. Using the heart of the horseshoe crab,
he demonstrated that the impulse from the heartbeat came from the
nerves and not from the muscles. Largely as a result of these two

researches, Carlson was called to the
University of Chicago, where he has
remained permanently.

At Chicago, his researches expanded
to the study of the ductless glands, digestion, and hunger. Carlson had been
startled by the prevalence of simple
goiter in the Midwest. He, and the students who soon flocked to work under
him, studied the workings of the thyroid
gland in both men and animals. Using
goiterous dogs, "pardoned" from the
pound by the mayor, Carlson and his
group showed that thyroid deficiency in
the mother resulted in thyroid deficiency in the pups. They demonstrated that removing the parathyroids from dogs resulted in a
lowering of the calcium level in their blood streams and in the
uncontrollable muscle twitchings of tetany. Studying the pancreas
in 1912, a student of Carlson's isolated a crude extract of insulin,
which mitigated the effects of diabetes in dogs, ten years before the
classical work of Banting and Best. Carlson also showed that removing the pancreas from a pregnant dog resulted in diabetes only after
the pups were born; insulin produced by the pancreases of the unborn
pups protected the mother until then.

In this period Carlson found a young Czech immigrant, whose
throat had been sealed as the result of an accident, who fed himself
through an opening in the stomach. This "second Alexis St. Martin,"
as Carlson called him, served as the subject of countless experiments
in the physiology of digestion, Carlson's most thoroughly explored
field. Here Carlson established a basic distinction between hunger
and appetite, showing that the first was independent of the second
and resulted from periodic contractions in the stomach itself. He

proved that the gastric juices flow even when the stomach is fasting and that their flow is rhythmical rather than dependent on the stimulus of food. He and his associates, who later included co-author Johnson, deliberately starved themselves and recorded the contractions of their own stomachs on rubber balloons which they swallowed. Carlson's research on digestion appeared in the book, *Control of Hunger in Health and Disease;* his entire scientific output totals almost two hundred and fifty papers.

Carlson's laboratory career was interrupted by World War I, in which he served as a lieutenant colonel in the sanitary corps. At the war's end he remained in Europe to aid Herbert Hoover in feeding starving children. These activities marked the start of a new and major phase in Carlson's career. The pioneering experimenter, the driving but stimulating teacher, turned an increasing amount of his energies to integrating science and daily life. He became as familiar on the lecture platform as in the laboratory. He denounced fascism and communism and worked for peace. He crusaded against quacks and antivivisectionists, winning their undying enmity, and campaigned against alcoholism, bulwarked by firsthand knowledge of its harmful effects on nutrition. He served on many official advisory boards, including the National Research Council and the Office of Scientific Research and Development.

Honors came thickly. He served as president of the American Association for the Advancement of Science, the American Physiological Society, the Association of American University Professors, the Federation of American Societies for Experimental Biology, the National Society for Medical Research, the Union of American Biological Societies, and the Institute of Medicine. He became a member of the National Academy of Sciences. He won honorary degrees from eight universities, including the rarely awarded honorary M.D. from the University of Lund in Sweden.

At the University of Chicago he became professor of physiology, chairman of the department, and Frank P. Hixon Distinguished Service Professor of Physiology. His use of scientific method entered all aspects of his own life; the phrase, "Vot iss de evidence?" became his hallmark. He called on physiology to help regulate his own physical health. Naturally enough, when he reached emeritus status, he turned to the study of the problems of aging, advocating an active and busy life for those past maturity.

The first relationship between Victor Johnson and Carlson was that of student and teacher. Born in Chicago on January 19, 1901, Johnson entered the University of Chicago and soon found himself assisting Carlson in research on hunger and the ductless glands. Johnson received his Bachelor's degree in 1926 and began teaching physiology two years before winning his Ph.D. From the very first, Johnson became, in Carlson's own words, "a great investigator and a great teacher." He collaborated with Carlson in the latter as well as the former, winning a $1,000 prize for excellence in undergraduate

teaching. He continued his own studies, receiving his M.D. degree in 1939. In research he concentrated upon the lymph system of the human body and then upon the study of blood and circulation. Among other things, he discovered that the products of fat digestion are injurious to red blood cells. Significant injury is avoided by the absorption of these fat-digestion products into the lymph vessels, which contain virtually no red blood cells. He also clarified the manner in which blood and air are brought together with maximum efficiency in the lungs. He has published many scientific papers and is co-author of the *Elements of Electrocardiographic Interpretation*, a standard guide for medical students and physicians.

In 1943, after becoming associate professor of physiology and dean of students in the University of Chicago's Division of the Biological Sciences, he was appointed secretary of the American Medical Association's Council on Medical Education and Hospitals, in which capacity he was mainly concerned with maintaining high standards of medical education in this country. He served by providing consultation advice to medical schools having difficulties, to universities contemplating the establishment or expansion of medical schools, and to hospitals engaged in the advanced education of physicians. Four years later Johnson became director of the Mayo Foundation for Medical Education and Research, with headquarters at the world-famed Mayo Clinic in Rochester, Minnesota, as well as professor of physiology at the University of Minnesota. Here he has been primarily engaged in the direction of the largest program in the world for the advanced training of and research by young physicians seeking to become specialists in some field of medicine.

Like Carlson, Johnson has also been active in the social applications of science and the scientific method. During World War II he served as a civilian consultant to both the Army and the Navy, to the United States Public Health Service, and to the Red Cross. On the national scene he is a member of the board of directors of the American Medical Education Foundation and adviser to the National Fund for Medical Education. These organizations raise and distribute funds to support medical schools. He continues as a member of the Council on Medical Education and Hospitals of the American Medical Association, which is concerned with medical education standards in the United States.

INDEX

Abdominal cavity, 222
  digestive organism, 266
  pressure fluctuations in, 249
Abscess, 120, 580
Absorption
  and diffusion, 250–51
  of food, 214, 268, 290, 302–5, 314
  and secretion, 272
  of water, 290, 296–97, 302–5
Accelerator of nerves of heart, 156–57, 431
Accommodation, 471–72
Acetoacetic acid, 315, 514
Acetone, 234, 514
Acetyl choline, 413, 431
  esterase of, 413
Achromatic vision, 480–83
Acid, 28
  in diabetes, 315, 514
  in gastric juice, 274
  neutralization, 258–59, 315
  and oxyhemoglobin dissociation, 255–57
  and respiration, 241
  in smoking, 354
  taste of, 496
  in urine, 357
Acid-base balance of blood, 258–59, 315
Acid control of pylorus, 295–96
Acid stomach, 305–6
Acidosis, 259
  in diabetes, 514
Acoustic labyrinth, 488–89
ACTH, 543, 566, 574
Action current; see Action potentials
Action potentials, 141
  in aortic depressor nerve, 163, 197
  in brain, 447
  in carotid sinus nerve, 198
  in glands, 142, 400
  of heart, 141, 148–49
  in muscle, 141
  in nerve, 386, 389, 395–96
  in skeletal muscle, 375
  in vagi from lungs, 238
Activation; see Stimulation
Adaptation in evolution, 2, 6, 603–5
Adaptation of rods in dark, 481
Addison, Thomas, 539
Addison's disease, 539–42
Adenosine triphosphate, 380
Adrenal cortex
  and hypophysis, 543, 574

Adrenal glands, 504, 535–43
  and hypophysis, 566
Adrenalin, 535
Adrenocorticotrophic hormone, 543, 566
Aerobic oxidation, 49
Afferent nerves, 157, 391–92, 404–6
  of blood pressure, 196–203
  of breathing, 236–39
  of heart rate, 161–64
  in reflexes, 418, 419
Afterdischarge, 426, 433
Age and heat production, 326–27
Agglutination
  of bacteria, 591
  of colloidal particles, 47
  of red blood cells, 88–90, 107–10
Agglutinins, 591
Agranular white blood cells, 115
Agranulocyte, 115
Air
  complemental, 232
  dead, 233
  in lungs, 231–33
  minimal, 232
  passageway for, 219, 293
  residual, 232
  supplemental, 232
  tidal, 232
Air hunger, 240
Air respired
  composition of, 233–34
  volume of, 231–33
Air sacs; see Alveoli
Albumin, 72
Alcohol, 31
  absorption of, 302
  excretion of, by lungs, 234
  hemolytic action of, 87
  use of, 350–53
"Alcoholics Anonymous," 353
Alcoholism, 352–53
Aldehydes in smoking, 354
Alimentary tract; see Digestive tract
Alkalies, 28
  on pyloric sphincter, 295–96
  taste of, 496
  in urine, 358
Alkaline reserve, 259
Alkalinity
  of bile, 277
  of intestinal juice, 277
  of pancreatic juice, 277

Allantois, 606, 608

All-or-none law, 388–90
    of heart, 140–41, 151
    of iron wire, 400
    of nerve, 140, 388–90, 394–95
    of skeletal muscle, 140–41, 373–75

Altitude and red blood cell formation, 96

Alveolar cells and secretion, 250

Alveoli, 219, 220, 221
    carbon dioxide in, 256
    gaseous exchange with blood, 249–51
    oxygen in, 249–51, 255–57

Ameba, 52, 64, 118, 580, 599–600, 614

Amebic dysentery, 580

Ameboid cells, 118–19

Ameboid movement, 118–19
    of leucocytes, 118–19, 587
    of macrophages, 587

Amines, 56, 278

Amino acids, 37–39
    absorption of, 304
    in diabetes, 514
    in digestion, 274, 276, 277
    essential, 341
    formula of, 317
    and protein synthesis, 315
    in proteins, 37–39, 340
    synthesis of, 48, 340–41

Amino group, 37

Ammonia, 72
    and acid-base balance, 315
    formation of, 315, 316
    in smoking, 354
    in urine, 357

Amnion, 606, 608, 617

Amniotic fluid, 606, 608, 617, 618

Amphibia
    breathing in, 217, 218, 224
    circulation in, 131–32
    effect of cold on, 330
    evolution of, 605–7
    eye development in, 612
    reproduction in, 605–7

Ampulla of semicircular canals, 493

Amylase, 273, 276

Amylon, 273

Amylopsin, 276

Anabolism, 48

Anaerobic oxidation, 49

Analogous structures, 585

Anelectrotonus, 396–97

Anemia, 97, 98–104
    from atom-bomb explosions, 124
    in embryo and infant, 609, 621–22
    and hemoglobin deficiency, 91
    in infants, 112–13
    pernicious, 102–4
    from Rh factor, 112–13
    from ulcers, 307

Anesthesia, 9–10, 351
    and gastrointestinal movements, 285
    hemolysis in, 87
    "nature's," 162
    and reflexes, 422, 424

Animal starch; see Glycogen

Anode, 396

Anterior chamber of eye, 469

Anthrax, 590–91

Antibiotics, 594–98

Antibodies, 69, 591–92
    in blood, 73

Anticoagulant, 77

Antigens, 591–92

Antiprothrombin, 79

Antithrombin, 79

Antitoxin, 591–92

Anus, 266, 267, 610

Anvil; see Incus

Aorta, 127, 166
    blood pressure in, 172, 174
    of embryo, 618, 619

Aortic body, 239

Aortic depressor nerve; see Depressor nerve

Ape-man of Java, 444

Apes, learning in, 455, 456

Aphasia, 451–52

Appendicitis, 308, 500

Appendix (veriform), 267, 307, 613

Appetite, 298–99, 349

Aqueous humor of eye, 469

Arteries, 127, 166–67
    carbon dioxide in, 256
    common carotid, 198
    elasticity of, 171–72
    external carotid, 198
    hardening of, 186, 208–9
    internal carotid, 198
    of kidneys, 366
    oxygen in, 256
    pressure in; see Blood pressure
    pulmonary, 127, 128

Arterioles, 166
    carbon dioxide action on, 179
    epinephrine action on, 180
    metabolite action on, 179
    pituitrin action on, 559
    pressure in; see Blood pressure
    smooth muscle of, 178
    tonic constriction of, 178

Arthritis, 543

Artificial respiration, 244–46

Aschner, Bernhard, 561

Association areas of brain, 448, 450–52

Asthma, 261, 537

Astigmatism, 475

Athlete's foot, 580

Atomic explosions, blood damage in, 124–25
Atoms, 23
    radioactive, 117
    "tagged," 43, 95
Atria of heart, 132, 144–46
Atrioventricular (A–V) bundle, 135, 145
Atrioventricular node, 135, 145
Atrioventricular valves, 132–33, 146, 147
Atropin, 283, 410–11, 476
Audition; *see* Hearing
Auditory canal, 484
Auditory nerves, 440, 486, 488
Auditory ossicles, 485
Auricles of heart; *see* Atria of heart
Automaticity
    of breathing, 142, 234
    of heart, 142–44
    of intestines and stomach, 294–95
Autonomic nervous system, 406–11
    and epinephrine, 537
    and myelin sheaths, 401–2
Axones, 16

Bacilli, 580
    of anthrax, 590
    of diphtheria, 580
    of pneumonia, 580
    of tetanus, 376
Back-pressure–arm-lift method of artificial
    respiration, 244–46
Bacon, Francis, 296, 431–32
Bacteria, 3, 50, 580–81; *see also* Infections
    agglutination of, 591
    anaerobic, 216
    in atom-bomb injury, 124
    barriers against invasion of, 582–87
    in colon, 278
    defenses against, 119–20, 221, 260–61, 582–
        87
    destruction of, by chemicals, 594–98
    in feces, 290
    fever produced by, 337
    in gallstones, 308
    hemolysis by, 87
    kinds of, 580
    and lymph nodes, 213–14, 588–89
    phagocytosis of, 119–20, 587
    useful, 581
Bacteriolysins, 591
Bainbridge reflex, 163–64, 175–76
Balance, sense of, 488–91
Balloon-manometer technique, 227–28, 249
    and digestive movements, 285, 286
    and hunger contractions, 299–300
    and swallowing, 286–87, 294
Banting, Frederick G., 511–12
Basal conditions, 325
Basal metabolism, 324, 325–30
    apparatus for, 323

in cretinism, 521
in hyperthyroidism, 518, 526–28
in hypothyroidism, 517–18, 526–28
and sleep, 460
Bases; *see* Alkalies
Basilar membrane, 486, 487–88
Basophilic white blood cells, 116, 117, 120
Bayliss, Sir William M., 508, 568
Bell, Sir Charles, 405
Bell and Magendie, law of, 403–6
Benzoic acid, 359
Benzol, 87, 99
Beriberi, 344–45, 348
Bernard, Claude, 65, 179, 311, 360, 507, 577
Best, Charles H., 512
Beta-hydroxybutyric acid, 514
Bicarbonate in carbon dioxide transport, 252,
    258
Bile, 275, 276
    and clotting of blood, 81
    secretion of, 282
    and vitamin K, 81
Bile duct, 94, 267
Bile pigment, 94
    in feces, 94, 290
    in pernicious anemia, 103
Bile salts, 275
Binocular vision, 477–79
Biochemistry, 5
Biological warfare, 598
Biology, 1–5
Biophysics, 5
Birds, 605, 606–7
Birth, phenomenon of, 615–21
Birth canal, 615, 617, 618
Birthmark, 609
Bladder; *see* Gall bladder; Urinary bladder
Bleeding; *see* Hemorrhage
Blind spot, 469, 480, 481
Blindness, 483
    night, 482
Blood (*see also* Blood glucose; Blood plasma;
        Clotting of blood)
    absorption of foods into, 304–5
    acid-base balance in, 258–59, 315
    alcohol in, 352
    alkali conservation in, 315
    amino acids in, 304
    atom-bomb damage of, 124–25
    bicarbonate in, 258–59
    buffers of, 258–59
    calcium and parathyroids, 532–33
    cancer of, 123
    carbon dioxide in, 251, 256, 257–59, 618–20
    carbonic acid in, 257
    chlorides and adrenal glands, 543

Blood—*Continued*
  composition of, 72–73
    in adrenal insufficiency, 543
    and breathing, 239–43
    regulation by kidneys, 357–58
  of embryo, 608, 610, 618–20
  epinephrine in, 535, 538
  fats, 72, 304–5, 314
  fibrinogen in, 71
  formed elements of, 65; *see also* Blood
      platelets; Red blood cells; White blood
      cells
  hormones in, 504–5
  hypophyseal hormones in, 564
  and internal environment, 65
  neutrality; *see* Blood, acid-base balance in
  oxygen in, 73, 255–57, 618–20
  oxygenated, 128, 131
  potassium and adrenal glands, 543
  proteins, 69–71
    and carbon dioxide transport, 258
    osmotic pressure of, 60, 70–71
  prothrombin in, 78
  reserves in spleen, 205
  Rh factor in, 110–13
  salts, 67–68, 73
  sex hormones in, 551–52, 555
  sodium and adrenal glands, 543
  specific gravity of, 66
  sugar; *see* Blood glucose
  transfusion of, 88–90, 104–15, 204
    reactions to, 105–6
  urine separation from, 359–60
  venous, 128
  viscosity of, 69, 191–92
Blood banks, 114–15
Blood cells; *see* Red blood cells; White blood
    cells
Blood clotting; *see* Clotting of blood
Blood flow, 168–76
  adjustments of, 176–78
  in aorta, 170–72
  in arteries, 170–72
  to brain, 162, 195, 202–3, 206
  breathing and, 231
  in capillaries, 169–72, 183, 186
  clotting and, 75
  and friction, 172–73, 190–92
  and gravity, 175, 249
  to kidneys, 366
  in muscle activity, 164, 175–76, 179–80,
      195, 206–8, 231, 261–62, 381–82
  and reflexes, 424
  resistance to, 172–74, 188, 190–92
  study of, 178–79
  and urine formation, 363–64
  variations in, 176–78
  in veins, 170, 174–76, 231, 249
  velocity of, 171, 173
Blood glucose, 68, 72, 312
  constancy of, 313, 574
  in diabetes, 513
  and epinephrine, 155, 537, 539, 571, 574
  and food ingestion, 311–13

  and hunger, 300–301
  and hypophysis, 575
  and insulin secretion, 573
  and urine secretion, 361–62, 364–65
Blood groups, 106–10
Blood plasma, 65, 66–74, 606; *see also* Blood
  a and *b* substances in, 108–10
  composition of, 72–73
  species differences in, 87–90
  viscosity of, 192
Blood platelets, 65, 116, 123–24
  in atom-bomb injury, 124
  and clotting, 76
  disintegration of, 76
  in hemophilia, 81–82
  prothrombin from, 78
  stability of, 82
Blood pressure, 182–206
  adjustments of, 188–206
  in adrenal insufficiency, 540
  aortic, 146–47, 172, 174
  and aortic depressor nerve, 197–203
  arterial, 174, 182–206
  arteriolar, 172, 174
  atrial, 172
  capillary, 172–74
  carbon dioxide and, 189, 194–95
  and cardiac output, 189
  and carotid sinus, 198–203
  diastolic, 174, 183, 185–88
  and friction, 172–73, 190–92
  gradient of, 172–74, 231
  and gravity, 202–3, 205–6
  in heart, 146–47
  and heart rate, 189
  after hemorrhage, 70, 204
  high, 209, 365–66, 463
  and kidney secretion, 363–64
  low, 209
  mean, 174, 183, 185
  measurement of, 186–88
  and peripheral resistance, 172–74, 190–203
  pulse, 174, 183, 185–88
  systolic, 174, 183, 185–88
  and vagus stimulation, 189
  and vasoconstriction, 189–203
  and vasodilation, 189–203
  venous, 172, 174, 206
  ventricular, 146–47
  and viscosity, 191–92
  and volume of blood, 188, 204–5
Blood sinuses, in placenta, 608
Blood types; *see* Blood groups
Blood vessels (*see also* Arteries; Arterioles;
    Capillaries; Veins)
  accessory, 578
  of allantois, 608
  carbon dioxide action on, 179
  chemical control of, 179–80
  constriction of, 176–78
  cross-sectional area of, 169–70
  of digestive tract, 270
  dilation of, 176–78

of embryo, 608, 610, 614, 618–20
epinephrine action on, 180
of heart, 135–36, 149
of kidney, 359–60, 365
nervous control of, 180–82, 192–94
occlusion of, 208
of placenta, 608, 618, 622
of rabbit's ear, 180–81
repair of, 80–81
of skin, 333
spasms of, 74, 149, 208
stopcocks in, 176–78
structure of, 166–68
tonus of, 180–81
of uterus, 622

Blood volume, 66, 210
and blood pressure, 204–5
in muscular exercise, 205, 262
regulation of, by kidneys, 358

Blue baby, 620

Blushing, 178, 196

Body-righting reflexes, 490–92; *see also* Posture

Body surface area and heat production, 325–26

Body temperature, 330–37
evolution of control of, 330–32
fluctuations in, 336–37, 460
in muscular exercise, 154
regulation of, 248, 330–37
and sleep, 460
and thyroids, 530

Body weight, 328–30, 354; *see also* Growth

Boils, 580, 587

Bone marrow, 92, 93
altitude and, 96
benzol poisoning of, 87
fatty, 92, 93
in hemorrhage, 99
in infant, 621
in infections, 123
injuries of, 99
and oxygen deficiency, 96–98
platelets from, 123
prothrombin from, 78
red, 92, 93
red blood cell production in, 92, 93
white blood cell production in, 117
yellow, 93

Bones, 16, 92, 93
calcium in, 16, 343, 534
of ear, 484, 485
growth of, 346, 556, 560–63
and parathyroids, 534
phosphorous in, 16, 343
of skeleton, 369, 605
and vitamin D, 346, 347

Bound water, 26

Bowel distress, 297

Bowman, Sir William, 359

Bowman's capsule, 359–60

Brain, 391–92, 442–65; *see also* Brain stem; Cerebellum; Cerebral cortex; Medulla oblongata; Midbrain; Thalamus
abnormal function of, 352, 454, 463–65
alcohol effects on, 351, 352
control of spinal cord by, 417, 442, 444
embryonic development of, 610, 611
evolution of, 417–18
experimental studies of, 447
gray matter of, 445
irritability of, 33
of lower animals, 418, 443–44
of prehistoric man, 444
segmental portion of, 439
significance of, 442–44
size of, 444
suprasegmental portion of, 442
surface markings on, 444–45, 446, 448
ventricles of, 439,445–46, 612
white matter of, 445

Brain stem, 438–42; *see also* Medulla oblongata; Midbrain; Thalamus
autonomic nerves from, 409
functions of, 439–42
reflex centers in, 439–42
salivation center in, 281
sleep center in, 460
structure of, 438–39
temperature regulation by, 335

Brain waves, 453–54

Bread mold, 603

Breathing (*see also* Respiration)
automatic rhythm of, 142, 234
and blood pressure, 189
carbon dioxide control of, 240–43, 262, 618–20
center; *see* Respiratory center
chemical control of, 239–43
and defecation, 247–48, 298
depth of, 238, 239, 243, 262
inhibition of
protective function of, 236, 237, 586
in swallowing, 236, 237, 292
by vagi nerves, 237–39
and lymph flow, 212, 231, 249
mechanics of, 222–31
modifications of, 246–49
in muscular exercise, 223–24, 231–33, 239–44, 261–62
nervous control of, 234–39
of newborn animal, 618
nonrespiratory functions of, 246–49
passages, 219, 293, 586, 610
rate of, 239–43, 262
recording of movements of, 227–28
reflexes of, 236–39, 248, 292–93
and straining, 247–48, 298
systems, evolution of, 217–18
in trained subject, 262–63
vagus nerves in, 237–39
and venous blood flow, 175–76, 231, 249
and vocalization, 246–47
voluntary control of, 242, 246–48

Bright's disease, 365–66

Bronchi, 219
Bronchitis, 260
Brown-Séquard, Charles E., 508, 540, 558
Buccal cavity; *see* Mouth
Buchner, Edward, 42

Calcium (*see also* Salts)
  in bones, 16, 343, 534
  and clotting of blood, 68, 77
  in diet, 343
  and heartbeat, 30, 143
  and irritability of cells, 30, 532–33, 534
  and parathyroids, 532–33
Caloric requirements in diet, 338
Calorie, large, 320
Calorimetry, 319–21
  indirect, 322–24
Cancer, 307–8, 601
  of the blood, 123
  and smoking, 354
  of thyroids, 529
Cannon, Walter B., 413, 537
Cannula, 179, 358
Capillaries, 50, 166, 167–68
  blood flow in, 169–72, 186
  in breathing systems, 218, 221
  discovery of, 129
  functions of, 167–68, 170
  of intestines, 312
  of kidneys, 360
  of liver, 312
  of lungs, 221
  lymphatic, 211–12
  of placenta, 608
  of uterus, 607
  water balance in, 59–61, 69–71
Carbohydrates, 31–34; *see also* Glucose; Glycogen; Starches
  in cells, 31–34
  combustion of, 33, 313–14, 314–15, 320–21
  in diet, 338–40
  digestion of, 272, 274, 276
  metabolism of, 311–14
    in adrenal insufficiency, 542
    in diabetes, 512–14
    and insulin, 512–14
  in milk, 621
  in nutrition, 338, 339–40, 621
  from proteins, 39–40, 317, 514
  respiratory quotient (R.Q.) of, 321–22
  storage of, 32–33, 35–36, 311–13, 507, 513–14
  transformed into fats, 35–36, 317–18
Carbon dioxide
  in air respired, 233–34, 256
  in alveoli, 250, 256
  in blood, 73, 251, 256, 257–59
  and blood pressure, 189, 194–95
  and blood vessels, 179
  and breathing, 240–41, 262, 618
  diffusion of, 52–53, 249–51
  in embryo, 606, 608, 620

  and heart, 143–44, 155–56
  indispensability of, 195, 242–43
  and oxyhemoglobin dissociation, 256
  production of
    in animals, 318–19
    in embryo, 606, 608
    by kidneys, 363
    measurement of, 320
    by muscle, 379–81
    by nerve tissue, 386
    in secretion, 271–72
  and respiratory center, 240–43
  in tissues, 256
  transportation in blood, 252, 257–59
  and vasoconstrictor center, 194–95
Carbon monoxide, 257, 354
Carbonic acid, 240, 257
Carboxyhemoglobin, 257
Cardiac cycle, 144–49
Cardiac output of blood, 151–64, 189
Cardiac reflexes, 160–64
Cardioaccelerator centers, 156, 157, 159
Cardioinhibitory center; *see* Vagus center
Cardioinhibitory nerves, 157–58, 159
Carotid body, 239
Carotid sinus, 163, 200
  and blood pressure, 197–203
  and breathing, 239
  and heart rate, 197–203
  nerve of, 197, 198
Cartilage, 15, 16, 219, 605
Castration, 558
Catabolism, 48
Catalysts, 40, 345
Cataract, 476
Catelectrotonus, 396
Cathartics, 297
Cathode, 396
Cell membranes, 12–13
  electrical changes on, 55–56
  fats in, 36, 55, 315
  permeability of, 54–57
Cell organs, 13
Cells (*see also* Protoplasm)
  carbon dioxide removal from, 251
  chemical composition of, 23–44
  destruction of
    by dilute salt solution, 59–60
    by fat solvents, 55, 86–87
    by foreign plasma, 87–90
  discovery of, 12
  division of, 93, 599–601, 607
  energy relations in, 49–50
  fat utilization by, 35, 314–15
  glucose utilization by, 313–14
  irritability of; *see* Irritability of cells
  kinds of, 14–17
  methods of study of, 17–21
  microscopic examination of, 17–18
  movement of materials through, 50–61

organization of, 14–17
oxygen consumption of, 24, 251, 259–60
protein utilization by, 315–16
secretory, 270–71
shrinkage of, by concentrated salt solution, 59
significance of, 11–12
specialized activities of, 270–72
structure of, 13
work done by
  in absorption, 303–4
  in excretion, 356, 363
Cellulose, 12
  in diet, 349
  in feces, 290
  indigestibility of, 278
Centers, reflex, 160–61, 418–19, 432
  for body-righting, 441
  for breathing, 235–36
  for heart rate, 157, 160
  sleep, 441, 460
  temperature-regulating, 335–36
  vasomotor, 193–94, 196
  vomiting, 441
Central excitatory state, 433–34
Central inhibitory state, 433–34
Central nervous system, 160, 392, 415–17;
  see also Brain; Brain stem; Cerebellum;
  Cerebral cortex; Cerebral hemispheres;
  Medulla oblongata; Midbrain; Spinal
  cord; Thalamus
  evolution of, 417–18
  excitation in, 428–31
  inhibition in, 428–31
  integrative action of, 415–17
  in muscle training, 263
  myelin sheaths in, 401
  neurilemma sheaths absent in, 401
  regeneration in, 403
  segmental portion of, 439
  suprasegmental portion of, 442
Centrifuging of blood, 66
Cephalic phase of gastric secretion, 284
Cephalin, 72, 77
Cerebellum, 448, 461–62
  localization of function in, 462
  nerve pathways to, 438
Cerebral cortex
  association areas of, 448, 450–52
  frontal lobes of, 446, 448
  in learning, 443, 458
  localization of function in, 446–53
  microscopic structure of, 446
  motor area of, 437, 447, 448–49
  and muscle tone, 442
  occipital lobes of, 446, 448, 458, 477–78
  parietal lobes of, 446, 448
  sensory areas of, 437, 448, 449–50
  temporal lobes of, 446, 448
Cerebral hemispheres, 335, 439
  removal of, 443–44
  structure of, 445–52

Cerebrospinal fluid, 445–46, 463
Chemical formulas, 24, 31
Chemical treatment of infections, 594–98
Chemistry, 23–25, 26
Chemoreceptors, 494
Chest; see Thorax
Childbirth, 615–21
  pain in, 500
  pituitrin in, 559–60
  straining in, 298
Chills, in fever, 337
Chlorides
  in adrenal insufficiency, 541, 543
  in diet, 343
Chloroform action on liver cells, 87
Chlorophyll, 13
Chloroplast, 13
Cholecystokinin, 282
Cholera bacteria, 580
Cholesterol, 72, 84, 308
Chorion, 606, 608, 617
Choroid coat of eye, 469
Chromosomes, 601, 602
Cilia, 584
Ciliary body, 469, 472
Ciliary motion, 221, 261, 583–86
Ciliary muscles, 471, 472
Ciliated cells in man (see also Hair cells)
  in ear, 487
  in respiratory system, 221, 260–61
  in smell receptors, 495, 497
  in taste buds, 495
Circulation (see also Blood flow; Blood pressure; Blood vessels; Vasoconstriction; Vasodilation)
  in Amphibia, 131–32
  in birds, 131–32
  through capillary walls, 61
  changes of, at birth, 618–19
  coronary, 136
  disorders of, 208–10
  in embryo, 618–20
  evolution of, 63–65
  in fish, 131
  lymphatic, 211–14
  in mammals, 131–32
  in placenta, 608
  proof of, 127–30
  pulmonary, 127, 128, 129, 168–69, 618–20
  rate of, 168–69
  and respiration, 261–63
  systemic, 129, 169
  in trained subject, 262–63
Circulation time, 168–69
Circulatory system
  in food distribution, 64–65, 266
  model of, 177
  structure of, 127–28
Citrate and clotting, 77

Closure, in learning, 456
Clot-inducing material; *see* Thrombokinase
Clotting of blood, 74–82
  antiprothrombin in, 79
  antithrombin in, 79
  and bile, 81
  in blood vessels, 76, 80, 208
  calcium in, 68, 77
  defective, 81–82, 348
  and defibrination, 78–79
  and epinephrine, 155, 537, 539
  fibrinogen in, 47, 71, 72, 77–78
  prevention of, 75, 76, 77, 78–79
  prothrombin in, 78
  reactions of, 78
  in repair of blood vessels, 80–81
  sol-to-gel change in, 47
  thrombin in, 77
  thrombokinase in, 77
  and vitamin K, 81, 348
Clotting time, 74
Clumping; *see* Agglutination
Coagulation; *see* Clotting of blood
Cocci, 580
Cochlea, 484, 485–87
Cochlear canal, 486
Cold (*see also* Body temperature)
  and caloric requirements, 338
  common, 260
  effects of, on various animals, 330
  and heartbeat, 144, 153–55
  and heat production, 324
  and muscle tone, 324, 332
  of nose, and taste, 494
  receptors, 335, 498
  reflex adjustments to, 182, 335
  and temperature regulation, 334, 335–36
Cold-blooded animals, 332
Cold sense, 498–99
  area in brain, 449
  nerve pathways of, 438
Cold spots in skin, 498
Colitis; *see* Bowel distress
Collip, J. B., 512
Colloid in thyroid gland, 516, 524
Colloidal particles, 44–45, 47
Colloidal suspension, 44–46, 74, 275
Colon, 266, 267, 268
  absorption of water from, 302
  bacteria in, 278
  excretion by, 290
  peristalsis in, 290
Color vision, 482–83
Coma, 352
Combustion; *see* Oxidations
Compounds, 23
Conditioned reflexes, 280, 284, 421, 457–58
Conducting system of heart; *see* Atrio-
  ventricular (A–V) bundle

Conduction of nerve impulse; *see* Nerve
  impulses
Cones of retina, 469, 480–83
Connective tissue, 15, 16, 19, 20, 121
Consciousness
  and cerebellum, 461
  and cerebral cortex, 447
  loss of; *see* Fainting
Constancy
  of basal metabolic rate, 325
  of blood glucose, 68, 313, 574
  of blood pressure, 202–3
  of body temperature, 330
  of heart rate, 163, 200–202
  of internal environment, 65
  of plasma composition, 67
  of red cell blood cell count, 98
  of white blood cell count, 121
Constant currents and nerve stimulation, 396
Constants, physiological, 68, 98, 203
Constipation, 296–97
Constriction
  of blood vessels; *see* Vasoconstriction
  of pupil, 472, 476–77
Contractions
  of skeletal muscle; *see* Muscle, skeletal;
    Muscle, smooth
  of uterus, 615–17, 620
Controls in experimentation, 8
Convergence of eyes, 479
Convolutions of brain surface, 444–45
Convulsions, 33, 69, 463, 515, 531
Copper, 101, 343
Copulation, 546, 554, 605
Corn protein, 341
Cornea, 469, 471, 472, 483
Corneal reflex, 468, 586
Corpus luteum, 547, 548
  hormone of, 552–53
  and hypophysis, 563, 564
  in pregnancy, 552–53
  in sex cycle, 552–53
Corresponding points, 478
Cortex
  of adrenal glands, 535, 536, 541–43
  cerebral; *see* Cerebral cortex
  of kidney, 358, 359
Corti, Alfonso, 486
Corti, organ of, 486–87
Cortical hormone of adrenal glands, 543
Cortisone, 543
Cough, 237, 248, 261, 586
Cranial nerves, 439–40
Craniosacral nerves, 409–10
Cranium, 368, 369
Crayfish, equilibrium organ of, 489–90
Creatin, 71

Creatine phosphate, 380
Creatinin, 71, 357
Crenation of red blood cells, 86
Cretinism, 520–21, 523, 527, 528
Crista of semicircular canals, 493
Crocodiles, 605
Cross-matching, in blood transfusions, 106
Crossed extension reflex, 420–21
Cupula of semicircular canals, 493
Curare, 372
Current of action; see Action potentials
Cytolysis, 59, 84, 87
Cytoplasm, 13, 47, 605

Dead space, 232–33
Deafness, 488
Deaminization of proteins, 315–16
Decerebrate rigidity, 442
Defecation, 247–48, 298
Defenses against infection, 119–20, 221, 260–
    61, 582–87
Defibrination of blood, 78–79
Deficiency diseases, 348
Degeneration of skeletal muscle, 384
Demarcation potential, 395
Dendrites, 16
Depressor nerve, 162–63, 196, 197, 239
Depth perception, 477–79
Descartes, René, 570
Diabetes insipidus, 560
Diabetes mellitus, 34, 277, 509–14
    acetoacetic acid in, 315
    acid metabolites in, 514
    carbohydrate combustion in, 513–14
    cause of, 515–16
    and hypophysis, 515–16, 574–75
    urine volume in, 364
Diaphragm, 222, 228, 267
    capillaries in, 168
    in defecation, 298
    movements of, 223, 229
    nerves of, 234–35
Diarrhea, 296–97
Diastole, 137, 138, 144, 146, 147; see also
    Heart, contraction of; Heartbeat
Diet, 338–49, 621
Differentiation of organs, 609–12
Diffusion, 51–53, 57, 600
    in absorption of food, 303, 312
    in capillaries, 167
    in lungs, 52–53, 249–51
    in nerve polarization, 397
    in placenta, 608
    respiratory gases, 52–53, 249–51, 256
Diffusion gradients, 250, 251, 253–55

Digestion
    chemical factors in, 265, 272–78; see also
        Bile; Gastric juice; Intestinal juice;
        Pancreatic juice; Saliva
    of foods, 272–78
    mechanical factors in, 265; see also Defeca-
        tion; Digestive movements; Swallow-
        ing
Digestive enzymes, 41, 273–77
Digestive glands, 266–70
    control of, 276, 278–85
    embryonic development of, 610
Digestive movements, 148, 285–90; see also
        Defecation; Swallowing
    control of, 291–98
    of intestines, 289–90
    of stomach, 287–89
Digestive tract
    absorption of foods from, 290, 302–5
    disorders of, 305–8
    embryonic development of, 610
    evolution of, 64, 266
    structure of, 266–70
Dilation
    of blood vessels; see Vasodilatation
    of pupil, 472, 476–77
Dinosaurs, 607
Diphtheria, 260
    bacteria of, 580
    and heart, 149
    immunization against, 593
    kidney damage in, 365
Disaccharides, 32
Disease, 10–11, 576–80, 623
Distance perception, 477–79
Diuresis, 364–65
Dizziness, 493
Donor of blood transfusion, 90, 105
Dorsal roots of nerves, 404–6
Drug action on heart nerves, 410–11
Drugs, 351, 354
Du Bois-Reymond, Emil, 392, 395
Duckbill, 607
Ductus arteriosus, 618–20
Duodenum, 266, 267, 283, 505
    secretin from, 280–82, 568, 572–73
    ulcers of, 306–7
Dwarfism, 520–21, 561–63

Ear, 484–94
    infections of, 260, 485
Ear stones, 489, 491
Eardrum, 484, 485
Earthworm, 217, 417, 418
Eck, Nikolai, 316
Eck fistula, 316
Ectoderm, 17, 609–10, 613
Edema, 70, 175

Effectors, 419

Efferent nerves, 156–57, 391–92, 404–6
  of blood vessels, 180–82, 192–93
  of breathing, 234–35
  of heart, 156–57, 409
  in reflexes, 419

Egg, 603, 605; *see also* Ova
  food in, 605
  iron in, 622
  land, of reptiles and birds, 606–7, 608
  shell of, 606
  yolk in, 605, 608

Eijkman, Christiaan, 344

Elasticity
  of blood vessels, 166
  of lungs, 221–22, 224, 226, 230
  of red blood cells, 84

Electrical activity, of brain, 453–54, 460

Electrical changes; *see* Action potentials

Electrical charges
  on cell membranes, 55–56
  on colloidal particles, 47–48
  in nerve, 68
  on nerve fibers, 393
  on skeletal muscle, 375

Electrocardiogram, 141, 148–49

Electrocardiograph, 148

Electroencephalograph, 454

Electrolytes, 28

Electrotonus, 396–97, 400

Elements, 23

Elimination, 355–56

Embolus, 81

Embryo, 605, 606–8

Embryonic membranes, 606–7, 617

Emergency functions
  of epinephrine, 155, 538–39
  of physiological processes, 577

Emmetropia, 474

Emotions
  and alcohol, 353
  and blood pressure, 196
  diseases of, 464
  and epinephrine secretion, 155, 538–39, 571–72
  and heart rate, 161
  and high blood pressure, 209
  and ulcer, 306

Emulsions, 275

Endocrine glands, 504–8
  adrenals, 505, 535–43, 566
  control of, 571–75
  duodenum, 505, 568
  hypophysis, 448, 505, 559–67, 574–75
  liver, 569
  pancreas, 505, 508–16
  parathyroids, 505, 530–35
  pineal body, 448, 505, 570
  pituitary gland, 505, 559–67
  sex glands, 505, 544–58

stomach, 568–69
study of, 506–8
thymus, 505, 569–70
thyroids, 505, 516–30, 566
transplantation of, 88–90

Endoderm, 17, 609–10, 611, 613

Endolymph, 486, 493

Endothelium, 167

Energy, 49–50
  and diet, 338
  in foods, 313, 318–21
  as stimulus, 467

Energy debt in muscle, 382

Energy liberation
  anaerobic, 50, 216
  in blood vessels, 171–72
  by cells, 49, 216
  in diabetes, 514
  by heart, 171–72
  in muscle contraction, 243, 375, 378–83
  in nerve impulse conduction, 389
  in respiration, 49, 216
  in secretion, 272, 363

Enteroception, 422

Enterokinase, 276

Environment
  adjustments to, 368
  internal; *see* Internal environment

Enzymes, 40–43
  for acetyl choline, 413
  in blood, 73
  in carbohydrate storage, 312–13
  in cells, 40–43
  digestive, 41, 273–77
  oxidative, 41, 260, 313

Eosinophilic white blood cells, 116, 117, 120

Epididymis, 544, 546

Epiglottis, 228, 247, 293

Epilepsy, 463

Epinephrine, 535–39
  and Addison's disease, 541
  on blood pressure, 189, 190
  on blood vessels, 180
  and carbohydrate storage, 313
  and clotting of blood, 155, 537, 539
  emergency function of, 155, 538–39
  on heart rate, 155
  nervous control of secretion of, 538, 571
  and sympathin, 413

Epithelium, 14–15
  ciliated, 584
  olfactory, 497
  respiratory, 218

Epsom salts, 297

Equations in chemistry, 24

Equilibrium, sense of, 488–94

Erection, 194, 196

Erepsin, 277

Ergometer, 319, 320

Ergosterol, 346

Errors of observation and reasoning, 7–8

Erythrocytes; *see* Red blood cells

Esophagus, 228, 266, 267, 293; *see also* Swallowing
  nervous control of, 291–94
  peristalsis in, 285–87, 293–94

Estrogen, 551, 552–54

Estrus, 549

Ether, 31, 87

Eustachian tube, 483, 613

Evolution, 6–7
  of body-temperature regulation, 330–32
  of breathing mechanisms, 217–18
  of circulation, 63–65
  of digestive tract, 64, 266
  and embryonic development, 613–15
  of heart, 131–32
  hunger and appetite in, 298
  of mammals, 607–8
  of membranes of embryo, 608
  of nervous system, 417–18
  sex factor in, 604–5
  of sexual reproduction, 605–8
  of vertebrates, 605–9

Excitation in central nervous system, 428–31; *see also* Stimulation

Excitatory substance, 431–34

Excitement; *see* Emotions

Excretion(s), 355–56
  of alcohol, 350
  by colon, 290
  in feces, 290
  by intestines, 265, 290
  by kidneys, 272, 357–58
  by liver, 355
  by lungs, 355
  via placenta, 608

Exercise; *see* Muscular exercise

Exophthalmos, 527, 528

Experimentation, 9–10, 623

Expiration, 233–24; *see also* Breathing
  active, 223
  initiation of, 238–39
  passive, 223
  pressure changes in, 225–31

Expiratory movement in straining, 247–48

Expiratory reserve volume, 232, 233

Extensor muscles, 370, 420–21

Exteroception, 422

Extracts of endocrine glands, 506

Eye, 467, 468–83
  embryonic development of, 612
  infections, 344, 483
  and vitamin A, 344

Eyelids, 468

Eyestrain, 474

Fainting, 162, 195, 202–3, 206

Fallopian tube; *see* Oviduct

Far point of vision, 474

Farsightedness, 474–76

Fasting, 35, 39
  and hunger, 299–300
  nitrogen imbalance in, 339
  protein combustion in, 39, 316

Fat cells, 15, 16–17

Fatigue
  in central nervous system, 425
  and epinephrine, 537, 539
  in muscle, 175, 243–44, 375, 379, 383
  and muscle tone, 459–60
  in nerve, 383
  and reflexes, 425
  and sleep, 459
  of smell sense, 497–98

Fatlike compounds, 36, 84

Fats, 34–36
  absorption of, 214, 304–5
  of blood, 72, 304–5, 314
  in bone marrow, 92, 93
  caloric value of, 320–21
  from carbohydrates, 35–36, 317–18
  in cell membranes, 36, 55, 315
  in cells, 34–36
  and cholecystokinin formation, 282
  combustion of, 35, 314–15, 320–21, 514
  in diet, 338, 339–40, 621
  digestion of, 275–77
  emulsification of, 275
  heat liberation from, 35, 320–21
  insulating effect of, 36, 314
  metabolism of, 314–15, 320–21, 514
  in milk, 621
  in nutrition, 338, 339–40, 621
  respiratory quotient of, 321–22
  solvents and red blood cells, 86–87
  storage of, 35, 314, 317–18
  transformation into carbohydrate, 36, 317
  vitamins in, 346–47

Fatty acids, 34–35
  from fat digestion, 276
  formula of, 317

Feathers in temperature regulation, 334

Feces, 265, 290
  bile pigment in, 94, 290
  in constipation, 297
  in diarrhea, 296

Femur, 92, 369

Fermentation, 50
  by bacteria in colon, 278
  and muscle contraction, 379
  by yeast, 42, 50

Fertilization, 544, 547, 548, 551, 603, 607

Fetus, 607

Fever, 336–37
  and heart rate, 155
  in infections, 122
  and metabolic rate, 328

Fibrillation of atria, 150

Fibrin, 74, 77–78

Fibrinogen, 47, 71, 72, 74, 77–78

Fibula, 369

Filtration, 57, 58, 60–61
in capillaries, 53–54, 70–71, 167
in urine formation, 360

Filtration pressure, 58
hemorrhage and, 70–71
and kidney secretion, 363–64

Fischer, Emil, 26

Fish
breathing of, 217, 218
circulatory system of, 131
effects of cold on, 330
in evolution, 613
reproduction in, 605

Flagellum, 584

Flatworm, 64, 614
alimentary system of, 266
in evolution, 613, 615
mesoderm in, 610
nervous system of, 417
vegetative reproduction in, 603
waste elimination in, 356

Fleming, Alexander, 595

Flexion reflex, 419, 420, 586

Flexor muscles, 370, 420

Florey, Howard, 596

Focusing of eye, 471

Follicle-stimulating hormone, 564

Follicles
of ovaries, 546, 547, 548
of thyroid glands, 516

Follicular fluid, 546, 547, 548

Follicular hormone, 551–52
and hypophysis, 564

Foods (see also Carbohydrates; Diet; Fats;
Nutrition; Proteins; Vitamins)
absorption of, 290, 302–5
alcohol, 350–51
caloric value of, 320–21
combustion of, 33, 35, 39–40, 313–14, 320–
21, 513–14
digestion of, 272–78
distribution of, to body cells, 63–65, 266
in egg, 605
energy in, 313, 318–21
heat liberated from, 318–21
and heat production, 324–25
interconversions of, 40, 316–18
metabolism of, 311–18, 514
passageway for, 267, 268, 293
from plants, 341
preserving of, 343
refinement of, 343
roughage in, 296
specific dynamic action of, 324–25, 332
vitamins in, 345, 348
in yolk sac, 606, 608

Foramen ovale, 618–20

Forebrain, 439

Forgetting, 455–56

Formed elements of blood, 65–66; see also
Blood platelets; Red blood cells; White
blood cells

Fovea centralis, 480, 481

Frederique, Léon, 577

Frog; see Amphibia

Frontal lobes; see Cerebral cortex

Fructose, 32

Fruit juices and vitamins, 343

Fundamental colors, 482–83

Fungus, infections, 580

"Funny bone," 391

Furfural, 354

Gall bladder, 267, 282

Gallstones, 94, 308

Gametes, 603

Gamma globulin in poliomyelitis, 593

Ganglion (pl. ganglia)
autonomic, 408
of cranial nerves, 439
of dorsal roots, 404–5, 407, 408
embryonic development of, 610
evolution of, 417–18
paravertebral, 408
synaptic conduction in, 434

Ganglionated chain, 408, 409

Gastric juice (see also Stomach)
acid in, 274
bacterial destruction by, 586–87
collection of, 272, 282–83
digestion by, 272, 274, 275
neutralization of, in intestine, 277
in pernicious anemia, 103–4

Gastric secretion
cephalic phase of, 284
chemical control of, 276, 283–84
gastric phase of, 284
reflex control of, 283–84

Gastrin, 276, 284, 569

Gastrointestinal tract; see Digestive tract

Gastrovascular cavity, 63–64, 266

Geiger counter, 43

Gel state, 45–47, 71, 74

Gelatin, 45, 341

Genes, 600–601, 602, 604, 609

Genital system; see Sex organs

Germ layers, 609–10

Gestation; see Pregnancy

Ghosts of red blood cells, 85, 87

Gigantism, 561–63

Gill arches, 614

Gill slits, 484, 613

Gills, 131, 217, 218, 613

Glands
action potentials in, 142, 400

digestive, 266–70, 610
ductless; *see* Endocrine glands
excitation of, 400
of internal secretion; *see* Endocrine glands
mammary, 554, 565, 566, 567, 605, 607
secretion by, 270–72
sweat, 355, 610
Glasses, 474–76
Globin, 83
Globulin, 72, 593
Glomerulus of kidney, 360
Glottis, 247; *see also* Vocal cords
in breathing, 223
in defecation, 247, 298
in straining, 247, 298
Glucose, 31, 32, 33; *see also* Carbohydrates;
    Sugar
of blood; *see* Blood glucose
combustion of, 313–14, 513–14
diuresis by, 364–65
formula of, 24, 317
in hepatic vein, 311
and irritability of cells, 33–34, 68–69, 313,
    515
in portal vein, 311
resorption of, from urine, 361
storage of; *see* Glycogen
in urine, 311, 364–65, 510, 513
Glycerol, 34–35, 276
Glycocoll, 359
Glycogen, 32, 72; *see also* Carbohydrates
from fats, 317–18
in muscle contraction, 243–44, 379–82
from proteins, 317
storage of, 33, 35, 311–13, 507, 513–14
Glycosuria, 513
Goiter, 517, 521–24, 525–26
Gonads; *see* Ovaries; Testes
Gonorrhea, 483
"Goose flesh," in the cold, 334
Grafts; *see* Transplantation
Granular white blood cells, 115
Granulocyte, 115
Grape sugar, 23–24; *see also* Glucose
Graphic methods in physiology, 136–38
Gravity
    and blood flow, 175, 182–83, 249
    and blood pressure, 202–3, 205–6
    in expiration, 223
    posture and, 489–92
    and sacculus of inner ear, 489–91
    and utriculus of inner ear, 489–91
    and varicose veins, 210
Gray matter
    of brain stem, 438–39
    of cerebellum, 461
    of cerebral hemispheres, 445
    of spinal cord, 404, 405, 436, 437
Growth, 2, 48, 600, 612
    compensatory, 578

energy for, 326, 338
and hypophysis, 560–63
nitrogen balance in, 338–39
proteins in, 338
salts in, 342–43
and sex organs, 556
of sex organs, 552, 563
and testis hormone, 556
and thymus, 569–70
and thyroid glands, 520–21, 528
vitamins and, 346, 348
Growth-promoting hormone, 562–63
Gull, Sir William, 516
Gull's disease, 516–17
Gullet; *see* Esophagus
Gustatory; *see* Taste
Gyri of cerebral cortex, 446

Habit, 453
Hair, 334, 610
Hair cells (*see also* Ciliated cells)
    of cochlea, 487–88
    of olfactory endings, 495, 497
    of sacculus, 489
    of semicircular canals, 493
    of small receptors, 495, 497
    of taste buds, 495
    of utriculus, 489
Hales, Stephen, 183, 185
Hammer; *see* Malleus
Harvey, Sir William, 127–30, 144
Hay fever, 261
Head posture, 491
Health and disease, 10–11, 576–80
Hearing, 448, 449, 484–88
Heart (*see also* Heart rate; Heartbeat)
    accelerator nerves of, 156–57, 412, 432
    all-or-none law of, 140–41, 151
    of Amphibia, 131–32
    atria of, 128, 132–33, 135
    automaticity of, 142–44
    of birds, 132
    chambers of, 128, 131, 132–33, 134
    conducting system of; *see* Atrioventricular
        (A-V) bundle
    contraction of, 144–46
    damage to, 149–50, 152, 154, 528
    disorders of, 149–50
    distension of, by blood, 152
    drug effects on, 410–11
    ejection of blood by, 151–64, 188, 189
    electrical changes in, 141, 148–49
    of embryo, 608, 618–20
    embryonic development of, 610
    energy liberation by, 171–72
    enlargement of, 150, 578
    evolutionary development of, 131–32
    excitation process in, 400
    failure of, 209
    of fish, 131
    irritability of, 138

Heart—*Continued*
  leakage of; *see* Valves, leakage of
  of mammals, 131, 132
  minute volume output of, 151
  muscle of, 15, 16, 133–34, 141
  nerves of, 134–35, 142, 156–64, 409, 412,
    431
  nutrition of, 135–36, 150
  output of blood per beat by, 151–53
  pacemakers of, 135, 144–45, 157
  pressure changes in, 146, 147
  properties of, 138–42
  reflexes of, 160–64, 199–203
  refractory period in, 138–40
  sinus node of; *see* Sinoatrial node of heart
  sinus venosus of, 132, 144
  sounds of, 146–47
  staircase phenomenon in, 374
  Starling's law of, 152–53, 175
  stroke volume output of, 152–53
  structure of, 131–36
  vagus nerve of, 157–58, 410–11, 432
  valves of, 132–33, 134, 135, 146, 147, 171
  ventricles of, 128, 131–32, 133–34, 135

Heart rate, 153–64
  and blood pressure, 189
  and carbon dioxide, 155–56
  and carotid sinus, 197–203
  chemical control of, 155–56
  constancy of, 163, 200–202
  and depressor nerve, 162–63, 197
  and emotions, 161
  epinephrine and, 155
  governor of, 163, 200–202
  in muscular exercise, 154, 155, 262, 263
  nervous control of, 156–64
  reflexes affecting, 160–64, 199–203
  slowing of; *see* Inhibition, of heart
  and smoking, 354
  temperature effects on, 144, 153–55
  and thyroxin, 156

Heartbeat, 144–46; *see also* Systole
  cause of, 142–44
  in muscular exercise, 153, 154, 155, 262, 263
  recording of, 136–38
  sequence of events in, 144–46
  strength of, 141, 152–53, 154, 176
  in trained subject, 263
  and urine formation, 363
  vagus effect on, 157–58, 410–11, 432

Heat (*see also* Body temperature; Heat loss
    from body; Heat production in body;
    Heat sense)
  conservation of, by fats, 36, 314
  enzyme destruction by, 40
  and heartbeat, 144, 153–55
  and metabolism, 328
  protein destruction by, 40
  receptors for, 335, 498–99
  reflex adjustments to, 333–37
  and respiration, 241
  source of, in animals, 318–19
  and temperature-regulating center, 335–36
  and vasomotor reflexes, 333–37

Heat loss from body, 325, 336
  in fever, 337
  and production of heat, 332
  regulation of, 333–37

Heat production in body, 318–28, 335
  and age, 326–27
  basal; *see* Basal metabolism
  and body size, 325–26
  factors modifying, 324–25, 332–33
  in hyperthyroidism, 526–27, 530
  in hypothyroidism, 517–20, 521
  and loss of heat, 332
  measurement of, 319–21
  in muscle, 333, 376, 382–83
  in nerve, 386
  and oxygen consumption, 322
  regulation of, 335–36
  and sex, 326–28

Heat sense
  area in brain, 449
  nerve pathways of, 438
  receptors of, 498

Heatstroke, 337

Helmholtz, H. L., 387, 487

Hematin, 83, 94

Hemaglobin, 83; *see also* Oxyhemoglobin
  amino acids in, 38
  and carbon dioxide transportation, 258
  and carbon monoxide, 257
  and copper, 101
  formula of, 37
  of infant, 621
  iron in, 37, 68, 343
  oxygen transportation by, 83, 252–57
  in pernicious anemia, 103
  quantity of, in blood, 91–92
  reduced form of, 252–53, 258
  synthesis in bone marrow, 93
  in urine, 252

Hemolysin, 87

Hemolysis, 84–90, 591

Hemophilia, 81–82

Hemorrhage
  anemia from, 99
  and blood pressure, 70, 204–5
  and blood volume, 204–5
  and bone marrow, 98, 99, 577
  in high blood pressure, 209
  saline injections in, 191–92
  and splenic contractions, 205
  and transfusion, 88, 204
  from ulcers, 307
  urine after, 358, 363
  water balance after, 70–71, 577

Heredity, 609; *see also* Chromosomes; Genes

Hexose-phosphate, 379

Hiccough, 248

Hindbrain, 439

Hippuric acid, 359

Histology, 17

Homology, 439, 585, 613

Homothermal, 332

Hormones
  of adrenal cortex, 541–43
  of adrenal medulla, 535–39
  adrenocorticotrophic, 543, 566
  in blood, 73
  of duodenum, 568
  and enzymes, 42
  functions of, 505–6
  growth-promoting, 562–63
  of hypophysis, 559–67
  interrelationships of, 574–75
  of ovary, 551–58
  of pancreas, 511–16
  of parathyroid glands, 533
  of pituitary gland, 559–67
  and reproductive cycle, 551–58, 563–65
  secretion of, 504–6
  of sex glands, 551–58
  similarity of, in different species, 88–89
  of stomach, 568–69
  of testes, 556–58
  of thyroid glands, 524–25

Humerus, 369

Hunger, 298–302

Hydra, 63–64, 92, 118, 266, 356, 417, 418, 610, 613, 614

Hydrochloric acid, 28, 78
  bacterial destruction by, 586–87
  in digestion, 274
  and pancreatic secretion, 280–82, 568
  and pyloric sphincter, 295–96
  and secretin, 280–82, 568

Hyperactivity of endocrine glands, 507; *see also under* specific glands

Hyperglycemia, 513

Hyperinsulinism, 515

Hypermetropia, 474

Hyperparathyroidism, 534

Hyperpituitarism, 560–63

Hypertension, 209, 366

Hyperthyroidism, 86, 525–29
  exophthalmos in, 527

Hypertonic solutions, 59

Hypoactivity of endocrine glands, 507; *see also under* specific glands

Hypoparathyroidism, 530–35

Hypophysis, 448, 505, 559–67, 621
  and adrenal cortex, 543, 574
  and diabetes, 515–16, 574–75
  and growth, 560–63
  and reproduction, 563–65, 621
  and sex hormones, 553, 557–58
  and thyroid, 525, 563, 566

Hypotension, 209

Hypothyroidism, 516–20, 526

Hypotonic solutions, 59

Ideas in learning, 456–57

Image formation, 470–72, 474–76

Immune bodies; *see* Antibodies

Immunity, 261, 581–82, 589–94
  active, 592–94
  passive, 593–94

Implantation, 607

Incus, 484, 485

Indigestion, 305–6

Induced currents, 398

Induction, self-, 398

Inductorium, 398

Infantile paralysis, 246

Infections, 579–80; *see also* Bacteria
  acquired resistance to, 589–94
  of digestive tract, 308
  of eye, 344, 483
  and immunity, 589–94
  of kidneys, 365, 366
  leucocytosis in, 122–23
  localization of, in body, 587–89
  of lungs, 260–61
  of middle ear, 260, 485, 488
  natural resistance to, 581–82
  of respiratory tract, 260–61
  treatment of, with chemicals, 594–98
  in vitamin deficiency, 344, 348

Infections
  white blood cells in, 119–20, 587

Inflammation, 119, 587

Inhibition
  of blood vessel tone; *see* Vasodilatation
  in central nervous system, 428–31
  of heart, 157–58, 161
  of muscle tone, 428–31
  nature of, 159–60, 181–82, 411–13
  reciprocal, 428–31
  in reflexes, 428–31

Inhibitory substance, 411–13, 431–34

Injury potential, 395

Inner ear, 484, 485–87

Innervation (*see also* Nerves)
  reciprocal, 423, 428–31

Insanity, 352, 463, 464

Insects
  breathing in, 217
  equilibrium in, 492
  hearing in, 487

Insight in learning, 455, 456

Inspiration, 223–24; *see also* Breathing
  nervous control of, 234–39
  pressure changes during, 225–31

Inspiratory reserve volume, 232–33

Insulin
  destruction of, by enzymes, 512
  discovery of, 511–12
  functions of, 312, 512–14
  and hypophysis, 574–75
  overdosage of, 515
  secretion of, 572, 573–74

Integration
  in animals, 415–17, 603–4
  chemical, 416, 506
  hormonal, 506
  by nervous system, 415–17
  in plants, 416–17
  of sensations, 451
  by temperature changes, 416
Intelligence, 450, 609; *see also* Mentality and
    thyroid gland
Interconversion of foods, 39–40, 316–18
Intercostal muscles, 223
Internal environment, 65, 68, 356, 358, 368,
    577
  glucose in, 33, 68, 313
  neutrality of, 259
  salts in, 30–31, 68, 533, 541
Interpretation
  of retinal images, 472–74
  of sensations, 450–51
  of symbols, 451–52
Intestinal juice
  digestive enzymes in, 276, 277
  secretion of, 284–85
Intestinal movements; *see* Digestive move-
    ments
Intestinal poisons, 278
Intestinal wall
  semipermeability of, 56, 304
  structure of, 267–70
Intestinal worms, 580
Intestines, 266, 267, 268; *see also* Colon
  automatic rhythmicity of, 294–95
  digestion in, 275–77, 312
  of embryo, 608
  excretion by, 265, 290
  nerves of, 294–95, 409
Intrapulmonic pressure, 225–26
Intrathoracic pressure, 226–31
Intrathoracic space, 226–27, 228, 229
Invertase, 277
Iodine
  in diet, 343
  and goiter, 522–24, 527
  in thyroid hormone, 68, 343, 524–25
Ionization, 27–28
Ions, 27–28
Iris, 469, 472
  functions of, 476–77
  nerves of, 409
Iron
  and anemia, 101, 609
  for blood donors, 115
  in diet, 101, 343, 621
  in enzymes, 40
  in hemoglobin, 37, 68, 343
  in liver, 94, 101, 621
Iron lung, 246
Irritability of cells
  and glucose, 33–34, 68–69, 313, 515

and insulin, 34, 515
and parathyroid glands, 532–33
and salts, 30–31, 33, 533, 534, 543
Islets of pancreas, 509
Isomers, 31
Isotonic salt solutions, 59
Isotopes, radioactive, 43
Isthmus of thyroid gland, 516

Jawless and limbless vertebrates, 605
Jaws, embryology of, 614
Jenner, Edward, 590
Joint sense, 467
Joints, 368–70

Kidneys, 271, 357–60; *see also* Urine
  aglomerular, 363
  ammonia formation in, 315–16
  and blood composition, 357–58
  blood volume regulation by, 358
  disorders of 70, 208, 365–66
  embryonic development of, 610, 612
  functions of, 357–58
  and high blood pressure, 208, 366
  threshold of, 311, 361–62
Knee jerk, 419–20, 428, 433
Koch, Robert H., 581
Kymograph, 136–38, 228

Labor; *see* Childbirth
Labyrinths, 441, 488–89; *see also* Cochlea;
    Sacculus; Semicircular canals; Utriculus
Lactase, 277
Lactates of blood, 72
Lactation, 565, 567, 621; *see also* Mammary
    glands
Lacteals, 305
Lactic acid
  and breathing, 240
  combustion of, 243, 380–81
  in muscle contraction, 379–82
Lactose, 277
Laking of red blood cells, 87
Land egg, 606–7, 608
Landsteiner, Karl, 107
Langerhans, Ernst R., 509
Langerhans, islets of, 509
Language, interpretation of, 451
Large intestine; *see* Colon; Intestines
Laryngitis, 260
Larynx, 218, 247, 267, 292–93
  nerves of, 236–37, 292
  in swallowing, 292–93
Lashley, Karl S., 451, 458
Latent period, 388
  in reflexes, 426, 432–33
  in skeletal muscle, 375, 376
  in smooth muscle, 384

Laughing, 246–47

Lavoisier, Antoine L., 318, 319, 324

Laxatives, 297

Learning, 454–58
  and association areas, 450
  and cerebral hemispheres, 443
  and sensations, 467–68
  and vision, 473–74

Learning curve, 455

Lens, 469, 472
  elasticity of, 471
  embryonic development of, 612
  image formation by, 470–72
  opacities of, 476
  vision corrected by, 474–76

Lens system of eye, 470–72, 474–76

Lesion, 437

Leucemia, 123

Leucocytes, 115–16; see also White blood cells
  ameboid motion by, 118–19, 587
  in infections, 118–20, 587
  phagocytosis by, 118–20, 587

Leucocytosis, 122–23

Leucopenia, 122

Life
  origin of, 3–4, 26, 29
  properties of, 2

Ligaments, 368

Ligatures, 75

Light
  receptors for, 480–83
  refraction of, 470–71, 474–76
  stimulation of retina by, 481–82
  ultraviolet, and vitamin D, 346

Light reflex, 441, 477

Lillie, Ralph, 399

Lillie's iron-wire model of nerve, 399–400, 402

Limes and vitamin C, 343, 348

Limulus (the king crab), 142–43

Lipase, 276

Lipins, 36; see also Fats

Lipoids, 36; see also Fats

Liver, 128, 266, 267, 276, 312
  in anemia, 102–3
  bile pigment excretion by, 94
  bile secretion by, 282
  deaminization in, 316
  in defense against infection, 588
  endocrine function of, 569
  fibrinogen production by, 71
  glycogen in, 32, 311–13, 507, 513, 514
  iron storage in, 94, 101, 621
  in pernicious anemia, 102–3
  prothrombin from, 78
  red blood cell destruction by, 92–93
  and specific dynamic action, 325
  stimulation by secretin, 282
  urea formation in, 316

Living and nonliving, 3–5

Loeb, Jacques, 532

Loewi, Otto, 411

Longevity and alcohol, 352

Ludwig, Karl F., 185

Lumen, 267, 269

Lungs, 218–22, 267; see also Breathing; Respiration
  air in, 231–33
  of Amphibia, 218
  blood platelets from, 124
  dead space in, 233
  elasticity of, 221–22, 224, 226, 230
  of embryo, 610, 618–20
  embryonic development of, 610
  of fish, 217, 218
  infections in, 260–61
  pressure changes in, 225–27
  vital capacity of, 232–33

Luteinizing hormone of hypophysis, 564

Lymph, 117, 211–12

Lymph capillaries, 211–12

Lymph duct, 211, 249

Lymph flow, 212, 231, 249

Lymph glands; see Lymph nodes

Lymph heart, 212

Lymph nodes, 117, 588, 610

Lymph vessels, 211–14, 226, 610
  of digestive tract, 270, 304–5
  and fat absorption, 304–5, 314
  in infections, 213, 588–89

Lymphatic tissue, 117

Lymphocytes, 19–20, 116, 117, 121, 213, 214; see also White blood cells

Lymphoid tissue, 117

Lysin, 341

McLeod, J. J., 511, 512

Macrophages, 587

Maculi of utriculus and sacculus, 490

Magendie, François, 405

Magnesium sulfate, 297

Malarial parasite, 87, 580

Malingering, 463

Malleus, 485

Malpighi, Marcello, 129

Maltose, 277

Mammals, 605, 607–8

Mammary glands, 554, 565, 567, 605, 607, 621

Manganese in diet, 343

Manometer
  mercury, 184–85
  water, 227, 228

Margins of safety, 540
  in defenses against disease, 577–78
  in diet, 342, 349
  in kidney tissue, 366

Mating, 605; *see also* Copulation

Maximow, A. A., 19

Mechanisms, concept of, 5–6

Medicine and physiology, 10–11

Medulla
  of adrenal glands, 535–39, 571
  of brain, 157, 448; *see also* Medulla oblongata
  of kidney, 358–59

Medulla oblongata, 157, 448
  automatic nerves from, 409
  cardiac centers in, 157, 160
  reflex centers in, 439–41
  respiratory center in, 235–36
  vasomotor centers in, 193–94, 201

Membrane theory of nerve conduction, 393–400

Membranes
  of cells, 13, 47, 54–57
  of embryo, 606–7, 617

Memory, 455–56

Menstruation, 548, 549
  cessation of, 555–56
  disorders of, 555–56
  and follicular hormone, 551
  red blood cell count in, 99

Mental diseases, 464; *see also* Insanity

Mentality and thyroid gland, 518, 519, 520, 521, 528

Mesentery, 267

Mesoderm, 17, 64, 610, 611, 613

Metabolism, 2, 48–50
  basal; *see* Basal metabolism
  of carbohydrates, 311–14, 513–14
  of fats, 314–15, 514
  measurement of, 319–21
  of nervous system, 425
  of proteins, 315–16
  rate of; *see* Basal metabolism
  of skeletal muscle; *see* Muscle, skeletal
  of smooth muscle, 385

Metabolites, effect of
  on blood vessels, 179
  on heart, 152, 155–56, 374
  on muscle contraction, 374

Metric system, 30

Micromanipulation of cells, 18

Microphages, 587

Micturition; *see* Urination

Midbrain, 439, 448
  autonomic nerves from, 409
  reflex centers in, 441

Middle ear, 260, 484–85, 488, 613

Milk
  food value of, 101, 621
  iron content of, 101, 621
  secretion of; *see* Lactation

Mineral oil, 297

Minkowski, Oscar, 509–10

Minot, G. R., 102

Minute volume output of heart, 151

Mitosis, 93, 214, 601

Molecules, 23

Monkeys, learning in, 455

Monocular vision, 478

Monocytes, 116, 120

Mono-iodoacetic acid, 379

Monosaccharides, 31

Motor area of cerebral cortex; *see* Cerebral cortex, motor area

Motor nerves, 391; *see also* Efferent nerves; Nerves

Motor units, 373–74

Mouse, heat production by, 325, 326

Mouth, 266, 267, 293, 610

Movements
  digestive; *see* Digestive movements
  sensations of, 492–94
  of skeleton, 368–71

Mucosa; *see* Mucous membrane

Mucous membrane, 268–69

Mucus, 268, 583

Müller, Johannes P., 387

Müller's law, 391

Murmurs of heart, 147

Murphy, W. P., 102

Muscle (*see also* Muscle, skeletal; Muscle, smooth)
  cardiac, 15, 16, 133
  ciliary, 471, 472
  of hairs, 334, 408
  of heart, 15, 16, 133–34
  of iris, 472
  pilomotor, 408
  of skin, 370, 613

Muscle, skeletal
  blood flow in, 164
  contractions of, 368–71, 371–83
    and alcohol, 352
    action potentials in, 375
    and cerebellum, 461–62
    chemistry of, 243–44, 378–83
    control of, by brain, 448–49
    efficiency of, 382
    electrical changes in, 141, 375
    glycogen in, 243–44, 379–82
    and heat production, 327, 333, 376, 382–83
    and lymph flow, 212, 231, 249
    modulation of, 423–24, 462
    reflex control of, 372
    sol-gel, transformations in, 47
    strength of, 141, 152, 373–75, 377, 378, 462
    tetanic, 376–78
    and venous blood flow, 174–76
    work done in, 375, 382
  tone of, 159, 378, 423
    and cerebellum, 462

and cold, 324, 333
exaggeration of, 441–42
and fatigue, 459–60
and inhibition, 428–31
nerve pathways for, 438
reflex control of, 423, 441–42
and sleep, 459–60
in temperature regulation, 333, 335, 336
Muscle, smooth, 14–15
in arterioles, 178
in blood vessels, 166
contraction of, 384–85
of digestive tract, 269
embryonic development of, 610
in lungs, 222
of skin, 334, 613
in spleen, 205
tone in, 385
Muscle lever, 136–37
Muscle-nerve preparation, 141, 383, 387
Muscle sense, 467, 500
area in brain, 448, 449
nerve pathways of, 438
and sleep, 459–60
Muscles, skeletal, 14, 15, 368–71, 371–83; see
    also Muscle, skeletal, contractions of;
    tone of; Muscular exercise
agonistic, 429
all-or-none law of, 140–41, 373–75
antagonistic, 370, 429, 430
antigravity, 423, 442
biceps, 370
blood flow to, 206–8, 262, 381–82; see also
    Muscular exercise, blood flow in
of breathing, 222–24
and cerebellum, 461, 462
chemistry of, 379–82
cold effects on, 324, 333
control of, by brain, 448–49
co-ordination of, 462
degeneration of, 384
of ear, 484
embryonic development of, 610
energy liberation in, 243, 375, 378–83
enlargement of, from exercise, 263, 578
excitation process in, 400
extensor, 370, 428–31
of eye, 448, 468
fatigue of, 175, 243–44, 375, 379, 381
fibers of, 371
flexor, 370, 428–31
glycogen in, 311
group action of, 370
insertion of, 369
intercostal, 223
irritability of, 372; see also Irritability of
    cells
load effects on, 374–75
motor units and, 373–74
movements of; see Muscle, skeletal, con-
    tractions of
of neck, in body-righting, 492
nerve centers of, 160
nerve pathways to, 437, 438

origin of, 369
oxygen debt in, 243–44, 263, 382
paralysis of, 209, 372, 405, 449
properties of, 371–75
receptors in, 419–20, 422, 500
reciprocal innervation of, 423, 428–31
refractory state in, 372
of ribs, 222, 223
spindles in, 500
structure of, 368–71
synergistic, 370
tension of; see Muscle, skeletal, tone of
tetanus of, 376–78
and training, 383–84, 578
triceps, 370
twitch of, 375–76, 385, 388
voluntary control of, 448–49
weakness of, in adrenal insufficiency, 541
Muscles, striated; see Muscles, skeletal
Muscular exercise
blood flow in, 175–76, 179–80, 195, 206–8,
    231, 261–62, 381–82
and body weight, 329
breathing in, 223–24, 231–33, 239, 241, 262
and heart action, 153, 155, 262, 263
and heat production, 324
muscle changes produced by, 383–84, 578
oxygen consumption in, 150–51, 243–44,
    380–82
oxygen debt in, 243–44, 263, 382
splenic contraction in, 205, 262
Muscular work and diet, 338
Mutations, 604
Myelin sheath of nerves, 400–403, 407
Myopia, 474
Myxedema, 517, 519, 528

Nasal cavity, 293
Nausea, 280, 493
Neanderthal man, 444
Near point of vision, 471
Nearsightedness, 474–76
Nerve (see also Nerves)
plate, 610, 611
roots, 404–6, 437
trunks, 157, 160, 390, 391
tube, 610, 611
Nerve cells, 15, 16
Nerve fibers, 16
afferent, 404–6
efferent, 404–6
motor, 404–6
postganglionic, 408
preganglionic, 408
sensory, 404–6
somatic, 407
Nerve impulses, 11, 159, 385–95, 395–400
rate of transmission of, 387, 402–3
rhythm of, in reflexes, 427, 433
volleys of, 390, 427
Nerve-muscle junction, 383

652

Nerve-muscle preparation, 141, 383, 387
Nerve network
  in gastrointestinal tract, 270, 294–95
  in heart, 135, 142
  in *Hydra*, 417, 418
Nerve pathways, 437
  in brain stem, 439
  in spinal cord, 435–38
  of vision, 477–78
Nerves (*see also under* specific nerves)
  activation of; *see* Stimulation, of nerve
  of adrenal glands, 538, 571
  afferent; *see* Afferent nerves
  all-or-none law in, 140, 388–90, 394–95
  auditory, 440
  autonomic, 401, 406–11
  of blood vessels, 166, 180–82, 192–93
  of breathing, 234–36
  of carotid sinus, 197, 198
  centers; *see* Centers, reflex
  central end of, 157
  cervical sympathetic, 179, 180
  conduction in; *see* Nerve impulses
  cranial, 439–40
  craniosacral, 409–10
  degeneration of, 403
  depressor, 162–63, 196, 197–203, 239
  of digestive tract, 270, 409
  effects of, by chemicals, 411–13
  efferent; *see* Efferent nerves
  electrical changes in, 386, 393–400
  electrotonus in, 396–97
  embryonic development of, 610, 611
  of esophagus, 291–94
  extrinsic, 134
  fatigue in, 383
  glossopharyngeal, 236, 291
  of heart, 134–35, 142, 156–64, 409, 412, 432
  intercostal, 234–35
  of intestines, 294–95, 409
  intrinsic, 135
  iron-wire model of, 399–400, 402
  Lillie's model of, 399–400, 402
  of lung, 235, 237–39
  messages; *see* Nerve impulses
  mixed, 404, 406, 408
  modulator, 197–203
  of muscles, 372
  olfactory, 280, 440, 494, 495, 497
  optic, 280, 440, 468, 469, 478
  of pancreatic islets, 572
  parasympathetic, 409–10
  peripheral end of, 157
  phrenic, 234–35
  polar stimulation of, 397–98
  pressor, 196, 197
  refractory period of, 390–91, 394
  regeneration of, 403
  of salivary reflex, 279–80, 281
  sheaths of, 400–403
  of smell; *see* Olfactory nerve
  spinal, 403–6
  stimulation of, 156, 385–93, 397–98
  of stomach, 294–95, 409

superior laryngeal, 236, 237, 292
  of swallowing, 291–94
  sympathetic, 407–9
  of taste, 494, 495
  thoracolumbar, 407–9
  of thyroid glands, 572
  trigeminal, 291
  vagus, 440
  vasoconstrictor, 180–82, 192–93
  vasodilator, 180–82
Nervous system; *see* Autonomic nervous system; Brain; Brain stem; Central nervous system; Cerebellum; Cerebral cortex; Cerebral hemispheres; Medulla oblongata; Midbrain; Nerves; Spinal Cord; Thalamus
Nervous tension
  and digestive tract, 305
  in hyperthyroidism, 527
Neural plate; *see* Nerve plate
Neural tube; *see* Nerve tube
Neurilemma sheath of nerves, 400–403
Neurohumeralism, 411–13
Neurones; *see* Nerves; Nerve cells
Neuroses, 464
Neutrality regulation; *see* Acid-base balance of blood
Neutrophilic white blood cells, 116
Nicotine, 354, 411
Nicotinic acid, 347
Night blindness, 482
Nitrogen
  in air, 234
  in blood, 73
  equilibrium, 338–40, 341
  excretion, 321, 338, 339
  in protein, 37, 321, 338
Nodal tissue of heart, 135, 145
Nonacoustic labyrinths, 489
Nourishment; *see* Diet; Nutrition
Nucleoproteins, 357
Nucleus
  of cell, 13, 605
  in cell division, 93, 602
  in central nervous system, 441
  and nerve-cell integrity, 403
  red, 442
  in red blood cells, 82–83, 93
  significance of, in cell, 18
Nutrition, 338–49; *see also* Diet
  and anemia, 99–102
  of embryo, 607, 619
  of heart, 135–36, 150
  of infant, 605, 607–8, 621–22
  of nerve, 402, 403
Nystagmus, 492

Occipital lobes; *see* Cerebral cortex
Odor; *see* Smell

Olfaction; *see* Smell

Olfactory nerve, 280, 440, 494, 495, 497

Olfactory tract, 497

Oliver, George, 536

Ontogeny, 615

Opsonins, 591

Optic nerve, 280, 440, 468, 469, 478

Optic pathways, 477–78

Oral cavity; *see* Mouth

Organization of living things, 2

Organs, 14, 416
    embryonic differentiation of, 609–12
    internal; *see* Viscera

Orientals, basal metabolism of, 325

Osmosis, 57–61
    in absorption of food, 303
    in capillaries, 59–61, 70, 167
    and cells, 59–60
    and hemolysis, 84–86
    in kidney function, 365
    and proteins of plasma, 60, 70
    and red blood cells, 85
    and salts, 59, 68, 84–85

Osmotic pressure, 58–61

Otitis media, 260, 588

Otoliths, 489, 490

Outer ear, 484–85

Ova, 544, 546–49, 554, 563; *see also* Egg

Ovaries, 505, 544–45, 546–56
    cyclic changes in, 546–49
    embryonic development of, 610
    functions of, 554, 621
    hormones of, 549–55, 664–65
    and hypophysis, 463–65
    in menstrual cycle, 549–55
    and thyroid glands, 518, 520

Oviduct, 546, 547, 607

Ovulation, 547, 548

Oxidations, 24; *see also* Oxygen consumption
    aerobic, 49
    of alcohol, 350
    anaerobic, 49
    in animals, 318–19
    of carbohydrates, 313–14, 320–21, 513–14
    in cells, 24, 49, 216, 259–60
    and energy liberation, 49, 216
    of fats, 314–15, 320–21, 514
    and heat production, 320–21, 332
    in hyperthyroidism, 518, 426–27, 528
    in hypothyroidism, 517, 518, 521, 528
    in muscle; *see* Oxygen consumption, of
        muscle
    of proteins, 315, 316, 320–21

Oxygen
    in air, 233–34, 256
    in alveoli, 249–51, 255–57
    in blood, 73, 255–57
    diffusion of, 52–53, 249–51
    -hemoglobin combination, 83, 252–57
    life without, 50

partial pressure of, 254–55
    tension, 254–55
    in tissues, 255–57
    transportation of, in blood, 83, 98, 252–57
    volume of, in blood, 256–57

Oxygen consumption (*see also* Oxidations)
    of central nervous system, 425
    of embryo, 606, 608
    and heat production, 322
    of infant, 618
    by kidneys, 363
    measurement of, 320, 321–22
    of muscle, 150–52, 243–44, 381–82
    of nerve, 386
    in secretion, 271–72

Oxygen debt, 243–44, 263, 382

Oxygen lack, effect of
    on bone marrow, 96–98
    on breathing, 239–40
    on red blood cell formation, 96–98

Oxygenation of blood, 131, 132, 249–51

Oxyhemoglobin, 252–53, 255–57, 258; *see
    also* Hemoglobin

Oxytocin, 559

Pain
    area of brain, 449–50
    and blood pressure, 197
    and breathing, 236
    in childbirth, 615–16, 621
    in ears in airplaines, 485
    functions of, 579–80
    nerve pathways of, 438
    receptors in skin, 499
    and reflexes, 421
    sense, 449–50, 498–99

Palate, 292

Pancreas, 266, 267, 281, 505
    and diabetes mellitus, 509–14
    digestive enzymes of, 275–77
    endocrine activity of, 508–16
    islet tissue of, 509–11
    stimulation by secretin, 276, 280–82, 568,
        573

Pancreatectomy, 509

Pancreatic juice, 275–77, 280–82, 573

Panting, 248

Parallax, 478–79

Paralysis; *see* Muscle, skeletal paralysis of

Paramecium, 52, 584, 585

Parasites, 4, 580–82

Parasympathetic nerves, 409–10

Parathormone, 533

Parathyroid glands, 505, 530–35

Paresis, general, 352

Parietal lobes; *see* Cerebral cortex

Parthenogenesis, 603

Parturition; *see* Childbirth

Pasteur, Louis, 580, 590–91

Pavlov, Ivan, P., 283, 457–58

Pavlov pouch, 283

Pediatrician, 622

Pekin man, 444

Pellagra, 347

Pelvis, 369
  of kidney, 358–59

Penicillin, 595–96

Penis, 546

Pepsin, 274, 275, 276,

Peptidase, 276, 277

Peptids, 276

Peptones in protein digestion, 274, 275

Perfusion of organs, 20

Pericardium, 127

Perilymph, 486

Peristalsis, 285–90
  in colon, 290
  in esophagus, 87, 291, 293–94
  in small intestines, 289–90
  in stomach, 285–89, 295

Peritoneal cavity, 267

Peritoneum, 267

Peritonitis, 307, 308

Permeability; *see* Semipermeability

Pernicious anemia, 102–4
  and vitamin B_{12}, 103–4

Perspiration; *see* Sweat

Phagocytes, 118–20, 587

Phagocytosis, 118–20, 587
  in defense against infections, 118–20, 587
  in liver, 92–93, 588
  in lymph nodes, 213, 588–89
  of red blood cells, 92–93, 98
  in spleen, 588
  by white blood cells, 118–20, 587

Pharyngitis, 260

Pharynx, 218, 266, 267, 293
  afferent nerves of, 236–291
  embryonic development of, 613
  in swallowing, 291
  and thirst, 301–2

Phenol in colon, 278

Phosphates, 73, 379–82, 534
  in bones, 16

Phosphatides, 72

Phospholipin, in red blood cell, 84

Phosphorus in diet, 343

Photosynthesis, 24, 44, 49, 216, 380

Phrenology, 446

Phylogeny, 585, 615

Physical sciences, 1

Physiological constants; *see* Constancy; Constants, physiological

Physiological salt solution, 59

Pig
  birth of, 617
  embryonic development of, 611, 615, 616
  heat production by, 326

Pilocarpine, 410

Piltdown man, 444

Pineal body, 448, 505, 570

Pinna, 484

Pitch discrimination, 487–88

Pitocin, 559

Pitressin, 559

Pituitary gland; *see* Hypophysis

Pituitrin, 559

Placenta, 606, 608, 618, 619, 621

Plants, reproduction in, 600, 603, 605

Plasma; *see* Blood plasma

Platelets; *see* Blood platelets

Pneumograph, 227

Pneumonia, 260, 580

Pneumothorax, 230

Poikilothermal, 332

Poiseuille, Jean L-M., 185

Poison, alcohol as, 351–53

Polarization of nerve, 393

Poliomyelitis, 246
  gamma globulin in, 593

Polymorphonuclear white blood cells, 115

Polysaccharides, 32

Portal circulation; *see* Portal vein

Portal vein, 128, 304, 311, 312, 316

Posterior chamber of eye, 469

Postganglionic nerve fibers, 408

Posture
  and blood pressure, 202–3, 206
  maintenance of, 423, 441–42, 489–92

Potassium
  in adrenal insufficiency, 541, 543
  and heartbeat, 30

Potential of action; *see* Action potentials

Powder fuse and nerve impulse, 388–89

Precipitins, 592

Preformation of embryo, 609

Preganglionic nerve fibers, 408

Pregnancy
  corpus luteum in, 547–48, 552–53
  duration of, 615
  follicles in, 548
  and hypophysis, 563
  nitrogen balance in, 338
  and ovaries, 550–51
  tests for, 564
  and vitamin E, 348

Premature birth, 618

Pressor nerves, 196

Pressure
  of blood; *see* Blood pressure

intrapulmonic, 225–26
intrathoracic, 226–30
partial, 254–55
Pressure sense, 498–99
    area in brain, 449
    nerve pathways of, 438
Progesterone, 552, 553, 554
Projection of retinal images, 472–74
Proprioception, 422, 459–60, 467
Prosecretin, 78, 281
Prostate gland, 544, 546, 557
Protective mechanisms; *see* Defenses against
    infection
Proteinase, 275–76
Proteins, 36–40
    adequate, 341
    of animal origin, 341–42
    and asthma, 261
    and basal metabolism, 324–25, 332
    caloric value of, 320–21
    in cells, 36–40
    colloidal nature of, 44–45
    combustion of, 315, 316, 320–21
    conversion of, to carbohydrates, 39–40,
        317, 318, 514
    deaminization of, 315–16
    in diet, 338–42, 621
    digestion of, 274, 275, 276, 277
    foreign, 591
    and growth, 338
    heat liberation from, 320–21
    and heat production, 324–25, 332
    inadequate, 341
    loss of, from body, 340
    metabolism of, 315–16
    nitrogen in, 37, 321, 338
    numbers of, 38–39
    in nutrition, 338–42, 621
    osmotic pressure of, 60, 70–71
    of plant origin, 341
    of plasma, 69–71
    respiratory quotient of, 321–22
    sensitivity to, 261
    size of molecule of, 37, 45
    sparers, 340, 350
    specific dynamic action of, 324–25, 332
    specificity of, 38, 88, 591
    storage of, 39–40, 316
    synthesis of, 39, 315, 338, 340–42
    and virus, 3–4
    and water balance, 60, 69–71
Proteoses in protein digestion, 274, 275
Prothrombin, 78, 81, 348
Protoplasm (*see also* Cells)
    chemical composition of, 23–44
    internal organization of, 44
    wear and tear of, 48, 315, 338
Protozoa, 580, 584
Pseudopods, 119, 588
Psychiatry, 465
Psychoses, 464

Ptyalin, 273
Pulmonary circulation, 127, 128, 129, 168–
    69, 618–20
Pulse wave in arteries, 171
Pupil of eye, 469, 472
    constriction of, 472, 476–77
    dilation of, 472, 476–77
    reflex control of, 441
Purkinje, Johannes, 135
Purkinje tissue of heart, 135
Pus, 120, 587, 596
Pyloric sphincter, 269–70
    control of, 295–96
    opening of, 287–89, 295–96
Pyramidal cells of cerebral cortex, 446
Pyramidal tract of spinal cord, 436

Rabbit, body temperature of, 331
Race and basal metabolism, 325
Radioactive chemicals, 43–44, 95, 117
Radioactive iodine, 529
Radius, 369
Ranvier, Louis, 400
Ranvier, nodes of, 400, 402
Rat, learning in, 458
Rays, 605
Réaumur, René A., 272
Recapitulation principle, 613–15
Receptors, 418, 419
    for chemical stimuli, 494–98
    distance, 467–68
    of equilibrium sense, 489, 491, 492–94
    for light, 480–83
    for movements, 492–94
    in muscles, 419–20, 422, 500
    for rotation sense, 492–94
    in sensation, 467
    in skin, 498
    for smell, 494, 495, 496
    for sound, 484, 485–87
    for taste, 495
    in tendons, 500
Recipients in blood transfusions, 90, 105
Reciprocal innervation, 423, 428–31
Recovery process
    in muscle, 379–82
    in nerve, 390–91
Rectum, 266, 267, 268
    emptying of, 247–48, 298
    nerves of, 409
Red blood cell counts, 90–91
    altitude effects on, 96
    and liver feeding, 102–3
    in menstruation, 99
    and oxygen lack, 96–98
Red blood cells, 65, 82–104
    A and B substances in, 107–10
    agglutination of, 88–90

Red blood cells—*Continued*
  in atom-bomb injury, 124
  in clotting of blood, 74–75
  constancy of numbers of, 92, 97–98
  crenation of, 86
  destruction of, 92–94
  electrical charge on, 55–56
  and foreign proteins, 591
  formation of, 92–95, 101, 103
  fragility of, 86, 103
  hemoglobin in, 83, 91–92
  hemolysis of, 84–90, 591
  immature, 82–83, 93
  of infant, 621
  life-cycle of, 90–98
  numbers of, 82; *see also* Red blood cell
    counts
  oxygen transportation by, 83, 251
  permeability of membrane of, 54, 86
  phagocytosis of, in liver, 92–93, 98
  stroma of, 83–84, 103
  structure of, 82–83
  transfusion of, 88–90, 104–15
  in transfusions, 104–15

Red blood corpuscles; *see* Red blood cells

Red bone marrow; *see* Bone marrow, red

Reduction, 24; *see also* Oxidations

Reflexes, 11, 157, 392, 418–24; *see also under*
    specific reflexes (e.g., Knee jerk, etc.)
  abolition of, 424, 435–36
  acquired, 421
  Bainbridge, 163–64, 175–76
  body-righting, 489–92; *see also* Posture
  centers for; *see* Centers, reflex
  conditioned, 280, 284, 421, 457–58
  conduction of nerve impulse in, 424–34
  continuous, 423
  corneal, 468, 586
  for eyelid closure, 468, 586
  of gastric secretion, 283–84
  of heart, 160–64
  inherited, 421
  innate, 421
  kinds of, 419–24
  learned, 421
  of muscle tone, 423, 441–42
  of pupillary change, 477
  respiratory, 236–39, 248, 292
  of salivary secretion, 11, 279–80, 281
  stretch, 422–23
  swallowing, 291–94
  unconditioned, 421
  variability of, 427–28, 433
  vasomotor, 182, 194, 196–203

Refraction of light, 470–71, 474–76

Refractory period
  absolute, 377
  of heart, 138–40
  in iron wire, 400
  of muscle, 372
  of nerve, 390–91, 394
  relative, 377

Relaxation
  of heart, 137, 144, 145
  of muscle, 375, 376

Repair processes, 19–20, 120–21

Repression, 464

Reproduction, 2, 544–58, 600
  and alcohol, 352
  of cells, 600
  cycle of, 546–49
  hormone control of, 549–56
  and hypophysis, 563–65
  and metabolism, 48
  in plants, 600, 603, 605
  sexual, 603–5
  by spores, 603
  vegetative, 601–2

Reptiles, 605
  Age of, 607
  circulation in, 132
  effects of cold on, 330
  egg of, 606–7, 608

Resistance to blood flow, 172–74, 188, 190–
    203
  external, 191
  internal, 191

Respiration, 24, 49, 216, 380; *see also* Breath-
    ing; Oxidations
  artificial, 244–46
  cellular, 259–60
  -circulation relationships, 261–63
  external, 218
  internal, 218

Respiratory center, 235–36
  carbon dioxide control of, 240–43, 262
  chemical control of, 239–43
  effects of oxygen lack on, 240–43
  rhythm of, 237–39
  voluntary control of, 246–48

Respiratory gases (*see also* Carbon dioxide;
    Oxygen)
  exchanges of, with blood, 249–51
  transportation of, 251–59

Respiratory quotient, 321–22

Retina, 469, 472
  corresponding points on, 478
  embryonic development of, 612
  receptors in, 480–83

Rh factor in blood, 110–13

Rhesus monkey, 111

Rheumatic fever, 150

Rhinitis, 260

Rhodopsin; *see* Visual purple

Rhythmical segmentation in intestine, 289–
    90

Riboflavin, 347, 348

Ribs, 368–69
  in breathing, 222–24
  muscles of, 234–35

Rickets, 346, 347, 348

Righting reflexes, 489–92; *see also* Posture

Ringer's solution, 60

Rods of retina, 469, 480–83
Rotation sense, 488, 492–94
Roughage in diet, 296, 349

Saccharose digestion, 277
Sacculus, 484, 488, 489–91
Salamanders, 605
Saliva, 11, 248, 273–74, 333–34
Salivary glands, 11, 266, 267, 273
    nerves of, 409
Salivary secretion
    by conditioned reflex, 280, 457–58
    control of, 11, 279–80, 281
    in panting, 248, 333–34
    and smell of food, 280
    and sight of food, 280
    and thought of food, 280
Salivation; see Salivary secretion
Salt solutions
    hypertonic, 59, 86
    hypotonic, 59, 84
    injections of, in hemorrhage, 191–92
    isotonic, 59
    physiological, 59
Salts
    absorption of, 302–3
    conduction of electricity by, 28
    in diet, 342–43, 621
    in growth, 342–43
    and heartbeat, 30, 143
    and irritability of cells, 30–31, 33, 533, 534,
        543
    in nutrition, 342–43, 621
    osmotic properties of, 58–60, 68, 84–86
    of plasma, 67–68
    of protoplasm, 30
    in urine, 357
Sanitation, 582
Sarcoplasm, 15
Scarlet fever, 260
    and heart, 149
Schafer, Edward A., 536
Scheiner, Christopher, 472
Scheiner's experiment, 472–73
Schleiden, Matthias J., 12
Schwann, Theodor, 12, 400
Schwann, sheath of, 400
Scientific method, 7–11
Sclera of eye, 469
Scrotal sac, 557
Scurvy, 343, 348
Sea water
    and body fluids, 29–30
    and plasma salts, 67
Seasickness, 493
"Second wind," 244
Secretin
    and bile secretion, 282

pancreatic, 276, 280–82, 568
    secretion of, 280–82, 573
Secretion, 250–51, 270–72, 356
    and absorption, 272, 303–4
    in defense against infections, 582–83
    of digestive juices; see under specific diges-
        tive glands
    granules, 271
    of hormones; see under specific hormones
    internal; see Endocrine glands; Hormones
    of sweat, 333–34, 335, 355
    of urine, 271, 357–65
Segmentation
    embryonic development of, 614
    evolution of, 614
    in nervous system, 417–18, 439, 461
Selection, evolutionary, 604–5
Selective permeability; see Semipermeability
Semen, 546
Semicircular canals, 484, 488, 492–94
Semilunar valves, 132, 133, 146, 147, 170, 544
Seminal duct, 544, 546, 557
Seminal vesicle, 544, 546, 557
Seminiferous tubule, 544, 545
Semipermeability, 55–57
    and fat solubility, 55, 315
    of intestinal wall, 304
    of nerve membrane, 393
    of red blood cell, 54
Sensations
    areas in brain for, 447, 448–50
    and dorsal roots, 404–6, 437
    gradations of intensity of, 390
    integration of, in brain, 451
    interpretation of, in brain, 450–51
    nerve pathways for, 436–37
    and sleep, 458–59
    from viscera, 450, 467, 499–500
    Weber's law of, 500–501
Sense organs; see Receptors
Senses, 467–68; see also under specific senses
Sensory nerves; see Afferent nerves
Serum, 74, 79
Sex (see also Ovaries; Pregnancy; Reproduc-
        tion; Testes)
    characters, secondary, 554, 556–57
    cycle, 546–49, 563–65
    evolution of, 603–5
    determination in embryo, 609
    glands, 505, 544–58
    and heat production, 326–28
    and hypophysis, 553, 557–58, 563–65, 566,
        567
    organs of, 544–58
    reproduction by, 544–58, 603–5
    and thyroid glands, 518, 520, 521, 528
    urge, 554, 558
Sharks, 605
Shell of egg, 606
Sherrington, Charles S., 433

Shivering, 324, 333, 337
Shock
  spinal, 435–36
  traumatic, 210
Sighing, 248
Signal record, 137, 138
Silver nitrate, 483
Singing, 246–47
Sinoatrial node of heart, 135
  and heartbeat, 144, 145
  nerve-tissue in, 142
  as pacemaker, 145, 154
  and reflex center compared, 431–32
Sinus node of heart; *see* Sinoatrial node of
  heart
Sinus venosus, 132, 144
Sinuses; *see* Blood sinuses
Sinusitis, 260
"606" (Salvarsan), 594
Skates, 605
Skeletal muscle; *see* Muscle, skeletal
Skeleton, 368–71, 605, 610
Skin
  in breathing, 217, 218
  in defense against infection, 582
  embryonic development of, 610, 613
  heat loss from, 182, 332, 333
  muscles of, 370, 613
  sensations from, 467, 498–99
  vessels in temperature regulation, 333, 335
Skull, 368, 369
Sleep, 454, 458–61
  brain center for, 441, 460
  metabolism in, 324
Small intestine; *see* Intestines
Smallpox, 590, 598
Smell, 450, 496–98
Smoking, tobacco, 353–54
Smooth muscle; *see* Muscle, smooth
Snakes, 605; *see also* Reptiles
Sneeze reflex, 248, 261
Snellen, Hermann, 475
Snellen's type, 475–76
Snoring, 248
Socket of eyeball, 468
Sodium
  in adrenal insufficiency, 541, 543
  and cause of heartbeat, 30, 143
  in diet, 343
Sodium chloride, 24, 68
Sol state, 45–47, 71, 74
Solar plexus blow, 162
Solutions, 26–27, 51, 59
Sound perception; *see* Hearing
Specialization of cells, 14
Specific dynamic action of foods, 324–25, 332

Speech, 246–47
  and alcohol, 352
  area of Broca, 452
  interpretation of, 451
  saliva and, 273
Sperm, 544, 545, 546, 584, 603, 605
  production of, 544, 545, 563
Spherical aberration, 477
Sphincters, 269, 286, 289, 294
Spinal column, 368, 369
Spinal cord, 156, 157, 201, 392; *see also* Central nervous system
  ascending tracts of, 437
  autonomic nerves from, 409
  control of, by brain, 435–36, 442, 444
  descending tracts of, 437
  embryonic development of, 610, 611
  evolution of, 417–18
  gray matter of, 404, 405, 436, 437
  intermediary neurones in, 435
  nerve centers in, 160
  pathways of, 435–38
  reflex functions of, 421, 424–34
  and roots of nerves, 403–6, 419
  vasomotor reflexes via, 194, 196
  white matter of, 404, 436
Spiral bacteria, 580
Spirella, 580
Splanchnic nerves, 294
Spleen, 128
  contractions of, 205, 262
  red blood cell destruction in, 93
Spores, 603, 604
Staircase phenomenon, 374
Stapes, 484, 485
Starches, 32, 273–74, 276; *see also* Carbohydrates
Starfish, 612
Starling, Ernest H., 152, 508, 568
Starling's law of the heart, 152–53, 175
Starvation; *see* Fasting
Statistics, 9
Steapsin 276
Stereoscopic vision, 479
Sterility
  and hypophysis, 563
  in hypothyroidism, 520
  and ovaries, 550
  and testes, 556
  from vitamin E lack, 348
Sternum, 222, 369
Sterols and vitamin D, 346
Stethoscope, 146, 187, 294
Stimulation (*see also* Irritability of cells)
  chemical, 391, 494
  conditioned, 457
  by energy changes, 467
  laws of, 392–93
  of Lillie's iron wire, 400

of muscle, 371, 372
nature of, 96–97
of nerve, 157, 385, 386–93, 397–98
unconditioned, 457

Stirrup; *see* Stapes

Stomach, 228, 266, 267; *see also* Gastric
  juice; Gastric secretion
  automatic rhythmicity of, 294–95
  contractions in hunger, 299–300
  digestion in, 274, 276
  emptying of, 287–89, 295–96
  endocrine activity of, 568
  as food reservoir, 295
  movements of, 285–86, 287–89, 296
  mucous membrane of, 283
  nerves of, 294–95, 409
  peristalsis in, 285–86, 287–89, 295
  in pernicious anemia, 103–4
  as sense organ, 300
  tone of, 385
  ulcer of, 99, 391

Stones of kidney, 365

Stools; *see* Feces

Storage; *see* Carbohydrates; Fats; Glycogen;
  Proteins

Straining, 247–48, 298, 617

Streptomycin, 597

Striated muscle, 14–15; *see also* Muscle,
  skeletal

Stroke, paralytic, 209

Stroke volume of heart, 151

Stroma of red blood cell, 83–84, 103

Strychnine, 436

Succus entericus, 277

Sugar (*see also* Carbohydrates; Glucose)
  absorption into blood, 304, 312
  of blood; *see* Blood glucose
  cane, 32, 277
  grape, 32
  malt, 277
  milk, 277
  synthesis of, 24, 26, 216

Sugars
  compound, 32
  double, 32, 273–74, 276, 277
  simple, 31, 276

Sulci of cerebral cortex, 446

Sulfa drugs, 594–95

Sulfanilamide, 595

Summation
  in muscle, 374, 377–78
  in nerve, 399
  in reflexes, 426–27, 433

Sunshine and vitamin D, 346

Sunstroke, 337

Survival value in evolution, 604–5

Swallowing, 286–87
  of air by frog, 224–25
  by cilia in frog, 585

nervous control of, 291–94
respiratory inhibition in, 236, 292

Sweat (*see also* Glands, sweat)
  composition of, 355
  quantity of, secreted, 355
  in temperature regulation, 333–34, 336

Swim bladder of fish, 218

Switzerland, goiter in, 524

Symbols, interpretation of, 451–52

Sympathetic nerves, 407–9

Sympathetic nervous system; *see* Autonomic
  nervous system

Sympathico-mimetic action of epinephrine,
  537

Sympathin, 413, 431, 538

Synapses
  in automatic ganglia, 407, 408, 410, 411
  chemicals liberated at, 431–34
  conduction across, 424, 425–26, 431–34
  in reflexes, 419, 424
  in thalamus, 441

Syphilis, 609
  aortic damage from, 210
  blood-vessel damage from, 80
  brain damage from, 352, 463
  heart-valve damage from, 147
  and muscle movements, 424
  and muscle sense, 438
  spinal-cord damage from, 424, 438
  treatment with chemicals, 594

Systemic circulation, 129, 169

Systems, 14

Systole, 137, 138, 144, 146, 147; *see also*
  Heart, contraction of; Heartbeat

Table salt, 23, 68

Tagged atoms, 43; *see also* Radioactive chemicals

Tail, in evolution, 613, 616

Tambour, 227

Tapeworm, 4

Taste, 495–96
  area in brain, 450
  saliva and, 273

Taste buds, 11, 279, 281, 495, 496

Taste pores, 495

Tears, 468, 583

Tectorial membrane, 486, 487

Teeth, 346, 534, 610

Temperature (*see also* Body temperature;
  Cold; Heat)
  effects on heart, 144, 153–55
  sense of, 498–99
  and skin blood vessels, 152, 333, 335
  and taste, 495

Temporal lobes; *see* Cerebral cortex

Tendons, sensations from, 500

Tensions of gases, 253–55

Testes, 505, 544–45, 556–58
    embryonic development of, 610
    endocrine activity of, 556–58
    functions of, 558
    hypophyseal control of, 563
    and thyroids, 518, 520, 528
Testicular hormone, 556–58
Tetanus in muscle, 376–78
Tetany, parathyroid, 531–33
Thalamus, 437, 448
    pain center in, 450
    reflex centers in, 441
    synapses in, 441
    temperature regulation by, 335–36
Thermoregulation; *see* Body temperature,
    regulation of
Thiamine, 347
Thirst, 301–2, 560
Thoracic cavity; *see* Thorax
Thoracic (lymph) duct, 249
Thorocolumbar nerves, 407–9
Thorax, 222, 223
    elasticity of, 223
    pressure changes in, 225–31
Threshold
    of kidney, 311, 361–62
    of smell, 494
    of stimulation, 140, 389, 392
    of taste, 494
Threshold substances in urine, 361–62
Thrombin, 77
Thrombokinase, 77
Thrombus, 80, 208
Thymus, 505, 569–70
Thyroglobulin, 524
Thyroid cartilage, 517
Thyroid glands, 516–30; *see also* Thyroid
    hormone
    abnormalities of, 149–50, 516–24, 525–30
    cancer of, 529
    and hypophysis, 525, 563, 566
    location of, 505, 516
    nervous control of, 572
    structure of, 516
Thyroid hormone, 524–25
    and basal metabolism, 328, 518
    and body weight, 328–30
    and growth, 520–21, 528
    and heart rate, 156
    and heat production, 328, 332–33
    and hypophysis, 563, 566
    iodine in, 68, 343, 524–25
    and sexual development, 518, 520, 521, 528
Thyroid-stimulating hormone, 525
Thyroxin, 524–25; *see also* Thyroid hormone
Tibia, 369
Tickle, sense of, 467
Time record, 137, 138

Tissue culture, 19–20, 74, 121
Tissue fluids, 60, 69–71, 211–12, 213
Tissue juices and blood clotting, 75–76
Tissues, kinds of, 14–17
Toads, 605
Tobacco, 353–54
Tobacco mosaic virus, 3–4
Tone; *see* Tonus
Tongue, in swallowing, 292, 293
Tonsils, 260
Tonus, 159
    of arterioles, 178
    of blood vessels, 180–82
    of carotid sinus nerve, 199, 201
    of depressor nerve, 200, 201
    of skeletal muscle; *see* Muscle, skeletal,
        tone of
    of vagus center, 158–59
    of vagus nerve, 199–200, 201
    of vasoconstrictor center, 194, 199
    of vasoconstrictor nerves, 192–94, 201
Touch sense, 498–99
    area in brain, 449
    nerve pathways of, 438
Toxin-antitoxin in diphtheria, 593
Trachea, 219, 220, 293
Tracheitis, 260
Transfusion of blood, 88–90, 104–15, 204;
    see aslo Blood, transfusions of
Transplantation, 88–90
    of pancreas, 572
    of retina, 612
    of testis, 90, 558
    of thyroid glands, 572
Transportation
    of carbon dioxide, 251, 257–59
    of food to body cells, 64–65, 266
    of oxygen, 83, 98, 252–57
Traumatic shock, 210
Trial and error in learning, 454–56
Trichinella, 120
Tripalmitin, 34, 314
Tropisms in white blood cells, 120
Trypsin, 276
Trypsinogen, 276
Tryptophane, 341
Tuberculosis, 260, 366, 540, 597–98
Tubules of kidney, 359
Tumors
    of bone, 99
    of brain, 463
    of digestive tract, 307
Turtle, 331, 605
Twitch of muscle, 375–76, 388
Tympanic canal, 486
Tympanic membrane, 485–86
Typhoid fever, 580, 582

Ulcer
of duodenum, 306–7
and emotional factors, 306
of stomach, 99, 306–7, 391, 500
and vagus nerves, 306

Ulna, 369

Umbilical cord, 606, 618

Universal donors, in blood transfusions, 109

Universal recipients, in blood transfusions, 109

Unstriated muscle; see Muscle, smooth

Urea, 71
in blood, 72, 361
combustion of, 321
excretion by kidneys, 364
formation of, 315–16
in sweat, 355
synthesis of, 25–26
in urine, 355, 357, 361
and urine secretion, 364

Ureter, 357, 358, 359

Urethra, 358, 359, 544, 546

Uric acid, 71
excretion of, 362
in urine, 357

Urinary bladder, 358, 359
embryonic development of, 610
emptying of, 247–48, 298, 356
nerves of, 409

Urinary tract, 358–60

Urination, 247–48, 298, 356

Urine (see also Kidneys)
of bladder, 361
of Bowman's capsule, 360
composition of, 357
filtration of, 360
follicular hormone in, 551
glucose in, 510
hemoglobin in, 252
hypophyseal hormones in, 564
progesterone in, 552–53
quantity secreted, 357, 364–65
secretion by kidneys, 271–72, 357–65, 560
voided, 361

Uterine tube; see Oviduct

Uterus, 547, 607, 611
chemical control of, 549–53, 554
contractions of, 615–17, 620
menstrual changes in, 549, 563
pituitrin action on, 559–60

Utility versus mechanism, 5–6

Utriculus, 484, 488, 489–91, 493

Vaccination, 590, 591, 592

Vagina, 547, 615

Vagus center, 157, 160
reflex stimulation of, 161–63, 199–203
tonic action of, 199–200

Vagus fibers
postganglionic, 410–11
preganglionic, 410–11

Vagus material, 412

Vagus nerve (pl. vagi)
and blood pressure, 189, 190
in breathing, 237–39
drug effects on, 410–11
and gastric secretion, 283–84
and heart, 157–58, 411, 432
of intestines, 294, 409
of stomach, 294, 409
tonic action of, 158–59, 200, 201
and ulcer, 306

Valves
of heart, 132–33, 134, 135, 146, 147, 171
leakage of, 147, 150, 186
in lymph vessels, 212
narrowing of openings of, 150
in veins, 164, 174

Variation and evolution, 604

Varicose veins, 210

Vascular system; see Blood vessels; Circulation

Vasoconstriction, 176–78; see also Blood flow; Blood pressure
and blood pressure, 190, 194
and epinephrine, 180
in exercise, 195, 207–8, 262
from kidney damage, 366
nerves of, 180–82, 192–93
by reflexes, 193–94, 196–203
of skin vessels in cold, 182, 333, 335
and smoking, 354
tonic action of, 192–94, 201
and volume of organs, 178

Vasoconstrictor center, 203
carbon dioxide action on, 194–95
in medulla oblongata, 193–94, 201
reflex inhibition of, 199–203
tonic action of, 194, 199

Vasodilatation, 176–78; see also Blood flow; Blood pressure
and blood pressure, 190, 193
by carbon dioxide, 179
via dorsal root fibers, 406
and epinephrine, 180
in exercise, 195, 207, 262
mechanism of, 181–82
metabolites and, 179
nerves of, 180–82
by reflexes, 193–94, 196–203
of skin vessels in heat, 182, 333, 335

Vasodilator center, 194, 203

Vasomotor centers, 193–94, 196

Vasomotor nerves, 180–82, 192–94

Vasomotor reflexes, 182, 194, 196–203

Vegetables, proteins in, 341–42

Vegetarian diet, 341–42

Vegetative reproduction, 601–2

Veins, 166–67
  blood flow in, 170, 174–76, 231, 249
  blood pressure in, 172, 174, 206
  carbon dioxide in, 256
  hepatic, 311
  oxygen in, 256
  portal, 128, 304, 311, 312, 316
  pulmonary, 127, 128
  subclavian, 226
  valves in, 164
  varicose, 210
  vena cava, 127, 128

Venoms, hemolysis by, 87

Ventilation, 234

Ventricles
  of brain, 439, 445–46, 612
  of heart, 131–32, 133–34, 135

Venules, 170

Vermis of cerebellum, 461

Vertebrae, 369

Vertebral column, 368, 369

Vertebrates, 605

Vessels; *see* Blood vessels

Vestibular canal, 485–86

Vestibular nerve, 489

Villi of intestine, 212, 268

Virus, filterable, 3–4, 260, 580

Viscera (*see also* Esophagus; Stomach; Lungs;
    etc.)
  nerves of, 406–11
  pain from, 499–500

Visceral nervous system; *see* Autonomic
    nervous system

Viscosity of blood, 69, 191–92

Vision, 468–83
  achromatic, 480–81
  area for, in brain, 448, 449
  color, 482–83
  components of, 469–70
  in dark, 481–82
  defects of, 474–76
  depth perception in, 477–79
  distance perception in, 477–79
  double, 478
  and equilibrium, 489, 491
  image formation in, 470–72
  nerve pathways for, 477–78

Visual purple, 481–82

Vitamin(s), 343–49
  A, 344, 348, 482
  and anemia, 101
  B, 345, 347
    and pernicious anemia, 103–4, 348
  B complex, 348
  C, 345, 346, 348, 622
  D, 346, 347, 622
  deficiencies, 81, 344, 347, 348, 622

  in diet, 343–48, 622
  E, 348
  and enzymes, 42–43
  and growth, 346, 348
  K, 81, 348
  table of, 348

Vitreous humor of eye, 469

Vocal cords, 247, 293; *see also* Glottis
  in breathing, 223, 224
  in defecation, 248, 298
  in straining, 248, 298

Vocalization; *see* Speech

Voice box; *see* Larynx

Volition, 452

Voluntary activity, 452

Volvox, 614

Vomiting, 298, 441, 493

Von Mering, Johann, 509–10

Warm-blooded animals, 332

Waste elimination, 63–65, 355–56, 608, 609

Water, 23
  absorption of, 290, 296–97, 302–5
  accumulation in tissues, 70, 175
  balance in capillaries, 53–54, 59–61, 69–
    71, 577
  in blood, 67, 541
  in diet, 342
  in evolution, 606
  properties of, 26–27
  in protoplasm, 26–29
  reabsorption from urine, 360–61
  in stomach, 295

Weaning, 621

Weber, Ernst, 501

Weber's law of sensation, 500–501

Whipple, G. H., 102

White blood cells, 19–20, 65, 115–23; *see
    also* Leucocytes; Lymphocytes
  agranular, 115
  ameboid movement of, 118–19, 587
  in atom-bomb injury, 124
  counting of, 115, 121–23
  fate of, 118, 121
  functions of, 20, 118–21
  granular, 115
  in infections, 118–20, 587
  kinds of, 115–16
  life-span of, 117–18
  origin of, 117

White matter
  of brain stem, 438–39
  of cerebellum, 461
  of cerebral hemispheres, 445
  of spinal cord, 404, 436

Windows, round and oval, in inner ear, 485–
    86

Windpipe; *see* Trachea
Wöhler, Friedrich, 25
Wound healing, 19–20, 121

X-ray studies of gastrointestinal movements, 285–86

Yawning, 248

Yeast, fermentation by, 42, 50
Yellow body; *see* Corpus luteum
Yolk,
  in egg, 605, 608
  sac, 606, 608

Zein as an inadequate protein, 341
Zygote, 604, 605, 607